T0205576

Interfacing Humans and Robots for Gait Assistance and Rehabilitation

Carlos A. Cifuentes • Marcela Múnera

Interfacing Humans and Robots for Gait Assistance and Rehabilitation

Carlos A. Cifuentes
Department of Biomedical Engineering
Colombian School of Engineering Julio Garavito
Bogota, Colombia

Marcela Múnera
Department of Biomedical Engineering
Colombian School of Engineering Julio Garavito
Bogota, Colombia

ISBN 978-3-030-79632-7 ISBN 978-3-030-79630-3 (eBook)
https://doi.org/10.1007/978-3-030-79630-3

This Springer imprint is published by the registered company Springer Nature Switzerland AG
The registered company address is: Gewerbestrasse 11, 6330 Cham, Switzerland

Preface

In the past few years, new technologies have emerged that improve the living conditions of people who have suffered motor impairments or amputations. Assistive and rehabilitation robotics, implemented either employing hardware adaptations or high-level control approaches, has led to the appearance of several promising applications that promote independence in subjects limited by their impairment.

The concepts presented in this book are explored for the first time in assistive and rehabilitation robotics, which is the combination of physical, cognitive, and social human-robot interaction to empower the process of gait rehabilitation to assist human mobility. This book aims to consolidate the methodologies, modules, and technologies implemented in lower-limb exoskeletons, smart walkers, and social robots when human gait assistance and rehabilitation are the primary targets. This book intends to provide a comprehensive discussion of the possibilities that arise when including robotics in the treatment of gait-related impairments.

The first part of the book is devoted to understanding the human gait and the generalities of the platforms previously mentioned. Several applications and special considerations for the development and modeling are reviewed to define the architectures and systems used in all those robotic devices. As the book progresses, the design of human-robot interfaces must rely on diverse modalities related to motion intention and generation of voluntary movement so that their users may experience an influence at a mechanical or neural level. In this sense, the design of human-robot interfaces (both physical and cognitive) is explored through (1) the characterization of different sensors and actuators and (2) the implementation of control strategies based on bio-inspired algorithms and models.

The discussion then moves to the cognitive realm of rehabilitation therapies, where evidence of including social robots in combination with other platforms is addressed. After that, a look into the process of assessment of the devices supports the need for a multidisciplinary team in rehabilitation. The strong impact of integrating all the robotic devices shown has in gait assistance and rehabilitation is then presented in studies with both clinicians and patients.

This book presents the combination of emergent technologies and robotics science, such as soft robotics, force control, novel sensing methods, brain-computer interfaces, serious games, automatic learning, and motion planning in healthcare applications. From the clinical perspective, this book presents case studies for testing and evaluating how those robots interact with humans, analyzing acceptance, perception, biomechanics factors, and physiological mechanisms of recovery during the robotic assistance or therapy.

This book will enable readers to understand clinical needs, technology, and science of human-robot interaction behind robotic devices for assistance and rehabilitation, along with evidence and the implications related to the implementation of those devices in real therapy and daily life applications.

This book hopes to facilitate a conversation where all the rehabilitation actors and all roles in the discipline feel comfortable participating and contribute to the design of better tools to empower clinicians during the rehabilitation process.

Bogota, Colombia Carlos A. Cifuentes
Bogota, Colombia Marcela Múnera
May 2021

Acknowledgment

This book is based on the research and work of the Center for Biomechatronics at the Colombian School of Engineering Julio Garavito. The work presented in this book has been carried out with financial support from the Colombian Ministry of Science, Technology, and Innovation (Minciencias, Colombia), which funds the AGoRA Project, entitled Development of an Adaptable Robotic Platform for Gait Rehabilitation and Assistance (Grant 801-2017). AGoRA project is also funded in part by the Colombian School of Engineering Julio Garavito.

The authors would like to thank the healthcare institutions, clinicians, and patients involved in this research. Likewise, we are grateful to the international collaborators, without whom this work would not have been possible. Remarkably, Center for Assistive Technology at Federal University of Espirito Santo (Brazil); the Neural and Cognitive Engineering Group; the Center of Automation and Robotics and the Neural Rehabilitation Group, Cajal Institute, at Spanish Council for Scientific Research (Spain); the Institute of Automatics at National University of San Juan (Argentina); Club de Leones Cruz del Sur Rehabilitation Center (Punta Arenas, Chile); Brain-Machine Interface Systems Lab at Miguel Hernández University of Elche (Spain); BruBotics at Vrije Universiteit Brussel (Belgium); Hocoma (Switzerland); the School of Engineering, Science and Technology at Universidad del Rosario (Colombia); the Institute for Physiotherapy at Zurich University of Applied Sciences (Switzerland); the Department of Electrical Engineering at University of Magallanes (Chile); the National Hospital for Paraplegics (Spain); the Department of mechanical and Aerospace Engineering at Syracuse University (USA); the Centre for Advanced Robotics at Queen Mary University of London (UK); EPF Graduate School of Engineering (France); Institut de Biomécanique Humaine Georges Charpak (France); and Department of Mechanical Engineering at University College London (UK).

Furthermore, we would like to thank several students who helped us develop and implement fantastic and challenging ideas for the AGoRA Project of the Biomedical Engineering Program (Agreement between the Colombian School of Engineering and the Universidad del Rosario).

Contents

List of Figures

List of Tables

Introduction to Robotics for Gait Assistance and Rehabilitation

1

Sergio D. Sierra M., Luis Arciniegas-Mayag, Margarita Bautista, Maria J. Pinto-Bernal, Nathalia Cespedes, Marcela Múnera, and Carlos A. Cifuentes

1.1 Human Gait

The ability to freely move and interact within a specific environment provides people with autonomy and independence during daily life activities. In this way, human mobility is one of the essential capacities for proper development and well-being. In general terms, human gait is a complex behavior that involves the musculoskeletal system, the nervous systems, and the cardio-respiratory system [1, 2]. Human gait requires the central nervous system (CNS) activation, the transmission of electrical signals to the muscles, muscular activation, and sensory information feedback [3]. Altogether, these systems enable gait initiation, planning, and execution, while being adapted to satisfy motivational and environmental demands of the individual [4]. The ability to walk is usually acquired at the first years of life [5]. During this stage, skills such as body weight balancing and upright standing are learned [6]. Once the individual learns to walk, this ability becomes spontaneous and unconscious, becoming an energy-efficient task [5, 6].

S. D. Sierra M. · L. Arciniegas-Mayag · M. Bautista · M. J. Pinto-Bernal · M. Múnera
C. A. Cifuentes (✉)
Biomedical Engineering Department of the Colombian School of Engineering Julio Garavito, Bogotá D.C., Colombia
e-mail: sergio.sierra@escuelaing.edu.co; luis.arciniegas@mail.escuelaing.edu.co; laura.bautista-a@mail.escuelaing.edu.co; maria.pinto@mail.escuelaing.edu.co; marcela.munera@escuelaing.edu.co; carlos.cifuentes@escuelaing.edu.co

N. Céspedes
Centre for Advanced Robotics at Queen Mary University of London, London, England
e-mail: n.cespedesgomez@qmul.ac.uk

C. A. Cifuentes, M. Múnera, *Interfacing Humans and Robots for Gait Assistance and Rehabilitation*, https://doi.org/10.1007/978-3-030-79630-3_1

1.1.1 The Gait Cycle

The complexity of the human gait has become a topic of major interest within the field of human movement sciences. It is a key component in the investigation of pathological gait patterns. Human gait is a cyclic activity that can be described as a series of discrete events [7]. The gait cycle is often defined as the period of time from the initial contact of one foot with the ground to the following occurrence of the same event with the same foot (it is also known as stride). The gait cycle consists of two major phases: stance phase, which cover 60% of the cycle and swing phase that covers the remaining 40% [8]. **Stance phase** is the term used to designate the entire period during which the foot is on the ground. Whereas **swing phase** is defined as the period when the foot is in the air and the limb advances in preparation for subsequent foot contact [9]. The precise duration of these gait cycle intervals varies with the person's cadence. The duration of both gait periods is shortened as gait velocity increases and becomes progressively greater as speed slows.

The gait cycle provides a means of correlating the simultaneous actions of the individual joints into patterns of total limb function and delineate in an orderly manner their specific biomechanical functions [10]. To this end, each pattern motion is related to a different functional demand that is why a more detailed breakdown of gait cycle is required, in such cases, four, six, or even eight different phases have been considered depending on the specific type of application. The most detailed gait cycle classification recognized by the literature is illustrated in Fig. 1.1 [10]. Stance phase consists of five events: heel strike (HS), Flat Foot (FF), midstance (MST), Heel-Off (HO), and preswing (PS). Swing phase, on the other hand, consists of the other three events: Toe-Off (TO), midswing (MSW), and terminal swing (TSW). Each gait phase has a functional objective and a critical pattern of selective synergistic motion to accomplish it [11].

Heel strike also known as initial contact begins when the foot strikes the ground and marks the beginning of stance phase [11]. It is essential to highlight that for individuals with some pathologies, heel contact may not occur; hence, the term initial contact is more appropriated to be used. **Flat foot** or loading response is the first period of double limb support defined from HS (0%) to approximately 8–12% of the gait cycle [10, 11]. During this phase, the body absorbs the impact of the foot. The foot is with its entire length in contact with the ground, and the body weight is fully transferred onto the stance limb that acts as a shock absorbent resulting in knee flexion, coincident with load acceptance and deceleration of the body. The period from HS trough FF enables the limb to accomplish a basic task denominated as weight acceptance. It is the most demanding task in the stride, because three functional patterns are needed, shock absorption, initial limb stability, and control of forward progression [10–12].

Midstance begins when the contralateral foot leaves the ground, initiating opposite limb swing phase, and ends at the instant when the body center of mass is decelerating until it is aligned over the forefoot. MST represents the first half of single limb support, which covers approximately 8–12% to 30%. Single limb

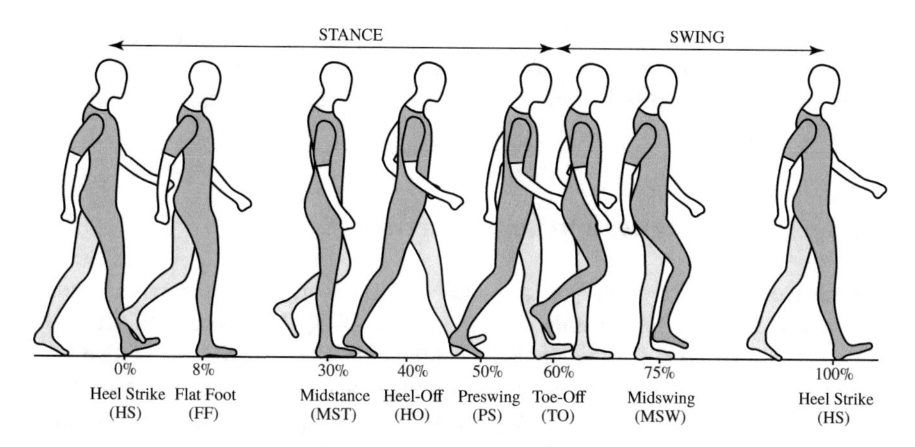

Fig. 1.1 Gait cycle is divided into seven phases. Illustrates seven periods that were summarized into one gait cycle based on 100% of the gait. One gait cycle can be described as a dynamic and continuous occurrence of seven phases from heel contact to the next heel contact. Different nomenclatures for the gait phases are used in the literature, this is the most used

support marks the period from MST through HO, when the opposite limb is in swing phase. During this period, one limb has the total responsibility for supporting body weight in both the sagittal and coronal planes. **Heel off** constitutes the second half of single limb support from 30 to 50% of the overall of the gait cycle [10–12]. This period begins at the time of heel leaves the ground and extends until the contralateral limb contacts the ground. During this event, the center of mass leads the forefoot and accelerates as it is falling forward towards the unsupported limb [11].

Preswing constitutes the last phase of the stance phase, i.e., it is the transition phase between stance phase and swing phase. During this event, the swinging leg acts as a compound pendulum [10, 13, 14]. The period of the pendulum is controlled by the mass moment of inertia. Variations in gait cadence are highly dependent on an individual's ability to alter the period of this pendulum. Hence, this period is associated with limb advancement [10]. Limb advancement begins in the final phase of stance and then continues through the entire swing period. It means that during limb advancement four gait phases are involved: preswing (end of stance), initial swing, mid swing, and terminal swing. This period has the purpose to meet the high demands of advancing the limb. Therefore preparatory posturing begins in stance. Then the limb swings through three postures as it lifts itself, advances and prepares for the next stance interval.

Toe off is a short period when the toes leave the ground like their name said. During this period, the stance limb is unloaded, and body weight is transferred onto the contralateral limb. It is a period of modulated acceleration that covers the time from 62 to 75% of the overall gait cycle and usually occupies one-third of the swing phase [10, 12]. **Midswing** begins at the moment the foot leaves the ground and continues until maximum knee flexion occurs, when the swinging limb is aligned with the contralateral limb, and ends when the swinging limb is in front of the stance limb. This period covers the middle third of swing phase from 75 to 87% of the overall gait cycle. **Terminal swing** is the last event of the swing phase from 87 to 100%, which initiates with vertical tibial alignment and continues until HS [10–12].

1.1.2 Gait Assistance and Rehabilitation

Despite the evident complexity of gait, individuals usually exhibit smooth, regular, stable, and repeated movements during walking [1, 3]. However, affectations to these systems could result in disorders or limitations to mobility [15]. Particularly, as the human gait gathers almost of the muscles of the body, as well as involves several cortical and sub-cortical structures, the training processes after physical and neurological injuries are usually challenging and long [5]. In this sense, the gait quality and locomotion capacities constitute important indicators of the overall health of an individual [16]. Thus, the presence of neurological alterations or musculoskeletal pathologies might lead to atypical gait patterns, weakness or loss of motor control [17].

The most common causes of gait impairments include neurological conditions (e.g., cerebrovascular accidents, spinal cord injuries, Parkinson disease, and Huntington disease), orthopedic disorders (e.g., osteoarthritis, skeletal deformities, and muscular dystrophy), and several medical conditions (e.g., coronary heart disease, respiratory insufficiency, and obesity) [2,15,18]. With aging, the risk factor of health conditions that affects well-being and overall autonomy is increased. This effect includes gait disorders [19]. Older adults often exhibit cardiovascular complications, musculoskeletal diseases, cancer, and impaired proprioceptive functions [20–22]. In general, these conditions result in cognitive and physical limitations and can cause the partial loss or degradation of the upper and lower-limbs' healthy functioning [23–25].

Physical rehabilitation aims at restoring people's movement and functioning affected by injuries, illness, or disability [26]. Gait rehabilitation and assistance therapies focus on providing, compensating, increasing, or re-training the lost locomotion capacities and the affected cognitive abilities of the individual [18]. In general terms, rehabilitation interventions seek to improve walking performance by (1) eliciting voluntary muscular activation in lower limbs, (2) increasing muscle strength and coordination, (3) recovering walking speed and endurance (i.e., usually accompanied with cardiovascular training), and (4) maximizing lower-limbs range of motion [27, 28]. In this sense, physical rehabilitation includes several techniques and approaches, ranging from overground and conventional gait training to robot-assisted and machine-based therapies [29, 30]. These strategies emphasize weight support, body mass propulsion, as well as on balance and postural control during movement [31].

Bearing in mind the above mentioned, this chapter presents an introductory overview of robotics for human gait rehabilitation and assistance. Advances in engineering and healthcare have led to the development of rehabilitation devices based on multiple robotic principles. Therefore, throughout this chapter, as well as in the remainder of this book, the focus will be on (1) Wearable Robotics, (2) Mobile Robotics, (3) Social Robotics, and (4) Combined Robotic Platforms with a particular emphasis on gait rehabilitation and assistance. Thus, the following sections present a definition of these fields of robotics, a description of how they can be applied to healthcare and physical training, and evidences of their use in rehabilitation therapies and gait assistance.

1.2 Wearable Robotics

The first type of device studied in this chapter used in gait assistance and rehabilitation is wearable robotic devices. This section introduces them, focusing on lower-limb exoskeletons. The following topics are presented: (1) a description of wearable robotics, (2) an introduction to lower-limb exoskeletons and their main objectives in the field of assistance and rehabilitation, (3) their evidence in gait rehabilitation scenarios, and finally (4) their evidence in gait assistance scenarios.

1.2.1 Defining Wearable Robotics

The robotic devices were used to support and develop some tasks that require different interaction levels with humans [32]. This Human–Robot Interaction (HRI) has been involved in activities where the robot has physical interaction with various parts of the human body [33]. Therefore, these robotic devices are being developed according to some standards that consider the user's safety when operating this device [34–36]. As a result, the definition of wearable robots emerges. As mentioned by *Pons* [37], "a wearable robot is person-oriented robots, that complement the limb primary movements or replace a human body limb". Currently, the development of wearable robot is applied to different areas according to the user's workspaces. For instance, applications where the user's body supports high loads for a long time [38,39], replacement of a human body limb [40], or robotic devices that complement or assist different human body movements [41–44].

As mentioned, the implementation of wearable robots involves the performance of a robotic device that supports and assists the user's limbs movements. In the first case, the wearable robots are used in clinical scenarios. The user wears the device in the therapy sessions to recover different primary motor functions to the human body. In the second case, the user cannot generate the required muscle activity to execute the limb's primary movements. For this reason, this device assists the limb's movements. The wearable robots focus on factors such as the shape, weight, and kinematic of the robotic device [37]. Each of these factors is directly dependent on the number the links and joints that provide kinematic compatibility [45, 46], the different types of physical interfaces to ensure a mechanical power direct transfer [47], and the type of actuation system that provides the backdriveability condition to the wearable robot [48–51]. Finally, the estimation of the physical Human–Robot

Interaction (pHRI) [52] allows the device to provide safety and comfort in support of different activities of daily living [53]. In this way, wearable robots are designed to apply the pHRI in various strategies depending on the user in developing different tasks in particular workspaces.

The assessment of the wearable is focused on critical aspects related to the functional performance applied on the interaction between the wearable robot and the user [54,55]; the user experience focused on the embodiment [56]; the wearable robot usability [57]; and the methodology that comprises the reproducibility of the experiments executed with the wearable robot controlled by the user, the acquisition and the facility in interpreting the results [58]. This device is assessed in two aspects related to the electrical and mechanical structure [36]. In this sense, some international organizations have raised several standards to guarantee user safety. For instance, the International Organization for Standardization (ISO) in collaboration with the Electrotechnical Commission (IEC) identifies the particular hazards presented in the Human–Robot Interaction. Where the IEC 62061 formulates a safety standard for robots and robotics personal care [34], the ISO 13482 "Robots and robotic devices—Safety requirements for personal care robots" [34], and the ISO 22523 "External limb prostheses and external orthoses—Requirements and test methods" [35] focused in the requirements and test methods for lower-limb exoskeletons.

Currently, there are various wearable robots that have fulfilled all the relevant factors mentioned, offering different services according to the population. Some of these devices are robotic prostheses, aimed at replacing the limb of a user who has suffered limb loss [59,60]; exoskeletons or orthoses focused on the industry whose purpose is to reduce the load perceived by the user [38,39]; and finally, exoskeletons concentrate on the rehabilitation and assistance of activities of daily living that currently have been developed for various joints of the human body [42,61,62]. The development of these exoskeletons includes the use of different instruments for data acquisition from the user. Finally, the development of different control strategies has allowed the user to participate in the therapy sessions or improve the user's quality of life with activity of daily living assistance.

1.2.2 Lower-Limb Exoskeletons

Lower-limb exoskeleton has been implemented for various applications related to increasing a person's motor capabilities. As mentioned by *Minchala et al.* defined a lower-limb exoskeleton as an anthropomorphic mechanical device that conforms to the user's anatomy. Where the movements generated by the device resemble various movements of the human body [63], therefore, its main objective is focused on three main aspects: increasing the strength of the human body [64–67], motor control rehabilitation of the human body [48, 68–71], and the assistance of the user's movements in various activities [42,61,71–73]. Lower-limb exoskeletons classification can be executed from various viewpoints starting with the different control strategies, workspace settings, tasks assisted by the device, and actuation systems. This book focuses on lower-limb exoskeletons developed for the rehabilitation of people suffering from gait-associated pathologies and gait assistance.

1.2.2.1 Lower-Limb Exoskeletons in Gait Rehabilitation

Lower-limb exoskeletons for gait rehabilitation are designed for the medical field where they are considered a tool for use in therapy sessions. The application of these devices fulfills two objectives. The first one is focused on supporting the recovery motor control [74] in the lower limbs by stimulating neuronal plasticity [75–77]. Second, the use of these robotic devices in rehabilitation decreases the therapist's workload to supervise the therapy session [68, 78–80]. In this way, the lower-limb exoskeletons support the user's movements and assist in the performance of the therapy sessions proposed by the therapist.

The exoskeleton for gait rehabilitation applied in indoor environments is integrated into a robotic platform complemented by bodyweight support [81, 82] and a treadmill [49, 83]. This lower-limb exoskeleton design provides an environmentally controlled working area in the development of the walking activity. Lower-limb exoskeleton is generally composed of different degrees of freedom (DoF), providing movement in the three human body planes. As a result, these rehabilitation platforms assimilate the user's lower-limb's natural movements during the walking activity. Additionally, each platform has a sensory interface to monitor the gait speed [83] and the gait cycle [84, 85], the joint's range of motion (RoM) [83], force/torque generated between the user and the wearable robot [50], the assistance level provided by the exoskeleton [81], and other kinetic and kinematic parameters to develop various control strategies, and to monitor the therapy session.

Different gait rehabilitation platforms use lower-limb exoskeletons to support the user's lower-limb movements in walking activity. For instance, *ALTACRO* (Vrije University, Belgium) comprised of the MACCEPA actuation system provides adaptable compliance during the walking activity using 12 DoF in Hip, knee, and ankle joints [49, 86]. *LOPES* (University of Twente, Netherlands) exoskeleton applies a bowden cable system actuation [78, 83, 87] to implement the "Robot in charge", "Patient in charge", and the "Therapist in charge" modalities [83]. As a result, the exoskeleton allows movements in the leg and a free 3-D translation of the pelvis [87]. *ALEX* (University of Delaware, USA) applies force fields to guide the user's foot trajectory into a prescribed gait pattern [82, 88]. Finally, a commercial rehabilitation such as *Lokomat* [68] is classified as a powered gait orthosis that can increase therapy session's intensity and reduce the therapist burden and healthcare cost [89]. Figure 1.2a presents a lower-limb exoskeleton schematic for gait rehabilitation for indoor environments.

On the other hand, the literature presents different lower-limb exoskeleton types of gait rehabilitation where the user executes the walking activity in outdoor environments using a lower-limb exoskeleton [90]. In this sense, the device's design changes compared to the exoskeletons mentioned in the last paragraph. The new design of these exoskeletons presents a portable design. This characteristic includes a portable power supply, processing module, and a sensory interface that the user

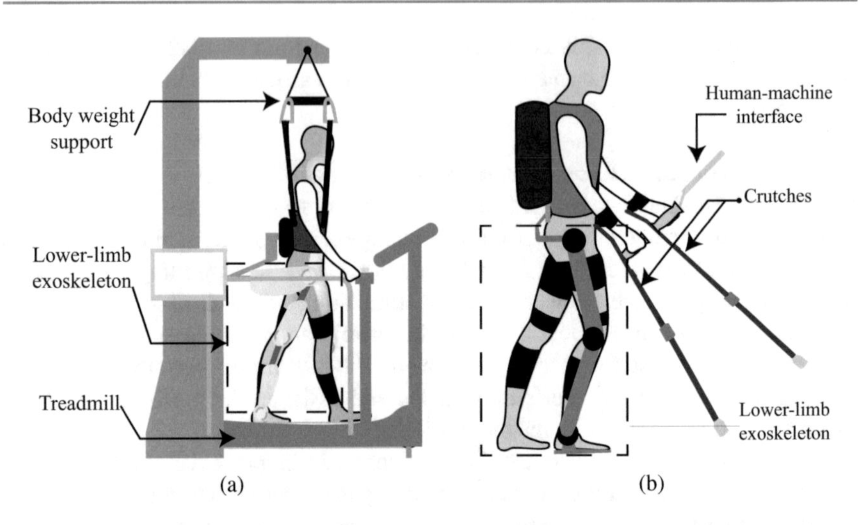

Fig. 1.2 Lower-limb exoskeleton schematics; (**a**) lower-limb exoskeleton for gait rehabilitation in indoor environments example. Generally, this platform comprises a bodyweight support, Treadmill, and a lower-limb exoskeleton designed with various DoF that covers the movements in different human body planes; (**b**) lower-limb exoskeleton for gait rehabilitation/assistance in outdoor environment example. This wearable robot is composed of a lower-limb exoskeleton that generally supports the sagittal plane movements, instrumented crutches, and a human–machine interface

could transport in various environments. Additionally, aspects such as the actuation system size [91], exoskeleton size [92], weight [74], and the kinematic configuration [45,65] are taken into account to improve the interaction between the wearable robot and the user [93]. To fulfill the main objective of rehabilitating the walking activity in outdoor environments, some of the above aspects, such as design considerations regarding the number of DoF, generally, focus on the human body sagittal plane. Finally, considering that these exoskeletons do not have a bodyweight support, in some cases, the lower-limb exoskeletons are complemented using walkers [94] and crutches [95] to provide support and stability during the walking activity.

Currently, lower-limb exoskeletons have been implemented in the rehabilitation of activities of daily living employing different methodologies. For instance, the *ALLOR* (Federal University of Espirito Santo, Brazil) is a unilateral knee exoskeleton that changes the impedance components using virtual damping according to a detected gait phase during the walking activity [69, 94]. *BioMot* (Future and Emerging Technologies (FET), Spain) is a bilateral lower-limb exoskeleton that uses a MACCEPA to change the system stiffness through the Hook law [70]. *BLEEX* exoskeleton (University of California Berkeley, USA) now known as *Ekso GT* was designed in the rehabilitation and assistance of stroke patients and spinal cord injury, respectively [71, 91, 96, 97]. Figure 1.2b shows an example of a lower-limb exoskeleton for the rehabilitation of activities of daily living.

The exoskeletons mentioned above are based on a rigid mechanical structure to transfer the calculated torques to the users' lower-limbs. However, the latest developments presented in the literature show different other wearable robot based on soft technology. In this field, the exosuits have been designed to fulfill the lower-limb exoskeleton purpose offering a lightweight device and an actuation system that generate the required torque profiles in the user's lower limbs. For instance, *XoSoft* (Fondazione Istituto Italiano di Tecnologia, Italy) and *Myosuit* (ETH Zurich, Switzerland) are exosuits designed for activity of daily living rehabilitation using cable-driven actuation to complement the lower-limb movements [48, 98–100]. As a result, implementing these exosuits in therapy allows different joint lower-limb movements and decreases the exoskeleton weight.

1.2.2.2 Lower-Limb Exoskeletons in Gait Assistance

Another field where the lower-limb exoskeletons have been applied is in gait assistance. Compared to the exoskeleton mentioned in Sect. 1.2.2.1, these lower-limb exoskeletons' primary objectives are to provide a high level of assistance in the human body's primary movements [90, 95]. Figure 1.2b shows examples of lower-limb exoskeleton design for this end. For this reason, the control strategies proposed in these devices have not considered the force/torque between the user and the wearable robot [42, 71, 101]. In this case, the user that wearers the lower-limb exoskeleton operates the wearable robot. Therefore, these devices propose various human–robot interfaces such as wrist-watch style controller [42, 61], external computers [73], smartphone app [74], among others. The user can mobilize in outdoor environments employing this category of lower-limb exoskeletons. In this sense, some characteristics of lower-limb exoskeletons for activity of daily living rehabilitation are applied in these exoskeletons. For instance, actuation system size, exoskeleton size, weight, and kinematic configuration are considered. Parameters such as kinematic the joint angular position, the joint angular velocity are estimated, and other kinematic and kinetic parameters are required for control strategies applied in these devices [42, 72, 91, 101].

Lower-limb exoskeletons for assistance have been presented in the literature and some of these devices are patented for commercialization. Various of these devices demonstrated their feasibility and performance *CYBATHLON* [102], often mentioned in the literature as the Paralympic Games, where the paraplegic users compete in the development of various activities of daily living. As an example *ReWalk* (ReWalk Robotics, United Kingdom) [61, 103], *REX* (REX Bionics, New Zealand) [42], *Indego* (Vanderbilt University, USA) [42], *VariLeg* (ETH Zurich, Switzerland) [57, 73, 104], *WalkON* (Sogang University, South Korea) [72] are highlighted in these applications focused on the assistance of walking activity, sitting/standing activities, ascending/descending stairs, and walking on sloping surfaces.

Table 1.1 shows some relevant characteristics of the exoskeletons for gait assistance and rehabilitation reviewed in this section. There were mentioned some lower-limb exoskeletons classified by the category (gait rehabilitation/gait assistance), degree of freedom for each limb, the bilateral/unilateral exoskeleton, and some activities of daily living that the exoskeletons provide a user's support and assistance.

Table 1.1 Lower-limb exoskeletons general description

Exoskeleton	Category	Degree of freedom	Main function	Activity of daily living
ALTACRO [49, 86]	Gait rehabilitation	12 DoF (hip, knee, and ankle).	Rehabilitation (Bilateral exoskeleton)	Walking
LOPES [78, 83, 87]	Gait rehabilitation	3 DoF (hip and ankle).	Rehabilitation (Bilateral exoskeleton)	Walking
ALEX [82, 88]	Gait rehabilitation	9 DoF (trunk, hip, knee, and ankle).	Rehabilitation (Unilateral exoskeleton)	Walking
ALLOR [69, 94]	Gait rehabilitation	3 DoF (hip, knee, and ankle).	Rehabilitation (Unilateral exoskeleton)	Walking, Sitting, Standing
Myosuit [48, 99, 100]	Gait rehabilitation	NA	Rehabilitation (Bilateral exoskeleton)	Walking, Sitting, Standing
BioMot [70]	Gait rehabilitation	3 DoF (hip, knee, and ankle).	Rehabilitation (Bilateral exoskeleton)	Walking, Sitting, Standing
ReWalk [61, 103]	Gait assistance	3 DoF (hip, knee, and ankle).	Assistance (Bilateral exoskeleton)	Walking, Sitting, Standing, Ascending stairs, Descending stairs
VariLeg [57, 73, 104]	Gait assistance	3 DoF (hip, knee, and ankle).	Assistance (Bilateral exoskeleton)	Walking, Sitting, Standing, Ascending stairs, Descending stairs
WalkOn [72]	Gait assistance	3 DoF (hip, knee, and ankle).	Assistance (Bilateral exoskeleton)	Walking, Sitting, Standing, Ascending stairs, Descending stairs
Anklebot [105]	Gait assistance	2 DoF (ankle).	Rehabilitation (Unilateral exoskeleton)	Walking

In conclusion, this section presented the general definition of a wearable robot and relevant aspects for their development. For example, lower-limb exoskeletons are the product of the development of wearable robots that have been developed as tools for rehabilitation and assistance for users with pathologies associated with the lower limbs. Some characteristics that identify exoskeletons designed for rehabilitation and assistance were shown. Finally, the exoskeletons developed in the last decades were mentioned, considering the difference between exoskeletons designed for the rehabilitation environment and the user's assistance environment.

1.3 Mobile Robotics

The second type of device used in gait assistance and rehabilitation therapies is mobile robotics. This section presents an introduction, concentrating on robotic or smart walkers. The following topics are presented: (1) a description of several mobile conventional devices, (2) a description of the smart walkers, from their overall structure to their functioning and interaction channels, (3) the evidence of smart walkers use in gait assistance and rehabilitation, with several applications in clinical scenarios, and finally (4) alternative mobile robots for gait rehabilitation, where other assistance platforms based on mobile robots will be showcased.

1.3.1 Defining Mobile Assistive Devices

Mobile devices for gait assistance often include ambulatory training tools and wheeled-based structures. For instance, these devices exploit the widely studied benefits of wheeled mobile robots, such as stability, balance, power autonomy, mechanical simplicity, among others. Mobile assistive devices intend to overcome physical limitations by maintaining or improving individuals' functioning and independence in clinical and everyday scenarios [106]. Moreover, these devices can be classified into conventional and robotic devices. The most common conventional mobile devices are manual wheelchairs, walking sticks or canes, walking frames, and rollators [107] (See Fig. 1.3).

On the one hand, manual wheelchairs are a standard solution when lower-limb locomotion capacities are entirely lost. In their most straightforward configuration, these devices consist of a chair fitted with wheels, where the user is entirely in charge of the device's propulsion [23]. The wheelchairs also provide proper fit and postural support based on each user's biomechanical and environmental requirements. On the other hand, the walking sticks or canes are simple assistive devices that aim to increase the patients' support base and improve their balance [108]. To provide

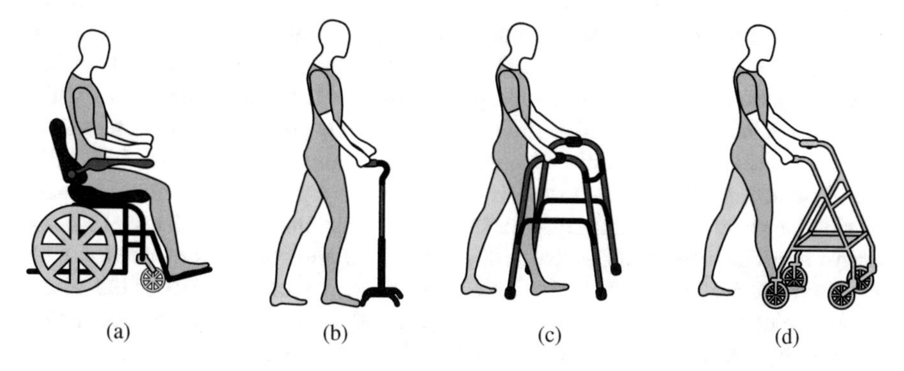

Fig. 1.3 Illustration of standard mobile conventional devices based on wheeled or ambulatory frames. (**a**) Manual wheelchairs. (**b**) Walking sticks. (**c**) Walking frames. (**d**) Rollators

weight support to the user, some variations include more than one support point to the floor. In this sense, multiple-legged canes or quad-canes are commonly found [108].

Finally, the walking frames or walkers are constituted by simple mechanical structures that improve overall balance, increase the users' base of support, enhance lateral stability, and provide partial weight-bearing [108, 109]. The walkers offer support and use the patients' remaining locomotion capability to move [109, 110]. These devices hold a rehabilitation potential as they encourage physical activity and social participation of people with mobility impairments [27]. Several types of conventional walkers can be found.

The standard walkers or walking frames are constituted by a four-legged frame with rubber tips that should simultaneously contact the floor [108, 110]. Although this configuration offers maximum stability, the patient must lift the frame and moved it forward during walking. In this sense, the use of standard walkers results in slower, and often abnormal gait patterns [108]. Moreover, this type of walker requires some degree of upper body strength and cognitive ability [110]. The front-wheeled walkers include wheels at the front legs making them more suitable for upper-limbs weakness [108]. These walkers reduce the risk for falling as lifting the device is not required, and it promotes forward displacements of the center of gravity [110]. The front-wheeled walkers facilitate more normal gait patterns and higher walking speeds [110]. The four-wheeled walker or rollators included wheels with pivot and rolling points requiring fewer users' effort [108, 110]. Since these walkers provide minor stability, they are often used by higher functioning patients with minimal weight-bearing requirements [108, 110]. These walkers are intended to be used in community scenarios during daily living activities, as they provide the most natural gait patterns and highest walking speeds [110]. These devices are usually equipped with shopping baskets, a resting seat and hand brakes [110]. These rollators are also found in three-legged or u-shaped configurations [109].

Conventional walkers are widely used by older adults or neurological patients with high independence levels [18]. However, several issues restrict their implementation in rehabilitation settings, complex scenarios, or patients requiring higher degrees of assistance [107]. On the one hand, several studies have reported that the conventional walking frame does not ensure enough safety during walking, since there is a considerable risk for falls [111, 112]. Moreover, the users' natural balance and energetic costs are often compromised with conventional walkers [107]. On the other hand, the user-walker interaction is entirely passive, so that the walker does not provide any additional physical and cognitive support [18, 107, 110]. For instance, patients with visual impairments may require assistance for safe navigation or guiding. Similarly, patients with reduced muscle capacity may require active assistance from the device [107, 113]. Other shortcomings associated with conventional walkers involve the inability of these devices to monitor the users' condition or to track the quality of users' gait. Additionally, these devices cannot provide any feedback to the users when there is an inappropriate interaction.

1.3.2 Smart Walkers

As previously stated, the walking frames or rollators exploit the patients' remaining locomotion capacities to provide gait assistance. They are usually prescribed for patients in need of assistance during functional daily living tasks [107]. Similarly, literature evidence shows that walker-assisted gait often elicits essential psychological benefits, including increased confidence and safety perception during ambulation [27, 107]. However, critics regarding the use of these devices often point out the lack of safety measures, the increased energy expenditure, and the inability to track users' progress and health status.

Bearing in mind the above mentioned, the possibility of improving and enhancing the functionalities of the conventional walkers through the inclusion of robotic technology has arisen. In general, there have been several research projects focused on creating robotic versions of canes, wheelchairs, and, particularly, walkers. The terms "robotic", "intelligent", or "smart walkers" (SWs) refer to those conventional walking frames and rollators that have been empowered with sensing interfaces, actuation technologies, and control strategies [27]. The inclusion of these technologies in the smart walkers allows providing more efficient and robust gait rehabilitation and assistance [27]. There are several benefits brought by the integration of technology and robotics, for instance (1) the execution of precise and repeatable tasks, (2) the implementation of intensive activities with programmable and measurable difficulty, (3) the online measurement of the performance and physiological state of the patients, (4) the implementation of more engaging rehabilitation environments through the use of virtual and augmented reality, (5) the assessment of the patients' rehabilitation progress, and (6) the reduction of the physical effort of the therapists [25, 28, 31, 114].

More specifically, thanks to the sensors, actuators, and control strategies implemented in the SWs, they are capable of multiple high-level functions that can be grouped in the following categories:

- **Biomechanical and Health Monitoring:** The smart walkers can gather information collected by their internal sensors and external sensors worn by the user. In this sense, the smart walkers can estimate biomechanical indicators from sensors such as Inertial Measurement Units (IMUs), Electromyography sensors (EMG), or sensors mounted on the device such as ultrasonic boards, Laser Rangefinders (LRF), cameras, among others [27, 107]. These sensors allow the estimation of gait spatiotemporal parameters such as cadence, speed, step length, stride length, gait symmetry, among others [115]. These parameters are relevant as they may help tracking and quantifying the users' rehabilitation progress. Moreover, control strategies can also be implemented using this information [113, 116, 117]. Finally, the patients also wear heart rate sensors, oxygen saturation sensors, and other physiological sensors to estimate the overall health status.
- **Estimation of Movement's Intention:** An essential functionality that the smart walkers' should provide is identifying the users' intentions to move. This func-

tionality allows the smart walkers to guarantee intuitive and natural interaction to their users. To this end, the smart walkers are often equipped with sensors on the forearm supports and handlebars to estimate the patients' intention of movement [110]. The most common approaches exploit the resulting force and torque exerted on the walker by the user to generate linear and angular reference velocities [107, 118]. Other techniques are based on cognitive interaction, where the user can control the smart walker without physically interacting with it, using voice commands or body gestures [18, 113].

- **Guidance and Navigation:** In the robotics field, navigation basically refers to a set of systems that allow a robot to safely move from one point to another within a specific environment [119]. In walker-assisted gait, navigation refers to safely guiding the users through different settings, while satisfying their social and motivational demands [107]. To this end, the smart walkers require obstacle detection and avoidance techniques, map building modules, autonomous localization algorithms, and path following strategies [107]. These functionalities can be implemented whether independently or in conjunction with user interaction modules, so that shared control strategies can be designed to regulate the users' role during therapy [27, 120]. In this way, guidance and navigation modules help with patients with cognitive requirements or visual impairments.
- **Safety Provision:** The smart walkers often assist the users in complex environments where dynamic situations might occur. Therefore, smart walkers equip redundant systems to react to hazardous situations rapidly. These modules commonly use rule-based algorithms, where different conditions and constraints are established to limit or stop the smart walker's motion [107]. For instance, these modules allow the device to be blocked in presence of stairs, glass walls, and dangerous walkways. Using distance sensors, these modules can detect obstacles in front of the device and limit the speed depending on the proximity to the obstacles [27, 120]. Moreover, smart walkers are often remotely monitored and controlled by healthcare professionals. Thus, they can limit or stop the device's movement in emergencies. Finally, along with these functionalities, the smart walkers also implement fall prevention strategies based on IMU sensors.
- **Feedback Strategies:** The smart walkers implement multiple communication channels to provide cognitive assistance or communicate relevant information to the users. For example, during navigation tasks in patients with visual impairment, the smart walkers use haptic and auditory feedback to guide them [118, 120]. Other implementations are based on visual feedback to let the users know their performance and status during the therapy [110, 121].
- **Remote control:** Gait rehabilitation therapies demand close accompanying of physiotherapists to provide postural corrections and therapy monitoring [55, 122]. In this sense, several smart walkers have implemented remote control strategies, so that the healthcare professionals can remotely assess the session data, override the device or control the smart walker's behavior [107, 123].

According to their end purpose, smart walkers can be found in multiple configurations depending on their actuators' type, the implemented sensory interfaces, the implemented functionalities and their mechanical structure. The smart walkers' propulsion interface can be classified into active, passive or hybrid devices. This classification simply suggests if the propulsion is entirely accomplished by the user (i.e., passive), aided by actuators on the device's wheels (i.e., active), or co-accomplished by the user and the device (i.e., hybrid). These topics will be further studied in the following sections and chapters.

1.3.3 Smart Walkers in Gait Assistance and Rehabilitation

Considering the above functionalities, the smart walkers have also proven helpful in gait assistance and rehabilitation therapies. Depending on each user's specific needs, these functionalities can be adjusted to provide targeted tasks to meet the users' clinical and personal requirements. In this sense, literature evidence shows that smart walkers have been successfully implemented in the following scenarios:

- **Smart walkers to provide stability and motion support:** Similar to the conventional type, the smart walkers can provide partial body weight support, physical stability, and balance in active [107, 124] or passive configurations [120, 125]. It is worth mentioning that smart walkers in passive modality require a considerable amount of postural control and walking ability by the user. Thus, passive smart walkers can provide motion support in later stages of rehabilitation or home-based scenarios. The smart walkers in active modality require less energetic input from the user and can offer automatic context-aware propulsion. Active smart walkers tend to be used in the early stages of rehabilitation and in clinical scenarios. In both cases, the ability to provide motion support makes the smart walkers useful for individuals with partial mobility loss, presenting different residual motor capacity levels [126]. Moreover, passive or active smart walkers help the patients gain or increase their independence during daily tasks and may positively impact on self-esteem and social interaction [27, 126].
- **Smart walkers to provide functional and daily tasks assistance:** The smart walkers are commonly characterized by forearm supports equipped with force sensors and pressure sensors, or devices to extract users' intentions to move (e.g., joysticks, voice recognition modules) [27, 121]. The information extracted from such sensors (e.g., force signals) can be converted into navigation or velocity commands to make the smart walker move according to the users' motivational demands [107, 118, 127]. This information is also used to detect undesired behaviors such as leaning more to one side of the walker, which is common in patients with hemiplegia or hemiparesis [128, 129]. In this sense, the smart walkers can cope with the users' intentions to assist them during their daily living activities. Moreover, several robotic walkers use this information to detect the choice to perform sit-to-stand or stand-to-sit transfers by the user [130, 131]. The strategies implemented to aid the user during these transfers range from

(1) simple activation of the braking system, to (2) active control of the smart walker in a forward direction to pull up the user, and (3) implementation of robotic supporting elements (e.g. forearm or chest support) to lift and guide the user [131]. Finally, smart walkers can also detect slopes, stairs or hazardous environments employing sensors such as laser rangefinders, ultrasonic sensors and cameras. With this information, the smart walkers can provide active propulsion to overcome rugged terrains such as slopes or ramps or completely stop their motion to avoid colliding with obstacles in the environment [107].

- **Smart walkers to provide guiding and safety:** As previously stated, smart walkers can perform safe navigation and obstacle detection in multiple environments. These functionalities can be used to guide people with visual or cognitive impairments. Different robotic walkers use auditory feedback to indicate the obstacles or the path to follow in guiding tasks [124]. Similarly, other proposals include haptic feedback by causing the device to vibrate when there is an obstacle nearby or making the device heavier when the user deviates from the route [118, 120]. Visual feedback has also been explored in the literature, which uses lights, screens or virtual reality to indicate the route [118, 132].
- **Smart walkers to monitor health status and gait quality:** The smart walkers can extract information from sensors worn by the users and onboard sensors [107]. This information is often processed to extract gait-related indicators such as speed, cadence, stride length, cycle duration, gait symmetry, among others [115]. The constant estimation of these indicators at every rehabilitation session allows the healthcare professionals to track the users' progress and assess their gait quality. Moreover, the smart walkers also receive information related to heart rate, skin impedance, oxygen saturation, among others. The data provided by these sensors is of great relevance since it allows to monitor the overall health status of the users and detect emergencies [27].

As presented in *Sierra et al.*, several smart walkers and on-going research projects involve the functionalities described above [107]. An updated and detailed description of some of these smart walkers is presented in Chap. 4. In the following section, some other devices based on mobile robots are presented.

1.3.4 Alternative Mobile Robots for Gait Rehabilitation

In case of total mobility impairment exhibited by the patient, several assistive devices based on mobile robots have been developed. Specifically, the conventional four-wheeled differential robots are commonly used to implement assistive devices such as robotic wheelchairs and autonomous vehicles for patients transportation (See Fig. 1.4).

Robotic or smart wheelchairs employ actuators, sensory interfaces and advance processing algorithms to provide easier and safer navigation [133]. Moreover, considering the patients' requirements, the robotic wheelchairs may include multimodal input interfaces, such as joysticks, touchscreens, voice recognition modules, image

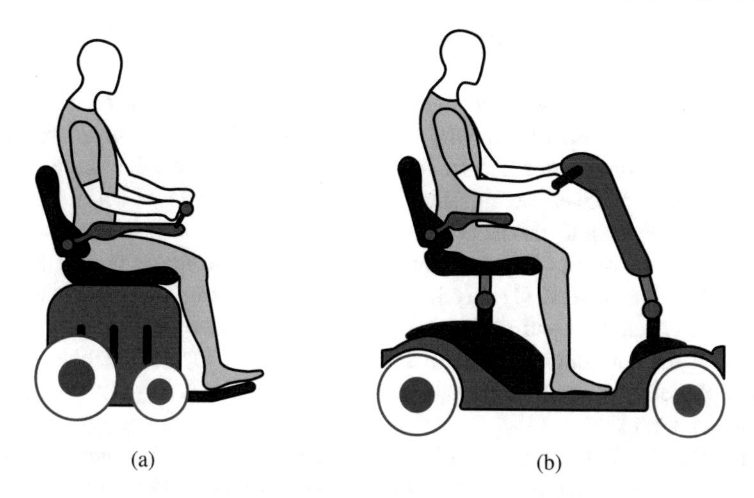

Fig. 1.4 (**a**) Illustration of a robotic wheelchair mounted on a four-wheeled mobile robot. (**b**) Illustration of an autonomous vehicle based on a robotic scooter

processing systems, and biosignals monitoring modules (e.g., electromyography and electroencephalography) [133].

Robotic scooters, standing vehicles and lifting robots are more commonly found around us [110]. These devices include several actuators and multimodal user interfaces to allow intuitive control and interaction. Moreover, these devices may be equipped with lifting mechanisms to provide sit-to-stand capabilities and partial body weight support [134].

At this point, the previous sections have highlighted and explained the benefits of employing wearable and mobile robotics in gait assistance and rehabilitation. However, there are other scenarios where robotics can foster physical training and gait assistance. In particular, patients can also benefit from social robots, in mobility and independence problems related to aging, cognitive syndromes, cardiac complications, depression, among others.

1.4 Social Robotics

In this section, a brief introduction to social robotics is presented. To achieve the complete understanding of social robots, this section provides: (1) the social robotics definition, where different definitions given to social robots are presented, (2) evidence of this type of robots in a healthcare context, and (3) applications of socially assistive robotics in gait assistance and rehabilitation.

1.4.1 Defining Social Robotics

Several definitions are given to social robotics through time. However, it is essential to clarify the meaning of *"social"*. This term represents two or more entities involved in the same context [135]. To illustrate a social robot's definition, different researchers' perspectives are currently used [136]. First, in 1999 *Duffy et al.* [135] defined a *social robot* as robots that can interact among more robots (*social robots*) or with humans (*societal robots*). The social robots according to these authors are composed of four layers: (1) physical layer, where the robot has a physic aspect within an environment and perform motor skills, (2) reactive layer to answer the stimulus of the environment through sensory interfaces, (3) deliberative layer that involves Beliefs-Desires Intention structure, and (4) the social layer in charge of the communication [135].

Afterward, in 2002 *Fong et al.* [137] define social robots as: *embodied agents* that are part of a heterogeneous group (robots or humans), and they can recognize and engage in social interactions. Moreover, *Fong et al.* established that social robots have environmental perceptions acquired from their own experience, i.e., they could learn a communicate within contexts. Thus, in 2004 *Bartneck and Forlizzi* [138] stated the *Design-centered Approach* of a social robot. Overall, they defined a social robot as an autonomous/semi-autonomous platform that interacts with humans following behavioral norms as in the human–human interaction. *Bartneck and Forlizzi* proposed two conditions to consider a platform as a social robot: (1) the robot has to be autonomous, it can interact cooperatively or non-cooperatively depending on the situation, (2) recognize human values, roles among others [138].

Finally, in 2018 *Breazeal et al.* [139] presented the sociable robot concept as a robot able to communicate and understand humans in social terms. In turn, human beings should be able to communicate and emphasize with *sociable robots*. In this case, *Breazeal et al.* established five fundamental characteristics to consider a robot sociable: (1) the embodiment in a situated manner (real or virtual), (2) has to have lifelike qualities, (3) the robot has to able of identifying the users, to create an interpersonal manner [139]. For instance, *Theory of mind* and *empathy* are essential to design human awareness. (4) The human has to understand the social robot's role, and (5) the robot has to learn social situations (by imitation or mimicry), to shape its history.

1.4.2 Social Robots in Healthcare

Socially Assistive Robots (SAR) are being tested and accepted in healthcare areas, such as rehabilitation and clinical assistance. Social care sectors beyond the traditional scope of surgical and rehabilitation robots are poised to become one of the most significant technological innovations of the twenty-first century [140]. Social robots have been developed and used in multiple clinical settings and home-based areas, providing physical, cognitive and social support, as well as coaching activities, such as exercise training, education, and monitoring [141].

In this field, researchers are focused on developing social robots to perform tasks with a high degree of autonomy, while holding natural interaction with the patients and the clinical staff [142]. This has enabled social robots to provide support in healthcare scenarios through social interaction. Social robots are attractive and engaging to a wide range of children. Children often perceive social robots as something in between a pet and a friend [143]. This makes social robots an interesting play therapy tool, where children can take safe risks to learn new skills and abilities [143].

According to literature evidence, SAR have been studied in mainly four healthcare scenarios with children [141]:

- **Social robots to help emotionally cope with illness:** They provide self-management tools as they help inform children about their medical conditions (e.g., cancer, diabetes) appropriately. Robots like *NAO* (SoftBank Robotics, France) are commonly used in this scenario, a significant emotional support tool for children [144, 145].

- **Social robots to support therapy or interventions in children with Autism Spectrum Disorder (ASD):** They have been used as active agents of reinforcement in semi-structured behavior for children with ASD. They increase eye contact [146, 147], improve and develop children's visual perspective-taking skills [148], enhance their joint attention [149], improve proprioception skills, improve in the production and recognition of children's facial emotions [150]. Robots like *NAO*, *CASTOR* (Colombian School of Engineering, Colombia) [151, 152], *Robonova* (Hitec Robotics, South Korea) [153], *Probo* (Vrije Universiteit Brussel, Belgium) [154], *Ono* (Ghent University, Belgium) [149], and *Kaspar* (University of Hertfordshire, UK) [148] are commonly used in ASD therapies.

- **Social robots to enhance well-being during inpatient stays:** They provide a positive effect on a child's well-being through distraction, engagement, or a positive impulse on communication [155–159]. Pet-like robots such as *Paro* (AIST, Japan) and *Huggable Bear* (MIT, USA), or humanoid robots such as *NAO* are commonly used in this scenario.

- **Social robots that provide distraction during a medical procedure** help decrease children's anxiety and stress levels [158, 160, 161]. The robot most used in this scenario is *NAO*, which has a promising potential as a distraction during a medical procedure, especially in vaccination.

Other compelling opportunities for social robots are in the context of eldercare. Regarding this population, their interaction includes educating, facilitating the older adults' communication and social connection with others, and assisting with adherence to care regimen through social support [162]. Specifically, in the adults and older population, SAR have been used in two healthcare scenarios mainly [141]:

- **Social robots as companions for older adults with disease:** They are widely used in mental health to aid and support loneliness, depression and anxiety [141]. Companion robots aim to enhance older people's health and psychological factors by providing companionship [163]. Robots like *Paro*, *NAO*, *Pepper* (Softbank Robotics, France), and *Buddy* (Blue Frog Robotics, France) are currently used for this purpose, especially in elderly care for dementia and physical rehabilitation [164].

- **Social robots as service robots in elderly care and well-being:** They are defined as assistive devices designed to support daily activities providing more independence to the users [163]. Service robots are mainly used to support older adults in home environments and healthcare centers. The most used robots in these areas are *HOBBIT* (Vienna University of Technology, Austria) [165], *RiSH* (Oklahoma State University, USA) [166], *Robot Era* (Project FP7/2007-2013) [167], and *RAMCIP* (Project EU Horizon 2020/643433) [168].
- **Social robots for exercising, coaching, and rehabilitation:** They act as companions or assistants in specific scenarios, with the aim of improving user performance or increasing motivation during certain tasks. Some scenarios in which these social robots have been shown to play a fundamental role are cardiac rehabilitation and physical training. In both cases, humanoid robots have been used due to their similarity to human beings. For example, *NAO* [169–172], as shown in Fig. 1.5, has shown positive results by increasing people's motivation to continue with their rehabilitation process. And *ROBOVIE* (Advanced Telecommunications Research Institute International, Japan) has shown positive results encouraging users to perform physical activity [173].

To summarizing, most of the studies reported positive effects of social robots within the older population. Nevertheless, some studies highlighted some issues and limitations about this technology [174], e.g., the novelty effect is decreased in some studies as the robots can become repetitive and predictable [175]. This issue can affect social interaction, which is reduced by time.

The studies identified in this section indicate that social robots can develop healthcare and well-being roles (i.e., assistance, companion, partner or coach/instructor). Social robots could be a valuable tool in healthcare personnel's repertoire to support children, adults, and the elderly in a medical environment that deals with stresses and loneliness [141]. Besides, in the children's population social robots might have the potential for engaging children, distracting, openness, develop their visual prospective-taking skills and decrease the level: of anger, fear, anxiety, and depression, among others. Such effects have also been noted for adults. The older under intervention with social robots, users showed improvements in social connections, communication, mood, and diminished loneliness, isolation, depression, and anxiety [141].

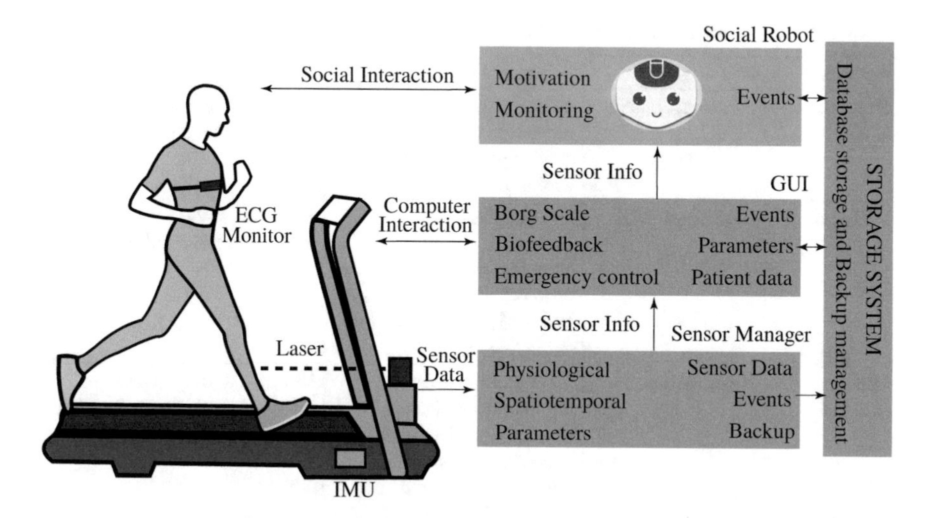

Fig. 1.5 Architecture of the integration of a *NAO* social robot in the cardiac rehabilitation setting. This humanoid robot acts as a companion, improving user performance and increasing motivation during therapy

1.4.3 Social Robots in Gait Assistance and Rehabilitation

As presented previously, social robots have been developed and assessed in healthcare areas, showing positive effects regarding motivation, adherence and engagement to the treatments. Furthermore, in gait rehabilitation and assistance, social robotics have to support procedures from different approaches (see Table 1.2).

For instance, to assist elderly individuals in gait rehabilitation *Scheidig et al.* [176] developed a *ROGER* robot platform. This platform was used to assist elderly patients after surgery in hip endoprosthetic. *ROGER*'s role was to lead the patients during the therapy session and measured their gait patterns. The results showed that the platform was reliable in performing the gait assessments. Based on these results, *Trinh et al.* [177] conducted a study in which the *ROGER* platform was validated in more detail with 20 elderly patients. They considered three evaluation aspects: robust collision avoidance, user-centered navigation, and reliable person perception. The results suggested that the patients felt safe and motivated using the robot. Besides, *ROGER* was a valuable tool for the clinical staff as they managed and controlled the sessions. However, in regards to navigation, the results showed that further improvements are still needed for fully autonomous training, such as increasing the positional accuracy of the 3D obstacle and person detection, especially during situations where the robot is very close to the user.

Similarly, *Piezzo et al.* [178] presented a *Pepper* robot's feasibility study to guide elderly individuals during walking. Qualitative study and gait analysis through the video analysis were performed to measure patients' response to the robot's role. Results showed that elderly patients trust the robot, and intrinsic motivation was

Table 1.2 Social robots for gait rehabilitation and assistance scenarios

Authors	Robot/type	Objectives	Robot's role
Piezzo et al. [178]	*Pepper*/Humanoid	To design and develop a humanoid robot to guide elderly patients during walking.	Support, monitor and motivate older patients during walking activities.
Buitrago et al. [181]	*NAO*/Humanoid	To use a SAR as an alternative and complementary method to promote the participation and motivation of children with Cerebral Palsy.	Support, provide feedback and encourage the children during the session.
Malik et al.[182]	*NAO*/Humanoid	To determine how a humanoid robot can be used as an assistive technology in specific therapy for children with Cerebral Palsy.	Motivate and keep the children's engagement during therapy.
Sheidig et al.[176]	*ROGER*/Humanoid Mobile	To validate the ROGER robot use in a real-world scenario.	Coach in therapies for gait rehabilitation. Assess the gait patient's performance.
Trinh et al.[177]	*ROGER*/Humanoid Mobile	To integrate real-time gait pattern analysis into a robotic application.	Support, monitor, give feedback, and employed different task-specific during the session.
Cespedes et al.[179, 180]	*NAO*/Humanoid	To design and evaluate a social human–robot interaction system to support conventional therapies with Lokomat.	Motivate, monitor the patient's performanc, and provide feedback during the treatment.

influenced positively. Furthermore, the robot also aids the patients to complete their goals during the sessions (e.g., walking from a goal point to another). Moreover, the *NAO* robot has also been used to assist elderly patients in gait rehabilitation. Such can be illustrated in the research carried out by *Cespedes et al.* [179]. A human–robot interaction system that integrates sensors to measure physiological variables (heart rate, cervical and thoracic inclinations) and the exertion level perceived (Borg scale) was developed. This system allowed collaborating with the therapist's tasks to decrease their work burden and motivate them to improve their self-performance through rehabilitation. The results suggest that the social robot has a positive and well-received effect on the robot regarding companionship, social interaction. Following this work, a long-term study showed that the robot's support improves the patients' physiological progress by reducing their unhealthy spinal posture time. Most patients described the platform as helpful and secure, with positive acceptance [180].

On the other hand, as far as assistance to children in gait rehabilitation is concerned, the most widely used social robot is also the *NAO* robot. An illustration of this is the research conducted by *Buitrago et al.* [181], which developed a motor learning therapeutic intervention using a social robot *NAO* for a child diagnosed with cerebral palsy. The researchers remark on the robot's capability to facilitate the child's persistence in walking and achieve the therapeutic objectives. Even they stated that after the fifth session, the children reached the proposed goal in the gait training. On the other hand, social robots are also used to support gait rehabilitation of patients who went under surgery procedures and need training during recovery phases. Therefore, *Malik et al.* [182] used the robot *NAO* in specific therapy for children with cerebral palsy. This consisted of four interactive scenarios in human–robot interaction based on the Gross Motor Functional Measure measurement items. However, the researchers reported a positive impact during therapy and effective engagement between child and robot. During the study, some mechanical errors occurred, e.g., the child did not understand what the robot said or answered the question correctly, but *NAO* did not detect the answer.

To summarizing, most of the studies reported positive effects of social robots to assist gait rehabilitation. They highlighted that patients feel more motivated, safe and comfortable during the session and, in turn, improved their rehabilitation performance during therapy. However, there are some issues and limitations about this technology regarding their configuration and operating setup in some gait rehabilitation scenarios.

1.5 Combined Platforms

The integration of different robotic devices, such as those mentioned throughout this chapter, has shown promising results in clinical applications, allowing to improve and accelerate the recovery of lost or diminished functions. As a result of this integration, this section presents an innovative concept called combined platforms, highlighting their use in gait assistance and rehabilitation. The following topics are presented: (1) a description of combined robotic platforms, (2) an overview of their deployments in healthcare, (3) the evidence of the use of these platforms in gait assistance and rehabilitation, with some conclusions resulting from their application in clinical scenarios, and finally (4) the general characteristics of existing combined platforms, where both electronic and mechanical characteristics will be described.

1.5.1 Defining Combined Robotic Platforms

As presented, several types of robotic devices, such as smart walkers, active joint orthosis, and lower and upper body exoskeletons, have been developed to provide rehabilitation, assistance or augmented physical capabilities in different scenarios (e.g. clinical, industry and military) [74]. Mainly, in the clinical setting, most of these devices focus on maintaining or improving an individual's functioning and independence to facilitate participation and to enhance overall well-being [183]. Besides, in some cases, they can prevent deterioration and secondary health conditions such as bowel or bladder problems, depression, overweight, obesity, among other things [184, 185].

Because of the above, these devices have been integrated or combined to enhance and maximize their effects in rehabilitation therapies [27], being of great help for gait assistance and rehabilitation. That is why integrating two or more two robotic assistive devices can be defined as combined platforms, which aim to overcome and compensate for the physical limitations of those who suffer from them, applying to both clinical and everyday scenarios [106].

Although few studies refer to combined platforms, this new approach seeks to introduce new and reliable technologies in the rehabilitation process [186], offering interesting advantages such as the possibility of automated and personalized treatments reducing the fatigue associated with repetitive and monotonous exercises [187]. In addition to its ability to integrate sensors that provide a quantitative estimate of recovery [183].

For example, since 2006, a small number of combined platforms have been developed primarily aimed at gait rehabilitation in patients with neurological disorders [27, 188–190], and the clinical evaluation of these motor impairments using measures such as joint range of motion, strength, muscle reflexes, muscle activity and coordination [191, 192]. Among the studies already published, the *AGoRA* combined platform (Adaptable Robotic for Gait Rehabilitation and Assistance), shown below (see Fig. 1.6), stands out for being a clear example of a novel and affordable combined platform composed of two assistive devices [27].

The combined platform concept is best exemplified in Fig. 1.6 with the *AGoRA* platform, composed of a lower-limb exoskeleton and a smart walker for mobility assistance and gait rehabilitation. Although this platform is one of the most recent examples, other similar combined platforms are found in the healthcare literature.

1.5.2 Combined Robotic Platforms in Healthcare

The deployment of combined platforms in the health area is considered a potential tool to improve the disabled and older population [193]. And even though robotic assistive devices are focused on helping humans, they will not replace humans' roles in the setting [194]. In other words, these platforms have been introduced into the healthcare field to complement conventional therapeutic interventions, becoming an alternative treatment to improve the quality of life of older adults, persons with physical limitations or disabilities and those around them [195, 196].

These combined platforms' clinical impact can be seen when the affected person can re-enter the workforce, reduce the burden on caregivers, and live at home, rather than in long-term care facilities. Some other benefits include prevention of medical complications and improved self-image and life satisfaction [193]. Therefore, in recent years, researchers have proposed and developed robotic platforms integrated by various robotic assistive devices. These efforts involved

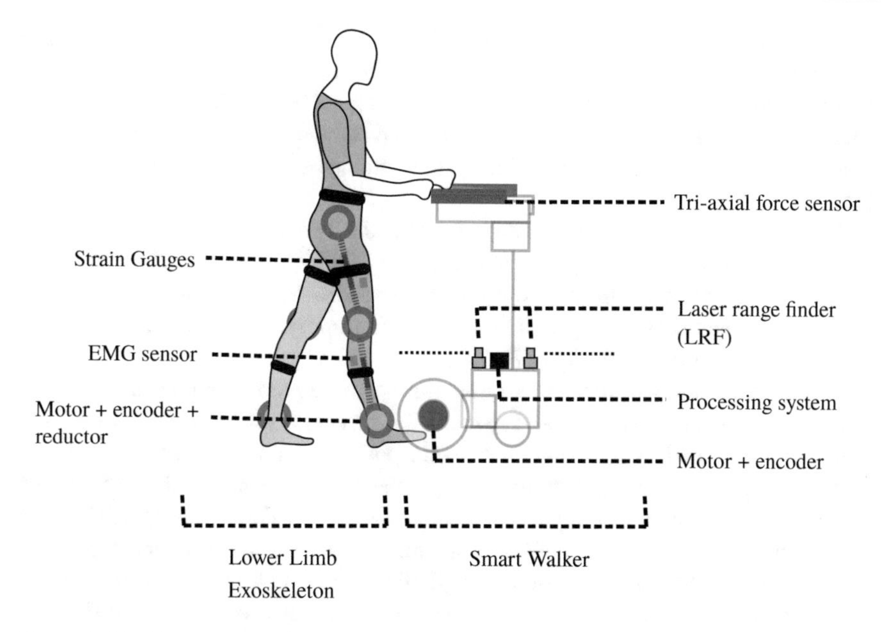

Fig. 1.6 *AGoRA* combined platform comprises a lower-limb exoskeleton and a smart walker oriented at gait rehabilitation of patients with mobility impairments

large and complex robotic systems, which evolved into more affordable systems that are both functional and aesthetically pleasing [193, 194].

Considering the above, most platforms today are integrated by an exoskeleton and a smart walker to assist people with reduced mobility. Next is a brief description of the existing combined platforms. Their use results are presented, allowing to offer a vision of the current situation of this new approach and the benefits that these can provide.

1.5.3 Combined Robotic Platforms in Gait Assistance and Rehabilitation

Combined platforms have shown positive effects in the area of rehabilitation and assistance in developing activities of daily living, helping to improve the independence and well-being of those who use them [81]. One of the first combined platforms developed was *EXPOS* [188]. This platform was proposed in 2006 by Sogang University as a viable solution to allow older people to live without the need for physical assistance from third parties. Furthermore, due to its ability to assist in the human body's movement, this platform has potential use in the gait rehabilitation of patients with neuromuscular diseases.

As a combined platform, *EXPOS* consists of a lower-limb exoskeleton, which is as light as possible for comfort, and a smart walker, which is heavy enough to keep the body in balance when held by the user [188, 197, 198]. Although this platform typically works on flat-controlled surfaces within the laboratory, it can be adapted for everyday use. As a qualitative result, after using *EXPOS* in the laboratory, a user stated that this platform helped him walk, sit, and stand quite well, demonstrating its viability for the assistance of the elderly and patients with neuromuscular diseases [188].

Nine years later, in 2015, researchers from the Federal University of Espírito Santo proposed a combined platform composed of a lower-limb exoskeleton and a smart walker. This platform makes it possible to re-establish neuromotor control in subjects with neurological injuries, in addition to improving mobility and safety of those who use it while walking [199, 200].

As a combined platform, the lower-limb exoskeleton provides an assistive torque that alleviates the load and reduces the muscular effort. The smart walker guides the user and helps maintain a stable posture. As a differential factor, the researchers proposed a new control strategy based on recognizing the intention of human movement by analyzing biomedical signals such as brain signals (EEG) and surface myoelectric signals (sEMG), which turned out to be valuable and applicable in future work. So far, this combined platform does not report quantitative or qualitative results, although it is recognized as an innovative robotic system that can be adjusted to fatigue situations or according to the evolution of the user's rehabilitation.

In 2017, researchers proposed the *CPWalker* platform aimed at gait training in children with cerebral palsy [189, 201]. As a combined platform, the *CPWalker* consists of a smart walker with bodyweight support and an exoskeleton for joint motion support [81], allowing the child to experience autonomous locomotion in a natural rehabilitation environment. Unlike the previous platforms presented, the use of *CPWalker* shows notable improvements in several physical skills as strength, stability, cadence, mean velocity, step length and symmetry in gait patterns [55]. All these improvements were achieved in the short term, so this research and its methodology could serve as an example for future clinical implementations of any robotic assistive device.

More recently, in 2018, Colombian researchers proposed and developed a new combined platform called *AGoRA*, aimed at gait assistance and rehabilitation [27, 202–204]. As shown in Fig. 1.6, *AGoRA* consists of a lower-limb exoskeleton with a bioinspired design [205], and a smart walker [202]. Although as a combined platform it has not been tested, both devices that compose it have already been assessed separately. For instance, stationary approaches such as the use of an ankle exoskeleton for motor recovery [206], the use of a lower-limb exoskeleton for knee rehabilitation [207], and their evaluations during walking have shown improvements in spatiotemporal and kinematic parameters, as well as their usability and performance through a natural interaction between users and devices [43, 46]. This platform will integrate the measurement of kinematic, physiological, and cognitive parameters to monitor the patient's condition, his evolutionary rehabilitation process and evaluate the effects of the platform's assistance. *AGoRA* platform presents interesting capabilities, such as the integration of the clinician the control loop

[208] and multimodal capabilities to enhance the navigation and the interaction with the environment [107]. In this sense, it is being studied the integration of cloud computing capabilities to empower the sensor processing and control strategies [209].

During that same year, Colombian researchers proposed a new combined platform that, unlike the previous ones, integrates a social robot and a robotic gait orthosis. Since then, the researchers have been developing innovative social robot interfaces to empower gait rehabilitation in patients with spinal cord injury or stroke [179, 180, 210]. This study is focused on studying the effectiveness of socially assistive robotics during gait training with Lokomat after suffering a neurological disease to maximize the probability of success of the rehabilitation. Different variables such as heart rate, spinal posture, spatiotemporal parameters, and perceived exertion are measured, providing a feedback mechanism through a socially assistive robot. As a result of the long-term study, two patients who suffered spinal cord injury have presented remarkable improvements in motor, cognitive and emotional processes [180].

All the platforms mentioned above have the main objective of being an effective tool for the assistance and rehabilitation of gait through robot-assisted therapy, especially in patients with neuromuscular disorders or also in older adults [211–214]. For this reason, these platforms have become a new alternative to help people with motor disabilities to recover or compensate for the loss of motor control, and re-establish their independence and well-being [215, 216].

In addition to sharing the same objective, these platforms have similarities at a structural level. Next, general aspects of their mechanical and electronic architecture will be delved into understanding their operation better.

1.5.4 General Features of Existing Combined Platforms

A combined platform's construction requires a high level of specialization in various scientific areas such as electronic control, mechanical design, and ergonomics [217]. The joint work of these different areas for the assembly of robotic devices for gait assistance and rehabilitation has advantages over conventional devices. It offers the user the opportunity to move autonomously, increasing his motivation and improving his physical capacities [7, 199, 218].

Some robotic devices such as exoskeletons and smart walkers, which are shown in Fig. 1.7, have presented promising results since they are equipped with different types of electronic sensors, such as kinematic sensors (angular position, speed, and acceleration) and kinetic (interaction force between the limb and each robotic device), which allow collecting relevant information to know the user's evolution or to control the robotic device [219–221]. This last idea has become an innovative feature implemented by some existing robotic devices, ensuring better controllability by analyzing biosignals such as EEG and sEMG for human motion intention detection [199, 218].

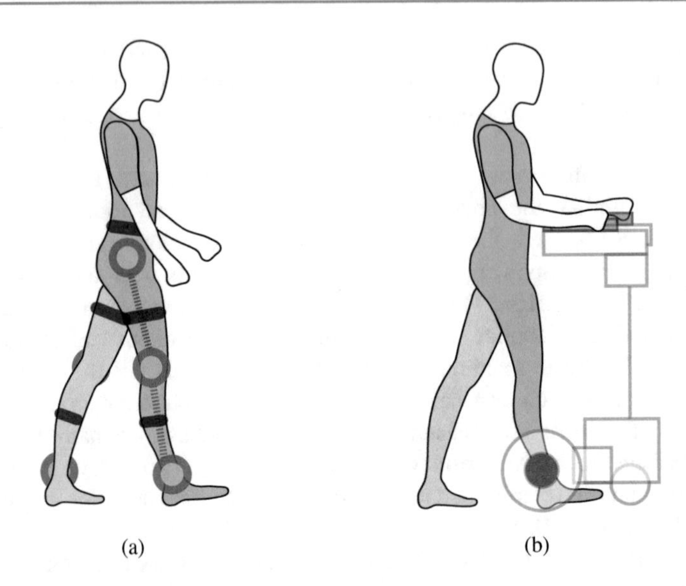

Fig. 1.7 Illustration of the most common robotic devices that make up the currently existing combined platforms. (**a**) Robotic lower-limb exoskeleton. (**b**) Smart robotic walker

That is why, considering the electronic operation and potential of these robotic devices, most of the existing combined platforms for mobility assistance and gait rehabilitation integrate an exoskeleton and a smart walker. On a mechanical level, as shown in Fig. 1.7a, exoskeletons are designed for the lower extremities, providing the ability to assist and support the gait process. Depending on the pathological conditions or the needs of each user, the exoskeletons can be unilateral or bilateral. Besides, the ergonomic design allows them to be adapted according to each user's anthropometry [219, 221].

On the other hand, most smart walkers, as shown in Fig. 1.7b, have been developed on mobile platforms, allowing to guarantee a gait pattern that is as natural as possible [110, 218, 222]. Furthermore, considering that its movement is carried out on wheels, less force must be exerted to move it, since no lifting is necessary [110, 218, 223, 224]. However, since the wheels can run freely, this type of walker requires from the user better control and a good balance [110, 218, 223–225].

These general mechanical and electronic aspects make integrating two or more than two robotic assistive devices a potential, innovative, and affordable solution to help people with motor disabilities regain or compensate for the loss of motor control. Although most existing combined platforms for gait rehabilitation integrate an exoskeleton and a robotic walker, rapid advances in the technology area will make it possible in the short term to innovate and integrate new platforms with different devices that are increasingly lighter, smarter, more sophisticated, and aesthetically pleasing.

1.6 Scope of the Book

This book is organized in 13 thematic chapters (**Chaps. 2–14**), addressing relevant robotic technologies that have been integrated into healthcare and physical training environments to provide gait assistance and rehabilitation. **Chapter 2** introduces the basic concepts of kinematic modelling, actuation systems, and sensing architectures of the different type of robotic devices here explained: lower-limb exoskeletons, social robots, and smart walkers. That chapter reviews the most common configurations for both commercial and research devices on the field.

Chapter 3 introduces the design process for lower-limb exoskeletons, from which the main modules and considerations of design can be extracted to analyze all other robotic devices presented. Through the definition of user-centered and device-centered features, the platforms are broken down into several modules that are addressed in the next chapters.

Similarly, **Chap. 4** explores the design of smart walkers. From physical structures to sensory interfaces and control strategies, that chapter presents the different smart walkers developed until now, and their main characteristics.

Chapter 5 describes some of the most relevant spatial and temporal indicators that are used to characterize human gait, and therefore, track and report a patient's rehabilitation progress, and detect anomalies in their gait patterns. That chapter also presents some wearable sensors that allow the acquisition of such indicators and methodologies that make use of them to develop high-level controllers in wearable and mobile robotic devices.

Chapter 6 focuses on the module of actuators, specifically in the area of flexible and soft actuators for the development of assistive robotic devices. It presents the steps involved in their characterization and the understanding of their capacities and limitations to correctly implement them in assistive and rehabilitation applications. The chapter finishes with concrete examples of robotics devices that include each type of actuation.

Chapter 7 centers in the field of flexible actuators and presents an overview of the variable stiffness actuators in terms of their principles, setups, and characteristics. The chapter introduces *T-FLEX*, an ankle exoskeleton based on this technology, and shows two preliminary case studies with healthy participants using this device.

Having the sensor and actuator modules explained. **Chap. 8** explores control strategies in lower-limb exoskeletons. Two controllers are developed for the *AGoRA* exoskeleton based on the principle of impedance, as methodologies for gait assistance. The controllers are implemented in two case studies.

Chapter 9 introduces the relevance of using the Brain–Computer Interface for neurorehabilitation and the stages in the universal design of a BCI system. That chapter also analyzes some of the most relevant works related to BCI-based control for lower-limb exoskeletons. Finally, a study case of a stroke survivor commanding *T-FLEX* through a BCI interface is presented.

Similar to **Chap. 8**, **Chap. 10** presents control strategies for smart walkers. These strategies contemplate smart walkers' interaction with both humans and the environment and allow smart walkers to respond to the user's movement intentions, to guide a user between two points or along a trajectory, among many others behaviors.

Chapter 11 explores the incorporation of socially assistive robotics to gait rehabilitation scenarios. That chapter provides an introduction to the benefits of including social interaction in rehabilitation processes, through relevant characteristics of social robotics and their role in patient-robot interfaces.

Chapter 12 introduces one of the new trends in this challenging field of rehabilitation robotics. Serious games in robot-assisted therapies are covered in terms of their interaction technology and feedback and incentive strategies. That chapter presents the development of *Jumping Guy*, a serious game for ankle rehabilitation therapy with *T-FLEX*.

After understanding of the different modules and strategies available in the robotic devices, **Chap. 13** addresses the methods, metrics, and equipment used to assess their performance in gait assistance and rehabilitation scenarios. It starts by explaining the basic concepts to characterize a motor skill and finishes with practical examples of assessments in the field.

Finally, **Chap. 14** is devoted to the compilation and evaluation of the experiences of clinicians using the rehabilitation robotics presented along the book. That chapter collects several techniques that can be used to analyze clinicians and patients opinions and to quantify this qualitative information. The book concludes with an analysis of the importance of using these techniques and the new opportunities that arise for robotics in rehabilitation and clinical programs.

References

1. C.L. Vaughan, Theories of bipedal walking: an odyssey. J. Biomech. **36**, 513–523 (2003)
2. W. Pirker, R. Katzenschlager, Gait disorders in adults and the elderly. Wien. klin. Wochenschr. **129**, 81–95 (2017)
3. L.M. Decker, F. Cignetti, N. Stergiou, Complexity and human gait. Rev. Andal. Med. Deport. **3**(1), 2–12 (2010)
4. A.S. Buchman, P.A. Boyle, S.E. Leurgans, L.L. Barnes, D.A. Bennett, Cognitive function is associated with the development of mobility impairments in community-dwelling elders. Am. J. Geriatr. Psychiatry **19**(6), 571–580 (2011)
5. É. Watelain, Human gait: from clinical gait analysis to diagnosis assistance. Mov. Sport Sci. **98**(4), 3 (2017)
6. K.E. Adolph, W.G. Cole, M. Komati, J.S. Garciaguirre, D. Badaly, J.M. Lingeman, G.L.Y. Chan, R.B. Sotsky, How do you learn to walk? Thousands of steps and dozens of falls per day. Psychol. Sci. **23**, 1387–1394 (2012)
7. A. Kuo, M. Donelan, Dynamic principles of gait and their clinical implications. Phys. Ther. **90**, 157–174 (2009)
8. Z.O. Abu-Faraj, G.F. Harris, P.A. Smith, S. Hassani, Human gait and clinical movement analysis, in *Wiley Encyclopedia of Electrical and Electronics Engineering* (Wiley, Hoboken, 2015), pp. 1–34
9. C. Vaughan, B. Davis, J. O'Connor, The three-dimensional and cyclic nature of gait, in *Dynamics of Human Gait*, vol. 168 (Kiboho Publishers, Cape Town, 1999), pp. 16–17

10. J. Perry, J.R. Davids et al., Gait analysis: normal and pathological function. J. Pediatr. Orthop. **12**(6), 815 (1992)
11. J.R. Gage, Gait analysis. An essential tool in the treatment of cerebral palsy. Clin. Orthop. Relat. Res. **288**, 126–134 (1993)
12. M.W. Whittle, Clinical gait analysis: a review. Hum. Mov. Sci. **15**(3), 369–387 (1996)
13. R. Hicks, S. Tashman, J. Cary, R. Altman, J. Gage, Swing phase control with knee friction in juvenile amputees. J. Orthop. Res. **3**(2), 198–201 (1985)
14. S. Tashman, R. Hicks, D. Jendrzejczyk, Evaluation of a prosthetic shank with variable inertial properties. Clin. Prosthet. Orthot. **9**(3), 23–28 (1985)
15. I. Carrera, H.A. Moreno, S. Sierra, A. Campos, M. Munera, C.A. Cifuentes, Technologies for therapy and assistance of lower limb disabilities: sit to stand and walking, in *Exoskeleton Robots for Rehabilitation and Healthcare Devices*, chap. 4 (Springer, Berlin, 2020), pp. 43–66
16. S. Schmid, K. Schweizer, J. Romkes, S. Lorenzetti, R. Brunner, Secondary gait deviations in patients with and without neurological involvement: a systematic review. Gait Posture **37**, 480–493 (2013).
17. M. Roberts, D. Mongeon, F. Prince, Biomechanical parameters for gait analysis: a systematic review of healthy human gait. Phys. Ther. Rehabil. **4**(1), 6 (2017)
18. C.A. Cifuentes, A. Frizera, *Human-Robot Interaction Strategies for Walker-Assisted Locomotion. Springer Tracts in Advanced Robotics*, vol. 115 (Springer International Publishing, Cham, 2016)
19. A.H. Snijders, B.P. van de Warrenburg, N. Giladi, B.R. Bloem, Neurological gait disorders in elderly people: clinical approach and classification. Lancet Neurol. **6**(1), 63–74 (2007)
20. R. Gheno, J. Cepparo, C. Rosca, A. Cotten, Musculoskeletal disorders in the elderly. J. Clin. Imaging Sci. **2**(3), 39 (2012)
21. MedlinePlus, Bone diseases (2016). https://medlineplus.gov/bonediseases.html
22. S.N. Nabili, C.P. Davis, Senior health: successful aging (2019)
23. Disabled World, Physical and Mobility Impairment: Information & News (2019). https://www.disabled-world.com/disability/types/mobility/
24. G. Sammer, T. Uhlmann, W. Unbehaun, A. Millonig, B. Mandl, J. Dangschat, R. Mayr, Identification of mobility-impaired persons and analysis of their travel behavior and needs. Transp. Res. Rec. J. Transp. Res. Board **2320**, 46–54 (2012)
25. T. Mikolajczyk, I. Ciobanu, D.I. Badea, A. Iliescu, S. Pizzamiglio, T. Schauer, T. Seel, P.L. Seiciu, D.L. Turner, M. Berteanu, Advanced technology for gait rehabilitation: an overview. Adv. Mech. Eng. **10**, 168781401878362 (2018)
26. National Health Service UK, Physiotherapy (2018). https://www.nhs.uk/conditions/physiotherapy/
27. S. Sierra, L. Arciniegas, F. Ballen-Moreno, D. Gomez-Vargas, M. Munera, C.A. Cifuentes, Adaptable robotic platform for gait rehabilitation and assistance: design concepts and applications, in *Exoskeleton Robots for Rehabilitation and Healthcare Devices* (Springer, Berlin, 2020), pp. 67–93
28. S.L. Chaparro-Cárdenas, A.A. Lozano-Guzmán, J.A. Ramirez-Bautista, A. Hernández-Zavala, A review in gait rehabilitation devices and applied control techniques. Disabil. Rehabil. Assist. Technol. **13**(8), 819–834 (2018)
29. R.A. States, E. Pappas, Y. Salem, Overground physical therapy gait training for chronic stroke patients with mobility deficits. Cochrane Database Syst. Rev. **2009**(3), CD006075 (2009)
30. A. Pollock, G. Baer, P. Campbell, P.L. Choo, A. Forster, J. Morris, V.M. Pomeroy, P. Langhorne, Physical rehabilitation approaches for the recovery of function and mobility following stroke. Cochrane Database Syst. Rev. **2014**(4), CD001920 (2014)
31. L.R. Sheffler, J. Chae, Technological advances in interventions to enhance poststroke gait. Phys. Med. Rehabil. Clin. N. Am. **24**, 305–323 (2013)
32. S. Haddadin, E. Croft, Physical human–robot interaction, in *Springer Handbook of Robotics* (Springer International Publishing, Cham, 2016), pp. 1835–1874
33. A.Q. Keemink, H. van der Kooij, A.H. Stienen, Admittance control for physical human–robot interaction. Int. J. Robot. Res. **37**, 1421–1444 (2018)

34. ISO 13482:2014(en) Robots and robotic devices — Safety requirements for personal care robots (2014). https://www.iso.org/obp/ui/#iso:std:iso:13482:ed-1:v1:en
35. ISO 22523:2006 External limb prostheses and external orthoses - Requirements and test methods (2006). https://www.iso.org/standard/37546.html
36. J.F. Veneman, Wearable robotics: challenges and trends, in *Challenges and Trends. Biosystems & Biorobotics* (2016), pp. 189–193
37. L. Bueno, F. Brunetti, A. Frizera, J.L. Pons, J.C. Moreno, E. Rocon, J.M. Carmena, E. Farella, L. Benini, Human–robot cognitive interaction, in *Wearable Robots* (Wiley, Chichester, 2008), pp. 87–125
38. E. Bionics, EVO (2020). https://eksobionics.com/ekso-evo/
39. Lockheed Martin, FORTIS (2014). https://www.lockheedmartin.com/en-us/products/exoskeleton-technologies/industrial.html
40. R. Alturkistani, A. Kavin, S. Devasahayam, R. Thomas, E.L. Colombini, C.A. Cifuentes, S. Homer-Vanniasinkam, H.A. Wurdemann, M. Moazen, Affordable passive 3D-printed prosthesis for persons with partial hand amputation. Prosthet. Orthot. Int. **44**(2), 92–98 (2020)
41. H. Lee, P.W. Ferguson, J. Rosen, Lower limb exoskeleton systems—overview, in *Wearable Robotics*, Chap. 11 (Academic, London, 2020)
42. R. Bionics, REX (2019). https://www.rexbionics.com/
43. D. Gomez-Vargas, F. Ballen-Moreno, P. Barria, R. Aguilar, J.M. Azorín, M. Múnera, C.A. Cifuentes, The actuation system of the ankle exoskeleton T-FLEX: first use experimental validation in people with stroke. Brain Sci. **11**(4), 412 (2021)
44. A. Peñas, J. Maldonado, O. Ramos, M. Munera, P. Barria, M. Moazen, H.A. Wurdemann, C.A. Cifuentes, Towards a fabric-based soft hand exoskeleton for various grasp taxonomies, in *The International Symposium on Wearable Robotics (WeRob2020) and WearRAcon Europe* (2020)
45. J. Beil, C. Marquardt, T. Asfour, Self-aligning exoskeleton hip joint: kinematic design with five revolute, three prismatic and one ball joint. IEEE Int. Conf. Rehabil. Robot. **2017**, 1349–1355 (2017)
46. F. Ballen-Moreno, C. Cifuentes, T. Provot, M. Bourgain, M. Munera, 3D relative motion assessment in lower-limb exoskeletons: a case of study with AGoRA exoskeleton, in *The International Symposium on Wearable Robotics (WeRob2020) and WearRAcon Europe* (2020)
47. A.F. Ruiz, A. Forner-Cordero, E. Rocon, J.L. Pons, Exoskeletons for rehabilitation and motor control, in *Proceedings of the First IEEE/RAS-EMBS International Conference on Biomedical Robotics and Biomechatronics, 2006, BioRob 2006*, vol. 2006 (2006), pp. 601–606
48. K. Schmidt, J.E. Duarte, M. Grimmer, A. Sancho-Puchades, H. Wei, C.S. Easthope, R. Riener, The Myosuit: bi-articular anti-gravity exosuit that reduces hip extensor activity in sitting transfers. Front. Neurorobot. **11**, 1–16 (2017)
49. P. Cherelle, V. Grosu, P. Beyl, A. Mathys, R. Van Ham, M. Van Damme, B. Vanderborght, D. Lefeber, The MACCEPA actuation system as torque actuator in the gait rehabilitation robot ALTACRO, in *2010 3rd IEEE RAS & EMBS International Conference on Biomedical Robotics and Biomechatronics* (IEEE, Piscataway, 2010), pp. 27–32
50. W. Huo, S. Mohammed, J.C. Moreno, Y. Amirat, Lower limb wearable robots for assistance and rehabilitation: a state of the art. IEEE Syst. J. **10**, 1–14 (2016)
51. M. Bortole, A. del Ama, E. Rocon, J.C. Moreno, F. Brunetti, J.L. Pons, A robotic exoskeleton for overground gait rehabilitation, in *2013 IEEE International Conference on Robotics and Automation* (IEEE, Piscataway, 2013), pp. 3356–3361
52. A. Schiele, F.C. van der Helm, Influence of attachment pressure and kinematic configuration on pHRI with wearable robots. Appl. Bionics Biomech. **6**(2), 157–173 (2009)
53. R. Alami, A. Albu-Schaeffer, A. Bicchi, R. Bischoff, R. Chatila, A. De Luca, A. De Santis, G. Giralt, J. Guiochet, G. Hirzinger, F. Ingrand, V. Lippiello, R. Mattone, D. Powell, S. Sen, B. Siciliano, G. Tonietti, L. Villani, Safe and dependable physical human-robot interaction in anthropic domains: state of the art and challenges, in *IEEE International Conference on Intelligent Robots and Systems* (2006)

54. E. Pirondini, M. Coscia, S. Marcheschi, G. Roas, F. Salsedo, A. Frisoli, M. Bergamasco, S. Micera, Evaluation of the effects of the Arm Light Exoskeleton on movement execution and muscle activities: a pilot study on healthy subjects. J. NeuroEng. Rehabil. **13**(1), 1–21 (2016)
55. L.F. Aycardi, C.A. Cifuentes, M. Múnera, C. Bayón, O. Ramírez, S. Lerma, A. Frizera, E. Rocon, Evaluation of biomechanical gait parameters of patients with Cerebral Palsy at three different levels of gait assistance using the CPWalker. J. NeuroEng. Rehabil. **16**, 15 (2019)
56. M. Pazzaglia, M. Molinari, The embodiment of assistive devices-from wheelchair to exoskeleton. Phys. Life Rev. **16**, 163–175 (2016)
57. J.T. Meyer, S.O. Schrade, O. Lambercy, R. Gassert, User-centered design and evaluation of physical interfaces for an exoskeleton for paraplegic users, in *2019 IEEE 16th International Conference on Rehabilitation Robotics (ICORR)* (2019), pp. 1159–1166
58. D. Torricelli, C. Rodriguez-Guerrero, J.F. Veneman, S. Crea, K. Briem, B. Lenggenhager, P. Beckerle, Benchmarking wearable robots: challenges and recommendations from functional, user experience, and methodological perspectives. Front. Robot. AI **7**, 168 (2020)
59. Ossur, Proprio foot. https://www.ossur.com/en-us/prosthetics/feet/proprio-foot
60. Ottobock, Empower (2020). https://www.ottobockus.com/products/empower-ankle/
61. A. Esquenazi, M. Talaty, A. Packel, M. Saulino, The ReWalk powered exoskeleton to restore ambulatory function to individuals with thoracic-level motor-complete spinal cord injury. Am. J. Phys. Med. Rehabil. **91**(11), 911–921 (2012)
62. Myoswiss, Myosuit (2020). https://myo.swiss/myosuit/
63. L.I. Minchala, F. Astudillo-Salinas, K. Palacio-Baus, A. Vazquez-Rodas, Mechatronic design of a lower limb exoskeleton, in *Design, Control and Applications of Mechatronic Systems in Engineering* (2017)
64. B.J. Makinson, Research and development prototype for machine augmentation of human strength and endurance. Hardiman I Project (1971)
65. H. Kim, Y.J. Shin, J. Kim, Kinematic-based locomotion mode recognition for power augmentation exoskeleton. Int. J. Adv. Robot. Syst. **14**(5), 1–14 (2017)
66. S. Galle, P. Malcolm, S.H. Collins, D. De Clercq, Reducing the metabolic cost of walking with an ankle exoskeleton: interaction between actuation timing and power. J. NeuroEng. Rehabil. **14**(1), 1–16 (2017)
67. L.M. Mooney, H.M. Herr, Biomechanical walking mechanisms underlying the metabolic reduction caused by an autonomous exoskeleton. J. NeuroEng. Rehabil. **13**(1), 1–12 (2016)
68. G. Colombo, M. Joerg, R. Scheier, V. Dietz, Treadmill training of paraplegic patients using a robotic orthosis. Percept. Mot. Skills **73**(1), 146 (2000)
69. A.C. Villa-Parra, D. Delisle-Rodriguez, T. Botelho, J.J.V. Mayor, A.L. Delis, R. Carelli, A. Frizera Neto, T.F. Bastos, Control of a robotic knee exoskeleton for assistance and rehabilitation based on motion intention from sEMG. Res. Biomed. Eng. **34**(3), 198–210 (2018)
70. T. Bacek, M. Moltedo, K. Langlois, G.A. Prieto, M.C. Sanchez-Villamañan, J. Gonzalez-Vargas, B. Vanderborght, D. Lefeber, J.C. Moreno, BioMot exoskeleton - towards a smart wearable robot for symbiotic human-robot interaction. IEEE Int. Conf. Rehabil. Robot. **2017**, 1666–1671 (2017)
71. E. Bionics, Ekso GT. https://eksobionics.com/
72. J. Choi, B. Na, P.-G. Jung, D.-w. Rha, K. Kong, Walkon suit: a medalist in the powered exoskeleton race of Cybathlon 2016. IEEE Robot. Autom. Mag. **24**(4), 75–86 (2017)
73. S.O. Schrade, K. Dätwyler, M. Stücheli, K. Studer, D.A. Türk, M. Meboldt, R. Gassert, O. Lambercy, Development of VariLeg, an exoskeleton with variable stiffness actuation: first results and user evaluation from the CYBATHLON 2016 Olivier Lambercy; Roger Gassert. J. NeuroEng. Rehabil. **15**(1), 1–18 (2018)
74. B. Chen, H. Ma, L.-Y. Qin, F. Gao, K.-M. Chan, S.-W. Law, L. Qin, W.-H. Liao, Recent developments and challenges of lower extremity exoskeletons. J. Orthop. Translat. **5**, 26–37 (2016)

75. J.D. Schaechter, Motor rehabilitation and brain plasticity after hemiparetic stroke. Prog. Neurobiol. **73**(1), 61–72 (2004)
76. G. Chen, C.K. Chan, Z. Guo, H. Yu, A review of lower extremity assistive robotic exoskeletons in rehabilitation therapy. Crit. Rev. Biomed. Eng. **41**(4–5), 343–363 (2013)
77. R.S. Calabró, A. Naro, M. Russo, P. Bramanti, L. Carioti, T. Balletta, A. Buda, A. Manuli, S. Filoni, A. Bramanti, Shaping neuroplasticity by using powered exoskeletons in patients with stroke: a randomized clinical trial. J. Neuroeng. Rehabil. **15**(1), 35 (2018)
78. B. Koopman, E.H. van Asseldonk, H. van der Kooij, Selective control of gait subtasks in robotic gait training: foot clearance support in stroke survivors with a powered exoskeleton. J. NeuroEng. Rehabil. **10**(1), 3 (2013)
79. L.D. da Silva, T.F. Pereira, V.R. Leithardt, L.O. Seman, C.A. Zeferino, Hybrid impedance-admittance control for upper limb exoskeleton using electromyography. Appl. Sci. (Switzerland) **10**(20), 1–19 (2020)
80. E. Fosch-Villaronga, B. Ozcan, The progressive intertwinement between design, human needs and the regulation of care technology: the case of lower-limb exoskeletons. Int. J. Soc. Robot. **12**, 959–972 (2020)
81. C. Bayon, R. Raya, S. Lerma Lara, O. Ramirez, J.I. Serrano, E. Rocon, Robotic therapies for children with cerebral palsy: a systematic review. Transl. Biomed. **7**(1), 44 (2016)
82. S. Banala, S.H. Kim, S. Agrawal, J. Scholz, Robot assisted gait training with active leg exoskeleton (ALEX). IEEE Trans. Neural Syst. Rehabil. Eng. **17**, 2–8 (2009)
83. R. Ekkelenkamp, J. Veneman, H. van der Kooij, LOPES: a lower extremity powered exoskeleton, in *Proceedings 2007 IEEE International Conference on Robotics and Automation* (IEEE, Piscataway, 2007), pp. 3132–3133
84. M.D. Sánchez-Manchola, M.J. Bernal, M. Munera, C.A. Cifuentes, Gait phase detection for lower-limb exoskeletons using foot motion data from a single inertial measurement unit in hemiparetic individuals. Sensors (Switzerland) **19**(13), 2988 (2019)
85. C. Di Natali, T. Poliero, M. Sposito, E. Graf, C. Bauer, C. Pauli, E. Bottenberg, A. De Eyto, L. O'sullivan, A.F. Hidalgo, D. Scherly, K.S. Stadler, D.G. Caldwell, J. Ortiz, Design and evaluation of a soft assistive lower limb exoskeleton. Robotica **37**(12), 2014–2034 (2019)
86. B. Brackx, V. Grosu, D. Lefeber, ALTACRO: design of the gait rehabilitation robot, in *Symposium on Robot-Assisted Gait Rehabilitation (ALTACRO)* (2013)
87. J.F. Veneman, R. Kruidhof, E.E. Hekman, R. Ekkelenkamp, E.H. Van Asseldonk, H. Van Der Kooij, Design and evaluation of the LOPES exoskeleton robot for interactive gait rehabilitation. IEEE Trans. Neural Syst. Rehabil. Eng. **15**(3), 379–386 (2007)
88. S.K. Banala, S.K. Agrawal, J. Scholz, Active leg exoskeleton (ALEX) for gait rehabilitation of motor-impaired patients, in *2007 IEEE 10th International Conference on Rehabilitation Robotics* (2007), pp. 401–407
89. C.R. Salvatore, A. Cacciola, F. Berté, A. Manuli, A. Leo, A. Bramanti, A. Naro, D. Milardi, P. Bramanti, Robotic gait rehabilitation and substitution devices in neurological disorders: where are we now?. Neurol. Sci. **37**(4), 503–514 (2016)
90. K. Han, J. Lee, W.K. Song, Application scenarios for assistive robots based on in-depth focus group interviews and clinical expert meetings, in *2013 44th International Symposium on Robotics, ISR 2013*, no. C (2013)
91. A.B. Zoss, H. Kazerooni, A. Chu, Biomechanical design of the Berkeley Lower Extremity Exoskeleton (BLEEX). IEEE/ASME Trans. Mechatron. **11**(2), 128–138 (2006)
92. C. Velandia, H. Celedon, D.A. Tibaduiza, C. Torres-Pinzon, J. Vitola, Design and control of an exoskeleton in rehabilitation tasks for lower limb, in *2016 21st Symposium on Signal Processing, Images and Artificial Vision, STSIVA 2016* (2016), pp. 1–6
93. M. Cempini, S.M.M. De Rossi, T. Lenzi, N. Vitiello, M.C. Carrozza, Self-alignment mechanisms for assistive wearable robots: a kinetostatic compatibility method. IEEE Trans. Robot. **29**(1), 236–250 (2013)
94. A.C. Villa-Parra, D. Delisle-Rodriguez, J.S. Lima, A. Frizera-Neto, T. Bastos, Knee impedance modulation to control an active orthosis using insole sensors. Sensors (Switzerland) **17**(12), 2751 (2017)

95. B. Chen, C.H. Zhong, X. Zhao, H. Ma, X. Guan, X. Li, F.Y. Liang, J.C.Y. Cheng, L. Qin, S.W. Law, W.H. Liao, A wearable exoskeleton suit for motion assistance to paralysed patients. J. Orthop. Translat. **11**, 7–18 (2017)
96. R. Steger, S.H. Kim, H. Kazerooni, Control scheme and networked control architecture for the Berkeley Lower Extremity Exoskeleton (BLEEX), in *Proceedings - IEEE International Conference on Robotics and Automation*, vol. 2006, May 2006, pp. 3469–3476
97. H. Kazerooni, R. Steger, L. Huang, Hybrid control of the Berkeley Lower Extremity Exoskeleton (BLEEX). Int. J. Rob. Res. **25**(5–6), 561–573 (2006)
98. T. Poliero, C. Di Natali, M. Sposito, J. Ortiz, E. Graf, C. Pauli, E. Bottenberg, A. De Eyto, D.G. Caldwell, Soft wearable device for lower limb assistance: assessment of an optimized energy efficient actuation prototype. *2018 IEEE International Conference on Soft Robotics, RoboSoft 2018* (2018), pp. 559–564
99. J.E. Duarte, K. Schmidt, R. Riener, J.E. Duarte, K. Schmidt, R. Riener, J.E. Duarte, K. Schmidt, R. Riener, The Myosuit : textile-powered mobility The Myosuit : Myosuit : textile-powered. IFAC-PapersOnLine **51**(34), 242–243 (2019)
100. F.L. Haufe, A.M. Kober, K. Schmidt, A. Sancho-puchades, J.E. Duarte, P. Wolf, R. Riener, User-driven walking assistance: first experimental results using the MyoSuit *, in *2019 IEEE 16th International Conference on Rehabilitation Robotics (ICORR)* (2019), pp. 944–949
101. Indego, Indego Exoskeleton (2012). http://www.indego.com/indego/us/en/home
102. C.E. Zurich, CYBATHLON (2020). https://cybathlon.ethz.ch/en
103. M. Talaty, A. Esquenazi, J.E. Briceno, Differentiating ability in users of the ReWalkTM powered exoskeleton: an analysis of walking kinematics. IEEE Int. Conf. Rehabil. Robot. **2013**, 6650469 (2013)
104. S.O. Schrade, Y. Nager, A.R. Wu, R. Gassert, A. Ijspeert, Bio-inspired control of joint torque and knee stiffness in a robotic lower limb exoskeleton using a central pattern generator. Int. Conf. Rehabil. Robot. **2017**, 1387–1394 (2017)
105. A. Roy, H.I. Krebs, S.L. Patterson, T.N. Judkins, I. Khanna, L.W. Forrester, R.M. Macko, N. Hogan, Measurement of human ankle stiffness using the anklebot, in *2007 IEEE 10th International Conference on Rehabilitation Robotics, ICORR'07*, vol. 00, no. c (2007), pp. 356–363
106. H. Bateni, B.E. Maki, Assistive devices for balance and mobility: Benefits, demands, and adverse consequences. Arch. Phys. Med. Rehabil. **86**(1), 134–145 (2005)
107. S.D. Sierra M., M. Garzón, M. Múnera, C.A. Cifuentes, Human–robot–environment interaction interface for smart walker assisted gait: AGoRA walker. Sensors **19**, 2897 (2019)
108. F.W. Van Hook, D. Demonbreun, B.D. Weiss, Ambulatory devices for chronic gait disorders in the elderly. Am. Fam. Physician **67**(8), 1717–1724 (2003)
109. R. Constantinescu, C. Leonard, C. Deeley, R. Kurlan, Assistive devices for gait in Parkinson's disease. Parkinsonism Relat. Disord. **13**, 133–138 (2007)
110. M. Martins, C. Santos, A. Frizera, R. Ceres, Assistive mobility devices focusing on smart walkers: classification and review. Rob. Auton. Syst. **60**, 548–562 (2012)
111. M. Amboni, P. Barone, J.M. Hausdorff, Cognitive contributions to gait and falls: evidence and implications. Mov. Disord. **28**, 1520–1533 (2013)
112. M. Montero-Odasso, J. Verghese, O. Beauchet, J.M. Hausdorff, Gait and cognition: a complementary approach to understanding brain function and the risk of falling. J. Am. Geriatr. Soc. **60**, 2127–2136 (2012)
113. W.M. Scheidegger, R.C. de Mello, S.D. Sierra M., M.F. Jimenez, M.C. Munera, C.A. Cifuentes, A. Frizera-Neto, A novel multimodal cognitive interaction for walker-assisted rehabilitation therapies, in *2019 IEEE 16th International Conference on Rehabilitation Robotics (ICORR)* (IEEE, Piscataway, 2019), pp. 905–910
114. J.-M. Belda-Lois, S. Mena-del Horno, I. Bermejo-Bosch, J.C. Moreno, J.L. Pons, D. Farina, M. Iosa, M. Molinari, F. Tamburella, A. Ramos, A. Caria, T. Solis-Escalante, C. Brunner, M. Rea, Rehabilitation of gait after stroke: a review towards a top-down approach. J. NeuroEng. Rehabil. **8**(1), 66 (2011)

115. A. Aguirre, S.D. Sierra M., M. Munera, C.A. Cifuentes, Online system for gait parameters estimation using a LRF sensor for assistive devices. IEEE Sens. J. 1 (2020). https://doi.org/10.1109/JSEN.2020.3028279
116. J. Ballesteros, C. Urdiales, A.B. Martinez, M. Tirado, Online estimation of rollator user condition using spatiotemporal gait parameters, in *International Conference on Intelligent Robots and Systems (IROS)* (2016), pp. 3180–3185
117. C.A. Cifuentes, C. Rodriguez, A. Frizera, T. Bastos, Sensor Fusion to control a robotic walker based on upper-limbs reaction forces and gait kinematics. in *5th IEEE RAS & EMBS International Conference on Biomedical Robotics and Biomechatronics* (2014), pp. 1098–1103
118. M.F. Jiménez, M. Monllor, A. Frizera, T. Bastos, F. Roberti, R. Carelli, Admittance controller with spatial modulation for assisted locomotion using a smart walker. J. Intell. Robot. Syst. **94**(7), 1–17 (2019)
119. S.F.R. Alves, J.M. Rosario, H. Ferasoli, L.K.A. Rincon, R.A.T. Yamasaki, Conceptual bases of robot navigation modeling, control and applications, in *Advances in Robot Navigation* (InTech, London, 2011)
120. A. Wachaja, P. Agarwal, M. Zink, M.R. Adame, K. Möller, W. Burgard, Navigating blind people with walking impairments using a smart walker. Auton. Robot. **41**, 555–573 (2017)
121. M. Martins, C. Santos, A. Frizera, R. Ceres, A review of the functionalities of smart walkers. Med. Eng. Phys. **37**, 917–928 (2015)
122. M. Belas Dos Santos, C. Barros de Oliveira, A. Dos Santos, C. Garabello Pires, V. Dylewski, R.M. Arida, A comparative study of conventional physiotherapy versus robot-assisted gait training associated to physiotherapy in individuals with ataxia after stroke. Behav. Neurol. **2018**, 2892065 (2018)
123. S.D. Sierra, M.F. Jimenez, M.C. Munera, A. Frizera-Neto, C.A. Cifuentes, Remote-operated multimodal interface for therapists during walker-assisted gait rehabilitation: a preliminary assessment, in *2019 14th ACM/IEEE International Conference on Human-Robot Interaction (HRI)* (IEEE, Piscataway, 2019), pp. 528–529
124. G.J. Lacey, D. Rodriguez-Losada, The evolution of Guido. IEEE Robot. Autom. Mag. **15**(4), 75–83 (2008)
125. C. Huang, G. Wasson, M. Alwan, P. Sheth, Shared navigational control and user intent detection in an intelligent walker, in *AAAI Fall 2005* (2005)
126. A.F. Neto, A. Elias, C.A. Cifuentes, C. Rodriguez, T. Bastos, R. Carelli, Smart walkers: advanced robotic human walking-aid systems, in *Intelligent Assistive Robots. Springer Tracts in Advanced Robotics* (Springer, Berlin, 2015), pp. 103–131
127. W.-H. Mou, M.-F. Chang, C.-K. Liao, Y.-H. Hsu, S.-H. Tseng, L.-C. Fu, Context-aware assisted interactive robotic walker for Parkinson's disease patients, in *2012 IEEE/RSJ International Conference on Intelligent Robots and Systems* (IEEE, Piscataway, 2012), pp. 329–334
128. T. Hellström, O. Lindahl, T. Bäcklund, M. Karlsson, P. Hohnloser, A. Bråndal, X. Hu, P. Wester, An intelligent rollator for mobility impaired persons, especially stroke patients. J. Med. Eng. Technol. **40**, 270–279 (2016)
129. K.-H. Cho, S. Pyo, G.-S. Shin, S.-D. Hong, S.-H. Lee, D. Lee, S. Song, G. Lee, A novel one arm motorized walker for hemiplegic stroke survivors: a feasibility study. BioMed. Eng. OnLine **17**, 14 (2018)
130. D. Chugo, T. Asawa, T. Kitamura, S. Jia, K. Takase, A moving control of a robotic walker for standing, walking and seating assistance, in *2008 IEEE International Conference on Robotics and Biomimetics* (IEEE, Piscataway, 2009), pp. 692–697
131. C. Werner, M. Geravand, P.Z. Korondi, A. Peer, J.M. Bauer, K. Hauer, Evaluating the sit-to-stand transfer assistance from a smart walker in older adults with motor impairments. Geriatr. Gerontol. Int. **20**, 312–316 (2020)
132. L. Palopoli, A. Argyros, J. Birchbauer, A. Colombo, D. Fontanelli, A. Legay, A. Garulli, A. Giannitrapani, D. Macii, F. Moro, P. Nazemzadeh, P. Padeleris, R. Passerone, G. Poier, D. Prattichizzo, T. Rizano, L. Rizzon, S. Scheggi, S. Sedwards, Navigation assistance and guidance of older adults across complex public spaces: the DALi approach. Intell. Serv. Robot. **8**, 77–92 (2015)

133. J. Leaman, H.M. La, A comprehensive review of smart wheelchairs: past, present, and future. IEEE Trans. Hum. Mach. Syst. **47**, 486–499 (2017)
134. R. Bostelman, J. Albus, Robotic patient lift and transfer, in *Service Robot Applications* (InTech, London, 2008)
135. B.R. Duffy, C.F.B. Rooney, G.M.P.O. Hare, R.P.S.O. Donoghue, What is a social robot? Computer 1–3 (1999)
136. F. Hegel, C. Muhl, B. Wrede, M. Hielscher-Fastabend, G. Sagerer, Understanding social robots, in *2009 Second International Conferences on Advances in Computer-Human Interactions*, pp. 169–174 (IEEE, Piscataway, 2009)
137. T. Fong, I. Nourbakhsh, K. Dautenhahn, A survey of socially interactive robots: concepts, design and applications. Des. Appl. Robot. Auton. Syst. **42**(3), 142–166 (2002)
138. C. Bartneck, J. Forlizzi, A design-centred framework for social human-robot interaction, in *RO-MAN* (2004), pp. 591–594
139. C. Breazeal, *Designing Sociable Robots* (The MIT Press, Cambridge, 2018)
140. T.S. Dahl, M.N.K. Boulos, Robots in health and social care: a complementary technology to home care and telehealthcare? Robotics **3**(1), 1–21 (2014)
141. C.A. Cifuentes, M.J. Pinto, N. Céspedes, M. Múnera, Social robots in therapy and care, in *Current Robotics Reports* (2020), pp. 1–16
142. T. Dahl, M. Kamel Boulos, Robots in health and social care: a complementary technology to home care and telehealthcare?, in *Robotics (MDPI, ISSN 2218-6581)*, vol. 3 (2014), pp. 1–21
143. I. Leite, C. Martinho, A. Paiva, Social robots for long-term interaction: a survey. Int. J. Soc. Robot. **5**(2), 291–308 (2013)
144. O.A. Henkemans, B.P. Bierman, J. Janssen, R. Looije, M.A. Neerincx, M.M. van Dooren, J.L. de Vries, G.J. van der Burg, S.D. Huisman, Design and evaluation of a personal robot playing a self-management education game with children with diabetes type 1. Int. J. Hum. Comput. Stud. **106**, 63–76 (2017)
145. R. Looije, M.A. Neerincx, J.K. Peters, O.A. Henkemans, Integrating robot support functions into varied activities at returning hospital visits: supporting child's self-management of diabetes. Int. J. Soc. Robot. **8**(4), 483–497 (2016)
146. S.S. Yun, J.S. Choi, S.K. Park, G.Y. Bong, H.J. Yoo, Social skills training for children with autism spectrum disorder using a robotic behavioral intervention system. Autism Res. **10**(7), 1306–1323 (2017)
147. Z. Zheng, H. Zhao, A.R. Swanson, A.S. Weitlauf, Z.E. Warren, N. Sarkar, Design, development, and evaluation of a noninvasive autonomous robot-mediated joint attention intervention system for young children with ASD. IEEE Trans. Hum. Mach. Syst. **48**(2), 125–135 (2018)
148. L.J. Wood, B. Robins, G. Lakatos, D.S. Syrdal, A. Zaraki, K. Dautenhahn, Developing a protocol and experimental setup for using a humanoid robot to assist children with autism to develop visual perspective taking skills. Paladyn **10**(1), 167–179 (2019)
149. A.A. Ramírez-Duque, T. Bastos, M. Munera, C.A. Cifuentes, A. Frizera-Neto, Robot-Assisted Intervention for children with special needs: a comparative assessment for autism screening. Robot. Auton. Syst. **127**, 103484 (2020)
150. W.C. So, M.K.Y. Wong, C.K.Y. Lam, W.Y. Lam, A.T.F. Chui, T.L. Lee, H.M. Ng, C.H. Chan, D.C.W. Fok, Using a social robot to teach gestural recognition and production in children with autism spectrum disorders. Disabil. Rehabil. Assist. Technol. **13**(6), 527–539 (2018)
151. D. Casas-Bocanegra, D. Gomez-Vargas, M.J. Pinto-Bernal, J. Maldonado, M. Munera, A. Villa-Moreno, M.F. Stoelen, T. Belpaeme, C.A. Cifuentes, An open-source social robot based on compliant soft robotics for therapy with children with ASD. Actuators **9**(3), 1–22 (2020)
152. A. Ramirez-Duque, L. Aycardi, A. Villa, M. Munera, T. Freire, T. Belpaeme, A. Frizera, C.A. Cifuentes, Collaborative and inclusive process with the autism community: a case study in Colombia about social robot design. Int. J. Soc. Robot. **13**(2), 153–167 (2021)
153. A. Peca, R. Simut, S. Pintea, B. Vanderborght, Are children with ASD more prone to test the intentions of the Robonova robot compared to a human? Int. J. Soc. Robot. **7**(5), 629–639 (2015)

154. B. Vanderborght, R. Simut, J. Saldien, C. Pop, A.S. Rusu, S. Pintea, D. Lefeber, D.O. David, Using the social robot probo as a social story telling agent for children with ASD. Interact. Stud. **13**(3), 348–372 (2012)
155. Y. Nakadoi, Usefulness of animal type robot assisted therapy for autism spectrum disorder in the child and adolescent psychiatric ward, in *New Frontiers in Artificial Intelligence*, ed. by M. Otake, S. Kurahashi, Y. Ota, K. Satoh, D. Bekki (Springer International Publishing, Cham, 2017), pp. 478–482
156. S. Jeong, D.E. Logan, M.S. Goodwin, S. Graca, B. O'Connell, H. Goodenough, L. Anderson, N. Stenquist, K. Fitzpatrick, M. Zisook, L. Plummer, C. Breazeal, P. Weinstock, A social robot to mitigate stress, anxiety, and pain in hospital pediatric care, in *ACM/IEEE International Conference on Human-Robot Interaction*, vol. 02, no. 1, March 2015, pp. 103–104
157. S. Jeong, C. Breazeal, D. Logan, P. Weinstock, Huggable: impact of embodiment on promoting verbal and physical engagement for young pediatric inpatients, in *RO-MAN 2017 - 26th IEEE International Symposium on Robot and Human Interactive Communication*, vol. 2017-Janua (2017), pp. 121–126
158. L.A. Jibb, K.A. Birnie, P.C. Nathan, T.N. Beran, V. Hum, J.C. Victor, J.N. Stinson, Using the MEDiPORT humanoid robot to reduce procedural pain and distress in children with cancer: a pilot randomized controlled trial. Pediatr. Blood Cancer **65**(9), e27242 (2018)
159. N. Céspedes Gómez, A.V. Calderon Echeverria, M. Munera, E. Rocon, C.A. Cifuentes, First interaction assessment between a social robot and children diagnosed with cerebral palsy in a rehabilitation context, in *Companion of the 2021 ACM/IEEE International Conference on Human-Robot Interaction*, HRI '21 Companion (Association for Computing Machinery, New York, 2021), pp. 484–488
160. T.N. Beran, A. Ramirez-Serrano, O.G. Vanderkooi, S. Kuhn, Humanoid robotics in health care: an exploration of children's and parents' emotional reactions. J. Health Psychol. **20**(7), 984–989 (2015)
161. S. Rossi, M. Larafa, M. Ruocco, Emotional and behavioural distraction by a social robot for children anxiety reduction during vaccination. Int. J. Soc. Robot. **12**(3), 765–777 (2020)
162. O. Korn, *Social Robots: Technological, Societal and Ethical Aspects of Human-Robot Interaction* (Springer, Berlin, 2019)
163. E. Martinez-Martin, A.P. del Pobil, Personal robot assistants for elderly care: an overview, in *Personal Assistants: Emerging Computational Technologies* (2018), pp. 77–91
164. W. Moyle, M. Bramble, C. Jones, J. Murfield, Care staff perceptions of a social robot called paro and a look-alike plush toy: a descriptive qualitative approach. Aging Ment. Health **22**(3), 330–335 (2018)
165. M. Bajones, D. Fischinger, A. Weiss, D. Wolf, M. Vincze, P. de la Puente, T. Körtner, M. Weninger, K. Papoutsakis, D. Michel et al., Hobbit: providing fall detection and prevention for the elderly in the real world. J. Robot. **2018** (2018). https://doi.org/10.1155/2018/1754657
166. H.M. Do, M. Pham, W. Sheng, D. Yang, M. Liu, Rish: a robot-integrated smart home for elderly care. Robot. Auton. Syst. **101**, 74–92 (2018)
167. F. Cavallo, R. Limosani, A. Manzi, M. Bonaccorsi, R. Esposito, M. Di Rocco, F. Pecora, G. Teti, A. Saffiotti, P. Dario, Development of a socially believable multi-robot solution from town to home. Cogn. Comput. **6**(4), 954–967 (2014)
168. G. Peleka, A. Kargakos, E. Skartados, I. Kostavelis, D. Giakoumis, I. Sarantopoulos, Z. Doulgeri, M. Foukarakis, M. Antona, S. Hirche et al., Ramcip-a service robot for MCI patients at home, in *2018 IEEE/RSJ International Conference on Intelligent Robots and Systems (IROS)* (IEEE, Piscataway, 2018), pp. 1–9
169. J. Casas, N. Cespedes, M. Múnera, C.A. Cifuentes, Chapter one - human-robot interaction for rehabilitation scenarios, in *Control Systems Design of Bio-Robotics and Bio-mechatronics with Advanced Applications*, ed. by A.T. Azar (Academic, London, 2020), pp. 1–31
170. J. Casas, E. Senft, L. Gutiérrez, M. Rincón-Rocancio, M. Munera, T. Belpaeme, C.A. Cifuentes, Social assistive robots: assessing the impact of a training assistant robot in cardiac rehabilitation. Int. J. Soc. Robot. 1–15 (2020). https://doi.org/10.1007/s12369-020-00708-y

171. J. Casas, N. Céspedes, C.A. Cifuentes G., L. Gutierrez, M. Rincón-Roncancio, M. Munera, Expectation vs. reality: attitudes towards a socially assistive robot in cardiac rehabilitation. Appl. Sci. **9**, 4651 (2019)
172. N. Céspedes, B. Irfan, E. Senft, C.A. Cifuentes, L.F. Gutierrez, M. Rincon-Roncancio, T. Belpaeme, M. Múnera, A socially assistive robot for long-term cardiac rehabilitation in the real world. Front. Neurorobot. **15**, 21 (2021)
173. D.J. Rea, S. Schneider, T. Kanda, "Is this all you can do? harder!": The effects of (im)polite robot encouragement on exercise effort, in *Proceedings of the 2021 ACM/IEEE International Conference on Human-Robot Interaction, HRI '21* (Association for Computing Machinery, New York, 2021), pp. 225–233
174. M.F. Damholdt, M. Nørskov, R. Yamazaki, R. Hakli, C.V. Hansen, C. Vestergaard, J. Seibt, Attitudinal change in elderly citizens toward social robots: the role of personality traits and beliefs about robot functionality. Front. Psychol. **6**, 1701 (2015)
175. D. Portugal, P. Alvito, E. Christodoulou, G. Samaras, J. Dias, A study on the deployment of a service robot in an elderly care center. Int. J. Soc. Robot. **11**(2), 317–341 (2019)
176. A. Scheidig, B. Jaeschke, B. Schuetz, T.Q. Trinh, A. Vorndran, A. Mayfarth, H.M. Gross, May I keep an eye on your training? gait assessment assisted by a mobile robot, in *IEEE International Conference on Rehabilitation Robotics*, vol. 2019, June 2019, pp. 701–708
177. T.Q. Trinh, A. Vorndran, B. Schuetz, B. Jaeschke, A. Mayfarth, A. Scheidig, H.-M. Gross, Autonomous mobile gait training robot for orthopedic rehabilitation in a clinical environment, in *2020 29th IEEE International Conference on Robot and Human Interactive Communication (RO-MAN)*, pp. 580–587 (IEEE, Piscataway, 2020)
178. C. Piezzo, K. Suzuki, Feasibility study of a socially assistive humanoid robot for Guiding elderly individuals during walking. Future Internet **9**(3), 30 (2017)
179. N. Céspedes, M. Munera, C. Gomez, C. Cifuentes G., Social human-robot interaction for gait rehabilitation. IEEE Trans. Neural Syst. Rehabil. Eng. **PP**, 1 (2020)
180. N. Céspedes, D. Raigoso, M. Munera, C.A. Cifuentes, Long-term social human-robot interaction for neurorehabilitation: robots as a tool to support gait therapy in the pandemic. Front. Neurorobot. **15**, 612034 (2021)
181. J.A. Buitrago, A.M. Bolaños, E. Caicedo Bravo, A motor learning therapeutic intervention for a child with cerebral palsy through a social assistive robot. Disabil. Rehabil. Assist. Technol. **15**(3), 357–362 (2020)
182. N.A. Malik, H. Yussof, F.A. Hanapiah, R.A.A. Rahman, H.H. Basri, Human-robot interaction for children with cerebral palsy: reflection and suggestion for interactive scenario design. Procedia Comput. Sci. **76**, 388–393 (2015)
183. J.C. Pulido, C. Suárez, J.C. González Dorado, A. Dueñas-Ruiz, P. Ferri, M. Encarnación, M. Sahuquillo, C. Echevarría, R. Vargas, P. Infante-Cossio, C. Luis, C. Parra Calderón, F. Fernández, A socially assistive robotic platform for upper-limb rehabilitation: a longitudinal study with pediatric patients. IEEE Robot. Autom. Mag. **26**, 24–39 (2019)
184. G. Krahn, D. Walker, R. Correa-de Araujo, Persons with disabilities as an unrecognized health disparity population. Am. J. Public Health **105**, e1–e9 (2015)
185. S. Gulley, E. Rasch, C. Bethell, A. Carle, B. Druss, A. Houtrow, A. Reichard, L. Chan, At the intersection of chronic disease, disability and health services research: a scoping literature review. Disabil. Health J. **11**, 192–203 (2018)
186. V. Haung, J. Krakauer, Robotic neurorehabilitation: a computational motor learning perspective. J. NeuroEng. Rehabil. **6**, 1–13 (2009)
187. M. Gilliaux, A. Renders, D. Dispa, D. Holvoet, J. Sapin, B. Dehez, C. Detrembleur, T. Lejeune, G. Stoquart, Upper limb robot-assisted therapy in cerebral palsy: a single-blind randomized controlled trial. Neurorehabil. Neural Repair **29**, 183–192 (2014)
188. K. Kong, D. Jeon, Design and control of an exoskeleton for the elderly and patients. IEEE/ASME Trans. Mechatron. **11**, 428–432 (2006)
189. C. Bayon, T. Martín Lorenzo, B. Moral, O. Ramírez, A. Moreno, S. Lerma Lara, I. Martínez, E. Rocon, A robot-based gait training therapy for pediatric population with cerebral palsy: goal setting, proposal and preliminary clinical implementation. J. NeuroEng. Rehabil. **15**, 69 (2018)

190. D. Delisle Rodriguez, A. Villa-Parra, J. Lima, L. Vargas-Valencia, A. Frizera, T. Bastos, Assessment of an assistive control approach applied in an active knee orthosis plus walker for post-stroke gait rehabilitation. Sensors **20**, 2452 (2020)
191. X. Giralt, L. Amigo, A. Casals, J. Amat, Robotic platform to evaluate the assistance and assessment on the rehabilitation loop, in *Converging Clinical and Engineering Research on Neurorehabilitation*, ed. by J.L. Pons, D. Torricelli, M. Pajaro. *Biosystems & Biorobotics*, vol. 1 (Springer, Berlin, 2013), pp. 1031–1035
192. S. Scott, S. Dukelow, Potential of robots as next-generation technology for clinical assessment of neurological disorders and upper-limb therapy. J. Rehabil. Res. Dev. **48**, 335–353 (2011)
193. S. Brose, D. Weber, B. Salatin, G. Grindle, H. Wang, J. Vazquez, R. Cooper, The role of assistive robotics in the lives of persons with disability. Am. J. Phys. Med. Rehabil. **89**, 509–521 (2010)
194. A.S. Prabuwono, K. Allehaibi, K. Kurnianingsih, Assistive robotic technology: a review. Comput. Eng. Appl. J. **6**, 71–78 (2017)
195. A. Meyer-Heim, H.J. van Hedel, Robot-assisted and computer-enhanced therapies for children with cerebral palsy: current state and clinical implementation. Semin. Pediatr. Neurol. **20**(2), 139–145 (2013)
196. D. Damiano, K. Alter, H. Chambers, New clinical and research trends in lower extremity management for ambulatory children with cerebral palsy. Phys. Med. Rehabil. Clin. N. Am. **20**, 469–491 (2009)
197. K. Kong, D. Jeon, Fuzzy control of a new tendon-driven exoskeletal power assistive device, in *Proceedings, 2005 IEEE/ASME International Conference on Advanced Intelligent Mechatronics* (IEEE, Piscataway, 2005), pp. 146–151
198. K. Kong, M. Tomizuka, H. Moon, B. Hwang, D. Jeon, Mechanical design and impedance compensation of SUBAR (Sogang University's Biomedical Assist Robot), in *2008 IEEE/ASME International Conference on Advanced Intelligent Mechatronics*, pp. 377–382 (IEEE, Piscataway, 2008)
199. A.C. Villa-Parra, D. Delisle-Rodríguez, A. López-Delis, T. Bastos-Filho, R. Sagaró, A. Frizera-Neto, Towards a robotic knee exoskeleton control based on human motion intention through EEG and sEMGsignals. Procedia Manuf. **3**, 1379–1386 (2015)
200. R. Sagaro Zamora, Proposal of an assisted-motion system for gait rehabilitation, in *International Workshop on Wearable Robotics WEROB2014*, Sept 2014
201. C. Bayón, O. Ramírez, J.I. Serrano, M. del Castillo, A. Moreno, J.M. Belda Lois, I. Martínez-Caballero, S. Lerma Lara, C. Cifuentes G., A. Frizera, E. Rocon, Development and evaluation of a novel robotic platform for gait rehabilitation in patients with cerebral palsy: CPwalker. Robot. Auton. Syst. **91**, 101–114 (2017)
202. S.D. Sierra M., J.F. Molina, D.A. Gomez, M.C. Munera, C.A. Cifuentes, Development of an interface for human-robot interaction on a robotic platform for gait assistance: AGoRA smart walker, in *2018 IEEE ANDESCON*, pp. 1–7 (IEEE, Piscataway, 2018)
203. M. Sánchez-Manchola, D. Gomez-Vargas, D. Casas-Bocanegra, M. Munera, C.A. Cifuentes, Development of a robotic lower-limb exoskeleton for gait rehabilitation: AGoRA exoskeleton, in *2018 IEEE ANDESCON, ANDESCON 2018 - Conference Proceedings* (2018), pp. 1–6
204. M. Sánchez-Manchola, D. Serrano, D. Gómez, F. Ballen, D. Casas, M. Munera, C.A. Cifuentes, T-FLEX: variable stiffness ankle-foot orthosis for gait assistance, in *Wearable Robotics: Challenges and Trends* (Springer, Cham, 2019), pp. 160–164
205. D. Casas, M. Gonzalez Rubio, M. Montoya, W. Sierra, L. Rodriguez, E. Rocon, C.A. Cifuentes, Bioinspired hip exoskeleton for enhanced physical interaction, in *Converging Clinical and Engineering Research on Neurorehabilitation II. Biosystems & Biorobotics*, ed. by J. Ibáñez, J. González-Vargas, J. Azorín, M. Akay, J. Pons (Springer, Cham, 2017), pp. 1497–1501
206. D. Gomez-Vargas, M.J. Pinto-Betnal, F. Ballén-Moreno, M. Múnera, C.A. Cifuentes, Therapy with t-flex ankle-exoskeleton for motor recovery: a case study with a stroke survivor, in *2020 8th IEEE RAS/EMBS International Conference for Biomedical Robotics and Biomechatronics (BioRob)* (2020), pp. 491–496

207. M.D. Sánchez-Manchola, L.J. Arciniegas Mayag, M. Munera, C.A. Garcia, Impedance-based backdrivability recovery of a lower-limb exoskeleton for knee rehabilitation, in *4th IEEE Colombian Conference on Automatic Control: Automatic Control as Key Support of Industrial Productivity, CCAC 2019 - Proceedings, Medellin* (2019), pp. 1–6
208. S.D. Sierra M., M.F. Jimenez, M.C. Múnera, T. Bastos, A. Frizera-Neto, C.A. Cifuentes, A therapist helping hand for walker-assisted gait rehabilitation: a pre-clinical assessment, in *2019 IEEE 4th Colombian Conference on Automatic Control (CCAC)* (2019), pp. 1–6
209. R.C. Mello, S.D. Sierra, M. Munera, C.A. Cifuentes, M.R.N. Ribeiro, A. Frizera-Neto, Cloud robotics experimentation testbeds: a cloud-based navigation case study, in *2019 IEEE 4th Colombian Conference on Automatic Control (CCAC)* (IEEE, Piscataway, 2019), pp. 1–6
210. M. Bautista, A. Garzón, S. Sierra Marín, M. Munera, C. Cifuentes G., Integration of a social robot in physical rehabilitation assisted by Lokomat, in *X Congreso Iberoamericano de Tecnologías de Apoyo a la Discapacidad IBERDISCAP2019, Buenos Aires* (2019), pp. 423–426
211. S. Hesse, A. Heß, C. Werner, N. Kabbert, R. Buschfort, Effect on arm function and cost of robot-assisted group therapy in subacute patients with stroke and a moderately to severely affected arm: a randomized controlled trial. Clin. Rehabil. **28**, 637–647 (2014)
212. A. Timmermans, R. Lemmens, M. Monfrance, R. Geers, W. Bakx, R. Smeets, H. Seelen, Effects of task-oriented robot training on arm function, activity, and quality of life in chronic stroke patients: a randomized controlled trial. J. Neuroeng. Rehabil. **11**, 45 (2014)
213. W.-w. Liao, C.-Y. Wu, Y.-W. Hsieh, K.-c. Lin, W.-Y. Chang, Effects of robot-assisted upper limb rehabilitation on daily function and real-world arm activity in patients with chronic stroke: a randomized controlled trial. Clin. Rehabil. **26**, 111–120 (2011)
214. J.-C. Fraile, J. Pérez-Turiel, E. Baeyens, P. Viñas, R. Alonso, A. Cuadrado, M. Franco, E. Parra Vidales, L. Ayuso, F. García-Bravo, F. Nieto, L. Lipsa, E2rebot: a robotic platform for upper limb rehabilitation in patients with neuromotor disability. Adv. Mech. Eng. **8**, 1–13 (2016)
215. P. Sale, M. Franceschini, A. Waldner, S. Hesse, Use of the robot assisted gait therapy in rehabilitation of patients with stroke and spinal cord injury. Eur. J. Phys. Rehabil. Med. **48**, 111–121 (2012)
216. C. Senanayake, S. Senanayake, Emerging robotics devices for therapeutic rehabilitation of the lower extremity, in *2009 IEEE/ASME International Conference on Advanced Intelligent Mechatronics* (IEEE, Piscataway, 2009), pp. 1142–1147
217. E. Tsardoulias, P. Mitkas, Robotic frameworks, architectures and middleware comparison. arXiv:1711.06842, Nov 2017
218. C. Valadao, E. Caldeira, T. Bastos-Filho, A. Frizera-Neto, R. Carelli, A new controller for a smart walker based on human-robot formation. Sensors **16**(7), (2016)
219. M. Sanchez-Villamanan, J. Gonzalez-Vargas, D. Torricelli, J. Moreno, J. Pons, Compliant lower limb exoskeletons: a comprehensive review on mechanical design principles. J. NeuroEng. Rehabil. **16**, 55 (2019)
220. Y. Long, Z.-j. Du, W. Wang, W. Dong, Development of a wearable exoskeleton rehabilitation system based on hybrid control mode. Int. J. Adv. Robot. Syst. **13**, 10 (2016)
221. A. Hasan, Designing and implementing an electronic system to control moving orthosis virtual mechanical system to emulate lower limb. Cogent Eng. **5**, 1456632 (2018)
222. E. Einbinder, T. Horrom, Smart walker: a tool for promoting mobility in elderly adults. J. Rehabil. Res. Dev. **47**, xiii–xv (2010)
223. S. Bradley, C. Hernandez, Geriatric assistive devices. Am. Fam. Physician **84**, 405–411 (2011)
224. A.D.C. Chan, J.R. Green, Smart rollator prototype, in *2008 IEEE International Workshop on Medical Measurements and Applications* (IEEE, Piscataway, 2008), pp. 97–100
225. G. Lacey, S.M. Namara, K.M. Dawson-Howe, Personal adaptive mobility aid for the infirm and elderly blind, in *Assistive Technology and Artificial Intelligence* (Springer, Berlin, 1998), pp. 211–220

Kinematics, Actuation, and Sensing Architectures for Rehabilitation and Assistive Robotics

2

Sergio D. Sierra M., Luis Arciniegas-Mayag, Orion Ramos, Juan Maldonado, Marcela Múnera, and Carlos A. Cifuentes

2.1 Introduction

In the process of analyzing and designing robotic technologies for rehabilitation and gait assistance, it is necessary to understand the different elements that describe and compose such a device. In this sense, this chapter presents the necessary tools to mathematically model rehabilitation devices such as lower-limb exoskeletons, social robots, and robotic walkers. Likewise, the concepts necessary to understand the actuation systems that allow the movement of these devices and the safe interaction with the users are presented. Finally, the most common sensing architectures reported in the literature are described, which allow these devices to acquire information from their internal systems, their environment, and the user.

2.2 Robotic Geometric and Kinematic Modeling

In robotics, kinematics studies the motion of a robot part concerning a reference system. In a manipulator robot, this system is usually chosen according to the task to be performed by the robot with respect to the base and the end effector. Geometric kinematic models essentially involve the relationship between the robot and its workspace, usually a Cartesian space. In this chapter the position of robot links will be considered in static situations only. To understand the complex geometry of a robot, one must add frames to the various parts of the mechanism and then describe

S. D. Sierra M. · L. Arciniegas-Mayag · O. Ramos · J. Maldonado · M. Múnera ·
C. A. Cifuentes (✉)
Biomedical Engineering Department of the Colombian School of Engineering Julio Garavito, Bogotá D.C., Colombia
e-mail: sergio.sierra@escuelaing.edu.co; luis.arciniegas@mail.escuelaing.edu.co; orion.ramos@mail.escuelaing.edu.co; juan.maldonado-me@mail.escuelaing.edu.co; marcela.munera@escuelaing.edu.co; carlos.cifuentes@escuelaing.edu.co

C. A. Cifuentes, M. Múnera, *Interfacing Humans and Robots for Gait Assistance and Rehabilitation*, https://doi.org/10.1007/978-3-030-79630-3_2

the relationships between these frames. The study of the kinematics of manipulator robots relates, among other things, to how the locations of these frames change as the mechanism articulates. In other words, either forward or inverse kinematics relates the end position of the end effector to the angles of the robot's joints.

This section will explain the difference between forward and inverse kinematics applied in robotics, develop the Denavit–Hartenberg methodology for the solution of forward kinematics, and present some examples with exoskeletons and humanoid robots that have different kinematic chains with different degrees of freedom.

2.2.1 Forward vs. Inverse Kinematics

The study of kinematics in robotics is based on calculating the parameters that define the positions of a specific part of the robot. These parameters can be joint angles or equations that depend only on the robot's dimensions. In manipulator robotics, there are two types of kinematics, forward kinematics and inverse kinematics. Each has a specific purpose and is calculated in different ways. Forward kinematics relates the final position of the kinematic chain concerning the angles of each joint. In other words, it finds the (X, Y, Z) position of the hand, head, or leg if the angles of each joint are known. On the other hand, inverse kinematics relates the angles of each joint to a desired (X_f, Y_f, Z_f) position. For example, in robots with arms, this type of kinematics is used when the robot is to point to something specific, so the inverse kinematics solution will find the necessary angles that must be placed in each motor or joint for the hand to point precisely to a defined position. Figure 2.1 shows the differences between the two types of kinematics graphically.

To solve the forward kinematics, the lengths of the kinematic chain's links must be known to use them with mathematical tools that finally model the robot's position. The most commonly used and standardized method to solve this kinematic problem is called Denavit–Hartenberg (DH) [1,2]. This method is based on finding 4 parameters $(\theta_i\ d_i\ \alpha_i\ a_i)$ per robot segment that relate the robot base to the final part of the kinematic chain (P_f). Usually this type of kinematics can be expressed with the Denavit–Hartenberg Table [2].

The most commonly used methods to find the inverse kinematics solution are the geometric method and the screw method. However, the second depends on

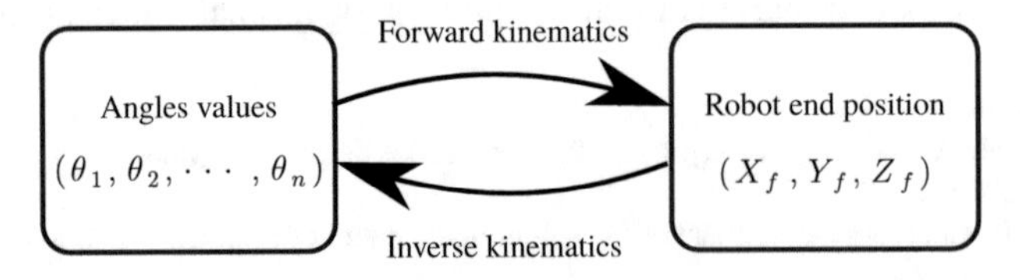

Fig. 2.1 Scheme that relates the angles of the joints and the positions of the links with the forward and inverse kinematics. If the angles of the joints are known, the final position can be calculated. If the final position is known, the angles required at each joint can be calculated

the previous calculation of the Denavit matrices [3]. The geometric method uses trigonometric rules to find angles of the triangles formed by the kinematic chain's links in a specific plane. Usually, functions such as arctan and decompositions of the laws of cosines or sines are used. The geometric method's solution will be different for each robot configuration, so it is more complex to calculate than forward kinematics.

2.2.2 Denavit–Hartenberg Convention

The relationship between the joints and the robot workspace is realized via geometric modeling. The most convenient way to perform this process is denoted as the Denavit–Hartenberg (DH) method. This approach focuses on forward kinematic computation only. The goal is to calculate the position and orientation of the end effector relative to the base as a function of the joint variables. This method consists of representing the robot's kinematic chain using a set of bodies connected by joints. These bodies are called links or segments. The joints or degrees of freedom (DoFs) are those that connect two links. The DH convention expresses the rotation and translation using a single homogeneous matrix of one DoF of the kinematic chain in mathematical terms.

The structure of the DH matrix is generally composed of the rotation and translation of the axes involved in the degree of freedom and the distance between them. The matrix will always have dimensions of 4×4 as follows:

$$H_n = \begin{bmatrix} R_{n(3\times3)} & t_{n(3\times1)} \\ 0_{1\times2} & 1 \end{bmatrix}, \tag{2.1}$$

where $R_{n(3\times3)}$ is the rotation matrix (3×3) that considers the rotation in x-, y-, and z-axis and $t_{n(3\times1)}$ is the translation vector with size (3×1), where each position of the vector is a translation on the x-, y- and z-axis.

The DH method lists the links starting from the arm's fixed base, called link 0. The first moving body is the link 1 and successively until reaching the free end of the arm (end effector), which is link n. The purpose of this is to locate the frames in each joint in the most appropriate way to obtain the DH parameters. The location of the frames according to the method can be summarized in the following six steps:

1. Identify the joint axes and mark the lines on them. The following steps consider two of these adjacent lines (on the i and $i+1$ axes).
2. Identify the common perpendicular line between them, or the point of intersection. This point will be the origin of the frame of the i_n link.
3. Assign the Z_i-axis in a manner that it points to the joint axis.
4. Assign the X_i-axis to be normal to the plane of Z_i at its intersection point.
5. Assign the Y_i-axis to complete the coordinate system of this frame according to the right-hand rule.
6. Perform these steps for the next joint until the joint n is reached.

Parameter	Name	Description
θ_i	Joint Angle	$X_i \xrightarrow[@Z_i]{\text{rotation,ccw}} X_{i+1}$
d_i	Joint Offset	$X_i \xrightarrow[@Z_i]{\text{distance}} X_{i+1}$
α_i	Twist Angle	$Z_i \xrightarrow[@X_i]{\text{rotation,ccw}} Z_{i+1}$
a_i	Link Length	$Z_i \xrightarrow[@X_i]{\text{distance}} Z_{i+1}$

Fig. 2.2 Description and calculation of the four DH parameters according to the axes and frames located in the system

After performing the location of all the frames in the kinematic chain, the four DH parameters ($\theta_i, d_i, \alpha_i, a_i$) are obtained for each joint. The way to obtain these parameters can be seen in Fig. 2.2. This way of obtaining the parameters is only functional if the frames were correctly placed according to the previous steps.

The angle θ_i will be the rotation in the Z_i-axis between the X axes of the links of the degree of freedom. In the particular case in which this angle depends on an actuator (motor), it is expressed as a variable angle on which the end effector's final position will depend. The parameter d_i is the distance between the two links of the degree of freedom around the Z_i-axis. If the actuator (linear motor) modifies this distance, it will be a variable parameter and not a constant value. The twist angle α_i is the angle formed between the Z axes of the degree of freedom around the X_i-axis. This value is usually constant, but as in the previous cases, if any actuator modifies it, it should be taken as a variable angle. Finally, the distance a_i is defined by the space generated between the Z_i and Z_{i+1} axes through the X_{i+1}-axis.

By identifying these parameters for each degree of freedom, it is possible to relate the base to the end effector depending on the variable angles and distances (which are modified by actuators). In other words, it will be possible to know the position and orientation of the end of the kinematic chain if the actual actuator's position that modifies angles or distances is known.

Based on this standard procedure to obtain the DH parameters, the following sections of the chapter will calculate the forward and inverse kinematics of the *AGoRA* lower-limb exoskeleton [4, 5] that has only two degrees of freedom in the same plane, and then calculate the kinematics of social robots where the motions are in the 3D Cartesian space. Finally, an approximation of the forward and inverse kinematics in mobile robots will be made.

2.2.3 Modeling Lower-Limb Exoskeletons

Each lower-limb exoskeletons presented in the literature show various kinematic models to mimic the human movements, where this parameter is estimated by applying the manipulator robot's concepts. The first step to define the kinematic parameter is presented through the DH convention explained in Sect. 2.2.2 where the joint angle θ_i, joint offset d_i, twist angle α_i, and the link length a_i are established. As an example, the *AGoRA* lower-limb exoskeleton is designed using these manipulator concepts. In this case, the *AGoRA* exoskeleton can express showing the DH convention of one limb. In this sense, the *AGoRA* exoskeleton is a 2 DoF robot where the joint angle is expressed by q_1 and q_2,l_1, and l_2 express the link lengths. Figure 2.3 shows the *AGoRA* exoskeleton identifying some DH parameters. Taking into account the parameters shown is defined the DH convention expressed in Table 2.1.

The DH convention shown in the table presents the *AGoRA* exoskeleton as a manipulator robot with 2 degrees of freedom (DoFs) focused on the movement generation in the sagittal plane. This exoskeleton is focused on the hip and knee joints of the patient's right limb.

In this sense, the *AGoRA* exoskeleton is presented in two homogeneous matrices shown in Eqs. 2.2 and 2.3

$$H_1 = \begin{bmatrix} \sin(q_1) & \cos(q_1) & 0 & l_1 \sin(q_1) \\ -\cos(q_1) & \sin(q_1) & 0 & -l_1 \cos(q_1) \\ 0 & 0 & 1 & 0 \\ 0 & 0 & 0 & 1 \end{bmatrix} \tag{2.2}$$

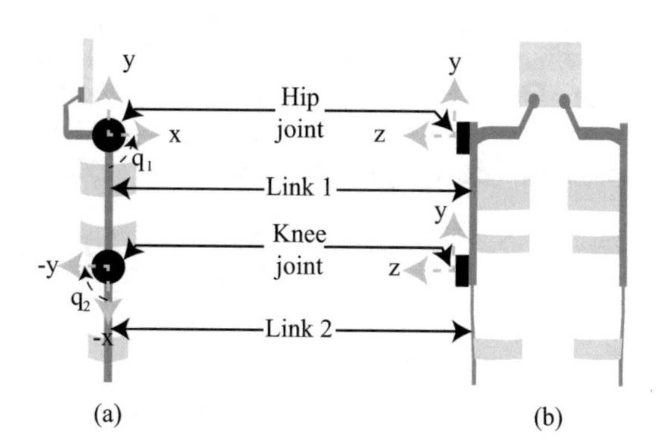

Fig. 2.3 AGoRA exoskeleton DH convention; (**a**) *AGoRA* exoskeleton's sagittal plane; (**b**) *AGoRA* exoskeleton's frontal plane

Table 2.1 *AGoRA* exoskeleton kinematic definition using the DH convention; θ_i is the revolute joint variable; d_i is the distance from the origin exoskeleton frame to the x-$axis$; a_1 presents the link's joint; α_i represents the x-$axis$ rotation

Joint	θ_i	d_i	a_i	α
1	q_1	0	l_1	0
2	q_2	0	l_2	0

$$H_2 = \begin{bmatrix} \sin(q_2) & \cos(q_2) & 0 & l_2 \sin(q_2) \\ -\cos(q_2) & \sin(q_2) & 0 & -l_2 \cos(q_2) \\ 0 & 0 & 1 & 0 \\ 0 & 0 & 0 & 1 \end{bmatrix}. \quad (2.3)$$

The *AGoRA* exoskeleton rotation and translation of each axis are described in three expressions. The hip and knee joint's rotation is presented in the *z-axis*. Consequently, each rotation has a translation value related to the link's length operated depending on the joint's rotation angle (q_1 and q_2). Equation 2.2 shows the rotation and translation of the shaft from joint 1 to joint 2. Equation 2.3 shows the rotation and translation of the shaft from joint 2 to the end effector. Finally, the multiplication of the matrix H_1 and H_2 (Eq. 2.4) provides information about the rotation and translation movements from the reference coordinate axis to the end-effector reference axis (Eq. 2.5). The values obtained in each of the equations are called homogeneous matrix transformations.

$$H_0^2 = H_1 H_2 \quad (2.4)$$

$$H_0^2 = \begin{bmatrix} -\cos(q_1+q_2) & \sin(q_1+q_2) & 0 & l_1 \sin(q_1) - l_2 \cos(q_1+q_2) \\ -\sin(q_1+q_2) & -\cos(q_1+q_2) & 0 & -l_1 \cos(q_1) - l_1 \sin(q_1+q_2) \\ 0 & 0 & 1 & 0 \\ 0 & 0 & 0 & 1 \end{bmatrix}. \quad (2.5)$$

Finally, the movement generation of an exoskeleton may be defined for each link geometrically. In that order, the forward kinematics of a manipulator robot provides end-effector coordinates information depending on the rotational information of each of the robot's links. In lower-limb exoskeletons, the end effector equals the distal location of the exoskeleton structure from the reference axis. For the *AGoRA* lower-limb exoskeleton, the forward kinematics of each link is expressed as follows:

$$Link_{1(x,y)} = \begin{bmatrix} l_{c1} \sin q_1 \\ -l_{c1} \cos q_1 \end{bmatrix} \quad (2.6)$$

$$Link_{2(x,y)} = \begin{bmatrix} l_1 \sin q_1 & l_{c2} \sin(q_1+q_2) \\ -l_1 \cos q_1 & -l_{c2} \cos(q_1+q_2) \end{bmatrix}. \quad (2.7)$$

Equation 2.6 shows the hip joint forward kinematic, where l_{c1} equals to the link 1 center of mass, and q_1 is the hip joint rotation angle. Equation 2.7 shows the knee joint forward kinematic, where l_1 is the link 1 length, and l_{c2} is the link 2 center of mass. Finally, q_2 is the knee joint rotation angle.

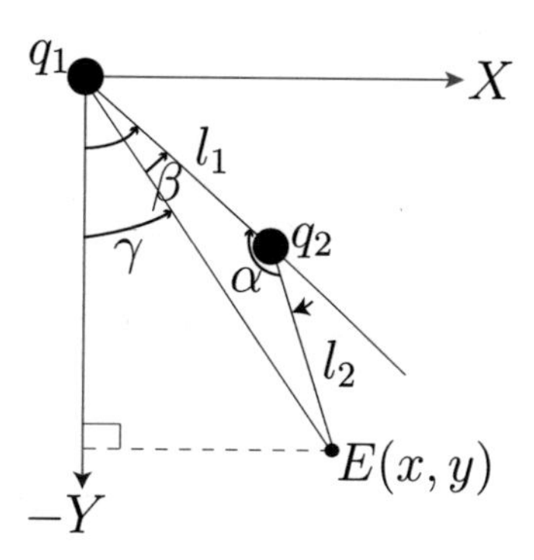

Fig. 2.4 Lower-limb *AGoRA* exoskeleton right limb, where l_1 and l_2 equal to the link 1 length and link 2 lengths, respectively; $E(x, y)$ is the end-effector position; q_1 and q_2 equal to the hip angle rotation and knee angle rotation, respectively

On the other hand, using the end-effector coordinates of the robot, it is possible to estimate the angles of each joint, which is named the inverse kinematics model. The estimation of q_1 and q_2 can be performed in two ways. The first one consists of solving these variables from Eqs. 2.6 and 2.7. The second method is performed geometrically. Figure 2.4 shows the development of this second method by positioning the *AGoRA* exoskeleton in a Cartesian plane and a point (x, y) where the end effector of the knee joint is located.

In the estimation of the q_1 and 1_2 values, q_2 is calculated as follows:

$$r^2 = x^2 + y^2 \tag{2.8}$$

$$r^2 = {l_1}^2 + {l_2}^2 - 2l_1l_2\cos\alpha \tag{2.9}$$

$$q_2 = \pi - \alpha. \tag{2.10}$$

Using Eq. 2.8 in Eq. 2.9 obtains $\cos\alpha$ value:

$$\cos\alpha = \frac{l_1^2 + l_2^2 - x^2 - y^2}{2l_1l_2}. \tag{2.11}$$

Applying Eq. 2.11 in Eq. 2.10 estimates the q_2 value:

$$q_2 = \arccos\frac{x^2 + y^2 - l_1 - l_2}{2l_1l_2}. \tag{2.12}$$

The estimation of q_1 value is performed as follows:

$$\gamma = q_1 - \beta, \tag{2.13}$$

where the $\tan\beta$ equals to

$$\tan \beta = \frac{-l_2 \sin q_2}{l_1 + l_2 \cos q_2} \tag{2.14}$$

$$\tan \gamma = \frac{x}{y}. \tag{2.15}$$

Applying tan in Eq. 2.13 replaces Eqs. 2.14 and 2.15 obtaining the following expression:

$$q_1 = \arctan\left(\frac{x}{y}\right) - \arctan\left(\frac{l_2 \sin q_2}{l_1 + l_2 \cos q_2}\right). \tag{2.16}$$

2.2.4 Modeling Social Robots

Once the kinematics of a single plane of motion device such as the *AGoRA* lower-limb exoskeleton is known, the kinematics of each of the moving parts of two social robots used for human–robot interaction will now be analyzed. The moving parts or kinematic chains to be analyzed of these robots in this section are defined as head, upper limb, and lower limb. The kinematic chain of the head only has one degree of freedom in the *CASTOR* robot [6, 7], the upper limb of the *CASTOR* robot has 3 degrees of freedom in different planes, and finally the lower limb of the *NAO* robot has 6 degrees of freedom. The following section will describe each of the degrees of freedom and elements needed to calculate the forward and inverse kinematics of the mentioned kinematic chains. The general architecture of the open-source *CASTOR* robot used for autism therapies and the *NAO* commercial robot from SoftBank Robotics [8] will be defined. It should be noted that a DoF is not necessarily located where the actuators or motors are placed, but is the point of motion as are the joints in the human body.

2.2.4.1 Modeling the Head and Upper Limb of the *CASTOR* Robot

In the case of the *CASTOR* robot, it has 14 degrees of freedom distributed into 12 active and 2 passive degrees of freedom. This means that this robot can change the position of the links of its system by using 12 actuators and has 2 joints to receive impacts or reject external disturbances. This system uses 7 servomotors (AX12, Dynamixel, Seoul Korea) to the kinematic chains of the head and upper limbs. The other 5 active joints deal with the gestures of the face. The definition of each degree of freedom of the robot is described as follows: 1 DOF for each elbow (i.e., flexo-extension movement), 2 DOFs per shoulder (i.e., flexo-extension and abduction movements), 1 DOF for head rotation movement, and 5 DOFs for facial expressions. Besides allowing deformation in the huggable structure and rejection of external stimulations, the robot incorporates 2 passive DOFs. The relevant joints for the kinematic analysis can be seen in Fig. 2.5.

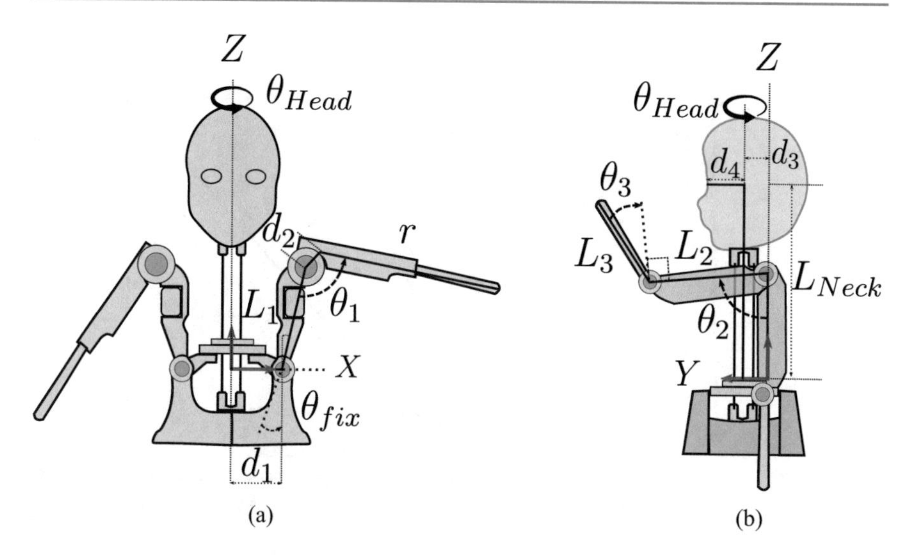

Fig. 2.5 Description of the angular and longitudinal variables involved in the *CASTOR* robot used to solve the forward and inverse kinematics. (**a**) Front view of the robot. (**b**) Side view of the robot

Table 2.2 DH parameters for the kinematic chain of the *CASTOR* robot head

$Matrix_i$	θ_i	d_i	a_i	α_i
1	0	0	d_3	0
2	θ_{Head}	L_{Neck}	0	0
3	0	0	d_4	0

By dividing this robot into two kinematic chains, head and upper limbs, it can be identified that the head part has one degree of freedom (θ_{head}) and the upper limbs have 3 degrees of freedom for each arm ($\theta_1, \theta_2, \theta_3$). To complement the general architecture of this robot, the system's overall dimensions must be defined. The *CASTOR* robot is 50 cm from base to head and 35.4 cm wide. Each arm of the robot has a total length of 35.6 cm [7].

2.2.4.2 The Head Kinematic Chain

As shown in Fig. 2.5 the kinematic chain of the head of the *CASTOR* robot only depends on one joint θ_{head} and the distances that are necessary to transport the base point (X_0, Y_0, Z_0) to the endpoint of the kinematic chain (X_f, Y_f, X_f), which in this case would be the eyes of the robot. Following the standard steps to complete the DH table [2], coordinate axes are established for each section of the robot neck kinematic chain, as seen in Fig. 2.6a. As can be seen, only the dimensions of the neck (L_{Neck}), the distance from the center to the neck (d_3), the joint rotation angle (θ_{Head}), and the distance from the neck to the eyes (d_4) are needed. From these coordinate axes, the DH steps are applied to complete Table 2.2.

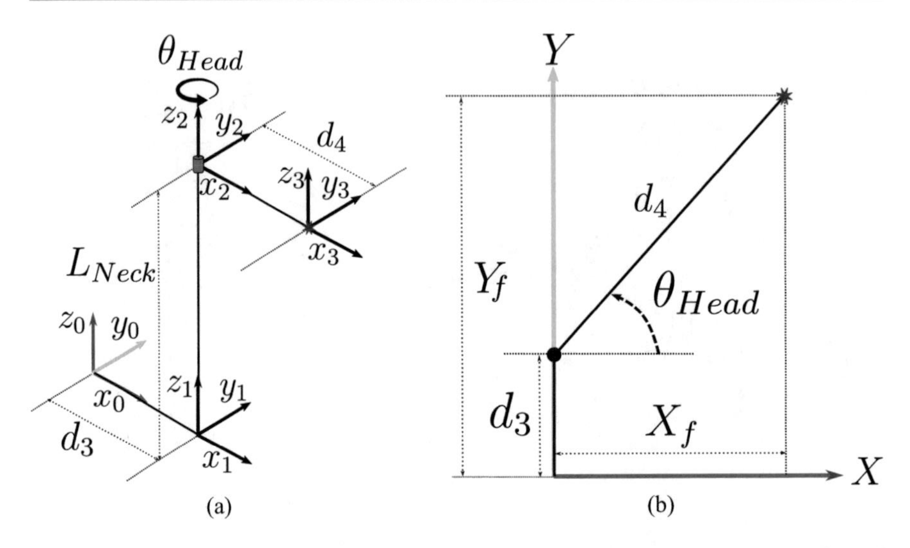

Fig. 2.6 (**a**) Kinematic chain of the *CASTOR* robot head with the coordinate axes located according to the Denavit–Hartenberg methodology. (**b**) View of the X–Y plane where the angle θ_{Head} is observed and calculated based on the triangle formed by the dimensions of the final position (X_f, Y_f) where the robot head should point. This calculation considers the separation distance d_3 between the base of the robot and the rotation point of the neck, since it can be different from zero

Once the parameters of Denavit have been obtained, the matrices of each row of the table must be calculated to relate the base of the robot with its endpoint. For each of the rows of Table 2.2, the matrix DH_i must be expressed according to Eq. 2.17, and multiply all these matrices. This last resulting matrix will be the one that relates all the spatial transformations that are necessary to apply to the origin to reach the final point.

$$DH_i = \begin{bmatrix} c\theta_i & -s\theta_i c\alpha_i & s\theta_i s\alpha_i & a_i c\theta_i \\ s\theta_i & c\theta_i c\alpha_i & -c\theta_i s\alpha_i & a_i s\theta_i \\ 0 & s\alpha_i & c\alpha_i & d_i \\ 0 & 0 & 0 & 1 \end{bmatrix}. \tag{2.17}$$

The three necessary matrices to perform the forward kinematics calculation are shown below in the *CASTOR* robot head, where Matrix 1 is defined as DH_1, Matrix 2 as DH_2, and Matrix 3 as DH_3.

$$DH_1 = \begin{bmatrix} 1 & 0 & 0 & d_3 \\ 0 & 1 & 0 & 0 \\ 0 & 0 & 1 & 0 \\ 0 & 0 & 0 & 1 \end{bmatrix} \tag{2.18}$$

$$DH_2 = \begin{bmatrix} \cos(\theta_{Head}) & -\sin(\theta_{Head}) & 0 & 0 \\ \sin(\theta_{Head}) & \cos(\theta_{Head}) & 0 & 0 \\ 0 & 0 & 1 & L_{Neck} \\ 0 & 0 & 0 & 1 \end{bmatrix} \tag{2.19}$$

$$DH_3 = \begin{bmatrix} 1 & 0 & 0 & d_4 \\ 0 & 1 & 0 & 0 \\ 0 & 0 & 1 & 0 \\ 0 & 0 & 0 & 1 \end{bmatrix}. \tag{2.20}$$

Therefore, the forward kinematics of the *CASTOR* robot head will be defined by the result of the matrix multiplication of R_t and P_0 (see Eq. 2.21), where R_t is the multiplication of the 3 matrices (Eqs. 2.18, 2.19, and 2.20) found in Table 2.2 and P_0 is the (X, Y, Z) coordinate vector of the robot base. The R_t matrix obtained for the CASTOR robot is shown in Eq. 2.22

$$P_f = R_t P_0 = \begin{bmatrix} x \\ y \\ z \\ 1 \end{bmatrix}, \tag{2.21}$$

where

$$R_t = DH_1 \cdot DH_2 \cdot DH_3 \,, \quad P_0 = \begin{bmatrix} 0 \\ 0 \\ 0 \\ 1 \end{bmatrix}$$

$$R_t = \begin{bmatrix} \cos(\theta_{Head}) & -\sin(\theta_{Head}) & 0 & d_3 + d_4\cos(\theta_{Head}) \\ \sin(\theta_{Head}) & \cos(\theta_{Head}) & 0 & d_4\sin(\theta_{Head}) \\ 0 & 0 & 1 & L_{Neck} \\ 0 & 0 & 0 & 1 \end{bmatrix}. \tag{2.22}$$

Solving Eq. 2.21 with the matrix already calculated allows obtaining the three equations that relate the final coordinates x_f, y_f, z_f with the angles of the kinematic chain, which means that the forward kinematics has been solved satisfactorily. In Eq. 2.23 the mathematical expressions of the solution of the forward kinematics of the *CASTOR* social robot neck can be seen.

$$\begin{bmatrix} x_f \\ y_f \\ z_f \\ 1 \end{bmatrix} = \begin{bmatrix} d_3 + d_4 \cos(\theta_{Head}) \\ d_4 \sin(\theta_{Head}) \\ L_{Neck} \\ 1 \end{bmatrix}. \tag{2.23}$$

The inverse kinematic calculation of the *CASTOR* robot head is performed by the geometrical method from the known robot dimensions. For the solution to this problem, it is necessary to define an endpoint for the kinematic chain. In this case, it would be the robots' view at the point (X_f, Y_f, Z_f). As this kinematic chain only consists of 1 DOF, it will only be necessary to find the equation that establishes the relation between the final point (X_f, Y_f, Z_f) and the angle that must be positioned in the neck θ_{Head}.

The top view is shown in Fig. 2.6b. The angle θ_{Head} can be identified in the triangle formed by dimensions X_f, Y_f, and d_3. The way to describe the angle by the geometric method, in this case, is to employ the tangent function since the dimensions of the opposite cathetus to the angle θ_{Head} are known. The mathematical expression that solves the inverse kinematics of the neck based on the known robot parameters is seen in Eq. 2.24.

$$\theta_{Head} = \tan^{-1}\left(\frac{X_f - d_3}{Y_f}\right). \tag{2.24}$$

2.2.4.3 The Upper-Limb Kinematic Chain

For the *CASTOR* robot arm, the kinematic calculation is more complex due to the number of degrees of freedom and their forms of motion. In Fig. 2.5 it can be seen how this kinematic chain depends on the three angles θ_1, θ_2, θ_3, the distances L_1, L_2, L_3, d_2, d_4, and the fixed angle of the passive degree of freedom of the robot θ_{fix}. Using the procedure for the DH parameter finding, Table 2.3 is made, which summarizes the forward kinematics of the left arm of the *CASTOR* robot.

Using these values the relationship between the angles $(\theta_1, \theta_2, \theta_3)$ and the robot hand, which is the endpoint of the arm, can be found. The final equation is given by Eq. 2.21, where R_t is the multiplication of the 7 DH matrices found in Table 2.3.

Table 2.3 DH parameters of the kinematic chain of the *CASTOR* robot arm obtained only by the known dimensions and angles of the robot

Matrix_i	θ_i	d_i	a_i	α_i
1	0	0	d_1	0
2	90	0	0	θ_{fix}
3	0	L_1	0	$180 - (\theta_{fix} + \theta_1)$
4	-90	0	$-d_2$	0
5	0	0	0	$-\theta_2$
6	0	L_2	0	$-(90 + \theta_3)$
7	0	L_3	0	0

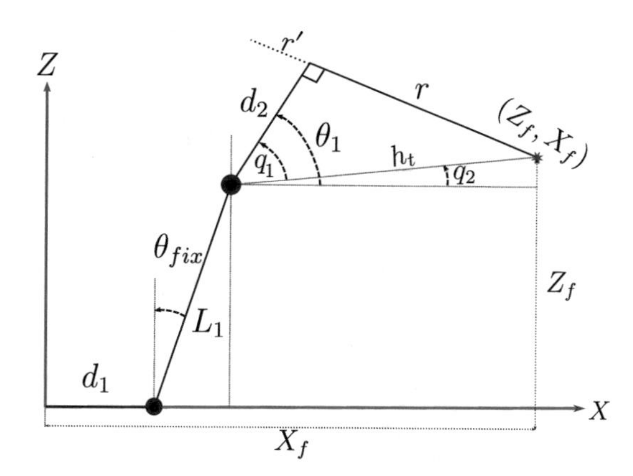

Fig. 2.7 $X - Z$ plane where the angle θ_1 and the actual dimensions needed to determine inverse kinematics equation of this first angle is seen

For the inverse kinematic calculation of this part of the robot it used the geometrical method starting from the known dimensions of the arm and the defined final position (X_f, Y_f, Z_f) to which the robot hand is to be pointed (see Fig. 2.5).

To calculate the first angle of the arm θ_1 the robot must be viewed from the $X-Z$ plane as seen in Fig. 2.7, since in this perspective all dimensions are known or can be calculated. As can be seen, the angle θ_1 is the sum of the two supplementary angles q_1 and q_2. So the mathematical part will be based on determining these angles based on the known dimensions.

The equation that solves the supplementary angle q_2 is seen in Eq. 2.25 and is based on the solution of the small right triangle that is formed with one leg on the X-axis and the other on the Z-axis. Simultaneously, the solution of the supplementary angle q_1 is calculated with the hypotenuse h_t of the small right triangle and the known leg dimension d_2 (see Eq. 2.26).

$$q_2 = \tan^{-1}\left(\frac{Z_f - L_1\cos\left(\theta_{fix}\right)}{X_f - d_1 - L_1\sin\left(\theta_{fix}\right)}\right) \tag{2.25}$$

$$q_1 = \cos^{-1}\left(\frac{d_2}{h_t}\right), \tag{2.26}$$

where

$$h_t = \sqrt{\left(X_f - d_1 - L_1\sin\left(\theta_{fix}\right)\right)^2 + \left(Z_f - L_1\cos\left(\theta_{fix}\right)\right)^2}.$$

As defined, the angle θ_1 is the sum of the two expressions of Eqs. 2.25 and 2.26. This expression, as can be seen, only depends on the known parameters based on the dimensions of the robot and the distances of the endpoint coordinates (X_f, Y_f, Z_f). Equation 2.27 shows the solution of angle θ_1 without mathematical reductions.

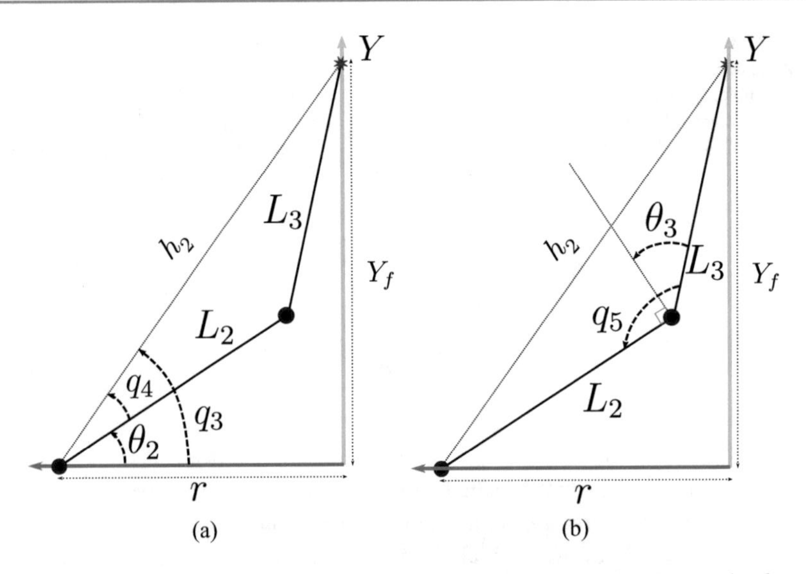

Fig. 2.8 The auxiliary $Y - r$ plane where the angles θ_2 θ_3 and the links L_2 and L_3 can be observed in their actual dimension necessary to determine the inverse kinematics equations of the angles of the joints 2 and 3. Part (**a**) shows the auxiliary variables for calculating the angle θ_2, and part (**b**) shows the auxiliary variables necessary to perform the calculation of the angle θ_3

$$\theta_1 = \cos^{-1}\left(\frac{d_2}{h_t}\right) + \tan^{-1}\left(\frac{Z_f - L_1\cos\left(\theta_{fix}\right)}{X_f - d_1 - L_1\sin\left(\theta_{fix}\right)}\right). \tag{2.27}$$

The angles θ_2 and θ_3 are calculated in the same way. In this case, the perspective of the $Y - r'$ plane as shown in Fig. 2.8 is used. This plane is the only one that allows seeing the dimensions of the links L_2 and L_3 in their actual size and where it is easier to find the geometric expressions to determine the angles of the remaining joints. The angle θ_2, as seen in Fig. 2.8a, is defined as the difference between the auxiliary angles q_3 and q_4, so it is necessary to find the geometric expressions from the triangles formed in the figure that describe these angles with the known dimensions.

The expression that defined the auxiliary angle q_3 (see Eq. 2.28) is calculated from the largest right triangle in Fig. 2.8a formed by the legs Y_f and r. Since the outer triangle is not a right triangle, the sine or cosine rule must be used to find the expression for the auxiliary angle q_4.

$$q_3 = \tan^{-1}\left(\frac{Y_f}{r}\right), \tag{2.28}$$

where

$$r = h_t \sin\left(q_1\right).$$

Since L_1 and L_2 are known from the dimensions of the robot and h_2 can be calculated since it is the hypotenuse of the right triangle formed by Y_f and r (see Fig. 2.8a), the cosine rule is used to relate the angle q_4 to the 3 dimensions of the sides of the triangle. In Eq. 2.29, the angle q_4 is shown cleared from the cosine rule.

$$q_4 = \cos^{-1}\left(\frac{L_3^2 - L_2^2 - h_2^2}{-2L_2h_2}\right), \tag{2.29}$$

where

$$h_2 = \sqrt{Y_f^2 + r^2}$$

.

Knowing the expressions of the supplementary angles of Eqs. 2.28 and 2.29, it is possible to express the angle θ_2 with only terms of known dimensions or final point coordinates. In Eq. 2.30, the inverse kinematics solution for the angle of joint 2 of the *CASTOR* robot arm is seen.

$$\theta_2 = \tan^{-1}\left(\frac{Y_f}{r}\right) - \cos^{-1}\left(\frac{L_3^2 - L_2^2 - h_2^2}{-2L_2h_2}\right). \tag{2.30}$$

The angle θ_3 is calculated in the same way, and the same plane as angle θ_2 (see Fig. 2.8b), but in this case, as the initial position of the robot forearm is at 90 degrees from the link L_2 (defined by the initial position), the general expression for this angle is seen in Eq. 2.31.

$$\theta_3 = 90 - q_5. \tag{2.31}$$

Computing the auxiliary angle q_5 employing the cosine rule (See Eq. 2.32) it is easy to complete the solution of the inverse kinematics of angle θ_3 of the kinematic chain of the *CASTOR* robot arm. In Eq. 2.33, the general solution of the last angle is presented.

$$q_5 = \cos^{-1}\left(\frac{h_2^2 - L_2^2 - L_3^2}{-2L_2L_3}\right) \tag{2.32}$$

$$\theta_3 = 90 - \cos^{-1}\left(\frac{h_2^2 - L_2^2 - L_3^2}{-2L_2L_3}\right). \tag{2.33}$$

Therefore, the equations modeling the inverse kinematics of the robot arm *CASTOR* can be summarized in Eq. 2.34. These inverse kinematics expressions and the forward kinematics calculations were used to develop a case study in which this robot helps children with autism through pre-configured gestures such as waving or pointing to their body parts to improve interaction with the environment [7].

$$\begin{aligned}\theta_1 =&\cos^{-1}\left(\frac{d_2}{h_t}\right)+\tan^{-1}\left(\frac{Z_f - L_1\cos\left(\theta_{fix}\right)}{X_f - d_1 - L_1\sin\left(\theta_{fix}\right)}\right)\\ \theta_2 =&\tan^{-1}\left(\frac{Y_f}{r}\right)-\cos^{-1}\left(\frac{L_3^2 - L_2^2 - h_2^2}{-2L_2h_2}\right)\\ \theta_3 =&90-\cos^{-1}\left(\frac{h_2^2 - L_2^2 - L_3^2}{-2L_2L_3}\right).\end{aligned} \tag{2.34}$$

Knowing the methodology of the kinematic calculation of the *CASTOR* robot, it is possible to perform analysis of more complex robots kinematically speaking. In the following section the kinematics calculation of the lower limb of the social robot *NAO* will be developed.

2.2.4.4 Modeling the Lower Limb of the *NAO* Robot

This social robot has 25 degrees of freedom to perform different tasks and movements divided as follows: in the head it has 2 DOFs, in each upper limb, this robot has 6 DOFs, and in the lower part it has 11 DOFs in total, distributed in each leg and the pelvis [9]. In this case, the 25 degrees of freedom of this robot are active, which means an independent motor modifies the system position for each joint. The movements of each degree of freedom are defined as follows: each arm has 2 DOFs at the shoulder, 2 DOFs at the elbow, 1 DOF at the wrist, and 1 additional DOF for the handgrip. The 2 DOFs on the head allow it to rotate about the yaw and pitch axes. For the lower kinematic chain, each leg has 2 DOFs at the ankle, 1 DOF at the knee, and 2 DOFs at the hip. The pelvis has a unique mechanism composed of two joints attached to each part of the hip. These joints are rotated at 45, facilitating control and reducing the system 1 degree of freedom less than other commercial robots [10]. Figure 2.9 shows the degrees of freedom (Fig. 2.9a) and the kinematic leg chain of the *NAO* robot (Fig. 2.9b).

As shown in Fig. 2.9b, the kinematic chain of the lower limbs consists of 6 DOFs that can be modified to adjust the position of each leg ($\theta_1 \cdots \theta_6$). Overall this robot has a height of 57.3 cm, a width of 27.5 cm, and each leg has a total length of 24.8 cm [8]. To finalize the calculations of the selected kinematic chains, it is necessary to develop the forward and inverse kinematics of the *NAO* robot leg. As already mentioned, this robot has 6 degrees of freedom that must be solved by the methods presented above. The solution of the kinematics is more complex and requires more advanced mathematics than the previous cases.

Based on the kinematic chain of the leg of this robot (see Fig. 2.9b), the DH table is generated, which collects the parameters that solve the forward kinematics of the lower limb. Table 2.4 presents 9 DH matrices that depend only on the dimensions of the leg and the angles of the motors.

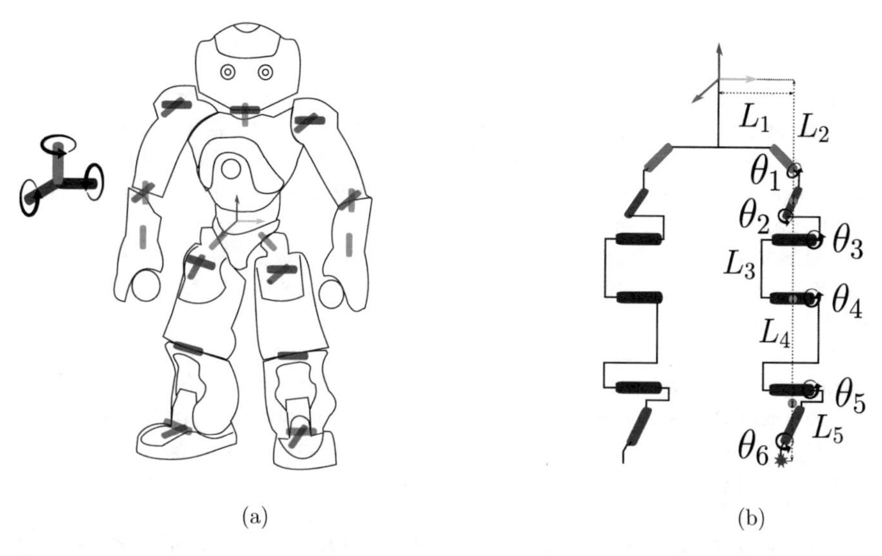

Fig. 2.9 The location of each degree of freedom of the *NAO* robot is shown in part (**a**). Part (**b**) shows the dimensions and angles involved in the kinematic chain of the robot leg

Table 2.4 DH parameters of the *NAO* robot leg kinematic chain obtained with only the known robot dimensions and known angles

Matrix_i	θ_i	d_i	a_i	α_i
1	90	0	0	45
2	90	L_1	0	180
3	θ_1	0	0	−90
4	$\theta_2 + 135$	0	0	−90
5	θ_3	0	L_2	0
6	θ_4	0	L_3	0
7	θ_5	0	0	90
8	θ_6	0	L_4	90
9	0	L_5	0	90

The solution of the inverse kinematics of the *NAO* robot leg is considered a high-level mathematical problem given that this 6 DOF kinematic chain generates a non-linear system that is quite complex to solve. To facilitate the calculations of this problem initially matrix transformations based on the forward kinematics are performed to reduce the complexity of the math. The complete calculation and solution of this kinematic chain can be found in [9, 11].

These same concepts of forward and inverse kinematics can be applied in mobile robotics, which aims to find the speeds of each wheel to reach a specific point (inverse kinematics) or according to the configuration of the moving wheels and their speeds to find out where the robot will reach in an estimated time (forward kinematics). The following section will explain how the modeling is performed in mobile robots.

2.2.5 Modeling Mobile Robots

As presented in previous sections, one of the main characteristics of robotic arms or manipulators, and thus crucial difference with mobile robots, is that they are fixed to a specific point and usually comprise of a single chain of actuated links [12]. In this sense, unlike the robotic arms in social robots or exoskeletons, which can move only in a specific workspace, mobile robots are capable of moving around freely and autonomously within a predefined environment [13]. This capability makes the mobile robots suitable for several applications, including gait assistance and rehabilitation.

Before delving into the modeling of these robots, it should be mentioned that the term "*mobile robots*" covers a wide variety of robots, including: (i) ground robots, where *wheeled mobile robots* and *legged mobile robots* are distinguished, (ii) aerial robots or *unmanned aerial vehicles*, (iii) aquatic robots or *autonomous underwater vehicles*, and (iv) hybrid robots, where mobile robots are equipped with one or more manipulators [13].

For the purpose of this book, the focus will be on wheeled mobile robots, since rehabilitation robots such as robotic wheelchairs, smart canes, and smart walkers can be labeled with this category. In general, these types of rehabilitation devices exhibit a locomotion configuration that is based on wheeled mechanical structures. The wheeled mobile robots are prevalent in both the industrial and rehabilitation concepts, given their low mechanical complexity and efficient energy consumption. In this regard, the following sections briefly present the basic concepts of kinematics and locomotion, focusing on smart walkers.

2.2.5.1 Wheeled Locomotion

The wheel is the most popular locomotion mechanism in mobile robotics, and it is also in robotic walkers. It has been demonstrated that implementing wheels provides outstanding efficiencies and simple mechanical structures [12]. Like most wheeled robots, the smart walkers are designed so that all wheels are always in ground contact. Thus, the smart walkers commonly include from three to four wheels to guarantee sufficient stable balance. Considering that patients frequently use smart walkers in clinical or indoor scenarios, no suspension systems are demanded. However, when interacting in uneven terrains, a suspension system might be required to maintain safe ground contact [12].

Considering that balance is rarely an issue in wheeled mobile robots, the focus is often on traction and stability, maneuverability, and control [12]. In this way, the design problem is commonly tackled by proposing an appropriate wheel design. Mainly, Page et al. reported that six types of wheels are often found in smart walkers. These wheel classes include: (i) fixed wheels, (ii) centered orientable wheels, (iii) off-centered wheels, (iv) Swedish wheels, (v) spherical wheels, and (vi) active split offset casters (ASOC) [14]. This report also stated that most of the smart walkers implement fixed wheels and off-centered wheels (i.e., caster wheels). The fixed wheels are commonly linked to the propulsion system, and the caster wheels provide

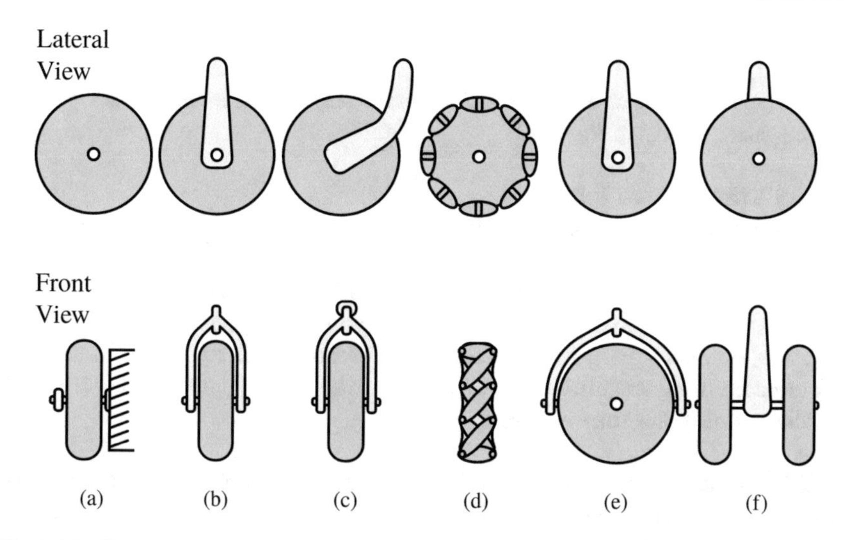

Fig. 2.10 Illustration of the primary wheel classes implemented in smart walkers. (**a**) Fixed wheel. (**b**) Centered orientable wheel or standard wheel. (**c**) Off-centered wheel or caster wheel. (**d**) Swedish wheel. (**e**) Spherical wheel. (**f**) Active split offset caster (ASOC) wheel

stability. Figure 2.10 shows these types of wheels, and a more detailed description of each wheel class can be found in [12, 14].

2.2.5.2 Wheel Configurations in Smart Walkers

Although the selection of the wheel's design is an essential issue, the geometrical arrangement and number of wheels attached to the robot's structure are directly related to the maneuverability, controllability, and stability of the platform. In general, any wheel configuration can be defined by the following elements: (i) the number and type of wheels, (ii) the wheels arrangement, and (iii) the locomotion type or actuated wheels. For instance, a particular robot might have two motorized fixed wheels in the rear and two caster wheels in front, whereas another robot might have two free wheels in the rear and one steered traction wheel in front.

As presented in [12, 13], rolling vehicles in industrial robotic applications can exhibit a wide range of wheel configurations, varying the elements mentioned above (i.e., type, number, and arrangement of wheels). Similarly, in [14], the authors reported the most common configurations in smart walkers. In general, the smart walkers can exhibit the following configurations:

- Two motorized fixed wheels in the rear and one free caster wheel in front [15]
- Two fixed wheels in the rear and one motorized caster wheel in front [16]
- Three motorized omnidirectional wheels [17]
- Two fixed wheels in the rear and two motorized and steered wheels in front [18, 19]
- Two motorized wheels in the rear and two free caster wheels in front [20]

- Two fixed wheels in the rear and two free caster wheels in front (i.e., passive device) [21]
- Two free caster wheels in the rear, two motorized wheels in the middle, and two free caster wheels in front [22]

2.2.5.3 Wheel Drive Types

The configurations mentioned above can also be classified in terms of the drive system that powers the wheels' locomotion. In general, wheeled mobile robots can be configured in six different drive types: (i) differential drive, (ii) tricycle, (iii) omnidirectional, (iv) synchro drive, (v) Ackerman steering, and (vi) skid steering [13]. For simplicity purposes, the focus in this book will be on the drive types that are used in the majority of smart walkers, which are: (i) differential drive, (ii) tricycle, and (iii) Ackerman steering.

- **Differential Drive:** This type of drive system consists of two fixed motorized wheels mounted on the left and right sides of the robot platform. The two wheels can be placed whether in front or in rear and are independently driven. To guarantee balance and stability, one, two, or more free castor wheels are used. Smart walkers such as *AGoRA Walker*, *UFES Smart Walker*, and the *ABSGo++* employ this drive type [15, 20, 22]. The motion possibilities in this configuration depend on the rotation speed and direction of the motorized wheels:
  - *Forward or backward motion:* The wheels rotate at the same speed, and depending on the direction, the structure moves straight forward or backward.
  - *Curved motion:* One wheel rotates faster than the other, making the structure follow a curved path. The slower wheel dictates the turning side.
  - *On-site turning motion:* The wheels rotate at the same speed in opposite directions, making the structure turn about the midpoint of the motorized wheels.
- **Tricycle:** This type of drive system has a single wheel that can be motorized, steered, or both [13]. Smart walkers such as *MARC* employ this drive type [16]. To guarantee balance and stability, two free fixed wheels are placed in the back. In this case, the motion possibilities work as follows:
  - *Forward motion*: The motorized wheel is in the middle position and driven at the desired speed.
  - *Curved motion*: The motorized wheel is positioned at a specific angle and driven at the desired speed.
  - *Circular motion*: The motorized wheel is at 90°, making the structure rotate in a circular path, whose center is the middle point between the rear wheels.
- **Ackerman steering:** This configuration describes the standard steering used in automobiles. Two linked motorized rear wheels and two linked steered front wheels characterize it. Smart walkers such as *GUIDO* and *c-Walker* use this drive type [18, 19]. There are only two motion possibilities:
  - *Forward or backward motion*: The rear wheels rotate at the same speed because they are linked, and the front wheels are not steered.

- *Curved motion*: The rear wheels rotated at the same speed and the front wheels are steered at the desired angle. The structure follows a curved motion with a minimum turning radius.

2.2.5.4 Mobile Robot Kinematics

Mobile robot kinematics describes the behavior of mobile robots considering their physical configuration in a defined workspace, the relations between their geometric parameters, and their mechanical constraints [12,23]. In this sense, this section seeks to describe such considerations when modeling robots like the smart walkers.

At this point, a key difference between mobile robots and manipulators or robotic arms arises, the position estimation. As explained in previous sections, the robot arms in social robots and the exoskeletons have one end that is treated as the fixed point or ground [12]. In this sense, the end effector's position can be instantaneously estimated by following the kinematic equations and measuring the position of each joint. However, with mobile robots the movement is not fixed, and the robot's motion must be integrated over time to estimate its position [12]. Thus, position estimation is a challenging task that cannot be achieved instantaneously.

With this in mind, understanding mobile robot kinematics addresses the problem of describing how each wheel contributes to the overall motion, as well as imposes motion constraints. Moreover, it also addresses the formulation of forward kinematic models that describe the robots' movement in terms of their geometry and wheels' behavior.

Representing Robot Position

The first step in deriving a kinematic model for a smart walker (or a mobile wheeled robot) is representing its position in a particular environment. To this end, the following assumptions are considered:

1. The smart walker is modeled as a rigid body on wheels.
2. The smart walker is only allowed to move in the horizontal plane.
3. The degrees of freedom of the wheels and internal joints of the smart walker are ignored.
4. The translational friction between the wheels and the ground at the point of contact is large enough so that the wheels do not experience translational slippage.
5. The rotational friction between the wheels and the ground at the point of contact is small so that the wheels can rotate.
6. The center of mass is located at the point of interest.

To set proper coordinate systems, two reference frames are defined, the global reference frame and the local reference frame. The global or inertial reference frame is located at the origin $O: \{X_I, Y_I\}$ of the horizontal plane. The local reference frame $\{X_R, Y_R\}$ is located at the robot chassis and defines the point of interest. According to this, only three elements are required to define the position or local frame of a smart walker, the X_R coordinate, the Y_R coordinate, and the orientation φ along with the Z_R or vertical axis. Thus, the pose (i.e., position and orientation) and velocities of a robot at the global reference frame are defined by Eqs. 2.35 and 2.36.

$$\xi_I = \left[x \; y \; \varphi\right]^T \tag{2.35}$$

$$\dot{\xi_I} = \left[\dot{x} \; \dot{y} \; \dot{\varphi}\right]^T . \tag{2.36}$$

Note that in Eqs. 2.35 and 2.36, the robot's position is defined by x and y, which represent any position in the global frame. Similarly, the angular difference between the global and local reference frames is represented by φ. As an illustration, Fig. 2.11 shows a representation of a smart walker in a given environment and the relationships between the global and local reference frames. In this case, the point of interest (i.e., local reference frame) is located at the user's estimated position.

This representation can be generalized employing the concept of *homogeneous transformations*. A *homogeneous transformation* describes the position and orientation of a solid body with respect to the global reference frame using a 4×4 transformation matrix **A**. This transformation matrix is described by Eq. 2.37.

$$\mathbf{A} = \begin{bmatrix} \mathbf{R} & \mathbf{p} \\ \mathbf{0} & 1 \end{bmatrix}, \tag{2.37}$$

where **p** is the position vector and **R** represents the rotation of the local frame with respect to the global frame. Rotations might occur along any axis. However, in mobile robots such as the smart walkers, the motion only occurs on a horizontal

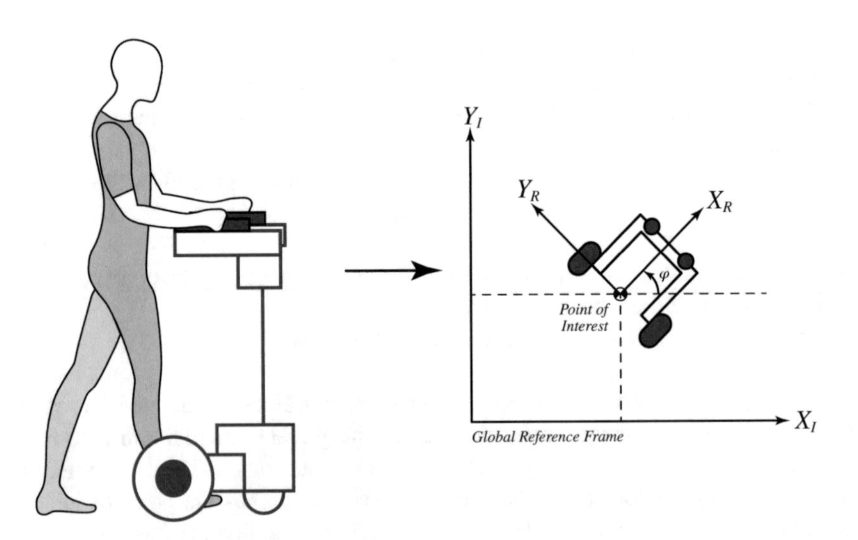

Fig. 2.11 Position representation of a smart walker in a given environment

plane. This also means that the rotations are only with respect to the vertical axis z. In this sense, the pose of a robot in the global reference frame, which is rotated about the vertical axis z, can be described as

$$\xi_I = \mathbf{R}_z \, \xi_R$$

$$\xi_I = \begin{bmatrix} \cos(\varphi) & -\sin(\varphi) & 0 \\ \sin(\varphi) & \cos(\varphi) & 0 \\ 0 & 0 & 1 \end{bmatrix} \begin{bmatrix} x_R \\ y_R \\ \varphi \end{bmatrix} \tag{2.38}$$

$$\xi_I = \begin{bmatrix} x_R \cos(\varphi) & -y_R \sin(\varphi) & 0 \\ x_R \sin(\varphi) & y_R \cos(\varphi) & 0 \\ 0 & 0 & \varphi \end{bmatrix}.$$

The orthogonal rotation matrix $\mathbf{R}_z$ is also useful to map motion along the axes of the global reference frame to motion along the axes of the local reference frame [12]. In particular, this mapping is a function of the pose of the robot as described in Eq. 2.39

$$\dot{\xi_R} = \mathbf{R}_z^{-1} \, \dot{\xi_I}$$

$$\dot{\xi_R} = \begin{bmatrix} \cos(\varphi) & \sin(\varphi) & 0 \\ -\sin(\varphi) & \cos(\varphi) & 0 \\ 0 & 0 & 1 \end{bmatrix} \begin{bmatrix} \dot{x_R} \\ \dot{y_R} \\ \dot{\varphi} \end{bmatrix}. \tag{2.39}$$

Nonholonomic Constraints

As described in [23], and without delving into the strict definition of nonholonomic constraints, mobile robots are systems that are subject to these constraints as they are under-actuated robots. When referring to *nonholomicity*, it is stated that if there is a difference between the number of degrees of freedom (n) of a robot and the independent motions (k) that the robot can produce, the *nonholomicity exists*. For instance, in smart walkers with differential drive configurations, only two wheels are actuated, that is, $k = 2$, whereas three degrees of freedom exist, that is, $n = 3$. Thus, there is one nonholonomic constraint (i.e., $n - k = 1$) [23].

In general, this nonholonomic constraint in mobile robotics often refers to: (1) the motion constraint of a disk that rolls on a plane without slipping or (2) the speed restriction in the robot's traverse direction [23]. This also means that the velocities $\dot{x}$, $\dot{y}$, and $\dot{\varphi}$ cannot take arbitrary values [23], and thus they are constrained as shown in Eq. 2.40.

$$\dot{x} \sin(\varphi) - \dot{y} \cos(\varphi) = 0. \tag{2.40}$$

Unicycle Kinematic Model

One of the most standard and straightforward kinematic models for wheeled mobile robots refers to the unicycle model. With this model, the robots are analyzed as

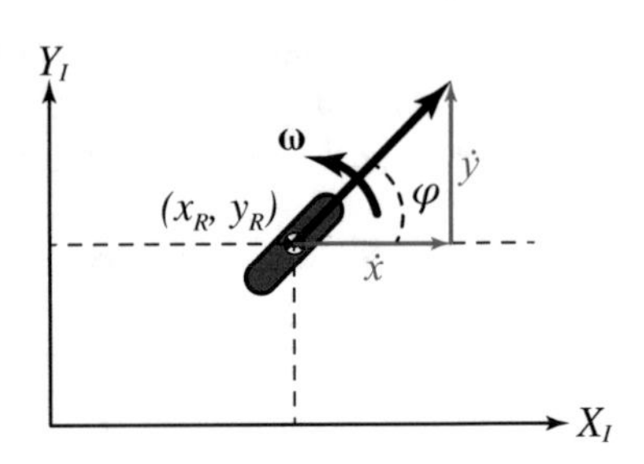

Fig. 2.12 Illustration of the unicycle kinematic model

if they were a simple conventional wheel rolling on a horizontal plane [23]. This configuration is illustrated in the global reference frame in Fig. 2.12.

The pose of the unicycle configuration can be described using the representation of Eq. 2.35, which is $\xi_I = [x_R, y_R, \varphi]^T$. This representation uses the position coordinates of the point of interest located at the ground contact of the wheel, and the orientation with respect to the x-axis [23]. Figure 2.12 also shows the linear velocity of the wheel $\boldsymbol{\mu}$ and the angular velocity about the vertical axis $\boldsymbol{\omega}$. Moreover, from Fig. 2.12 the motions along x-axis and y-axis can be used to describe the unicycle model as follows:

$$\dot{\xi}_I = \begin{bmatrix} \dot{x_R} \\ \dot{y_R} \\ \dot{\varphi} \end{bmatrix} = \begin{bmatrix} \mu \cos(\varphi) \\ \mu \sin(\varphi) \\ \omega \end{bmatrix} = \begin{bmatrix} \cos(\varphi) \\ \sin(\varphi) \\ 0 \end{bmatrix} \mu + \begin{bmatrix} 0 \\ 0 \\ 1 \end{bmatrix} \omega. \tag{2.41}$$

Note that the nonholonomic constraint can be derived from $\dot{x_R}$ and $\dot{y_R}$ by eliminating μ. Moreover, considering the unicycle model presented in Eq. 2.41, and assuming that the linear and angular velocities are the joints of the system, the Jacobian matrix of the system is [23]:

$$\mathbf{J} = \begin{bmatrix} \cos(\varphi) & 0 \\ \sin(\varphi) & 0 \\ 0 & 1 \end{bmatrix}. \tag{2.42}$$

Displaced Kinematic Model

Another interesting kinematic model arises when the point of interest is displaced to a new point in the robot's front. This model adds little complexity to the previous unicycle model, while functional, for instance, in path following tasks. Figure 2.13 illustrates the new location of the point of interest. As described in [24], although the displaced point of interest can have velocities in any direction, the robot is still considered as a nonholonomic robot.

Moreover, from Fig. 2.13 it can be observed that the x and y coordinates of the previous unicycle model are displaced by Δx and Δy, whereas the orientation of the robot is still about to vertical axis located in the middle of the wheels. In this sense, the pose of the robot is now defined by Eq. 2.43.

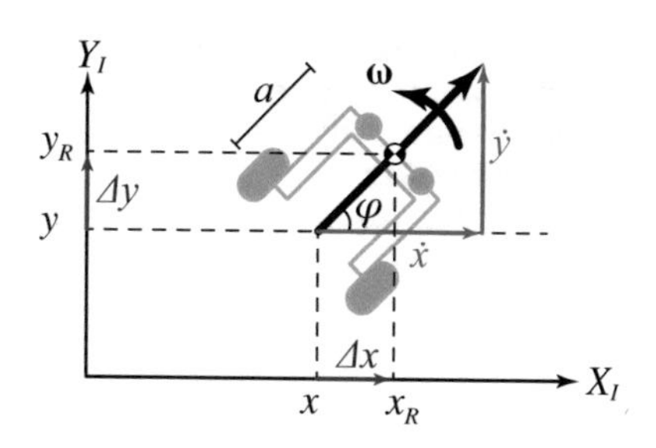

Fig. 2.13 Illustration of the displaced kinematic model

$$\xi_I = \begin{bmatrix} x + \Delta x \\ y + \Delta y \\ \varphi \end{bmatrix}. \tag{2.43}$$

Considering that the distance between the middle of the wheels and the new point of interest is a, the displacements Δx and Δy can be described as follows:

$$\xi_I = \begin{bmatrix} x + a\ \cos(\varphi) \\ y + a\ \sin(\varphi) \\ \varphi \end{bmatrix}. \tag{2.44}$$

To obtain the final displaced kinematic model, the motions along x_R and y_R are required. Thus, taking derivatives at both sides of Eq. 2.44, the following model is obtained:

$$\begin{aligned} \dot{\xi}_I &= \begin{bmatrix} \dot{x} - a\ \dot{\varphi}\ \sin(\varphi) \\ \dot{y} + a\ \dot{\varphi}\ \cos(\varphi) \\ \dot{\varphi} \end{bmatrix} = \begin{bmatrix} \mu\ \cos(\varphi) - a\ \omega\ \sin(\varphi) \\ \mu\ \sin(\varphi) + a\ \omega\ \cos(\varphi) \\ \omega \end{bmatrix} \\ \dot{\xi}_I &= \begin{bmatrix} \cos(\varphi) \\ \sin(\varphi) \\ 0 \end{bmatrix} \mu + \begin{bmatrix} -a\ \sin(\varphi) \\ a\ \cos(\varphi) \\ 1 \end{bmatrix} \omega. \end{aligned} \tag{2.45}$$

In this case, assuming the linear and angular velocities as the system's action variables, the Jacobian matrix of the system is

$$\mathbf{J} = \begin{bmatrix} \cos(\varphi) & -a\ \sin(\varphi) \\ \sin(\varphi) & a\ \cos(\varphi) \\ 0 & 1 \end{bmatrix}. \tag{2.46}$$

Differential Drive Kinematic Model

Considering that the most of smart walkers in the literature employ a differential drive locomotion, it is of great relevance to obtain a kinematic based on the independent speed of the actuated wheels. Figure 2.14 illustrates the geometry and

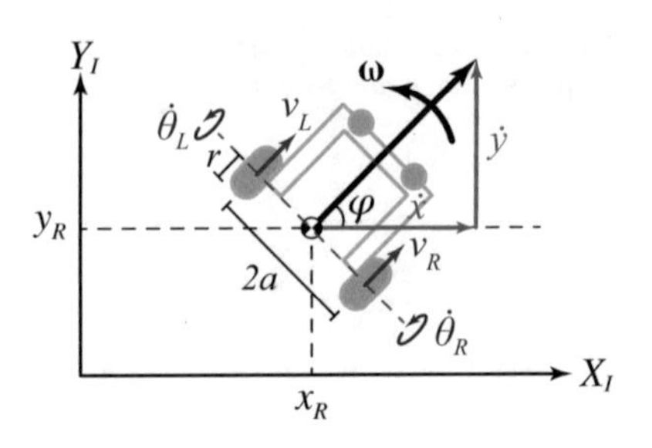

Fig. 2.14 Illustration of kinematic and geometrical parameters for the differential drive configuration

kinematic parameters of a differential smart walker with two actuated wheels in the rear.

As shown in Fig. 2.14, the angular speeds of the left and right wheels are $\dot{\theta_L}$ and $\dot{\theta_R}$, respectively. Similarly, the linear velocities of the left and right wheels are v_L and v_R, respectively. Moreover, the point of interest in this formulation is the same as in the unicycle, i.e., in the midpoint between the wheels, so that the linear and angular velocities of the smart walker are also μ and ω, respectively. In this sense, the velocities of each wheel can be described by Eq. 2.47 [23].

$$\begin{aligned} v_L &= \mu - a\dot{\omega}, \\ v_R &= \mu + a\dot{\omega}. \end{aligned} \tag{2.47}$$

From Eq. 2.47, v_R and v_L are added and subtracted to obtain

$$\begin{aligned} \mu &= \frac{1}{2}(v_R + v_L), \\ 2a\omega &= v_R - v_L. \end{aligned} \tag{2.48}$$

Considering the non-slippage condition previously defined, the linear velocities of the wheels can also be defined as $v_R = r\theta_R$ and $v_L = r\theta_L$, respectively. In this way, using the definition of $\dot{x_R}$ and $\dot{y_R}$ described by Eq. 2.41, the kinematic model for the differential drive configuration can be derived from Eq. 2.48 as follows:

$$\begin{aligned} \dot{\xi_I} &= \begin{bmatrix} \dot{x_R} \\ \dot{y_R} \\ \dot{\varphi} \end{bmatrix} = \begin{bmatrix} \frac{r}{2}(\dot{\theta_R}\cos(\varphi) + \dot{\theta_L}\cos(\varphi)) \\ \frac{r}{2}(\dot{\theta_R}\sin(\varphi) + \dot{\theta_L}\sin(\varphi)) \\ \frac{r}{2a}((\dot{\theta_R} - (\dot{\theta_L}) \end{bmatrix} \\ \dot{\xi_I} &= \begin{bmatrix} \frac{r}{2}\cos(\varphi) \\ \frac{r}{2}\sin(\varphi) \\ \frac{r}{2a} \end{bmatrix} \dot{\theta_R} + \begin{bmatrix} \frac{r}{2}\cos(\varphi) \\ \frac{r}{2}\sin(\varphi) \\ \frac{-r}{2a} \end{bmatrix} \dot{\theta_L}. \end{aligned} \tag{2.49}$$

In this case, assuming the linear velocities of the wheels as the system's joints, the Jacobian matrix is

$$\mathbf{J} = \begin{bmatrix} \frac{r}{2}\cos(\varphi) & \frac{r}{2}\cos(\varphi) \\ \frac{r}{2}\sin(\varphi) & \frac{r}{2}\sin(\varphi) \\ \frac{r}{2a} & \frac{-r}{2a} \end{bmatrix}. \tag{2.50}$$

2.3 Robotic Actuation Systems

An actuation system means a system that uses a type of energy at its input. As a result, it is generated mechanical energy, and the systems that provide this function are named transducers. These actuation systems are used for various workspaces and different applications. In this particular case, the rehabilitation and assistance applied robotics mentioned in Chap. 1 contemplates using actuation systems in robots whose working environment is shared with humans. Therefore, the implementation of different actuation systems is conditioned by the collaboration level that each of these robotic devices provides to the human. For this purpose, factors such as the interaction of forces presented between the human and the robotic device, guaranteeing the user's safety during the interaction with the robotic device, protecting the robotic device during the development of various tasks, and evaluating the similarity of the robotic device's movements compared to the human movements. In this sense, the purpose of this section is to show the different actuation systems used in devices designed for clinical environments. This section will focus on this topic in the following order: (i) actuation systems in lower-limb exoskeletons; (ii) actuation systems in social robotics; and (iii) actuation systems in robotic walkers.

2.3.1 Actuation Systems in Lower-Limb Exoskeletons

An essential factor in developing lower-limb exoskeletons is the implementation of an actuation system to transmit a force to the user's joints using an energy source. Currently, several proposals can be found that fulfill this purpose. Some of these include electric actuation systems, hydraulic actuation systems, and pneumatic actuation systems. Besides, various power transmission mechanisms complement these actuation systems. The main purpose of this section is to provide an actuation system general description mentioned. Additionally, some of the transmission mechanisms are usually mentioned in the literature and applied in the design and development of lower-limb exoskeletons.

2.3.1.1 Electric Actuation System

The electric actuation system is one of the most widely used applications related to the development of lower-limb exoskeletons [25]. This system uses electrical signals as a power source. In this case, electrical signals are converted into mechanical energy using DC electric motors. This actuation system can generate the torque profiles required to generate movement in the user's joints. However, the use of the electric actuation system has some disadvantages. For example, the actuation system's size and weight depend on the torque profiles that it can generate [26], the cost of this system is higher compared to other actuation systems (hydraulic and pneumatic) [27, 28], and it does not present a low impedance in comparison to the pneumatic actuators (back drivable device) [29, 30]. Some of these disadvantages can be compensated by using a sensory interface and various types of mechanical energy transmission. However, this increases the cost of the system. Finally, this system is used to implement position and velocity control in the user's joints [26]. Some examples of the use of this system will be shown in Sect. 2.3.1.4.

2.3.1.2 Hydraulic Actuation System

Other actuation systems have been used to develop active joints in lower-limb exoskeletons, such as hydraulic actuation systems. These systems are composed of a fluid or liquid (oil, fuel) with high system viscosity [31, 32]. The primary purpose of this system is focused on the generation of a force/torque that would generate the joint's movement [32]. One of the advantages in its implementation is that it has a high power-to-weight ratio and the ability to move objects at low speeds and operate at a constant pressure without requiring large amounts of additional energy [33]. Another property of its use is the zero impedance to the joint. This means that this actuation system will allow the movement of the joint without increasing the user's metabolic cost [34]. Finally, this system does not require that the complete structure of the actuation system be located in the joint of the device. Therefore, much of the actuation system can be located in a section where it will not suffer damage caused by the joint movement [34]. On the other hand, this actuation system is not efficient for implementing a position controller [32]. As mentioned, this system is implemented to perform force/torque based control strategies. As an example, Fig. 2.15 shows a schematic of the location of part of the hydraulic actuation system in a lower-limb exoskeleton joint.

The implementation of this actuation system in lower-limb exoskeletons has been relevant in some commercially distributed devices and others that have demonstrated their effectiveness in various competitions. Exoskeletons such as *BLEEX* (University of California Berkeley, USA) [35] apply this principle to complement hip, knee, and ankle movements in the sagittal plane, this design contributed to the *BLEEX* exoskeleton to operate at a speed of 1.3 m/s [36]. *CPWalker*, a system comprised of a body weight support (BWS), robotic walker,

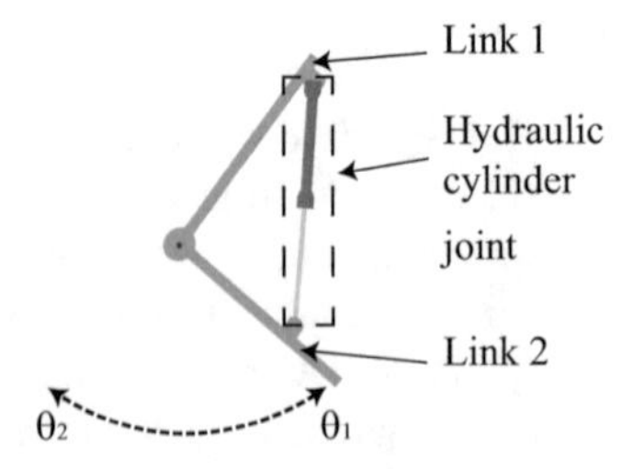

Fig. 2.15 Part of the hydraulic actuation system (hydraulic cylinder) is coupled between the two links. Subsequently, a stiffness variation applied in the hydraulic cylinder could allow various movements in the user's limbs

and lower-limb exoskeleton, uses a hydraulic pump in the system used for the BWS to lift the user's body [37]. Other exoskeletons such as those mentioned in [38] apply this actuation system at the hip and knee; thus, thrust force values from 5 N to 2 KN are obtained. An *ElectroHydraulic AFO* (EHO) (Université Laval, Canada) is considered a back drivable system given its implementation of hydraulic actuation systems.

2.3.1.3 Pneumatic System Actuator

The electrical actuation system has been applied for various applications. However, the acquisition of these actuation systems requires a high cost. For this reason, other alternatives are investigated to fulfill the electrical actuation system function. In this sense, the pneumatic actuation system has been adapted to these applications related to the exoskeleton design. Its performance is based on compressed air as an energy source to generate calculated pressures in the system presented in [39]. As a result, calculated force or torque is applied to the exoskeleton's joints. This actuation system shares some characteristics of hydraulic actuation systems, such as low effectiveness in developing a velocity or position control [39,40]. Additionally, it performs a non-linear response due to the element that it uses to generate energy. On the other hand, this system provides higher efficiency in implementing control strategies with force/torque [41]. Some of the advantages of this actuation system are the low cost, the facility for coupling in the mechanical structure, a lightweight structure [26, 42], and low difficulty for the system maintenance [42]. Some disadvantages of the system are the low effectiveness of implementing position or velocity control [42] and the actuation system not efficient in power transmission that is not used to generate fast responses [40]. Finally, the actuator behaves like a damping system that varies the stiffness of the joint and avoids abrupt movements of the system [40].

Previously, this actuation system was not considered for lower-limb exoskeleton actuation systems due to limited precision and accuracy in some cases. Therefore, they were implemented on stationary platforms. On the other hand, the performance of these systems could depend on the type of control applied on the platforms. In this sense, several lower-limb exoskeletons can be found that use this type of actuation system. For example, *ALEX* (University of Delaware, USA) uses a Pneumatic Muscle Actuator (PMA) to actuate the sagittal plane's hip and knee joints. As a result, an assist-as-needed (AAN) control system was developed, which adjusts the assist level of the device by evaluating the angular position of the joint [43]. Lower-limb exoskeletons for activities of daily living assistance such as *WalkOn* use pneumatic cylinders for knee joint actuation in the sagittal plane, obtaining 18 m/min of gait speed [44]. Likewise, these actuators are applied in the development of soft robots, where pneumatic muscles are placed to complement the muscle contraction for the generation of dorsi-plantarflexion and inversion–eversion movements in [45]. On the other hand, using these devices can also reduce the metabolic cost of a healthy user using an ankle exoskeleton during gait with a pneumatic actuation system [46].

2.3.1.4 Transmission Mechanisms

Lower-limb exoskeleton joints have been actuated by various transmission mechanisms; these mechanisms fulfill the objective of applying the assistive torques for the hip, knee, and ankle joints of the lower limb. Several actuation systems have been developed to support the user's movements and provide assistance in the primary movements of the lower limbs of human body. The following are the main categories:

- **Stiff Actuators:**
  The implementation of electric motors in the center of rotation (CR) of the joints of a lower-limb exoskeleton has been applied in the development of exoskeletons focused on activity of daily living assistance. This type of system has been used to execute predefined trajectories [47] applying a position controller. For instance, *ReWalk* (ReWalk Robotics, United Kingdom) [48], *Ekso* (University of California Berkeley, USA) [49], and *Indego* (Vanderbilt University, USA) [50] use this type of methodology for the execution of various activities of daily living with predefined trajectories. Usually, this actuation system focuses on the assistance of the hip, knee, and ankle joints in the sagittal plane, to provide primary movements of the human body in activities of daily living and walking activity. Figure 2.16 shows an example of a stiff actuator implemented in the lower-limb exoskeleton joint.

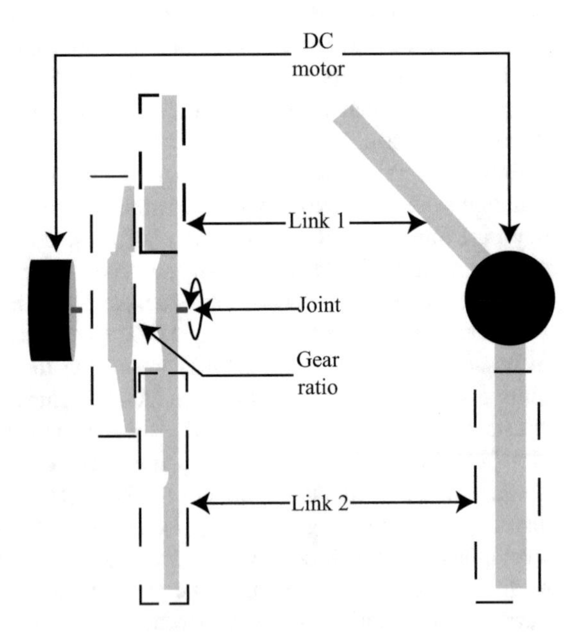

Fig. 2.16 Stiff actuator system schematic, the DC motor is coupled to the gearbox ratio. Subsequently, this actuation system is located directly in the exoskeleton's joint

This transmission mechanism uses a DC electric motor, and the gearbox magnifies the torque generated by the motor as follows:

$$\tau_j = \tau_n \times n_{gr}, \tag{2.51}$$

where τ_j is the joint torque, τ_n is equal to the motor nominal torque, and N_{gr} is the gearbox ratio. In general, exoskeletons with stiff actuators are developed for spinal cord injury users. Therefore, these are designed as portable and wearable devices, for prolonged use. These devices are complemented with a Graphical User Interface that provides the user with the capability to control the activities developed using the lower-limb exoskeleton [44, 48, 51]. On the other hand, this stiff actuator can be used in control strategies aimed in the physical Human–Robot Interaction (pHRI) using sensors that estimate the user's kinetic parameters and simulate a back drivable device. This actuation system will allow the generation of required torques to the user's lower-limb movements [4, 52, 53]. The disadvantage of using gearboxes is based on the reduction of the angular velocity of the actuation system by dividing the value of the motor nominal speed by the gearbox ratio. The implementation of DC motors with gearboxes is adapted to the joints to generate the desired angular positions in the walking activity, assisting the lower extremities [44, 54, 55]. On the other hand, the reduction ratio of the gearboxes limits the movements of the joint performed by the user, so this actuation system is used in lower-limb exoskeletons for people with spinal cord injury.

- **Compliant Actuators**
  The series elastic actuators (SEAs), for example, are presented to a shock tolerance actuation system [56]; this main characteristic focuses on uncoupling the joint to the DC motor and the gearbox ratio using an elastic element [57]. As a consequence, the SEA properties include a low impedance and low friction. Generally, it is implemented in the development of force controller; impedance–admittance controller is applied in devices to take advantage of the physical Human–Robot Interaction (pHRI) [41]. The SEAs respond to the replacement of high precision position controllers in manipulator robots that interact with users according to the workspace [58]. In the last years, various compliant actuators have been presented in the literature. This section is defined as a mechanically adjustable compliance and controllable equilibrium position actuator (MACCEPA) [59]. This actuator was used to develop some lower-limb exoskeletons [51,60,61] aimed at rehabilitating activities of daily living applying the pHRI definition.

  The MACCEPA is an actuation system focused on including the user's motion intention in the control strategies. Its development is based on a system mass–spring system, which exercises a pretension on the spring varying the stiffness of the joint for the development of various activities. Figure 2.17 shows the concept of the MACCEPA system presented in [59]. With radius (r), the disc rotates to

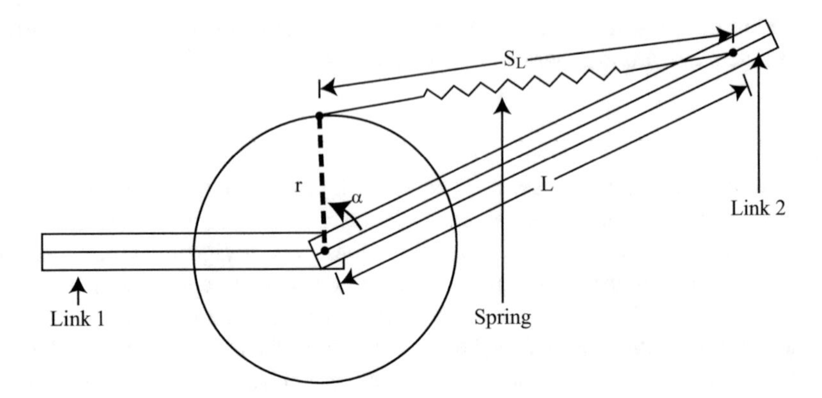

Fig. 2.17 MACCEPA definition, where L is the length of the starting point of the joint to the point where the spring is attached; S_L is equal to the length of the pretensioned spring; r is equal to the disc radius that provides the pretension to the system

vary the spring pretension, which provides stiffness to the system according to the activity to be performed.

The torque generated by this system is expressed by the equation shown below:

$$\tau = krL \times \sin\alpha \times \left(1 + \frac{P - |L - r|}{\sqrt{r^2 + L^2 - 2rL \times \cos\alpha}}\right), \tag{2.52}$$

where k equals to the system's elasticity constant, r is the disc radius that changes the pretension of the spring, L is the length from the disc location to the point where the end of the spring is attached, and P corresponds to the extension of the spring caused by the pretension [59]. So the MACCEPA actuator adjusts the rotation of the disc to generate pretension in the joint; this allows generating different torque profiles depending on the value of α and the elasticity constant of the spring. The actuation system has been characterized as a back drivable device. Likewise, the variation of the spring pretension can turn it into a non-back drivable system, capable of generating 70 Nm for the hip and knee joints [60].

Exoskeletons such as *ALTACRO* (Vrije University, Belgium) [60] implement the MACCEPA for gait rehabilitation by testing different levels of joint stiffness and the effects that this configuration has on gait [59]. Wearable lower-limb exoskeletons such as *Biomot* (Future and Emerging Technologies (FET), Spain) [61] apply this actuation system to the hip and knee joints. Proposing an exoskeleton focused on the assistance of these joints in the sagittal plane.

- **Cable-Driven Actuators:**
  The actuation systems presented in the literature for lower-limb exoskeletons were focused on the generation of the necessary torques for hip, knee, and in some cases ankle joints. On the other hand, some factors such as the actuation

system location, dependence on a rigid structure, weight, among others generate limitations for the movement of these joints. For this reason, some systems were developed, such as the driven cables, which present an actuation mechanism focused on the implementation of a mass–spring–damper system.

These actuation systems offer the capability to generate movements in the hip, knee, and ankle joints, without presenting constraints associated with the exoskeleton mechanical structure [62]. The implementation of this actuation system contributes to the pHRI, where the main objective is the user's participation in the development of activities of daily living. Their location is distributed in a way that they are not located directly in the lower-limb joints. This characteristic decreases the weight perceived by the user's joints [26, 63]. Additionally, the weight of these joints is considered lower than the implementation of electric motors adapted with gearboxes. Finally, the implementation of these systems does not require a rigid structure [64].

As an example, LOwer extremity Powered ExoSkeleton (*LOPES*) (University of Twente, Netherlands) [62] makes use of this system, proposing a series elastic actuator system using Bowden cables. The LOPES actuation system includes the use of Bowden cables, complemented with the implementation of springs for the lower-limb joints. This system was focused on the variation of the pretension of the springs, for the generation of assistive torques for the activity of the march. Factors such as wear and friction of the cables with the system structure are relevant for developing a closed-loop control [62]. As an example, Fig. 2.18a shows a schematic of this system. An Exosuit *Myosuit* (ETH Zurich, Switzerland) [65] is a wearable device that does not have a rigid structure, and implements driven cable actuators to assist the sitting-to-standing activity and walking activity. The *Myosuit* actuation system allows the user to generate voluntary movements monitored through loads cells and IMU sensors to support the motion intention generated by the user who suffers muscle weakness. Figure 2.18b shows a schematic of this exosuit in the user's lower limbs.

In conclusion, the actuation systems shown in this section are some of the relevant actuation mechanisms developed for the lower-limb exoskeletons most mentioned in the literature. Each of these actuation systems was aimed to use the pHRI in the control strategies developed according to the robot workspaces. In this case, the lower-limb exoskeletons mentioned are designed for the rehabilitation and assistance of the relevant activities of daily living performed by a user who suffers from a neurological disease. Each of the mentioned actuation systems presents some advantages and disadvantages depending on the requirements in maximum torque, maximum angular velocity, actuation system's weight, usability, and the required control strategies in developing various activities. Finally, these factors will affect the support and assist the hip, the knee, and the ankle joint in various primary movements of the human body.

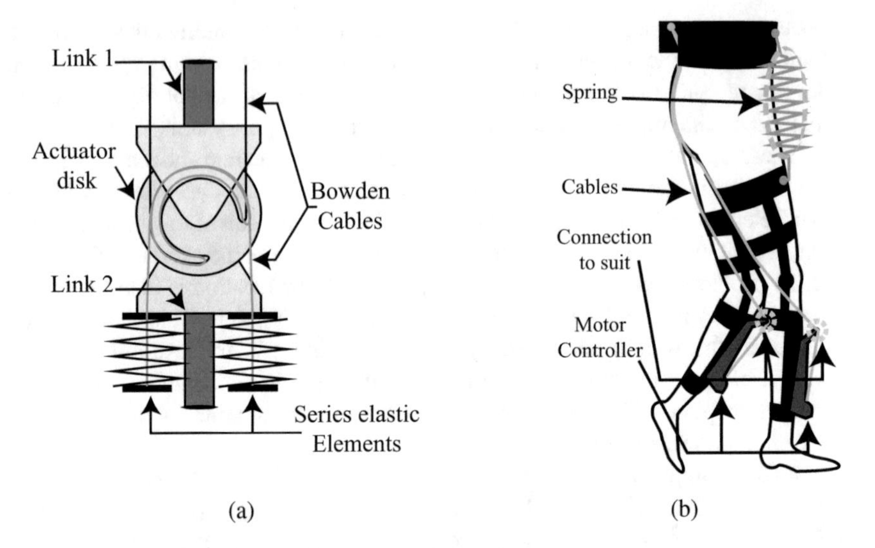

Fig. 2.18 Cable-driven actuators schematic of lower-limb exoskeletons developing activities of daily living; (**a**) actuation system schematic of the LOPES exoskeleton developed by Bowden cables located in lower-limb joints; (**b**) exosuit named *Myosuit* is considered a wearable exoskeleton that applies soft robotics to support various activities of daily living using driven cables actuation system in the lower-limb joints

2.3.2 Actuation Systems in Social Robots

As mentioned in the previous section (Sect. 2.3.1) there are different types of actuation for lower-limb exoskeletons, within which are pneumatic, hydraulic, or electric actuation systems. In Social Assistive Robots (SAR), different actuation systems based on electric actuation systems have been implemented, as shown in Table 2.5.

The table shows the most common types of actuation systems on social robots. In this case, most of the robots are actuated by stiff actuation systems and a few by an actuation system based on series elastic actuators (SEAs). The stiff actuation systems are composed of DC motors, and a gearbox that increases the torque generated by the motor. On the other hand, actuation systems based on SEA consist of an elastic element between the DC motor and the gearbox. This elastic element provides benefits such as shock tolerance, lower reflected inertia, and minor damage to the environment [82].

Most of the SAR contain many degrees of freedom because they attempt to generate realistic biological movements. The *NAO* robot is one of the robots with the most degrees of freedom allowing it to generate movements such as walking and dancing. Other robots, such as *CASTOR*, *KASPAR*, *Huggable Bear*, and *Paro*, have less complex actuation systems but perform limb movements or generate facial expressions.

Table 2.5 Actuation systems in Social Assistive Robots

Robot	Purpose	Actuation system	Degrees of freedom
CASTOR [7]	Autism spectrum disorder	SEA	14
Keepon [66]	Autism spectrum disorder	Stiff	4
Probo [67,68]	Autism spectrum disorder	SEA	20
KASPAR [69,70]	Autism spectrum disorder	Stiff	17
Pleo [71,72]	Support for well-being	Stiff	14
TIAGo [73]	Elderly care	Stiff	12
Pepper [74]	Elderly care	Stiff	19
HOBBIT [75]	Elderly care	Stiff	5
Huggable Bear [76]	Autism spectrum disorder	Stiff	8
NAO [77–79]	Autism spectrum disorder	Stiff	25
Paro [80,81]	Autism spectrum disorder	Stiff	7

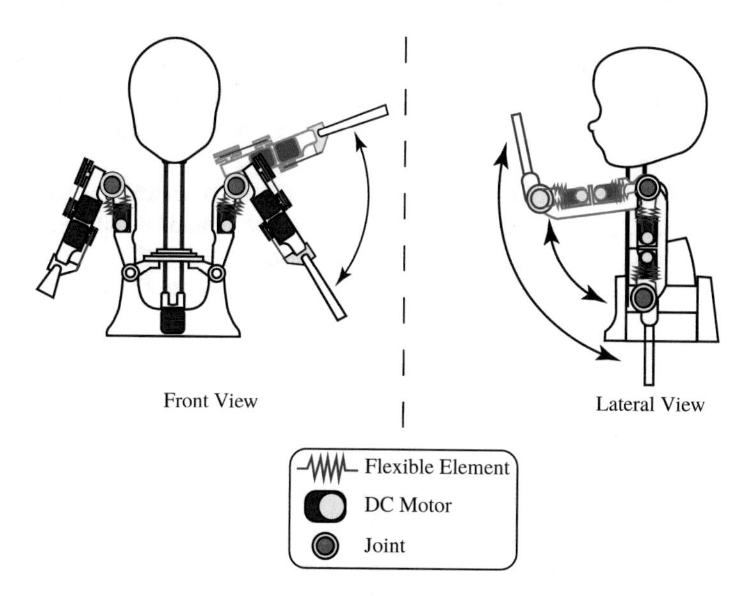

Fig. 2.19 Series elastic actuators (SEAs) in *CASTOR* robot improve the interaction and safeguard the structure to provide the ability to interact physically and socially with CwASD during therapy

The robots focused on therapies with children with autism spectrum disorder (CwASD) such as *CASTOR* and *Probo* are the only ones based on a SEA actuation system. This is for the type of actuation, and SEA allows the execution of therapies focused on physical interaction, necessary for the development of CwASD, without representing a risk for both the child and the robot [83]. Figure 2.19 shows the actuation system in *CASTOR* Robot, an example of SEA in SAR.

2.3.3 Actuation Systems in Smart Walkers

In case of mobile wheeled robots such as the smart walkers, the actuation systems are commonly more straightforward than the actuators employed by social robots and exoskeletons. In general, the actuation systems in smart walkers can be categorized according to their purpose: (1) actuation systems for motion and (2) actuation systems to interact with the user. As shown in Fig. 2.20, there are several actuation interfaces that can be equipped in smart walkers. Depending on the design requirements or rehabilitation purpose, these, more than one of these, actuation systems can be used.

2.3.3.1 Actuation Systems for Motion

Most of the robotic walkers reported in the literature implement differential drive systems, in which only two motorized wheels are used [15,20,22,84]. Usually, each of the actuated wheels is equipped with a brushed DC electric motor coupled to a gearbox to increase the delivered torque by the motor. However, the most current developments in robotic walkers migrate to brushless DC motors that are directly integrated into the wheels (i.e., BLDC hub), such as the wheel motors equipped in electric scooters.

Although the literature lacks in reporting the torque characteristics required for a walker-assisted gait application, the generated torque should be sufficient to move the robot chassis, while a person partially supports its weight on it. In this way, when selecting or designing these types of actuation system, it is helpful to consider that:

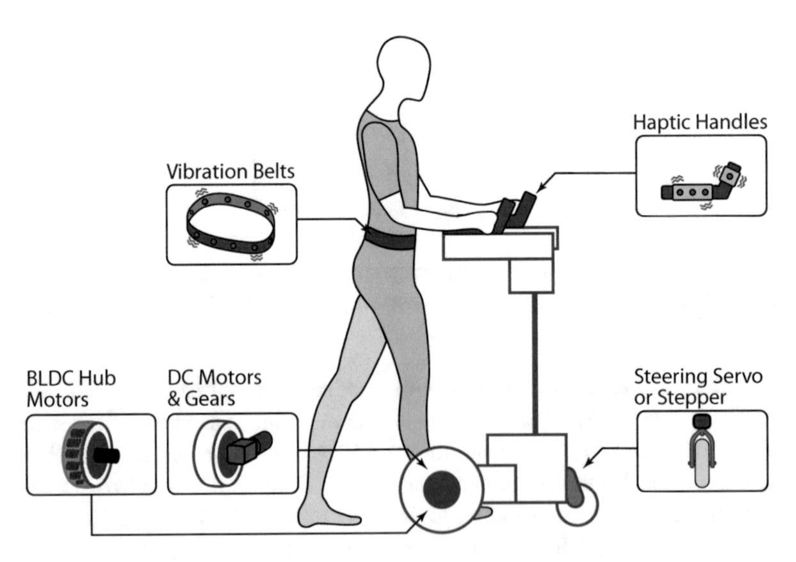

Fig. 2.20 Illustration of common actuation interfaces in smart walkers

1. The smart walker should support an average payload from 60 kg to 90 kg, based on the average weight of an adult male.[1]
2. The smart walker should be able to provide an average linear speed of 1.5 m/s, based on the average walking speed of a healthy adult.[2]

In the tricycle drive and Ackerman steering configurations, the actuation systems are frequently located in front to provide the steering of the device [16, 18, 19]. In such configurations, the actuation is often accomplished by electrical stepper motors or servomotors. Once again, even though the electrical characteristics have not been appropriately benchmarked in the literature, the actuation system should be able to steer the device considering the friction of the wheel and the weight of the smart walker while supporting a user.

Finally, electromechanical brakes or braking systems are often implemented in smart walkers. Although these devices do not provide motion during gait assistance, they are generally used to limit movement in hazardous situations or guidance and cognitive support tasks.

2.3.3.2 Actuation Systems to Interact with the User

There are several ways to communicate information with the user in rehabilitation and assistance tasks in smart walkers. One of these communication channels is achieved using actuators, whether worn by the user or mounted on the device, to provide haptic feedback to the user [21]. In these scenarios, small DC motors or vibration motors are often equipped, through which information such as the presence of obstacles, a dangerous situation, or the path to be followed can be indicated to the user. For instance, vibration belts and haptic handlebars have been reported in the literature [21, 85].

2.4 Robotic Sensory Architectures

One of the most important tasks in assistive and rehabilitation robots is the ability to acquire information about the users and the surrounding environment. This is commonly accomplished by sensing devices that are capable of measuring physical magnitudes and converting them into electrical signals. In this section, the most common sensors used in lower-limb exoskeletons, social robots, and smart walkers are presented. The particular selection of sensors and input devices in these devices is referred to as sensory architectures.

[1] https://www.cdc.gov/nchs/fastats/body-measurements.htm.

[2] https://www.sciencedirect.com/topics/medicine-and-dentistry/walking-speed.

2.4.1 Sensory Architecture of Lower-Limb Exoskeletons

The implementation of various sensory interfaces is developed with many objectives. First, the assessment of kinematic parameters for the gait pattern provides information to evaluate the development of this task by the health-care professional. On the other hand, the acquisition of this data is also processed to apply different control strategies implemented for the rehabilitation or assistance methods for the walking activity. These methods contribute to the therapy sessions an increase in patient compliance in developing therapy sessions promoting neural plasticity. This section mentions various sensors aimed at the acquisition of kinematic and kinetic parameters; these are relevant for the control strategies implemented in the lower-limb exoskeletons.

2.4.1.1 Kinematic Parameters

Different parameters are required in developing activities of daily living. As an example, the displacement, velocities, and acceleration of the exoskeleton joint are estimated. In this sense, this section shows some sensors that are used in the estimation of these parameters:

- **Magnetic/incremental encoders:**
  This sensor is used to estimate the number of motor shaft turns. In this case, the encoders are coupled in the center of rotation of each joint of the exoskeleton. As a result, the angular position of the joints in various planes of the human body is obtained. Therefore, this sensor is implemented in rehabilitation and assistive exoskeletons.

  Some examples of their application are presented in lower-limb rehabilitation exoskeletons such as *BioMot* (Future and Emerging Technologies (FET), Spain) [61], *ALEX* (University of Delaware, USA) [86], and *E_ROWA* [87] that monitor the joint angular position in the sagittal plane using it as an input to control strategies based on activities of daily living rehabilitation. On the other hand, robotic orthoses are instrumented to observe the behavior of a specific joint in various planes of the human body. Such is the case of *Ankleboot* (Massachusetts Institute of Technology, USA), which monitors ankle plantarflexion–dorsiflexion and inversion–eversion movements [88]. Finally, lower-limb exoskeletons for activities of daily living assistance such as *ReWalk* (ReWalk Robotics, United Kingdom) [48], *REX* (REX Bionics, New Zealand) [89], *VariLeg* (ETH Zurich, Switzerland) [51], *Indego* (Vanderbilt University, USA) [50], *HAL* (Tsukuba University/Cyberdyne, Japan) exoskeleton [90] use these sensors to measure the angular position of the hip, knee, and ankle joints in the sagittal plane in generating control strategies that provide 100% assistance to the user's joints.
- **Inertial Measurement Unit (IMU):**
  The acquisition of different human body parameters requires the user's instrumentation with non-invasive sensors, where the sensor placement time is a relevant factor to use with a lower-limb exoskeleton. For these requirements,

MEMS sensors are used. For instance, the Inertial Measurement Unit (IMU) sensors are commonly mainly implemented for these applications because of the size, the patient instrumentation time, and the parameters provided by the IMU sensor [91]. Generally, the IMU sensor is composed of a gyroscope, an accelerometer, and a magnetometer that provide angular velocity, linear acceleration, and magnetic field strength in different axes [92, 93]. The IMU sensor is located in various parts of the user's human body for the activity of daily living recognition using machine learning methods to process the IMU data [92, 94–96]. Into the IMU implementation advantages is taken into account the sensor placement time, the sensor size, and the parameters that can be estimated.

2.4.1.2 Kinetic Parameters

The interaction forces between the user and the lower-limb exoskeleton are estimated for the development control strategies focused on the pHRI. On the other hand, some methodologies to use the human body electrical signal are implemented to control the voluntary movements of the user's lower limb using a lower-limb exoskeleton. The main goal of the acquisition and processing of these parameters is to increase the patient's participation in the therapy sessions, where each signal is rendered in terms of position, velocity, or acceleration values. Subsequently, the lower-limb actuation system is commanded through these values.

- **Force sensor:** Currently, several sensors are implemented. Strain gauge sensors are located in the lower-limb exoskeleton mechanical structure [52, 97]; this allows to estimate if the interaction is in terms of force/torque. As an example, *AGoRA* exoskeleton (Colombian School of Engineering Julio Garavito, Colombia) [4, 53], *ALLOR* (Federal University of Espirito Santo, Brazil) [97], and *CPWalker* [98] apply this method in the estimation of the exerted forces and the development of different control strategies using this parameter as an input system.

The acquisition of kinetic and kinematic parameters was aimed to monitor and use an input signal in various control strategies. Generally, the sensor modules showed in this section are integrated into lower-limb exoskeletons for rehabilitation and assistance workspaces. Remarkably, the mentioned sensors are non-invasive sensors and the acquired outcomes provide various parameters that allow the healthcare professional to assess the development of activities of daily living.

2.4.2 Sensory Architecture of Social Robots

Sensory systems provide robots with the ability to sense and receive information from the surrounding environment. Navigation sensors such as lasers or ultrasonic sensors give the possibility to plan a trajectory or prevent the robot from colliding. Other robots include pressure sensors, tactile sensors, and microphones that allow better interaction with the patient/user. Also, robots include cameras that serve as

Table 2.6 Sensory systems in Social Assistive Robots

Robot	Sensors
CASTOR [7]	Tactile sensors
Keepon [66]	Tactile sensors, cameras, and microphone
Probo [67,68]	Tactile sensors, camera, and microphone
KASPAR [69,70]	Cameras and force sensors
Pleo [71,72]	Tactile sensors, camera, microphones, infrared sensors, temperature sensor, ground foot sensors, and orientation and motion sensors
TIAGo [73]	Microphones, force/torque sensor, laser, RGB-D camera, sonar sensor, and Inertial Measurement Unit (IMU)
Pepper [74]	Tactile sensors, cameras, microphones, 3D sensor, gyroscope, sonar sensors, lasers, and Bumper sensors
HOBBIT [75]	Lasers, depth camera, and RGB-D camera
Huggable Bear [76]	Cameras, force sensors, temperature sensors, electric field sensors, and temperature sensors
NAO [77–79]	Tactile sensors, microphones, OmniVision cameras, pressure sensors, inertial sensor, sonar rangefinder, and infrared sensors
Paro [80,81]	Tactile sensors, microphones, temperature sensors, and light sensors

the robot's eyes. This provides the ability to detect facial expressions, gestures, and distances. Table 2.6 shows different sensory systems implemented in social robots.

Most robots feature touch sensors, microphones, and cameras. On the other hand, more complex robots such as Pleo, TIAGo, Pepper, and *NAO* implement a variety of sensors that, as mentioned above, give the robot a better perception of the environment.

2.4.3 Sensory Architectures of Smart Walkers

As in mobile robots, the sensing devices equipped in smart walkers cover a wide range of sensors, ranging from devices to acquire information from the environment to devices to obtain information from the user. In this sense, this section briefly describes the most common sensors employed to extract information about the user's state and intentions. Moreover, considering that the smart walkers move around indoor environments, they require sensing devices to estimate their motion and global position, as well as sensors to overcome unforeseen environmental characteristics [5,22].

2.4.3.1 Sensing Loops in Smart Walkers

The sensors mounted in the smart walkers and the sensors worn by the users can be classified in terms of the source of information delivered by the sensing devices.

- **Proprioceptive Sensors:** These devices include those sensors that estimate information related to the internal systems of the smart walker. Commonly,

proprioceptive sensors in smart walkers are used to estimate the velocity, position, and orientation in a given environment, as well as, to estimate the battery level, internal temperature, brake's state, among others.
- **Exteroceptive Sensors:** These sensors are used to obtain information related to the environment and the user, and thus they are further classified into two categories or sensing interfaces:

1. *Human–Robot Interface (HRi)*: This interface refers to the communication channel between the user and the robot, and therefore to the sensors that acquire information from the user. With the proper implementation of this interface, the smart walker is capable of: (1) the recognition of the interaction forces between the user and the platform, (2) the estimation of the user's navigation commands or intentions of movement, (3) the detection of the user's presence and proper support on the walker, (4) the estimation of the user's gait parameters and biomechanical indicators, and (5) the monitoring of the user's overall health state [22].
2. *Robot–Environment Interface (REi)*: This interface refers to the sensing devices that provide information from the environment and surrounding objects. The sensors for environment sensing allow: (1) the building and autonomous update of the environment map, (2) the autonomous localization of the device in the environment, (3) the detection of surrounding obstacles and people, and (4) the detection of hazardous situations.

2.4.3.2 Common Sensors in Smart Walkers

The smart walkers can be equipped with a wide range of sensors for multiple purposes. For instance, Table 2.7 describes some of the sensors found in smart walkers, classifying them by typical use and classification. Regarding the classification of these sensors, they can be whether proprioceptive or exteroceptive. Those sensors classified only as exteroceptive are devices that can be used in both HRi or REi. An extensive description of some of the sensors presented in Table 2.7 can be found in [12].

In addition to the above, to analyze the sensory interfaces of smart walkers reported in the literature, Table 2.8 summarizes some of the most notable smart walkers, describing their main functionalities and sensory interfaces. For instance, the CO-Operative Locomotion Aide (COOL Aide) is a three-wheeled passive SW intended to assist the elderly with routine walking tasks [103]. It includes mapping and obstacle detection systems, as well as navigation and guidance algorithms. Additionally, it is equipped with force sensors on its handlebars and a laser rangefinder (LRF) to estimate the user's desired direction to turn [5]. Other passive walkers, such as those presented in [16,21], include navigation and guidance algorithms based on haptic feedback systems and laser-based mapping.

Regarding active smart walkers, multiple sensory interfaces have been implemented [15,17,18,20,99–101,104]. These interfaces are equipped with sensors such

Table 2.7 Classification of common sensors used in mobile robotics applications, focusing on walker-assisted gait

Typical use	Sensor	Classification
Tactile or physical interaction sensors	Bumpers	Exteroceptive—REi
	Contact switches	Exteroceptive—REi
User estimation	Joysticks	Exteroceptive—HRi
	Potentiometers	Exteroceptive—HRi
	Microphones	Exteroceptive—HRi
	Temperature sensors	Exteroceptive—HRi
	Hearth rate sensors	Exteroceptive—HRi
	Laser rangefinders	Exteroceptive—HRi
	Ultrasonic sensors	Exteroceptive—HRi
	Force sensors or load cells	Exteroceptive—HRi
	Plastic optical fiber (POF) sensors	Exteroceptive—HRi
	Inertial Measurement Units (IMUs)	Exteroceptive—HRi
	Foot and hand pressure sensors	Exteroceptive—HRi
Wheels and motors sensors	Encoders (all classes)	Proprioceptive
	Potentiometers	Proprioceptive
Orientation or localization sensors	IMUs	Proprioceptive
	Gyroscopes	Proprioceptive
	RFID readers	Exteroceptive—REi
Motion sensors	Accelerometers	Proprioceptive
	Encoders (all classes)	Proprioceptive
Ranging and obstacle sensing	Laser rangefinders	Exteroceptive—REi
	2D Light Detection and Ranging (LiDAR)	Exteroceptive—REi
	3D LiDAR	Exteroceptive—REi
	Ultrasonic sensors	Exteroceptive—REi
	Reflective sensors	Exteroceptive—REi
Vision sensors	CCD/CMOS cameras	Exteroceptive
	Depth cameras	Exteroceptive

Table 2.8 Related works involving smart walkers that integrate interfaces for Human–Robot–Environment Interaction

Walker	Sensory interface	Main functionalities
GUIDO [18]	– Force sensors – LRF – Sonars – Encoders	– Autonomous navigation – Detection of user's intentions – Sound feedback
XR4000 [99]	– Force sensors – LRF – Sonars – Infrared sensors – Encoders	– Autonomous navigation – Detection of user's intentions
ASBGo++ [20]	– Force sensors – LRF – Sonar – Infrared sensors – Camera – Encoders	– Autonomous navigation – Detection of user's intentions – Gait monitoring – User position feedback
JARoW [17]	– Infrared sensors – Encoders – LRFs	– User's position estimation and prediction – Obstacle avoidance
UFES [15]	– Force sensors – LRF – IMUs – Encoders	– Path following – Obstacle avoidance – Detection of user's intentions – Gait monitoring
PAMM [100]	– Force sensors – Sonars – Camera – Encoders	– Autonomous navigation – Health monitoring
MOBOT	– Force sensors – LRFs – Cameras – Kinect sensors – Microphones	– Autonomous navigation – Detection of user's intentions – Speech and gesture recognition – Body pose estimation – Gait Analyzer
CAIROW [101]	– Force sensors – LRFs	– Environment analyzer – Force analyzer – Gait analyzer
ISR-AIWALKER [102]	– Force sensors – Kinect sensor – Encoders – RGB-D Camera	– Detection of user's intention – Gripping recognition – Gait analyzer – Autonomous navigation

(continued)

Table 2.8 (continued)

Walker	Sensory interface	Main functionalities
COOL Aide [103]	– Force sensors – LRF – Encoders	– Autonomous navigation – Detection of user's intentions
Wachaja et al. [21]	– LRF – Tilting LRF	– 3D Mapping and localization – Obstacle avoidance – Vibrotactile feedback
MARC [16]	– Sonars – Infrared sensors – Encoders	– Path following – Obstacle avoidance
c-Walker [19]	– Kinect like sensor – RFID reader – IMU – Camera – Encoders	– Autonomous navigation – People detection and tracking – Guidance

as encoders, IMUs, and LRFs to provide navigation, guidance, and user interaction. Moreover, these smart walkers are also equipped with sensing technologies to estimate the user's intentions to move, based on gait analysis systems and rule-based algorithms.

2.5 Conclusions

Several aspects define a gait rehabilitation and assistance device. In this sense, this chapter introduced concepts about the kinematic modeling, actuation systems, and sensing architectures of lower-limb exoskeletons, social robots, and smart walkers. Different methodologies, such as forward and inverse kinematics, DH convention, and homogeneous transformations were explained, regarding the kinematic modeling.

In terms of the actuation mechanisms of these rehabilitation robots, a concise description of the actuators employed was presented. In particular, several insights were given to make an optimal selection of an appropriate actuation system depending on the task to be performed with the devices. Additionally, factors such as the actuation system cost, sensory interface for the implementation of control strategies, weight, and mechatronic integration must be considered.

Finally, the third aspect corresponds to the sensory interface that estimates the kinematic and kinetic parameters of the devices, and allows them to acquire internal and external information. Those presented in this chapter provide the input parameters for various control strategies and the required information to guarantee a safe and natural interaction in clinical and everyday scenarios.

References

1. M.W. Spong, S. Hutchinson, M. Vidyasagar, Forward kinematics: the Denavit-Hartenberg convention. *Robot Dynamics and Control*, 2nd edn. (Wiley, New York, NY, USA, 2004), pp. 57–82
2. A.A. Hayat, R.G. Chittawadigi, A.D. Udai, S.K. Saha, Identification of Denavit-Hartenberg parameters of an industrial robot, in *Proceedings of Conference on Advances in Robotics*, pp. 1–6 (2013)
3. Q. Chen, S. Zhu, X. Zhang, Improved inverse kinematics algorithm using screw theory for a six-DOF robot manipulator. Int. J. Adv. Robot. Syst. **12**(10), 140 (2015)
4. M. Sánchez-Manchola, D. Gomez-Vargas, D. Casas-Bocanegra, M. Munera, C.A. Cifuentes, Development of a robotic lower-limb exoskeleton for gait rehabilitation: AGoRA exoskeleton, in *2018 IEEE ANDESCON, ANDESCON 2018 - Conference Proceedings*, pp. 1–6 (2018)
5. S.D. Sierra M., L. Arciniegas, F. Ballen-Moreno, D. Gomez-Vargas, M. Munera, C.A. Cifuentes, Adaptable robotic platform for gait rehabilitation and assistance: design concepts and applications, in *Exoskeleton Robots for Rehabilitation and Healthcare Devices* (Springer, 2020), pp. 67–93
6. A.A. Ramírez-Duque, L.F. Aycardi, A. Villa, M. Munera, T. Bastos, T. Belpaeme, A. Frizera-Neto, C.A. Cifuentes, Collaborative and inclusive process with the autism community: a case study in Colombia about social robot design. Int. J. Soc. Robot., 1–15 (2020)
7. D. Casas-Bocanegra, D. Gomez-Vargas, M.J. Pinto-Bernal, J. Maldonado, M. Munera, A. Villa-Moreno, M.F. Stoelen, T. Belpaeme, C.A. Cifuentes, An open-source social robot based on compliant soft robotics for therapy with children with ASD. Actuators **9**(3), 1–22 (2020)
8. NAO Technical overview—NAO Software 1.14.5 documentation (2020)
9. N. Kofinas, E. Orfanoudakis, M.G. Lagoudakis, Complete analytical forward and inverse kinematics for the NAO humanoid robot. J. Intell. Robot. Syst. **77**(2), 251–264 (2015)
10. D. Gouaillier, V. Hugel, P. Blazevic, C. Kilner, J. Monceaux, P. Lafourcade, B. Marnier, J. Serre, B. Maisonnier, Mechatronic design of NAO humanoid, in *2009 IEEE International Conference on Robotics and Automation* (IEEE, 2009), pp. 769–774
11. N. Kofinas, E. Orfanoudakis, M.G. Lagoudakis, Complete analytical inverse kinematics for NAO, in *2013 13th International Conference on Autonomous Robot Systems* (IEEE, 2013), pp. 1–6
12. R. Siegwart, I.R. Nourbakhsh, D. Scaramuzza, *Introduction to Autonomous Mobile Robots*, 2nd edn. (MIT Press, 2011)
13. S.G. Tzafestas, Mobile robots, in *Introduction to Mobile Robot Control* (Elsevier, 2014), pp. 1–29
14. S. Page, L. Saint-Bauzel, P. Rumeau, V. Pasqui, Smart walkers: an application-oriented review. Robotica **35**, 1243–1262 (2017)
15. M.F. Jiménez, M. Monllor, A. Frizera, T. Bastos, F. Roberti, R. Carelli, Admittance controller with spatial modulation for assisted locomotion using a smart walker. J. Intell. Robot. Syst., 1 (2018)
16. G. Wasson, J. Gunderson, S. Graves, R. Felder, An assistive robotic agent for pedestrian mobility, in *Proceedings of the Fifth International Conference on Autonomous Agents - AGENTS '01* (ACM Press, New York, USA, 2001), pp. 169–173
17. G. Lee, E.J. Jung, T. Ohnuma, N.Y. Chong, B.J. Yi, JAIST Robotic Walker control based on a two-layered Kalman filter, in *Proceedings - IEEE International Conference on Robotics and Automation*, pp. 3682–3687 (2011)
18. G.J. Lacey, D. Rodriguez-Losada, The evolution of Guido. IEEE Robot. Autom. Mag. **15**(4), 75–83 (2008)
19. L. Palopoli, A. Argyros, J. Birchbauer, A. Colombo, D. Fontanelli, A. Legay, A. Garulli, A. Giannitrapani, D. Macii, F. Moro, P. Nazemzadeh, P. Padeleris, R. Passerone, G. Poier, D. Prattichizzo, T. Rizano, L. Rizzon, S. Scheggi, S. Sedwards, Navigation assistance and guidance of older adults across complex public spaces: the DALi approach. Intell. Serv. Robot. **8**, 77–92 (2015)

20. J. Alves, E. Seabra, I. Caetano, C.P. Santos, Overview of the ASBGo++ smart walker, in *2017 IEEE 5th Portuguese Meeting on Bioengineering (ENBENG)* (IEEE, 2017), pp. 1–4
21. A. Wachaja, P. Agarwal, M. Zink, M.R. Adame, K. Möller, W. Burgard, Navigating blind people with walking impairments using a smart walker. Autonomous Robots **41**, 555–573 (2017)
22. S.D. Sierra M., M. Garzón, M. Múnera, C.A. Cifuentes, Human–Robot–environment interaction interface for smart walker assisted gait: AGoRA walker. Sensors **19**, 2897 (2019)
23. S.G. Tzafestas, Mobile robot kinematics, in *Introduction to Mobile Robot Control* (Elsevier, 2014), pp. 31–67
24. C.Z. Resende, R. Carelli, T.F. Bastos-Filho, M. Sarcinelli-Filho, A new positioning and path following controller for unicycle mobile robots, in *2013 16th International Conference on Advanced Robotics (ICAR)* (IEEE, 2013), pp. 1–6
25. M.R. Islam, B. Brahmi, T. Ahmed, M. Assad-Uz-Zaman, M.H. Rahman, *Exoskeletons in Upper Limb Rehabilitation: A Review to Find Key Challenges to Improve Functionality* (Elsevier Inc., 2020)
26. A.J. Young, D.P. Ferris, State of the art and future directions for lower limb robotic exoskeletons. IEEE Trans. Neural Syst. Rehab. Eng. **25**(2), 171–182 (2017)
27. P. Zhang, *Transducers and Valves*, 1st edn. (Peng Zhang, 2010)
28. J. Varghese, V.M. Akhil, P.K. Rajendrakumar, K.S. Sivanandan, A rotary pneumatic actuator for the actuation of the exoskeleton knee joint. Theor. Appl. Mech. Lett. **7**(4), 222–230 (2017)
29. R.J. Varghese, D. Freer, F. Deligianni, J. Liu, G.-Z. Yang, Wearable robotics for upper-limb rehabilitation and assistance, in *Wearable Technology in Medicine and Health Care* (Elsevier, 2018), pp. 23–69
30. T. Sénac, A. Lelevé, R. Moreau, C. Novales, L. Nouaille, M.T. Pham, P. Vieyres, A review of pneumatic actuators used for the design of medical simulators and medical tools. Multimodal Technol. Interact. **3**(3), 1–22 (2019)
31. R. Bishop, *The Mechatronics Handbook, Second Edition - 2 Volume Set*. Mechatronics Handbook 2e (Taylor & Francis, 2002)
32. J. Segil, *Handbook of Biomechatronics* (Academic Press, 2019)
33. I. Sardellitti, E. Cattin, S. Roccella, F. Vecchi, M.C. Carrozza, P. Dario, P.K. Artemiadis, K.J. Kyriakopoulos, Description, characterization and assessment of a bio-inspired shoulder joint-first link robot for neuro-robotic applications, in *Proceedings of the First IEEE/RAS-EMBS International Conference on Biomedical Robotics and Biomechatronics, 2006, BioRob 2006*, vol. 2006, no. February 2001, pp. 112–117 (2006)
34. J. Zhu, Y. Wang, J. Jiang, B. Sun, H. Cao, Unidirectional variable stiffness hydraulic actuator for load-carrying knee exoskeleton. Int. J. Adv. Robot. Syst. **14**(1), 1–12 (2017)
35. H. Kazerooni, R. Steger, L. Huang, Hybrid control of the Berkeley Lower Extremity Exoskeleton (BLEEX). Int. J. Robot. Res. **25**(5-6), 561–573 (2006)
36. A.B. Zoss, H. Kazerooni, A. Chu, Biomechanical design of the Berkeley Lower Extremity Exoskeleton (BLEEX). IEEE/ASME Trans. Mechatron. **11**(2), 128–138 (2006)
37. C. Bayón, O. Ramírez, J.I. Serrano, M.D. Castillo, A. Pérez-Somarriba, J.M. Belda-Lois, I. Martínez-Caballero, S. Lerma-Lara, C. Cifuentes, A. Frizera, E. Rocon, Development and evaluation of a novel robotic platform for gait rehabilitation in patients with Cerebral Palsy: CPWalker. Robot. Autonom. Syst. **91**, 101–114 (2017)
38. H. Kim, Y.J. Shin, J. Kim, Kinematic-based locomotion mode recognition for power augmentation exoskeleton. Int. J. Adv. Robot. Syst. **14**(5), 1–14 (2017)
39. E.F. Kececi, Actuators, in *Mechatronic Components* (Elsevier, 2019), pp. 145–154
40. D. GRAY, Chapter 7 - actuators, in *Centralized and Automatic Controls in Ships,* ed. by D. GRAY (Pergamon, 1966), pp. 44–54
41. S. Arumugom S. Muthuraman, V. Ponselvan, Modeling and application of series elastic actuators for force control multi legged robots. J. Comput. **1**(1), 26–33 (2009)
42. M.A. Gull, S. Bai, T. Bak, A review on design of upper limb exoskeletons. Robotics **9**(1), 1–35 (2020)

43. S. Hussain, P.K. Jamwal, M.H. Ghayesh, S.Q. Xie, Assist-as-needed control of an intrinsically compliant robotic gait training orthosis. IEEE Trans. Ind. Electron. **64**(2), 1675–1685 (2017)
44. J. Choi, B. Na, P.-G. Jung, D.-w. Rha, K. Kong, WalkON suit. IEEE Robot. Autom. Mag. no. November, 75–86 (2017)
45. Y.-L. Park, B.-r. Chen, N. O. Pérez-Arancibia, D. Young, L. Stirling, R.J. Wood, E.C. Goldfield, R. Nagpal, Design and control of a bio-inspired soft wearable robotic device for ankle–foot rehabilitation. Bioinspiration Biomimetics **9**, 016007 (2014)
46. S. Galle, P. Malcolm, S.H. Collins, D. De Clercq, Reducing the metabolic cost of walking with an ankle exoskeleton: interaction between actuation timing and power. J. NeuroEng. Rehab. **14**(1), 1–16 (2017)
47. H. Lee, P.W. Ferguson, J. Rosen, *Chapter 11. Lower Limb Exoskeleton Systems—Overview* (INC, 2020)
48. A. Esquenazi, M. Talaty, A. Packel, M. Saulino, The ReWalk powered exoskeleton to restore ambulatory function to individuals with thoracic-level. Am. J. Phys. Med. Rehab., 911–921 (2012)
49. E. Bionics, Ekso GT (2020)
50. Indego, Indego Exoskeleton (2020)
51. S.O. Schrade, K. Dätwyler, M. Stücheli, K. Studer, D.A. Türk, M. Meboldt, R. Gassert, O. Lambercy, Development of VariLeg, an exoskeleton with variable stiffness actuation: First results and user evaluation from the CYBATHLON 2016 Olivier Lambercy; Roger Gassert. J. NeuroEng. Rehab. **15**(1), 1–18 (2018)
52. C. Bayón, O. Ramírez, F. Mollà, J. Serrano, M. Del Castillo, J. Belda-Lois, R. Poveda, R. Raya, T. Martín Lorenzo, I. Martínez Caballero, S. Lerma Lara, C. Cifuentes, A. Frizera, E. Rocon, CPWalker, robotic platform for gait rehabilitation and training in patients with cerebral palsy. IEEE Trans. Neural Syst. Rehab. Eng. (Under Review), 3736–3741 (2015)
53. M.D. Sánchez-Manchola, L.J. Arciniegas Mayag, M. Munera, C.A. Garcia, Impedance-based backdrivability recovery of a lower-limb exoskeleton for knee rehabilitation, in *4th IEEE Colombian Conference on Automatic Control: Automatic Control as Key Support of Industrial Productivity, CCAC 2019 - Proceedings* (Medellin, Colombia, 2019), pp. 1–6
54. M. Talaty, A. Esquenazi, J.E. Briceno, Differentiating ability in users of the ReWalkTM powered exoskeleton: An analysis of walking kinematics. *IEEE International Conference on Rehabilitation Robotics* (2013)
55. A. Ramanujam, C.M. Cirnigliaro, E. Garbarini, P. Asselin, R. Pilkar, G.F. Forrest, Neuromechanical adaptations during a robotic powered exoskeleton assisted walking session. J. Spinal Cord Med. **41**(5), 518–528 (2018)
56. W. Zou, N. Yu, Modeling and control of a cable-driven series elastic actuator (2018)
57. W.M. Dos Santos, A.A.G. Siqueira, *Impedance Control of a Rotary Series Elastic Actuator for Knee Rehabilitation*, vol. 19 (IFAC, 2014)
58. K. Isik, S. He, J. Ho, L. Sentis, Re-engineering a high performance electrical series elastic actuator for low-cost industrial applications. Actuators, 1–16 (2017)
59. R. Van Ham, B. Vanderborght, M. Van Damme, B. Verrelst, D. Lefeber, MACCEPA, the mechanically adjustable compliance and controllable equilibrium position actuator: Design and implementation in a biped robot. Robot. Autonom. Syst. **55**, 761–768 (2007)
60. P. Cherelle, V. Grosu, P. Beyl, A. Mathys, R. Van Ham, M. Van Damme, B. Vanderborght, D. Lefeber, The MACCEPA actuation system as torque actuator in the gait rehabilitation robot ALTACRO, in *2010 3rd IEEE RAS & EMBS International Conference on Biomedical Robotics and Biomechatronics* (IEEE, 2010), pp. 27–32
61. T. Bacek, M. Moltedo, K. Langlois, G.A. Prieto, M.C. Sanchez-Villamañan, J. Gonzalez-Vargas, B. Vanderborght, D. Lefeber, J.C. Moreno, BioMot exoskeleton - Towards a smart wearable robot for symbiotic human-robot interaction, in *IEEE International Conference on Rehabilitation Robotics*, pp. 1666–1671 (2017)
62. J.F. Veneman, R. Ekkelenkamp, R. Kruidhof, F.C. Van Der Helm, H. Van Der Kooij, A series elastic- and Bowden-cable-based actuation system for use as torque actuator in exoskeleton-type robots. Int. J. Robot. Res. **25**(3), 261–281 (2006)

63. J.C. Perry, J. Rosen, S. Burns, Upper-limb powered exoskeleton design. IEEE/ASME Trans. Mechatron. **12**(4), 408–417 (2007)
64. L.M. Mooney, H.M. Herr, Biomechanical walking mechanisms underlying the metabolic reduction caused by an autonomous exoskeleton. J. NeuroEng. Rehab. **13**(1), 1–12 (2016)
65. J.E. Duarte, K. Schmidt, R. Riener, J.E. Duarte, K. Schmidt, R. Riener, J.E. Duarte, K. Schmidt, R. Riener, The Myosuit : textile-powered mobility The Myosuit : textile-powered mobility. IFAC-PapersOnLine **51**(34), 242–243 (2019)
66. H. Kozima, M.P. Michalowski, C. Nakagawa, Keepon: A playful robot for research, therapy, entertainment. Int. J. Soc. Robot. **1**(1), 3–18 (2009)
67. J. Saldien, K. Goris, S. Yilmazyildiz, W. Verhelst, D. Lefeber, On the design of the huggable robot Probo. J. Phys. Agents **2**(2), 3–11 (2008)
68. K. Goris, J. Saldien, B. Vanderborght, D. Lefeber, Mechanical design of the huggable robot Probo. Int. J. Humanoid Robot. **8**(3), 481–511 (2011)
69. K. Dautenhahn, C.L. Nehaniv, M.L. Walters, B. Robins, H. Kose-Bagci, N.A. Mirza, M. Blow, KASPAR - a minimally expressive humanoid robot for human-robot interaction research. Appl. Bionics Biomech. **6**(3-4), 369–397 (2009)
70. S. Costa, H. Lehmann, K. Dautenhahn, B. Robins, F. Soares, Using a humanoid robot to elicit body awareness and appropriate physical interaction in children with autism. Int. J. Soc. Robot. **7**(2), 265–278 (2015)
71. Y. Fernaeus, M. Håkansson, M. Jacobsson, S. Ljungblad, How do you play with a robotic toy animal? A long-term study of Pleo, in *Proceedings of IDC2010: The 9th International Conference on Interaction Design and Children*, pp. 39–48 (2010)
72. J. Dimas, I. Leite, A. Pereira, P. Cuba, R. Prada, A. Paiva, Pervasive Pleo: Long-term attachment with artificial pets, in *Workshop on Playful Experiences in Mobile HCI* (2010)
73. S. Coşar, M. Fernandez-Carmona, R. Agrigoroaie, J. Pages, F. Ferland, F. Zhao, S. Yue, N. Bellotto, A. Tapus, ENRICHME: Perception and interaction of an assistive robot for the elderly at home. Int. J. Soc. Robot. **12**(3), 779–805 (2020)
74. M. Eskenazi, L. Devillers, J. Mariani, (eds.), *Advanced Social Interaction with Agents*, vol. 510 of *Lecture Notes in Electrical Engineering* (Springer International Publishing, Cham, 2019)
75. D. Fischinger, P. Einramhof, K. Papoutsakis, W. Wohlkinger, P. Mayer, P. Panek, S. Hofmann, T. Koertner, A. Weiss, A. Argyros, M. Vincze, Hobbit, a care robot supporting independent living at home: First prototype and lessons learned. Robot. Autonom. Syst. **75**, 60–78 (2016)
76. W.D. Stiehl, J. Lieberman, C. Breazeal, L. Basel, R. Cooper, H. Knight, L. Lalla, A. Maymin, S. Purchase, The Huggable: A therapeutic robotic companion for relational, affective touch, in *2006 3rd IEEE Consumer Communications and Networking Conference, CCNC 2006*, vol. 2, pp. 1290–1291 (2006)
77. A. Di Nuovo, D. Conti, G. Trubia, S. Buono, S. Di Nuovo, Deep learning systems for estimating visual attention in robot-assisted therapy of children with autism and intellectual disability. Robotics **7**(2), 25 (2018)
78. O.A. Henkemans, B.P. Bierman, J. Janssen, R. Looije, M.A. Neerincx, M.M. van Dooren, J.L. de Vries, G.J. van der Burg, S.D. Huisman, Design and evaluation of a personal robot playing a self-management education game with children with diabetes type 1. Int. J. Human Comput. Stud. **106**, no. January, 63–76 (2017)
79. Z. Shen, Y. Wu, Investigation of practical use of humanoid robots in elderly care centres, in *HAI 2016 - Proceedings of the 4th International Conference on Human Agent Interaction*, pp. 63–66 (2016)
80. T. Washio, A. Sakurai, K. Nakajima, H. Takeda, S. Tojo, M. Yokoo, (eds.), *New Frontiers in Artificial Intelligence*, vol. 4012 of *Lecture Notes in Computer Science* (Springer, Berlin, Heidelberg, 2006)

81. T. Shibata, T. Mitsui, K. Wada, A. Touda, T. Kumasaka, K. Tagami, K. Tanie, Mental commit robot and its application to therapy of children, in *IEEE/ASME International Conference on Advanced Intelligent Mechatronics, AIM*, vol. 2, no. July, pp. 1053–1058 (2001)
82. G.A. Pratt, M.M. Williamson, Series elastic actuators, in *IEEE International Conference on Intelligent Robots and Systems*, vol. 1, pp. 399–406 (1995)
83. J.J. Cabibihan, H. Javed, M. Ang, S.M. Aljunied, Why robots? A survey on the roles and benefits of social robots in the therapy of children with autism. Int. J. Soc. Robot. **5**(4), 593–618 (2013)
84. T. Mikolajczyk, I. Ciobanu, D.I. Badea, A. Iliescu, S. Pizzamiglio, T. Schauer, T. Seel, P.L. Seiciu, D.L. Turner, M. Berteanu, Advanced technology for gait rehabilitation: An overview. Adv. Mech. Eng. **10**, 1–19 (2018)
85. A. Trujillo-León, W. Bachta, J. Castellanos-Ramos, F. Vidal-Verdú, Assistive handlebar based on tactile sensors: Control inputs and human factors. Sensors **18**, 2471 (2018)
86. S.S. Banala, S.K. Agrawal, Active leg exoskeleton (ALEX) for gait rehabilitation of motor-impaired patients, in *IEEE Int Conf Rehabil Robot 2007, 401–7, Rehabilitation Robotics*, vol. 00, no. c (2007)
87. W. Huo, S. Mohammed, Y. Amirat, K. Kong, Fast gait mode detection and assistive torque control of an exoskeletal robotic orthosis for walking assistance. IEEE Trans. Robot. **34**(4), 1035–1052 (2018)
88. A. Roy, H.I. Krebs, S.L. Patterson, T.N. Judkins, I. Khanna, L.W. Forrester, R.M. Macko, N. Hogan, Measurement of human ankle stiffness using the Anklebot, in *2007 IEEE 10th International Conference on Rehabilitation Robotics, ICORR'07*, vol. 00, no. c, pp. 356–363 (2007)
89. R. Bionics, REX (2019)
90. H. Kawamoto, S. Lee, S. Kanbe, Y. Sankai, Power assist method for HAL-3 using EMG-based feedback controller, in *SMC'03 Conference Proceedings. 2003 IEEE International Conference on Systems, Man and Cybernetics. Conference Theme - System Security and Assurance (Cat. No.03CH37483)*, vol. 2 (IEEE, 2003), pp. 1648–1653
91. M.N. Victorino, X. Jiang, C. Menon, *Wearable Technologies and Force Myography for Healthcare* (Elsevier Inc., 2018)
92. H. Nguyen, K. Lebel, S. Bogard, E. Goubault, P. Boissy, C. Duval, Using inertial sensors to automatically detect and segment activities of daily living in people with Parkinson's disease. IEEE Trans. Neural Syst. Rehab. Eng. **26**(1), 197–204 (2018)
93. J.C. Moreno, F. Brunetti, E. Navarro, A. Forner-Cordero, J.L. Pons, Analysis of the human interaction with a wearable lower-limb exoskeleton. Appl. Bion. Biomech. **6**(2), 245–256 (2009)
94. M. Sánchez-Manchola, M.J. Bernal, M. Munera, C.A. Cifuentes, Gait phase detection for lower-limb exoskeletons using foot motion data from a single inertial measurement unit in hemiparetic individuals. Sensors (Switzerland) **19**(13), 2988 (2019)
95. M. Tariq, H. Majeed, M.O. Beg, F.A. Khan, A. Derhab, Accurate detection of sitting posture activities in a secure IoT based assisted living environment. Future Gener. Comput. Syst. **92**, 745–757 (2019)
96. B. Chen, C.H. Zhong, X. Zhao, H. Ma, X. Guan, X. Li, F.Y. Liang, J.C.Y. Cheng, L. Qin, S.W. Law, W.H. Liao, A wearable exoskeleton suit for motion assistance to paralysed patients. J. Orthopaedic Transl. **11**, no. March, 7–18 (2017)
97. A.C. Villa-Parra, D. Delisle-Rodriguez, J.S. Lima, A. Frizera-Neto, T. Bastos, Knee impedance modulation to control an active orthosis using insole sensors. Sensors (Switzerland) **17**(12), 2751 (2017)
98. C. Bayon, O. Ramirez, M. Del Castillo, J. Serrano, R. Raya, J. Belda-Lois, R. Poveda, F. Molla, T. Martin, I. Martinez, S. Lerma Lara, E. Rocon, CPWalker: Robotic platform for gait rehabilitation in patients with Cerebral Palsy, in *2016 IEEE International Conference on Robotics and Automation (ICRA)* (IEEE, 2016), pp. 3736–3741

99. A. Morris, R. Donamukkala, A. Kapuria, A. Steinfeld, J. Matthews, J. Dunbar-Jacob, S. Thrun, A robotic walker that provides guidance, in *2003 IEEE International Conference on Robotics and Automation (Cat. No.03CH37422)*, vol. 1, pp. 25–30 (2003)
100. M. Spenko, H. Yu, S. Dubowsky, Robotic personal aids for mobility and monitoring for the elderly. IEEE Trans. Neural Syst. Rehab. Eng. **14**(3), 344–351 (2006)
101. W.-H. Mou, M.-F. Chang, C.-K. Liao, Y.-H. Hsu, S.-H. Tseng, L.-C. Fu, Context-aware assisted interactive robotic walker for Parkinson's disease patients, in *2012 IEEE/RSJ International Conference on Intelligent Robots and Systems* (IEEE, 2012), pp. 329–334
102. J. Paulo, P. Peixoto, U.J. Nunes, ISR-AIWALKER: Robotic walker for intuitive and safe mobility assistance and gait analysis. IEEE Trans. Human Mach. Syst. **47**(6), 1110–1122 (2017)
103. C. Huang, G. Wasson, M. Alwan, P. Sheth, Shared navigational control and user intent detection in an intelligent walker, in *AAAI Fall 2005*, no. April (2005)
104. E. Efthimiou, S.E. Fotinea, T. Goulas, M. Koutsombogera, P. Karioris, A. Vacalopoulou, I. Rodomagoulakis, P. Maragos, C. Tzafestas, V. Pitsikalis, Y. Koumpouros, A. Karavasili, P. Siavelis, F. Koureta, D. Alexopoulou, The MOBOT rollator human-robot interaction model and user evaluation process, in *2016 IEEE Symposium Series on Computational Intelligence, SSCI 2016* (2017)

3 Fundamentals for the Design of Lower-Limb Exoskeletons

Felipe Ballen-Moreno, Daniel Gomez-Vargas, Kevin Langlois, Jan Veneman, Carlos A. Cifuentes, and Marcela Múnera

3.1 Introduction

The development of gait rehabilitation devices intends to enhance user's well-being, involving multiple engineering disciplines such as, mechanical, biomedical, and electronic engineering. Across these fields, the design process entailed common questions (i.e., what, who, where, when, why, how) along each step to achieve a device's prototype [1]. Each question provides meaningful information to discern the device's features, materials, functionalities, and geometry shape. Moreover, it helps to define the application's scope, including the design features that will be further addressed [2]. This chapter will be focused on the development of lower-limb exoskeletons and its design features. In this sense, design questions ease establishing the device's guidelines that are defined and addressed among the lower-

F. Ballen-Moreno · C. A. Cifuentes (✉) · M. Múnera
Biomedical Engineering Department of the Colombian School of Engineering Julio Garavito, Bogotá, Colombia
e-mail: felipe.ballen@mail.escuelaing.edu.co; carlos.cifuentes@escuelaing.edu.co; marcela.munera@escuelaing.edu.co

D. Gomez-Vargas
Biomedical Engineering Department of the Colombian School of Engineering Julio Garavito, Bogotá D.c., Colombia

Institute of Automatics, National University of San Juan, San Juan, Argentina
e-mail: daniel.gomez-v@mail.escuelaing.edu.co

K. Langlois
Robotics & Multibody Mechanics Research Group, Department of Mechanical Engineering, Vrije Universiteit Brussel, Elsene, Belgium
e-mail: kevin.langlois@vub.be

J. Veneman
Hocoma AG, Volketswil, Switzerland
e-mail: jan.veneman@gmail.com

C. A. Cifuentes, M. Múnera, *Interfacing Humans and Robots for Gait Assistance and Rehabilitation*, https://doi.org/10.1007/978-3-030-79630-3_3

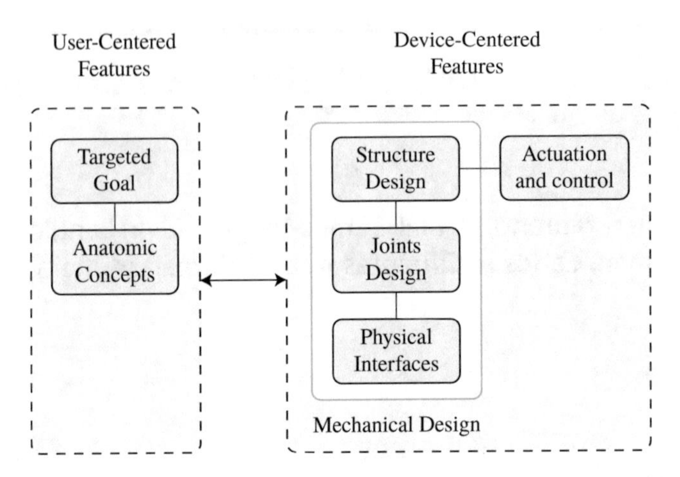

Fig. 3.1 Features to consider in the design process of a lower-limb exoskeleton. The design features are divided into two groups: user-centered features and device-centered features, showing the principal features per group

limb exoskeleton's design features, shown in Fig. 3.1. Each one is interlinked to ensure the lower-limb exoskeleton's main objectives for the user, either clinical or work-related.

As shown in Fig. 3.1, the design process presented in this chapter has two principal groups of device features such as features centered on the user and features focused on the device. User-centered features are the first step to consider along the design process. These features are separated into anatomic concepts and the targeted goal. On the one hand, the anatomic concepts determine the joint and limb characteristics used for the exoskeleton's design. On the other hand, the targeted goal is defined as the device's activity or activities to perform using the exoskeleton, delimiting different features' capabilities. Other design features have a bearing on the targeted goal, defining technical and aesthetic outcomes. At this point, the previous design features are related to the goal and features of the user. According to this information, the remaining features are aimed at the lower-limb exoskeleton.

The user-centered features define the device's attributes through four devices' features which are divided into mechanical structure, joints, physical interfaces, and actuation and control strategies. In this part of the design process, the mechanical design determines three out of four of the device-centered features, as is shown in Fig. 3.1. The device's design process could be addressed by a bioinspired or biomimetic approach to ensure these attributes.

These design processes are applied in two robotic fields divided into rigid and soft robotics [3]. A rigid or soft device's prototype can be achieved by an innovative design process known as bioinspired and biomimetic design. The difference between them relies on the materials and modeling applied to the prototype. Each process addresses the design features by understanding the inspiration source and how it is deployed. *Nagel* et al. [4] established an equivalence between engineering-to-biology functions, which can solve a specific problem through bioinspired design. Similarly, the biomimetic design also understands that biological functions through the design's prototype emulate biological behavior [5].

In this sense, this chapter describes each design feature involved in a lower-limb exoskeleton, focusing on user-centered and device-centered features. The primary source of information to establish multiple parameters for the exoskeleton relies on the user's features divided into the targeted goal (i.e., applications) and the anatomic concepts (i.e., anthropometric measurements, human body planes, and joint kinematics). The lower-limb exoskeleton's applications are addressed to analyze different scenarios and primary user's objectives providing the targeted goals. The anatomic concepts (e.g., anthropometric measurements and joint kinematics) aimed at hip, knee, and ankle joints. Then, device-based features are explained through developments presented in the literature.

3.2 User-Centered Features

The needs and anatomic concepts of the users are the principal sources to answer the design process's common questions. These sources are divided into targeted goals or applications of the lower-limb exoskeleton, and the anatomic concepts in that detail the principal human features to design a lower-limb exoskeleton. As shown in Fig. 3.1, the targeted goal defined the areas in which the lower-limb exoskeletons are deployed such as, augmentation, rehabilitation, and assistance. Moreover, the anatomic concepts will provide the human guidelines for the exoskeleton.

3.2.1 Targeted Goal Focused on the Applications

As mentioned before, the applications for lower-limb exoskeletons are mainly oriented to augmentation, rehabilitation, and assistance. Augmentation exoskeletons are commonly deployed in military and industry scenarios to enhance soldiers' or workers' capabilities (e.g., strength, fatigue, load carrying) [6]. On the other hand, rehabilitation and assistive devices focus on populations that suffer from neurological injuries or work-related accidents (i.e., spinal cord injury, stroke, cerebral palsy) [7]. Robotic aid has been divided into execution scenarios, such as overground tasks or treadmill-based training [8]. The lower-limb exoskeleton can include different functionalities and approaches to interact with the clinician and the user (i.e., patient, soldier, or worker), according to its scenario's execution. In this sense, populations' needs and activities lead to guidelines defined as design features for the robotic devices [2].

According to the application, these exoskeletons rely on multiple components that provide energy to the user [9]. In terms of energy, each application requires an amount of energy to perform a specific task. For instance, augmentation deploys more energy to the user than other applications (e.g., robot-assisted therapy or robotic aid for activities of daily living) where the user's enhancement is several times greater than the user's baseline. These other applications, as the rehabilitation and assistance, requires near to the user's baseline, considering the device's and user's weight.

3.2.1.1 Exoskeletons for Human Augmentation

The lower-limb exoskeleton's development for augmentation applications has been widely used in different scenarios such as, military, industry, and health to provide sufficient support and induce a reduction in user effort [10]. In this application, lower-limb exoskeletons are deployed to improve the user's walking capacities, or support a high load of military equipment. One of the main recognized exoskeletons is the *Berkeley Lower-Extremity Exoskeleton* (*BLEEX*, University of California, USA), which combines robotic force capabilities through two motorized anthropomorphic legs and a backpack frame-like in which heavy loads, up to 75 Kg, can be added. Without pre-programmed movements, the study results show that the kinematics and dynamics of *BLEEX* do not entirely match the user. Besides, its weight and size are aspects that limit its efficiency [11]. Another primary device is known as *HULC* (Human Universal Load Carrier, Ekso Bionics, USA), a full-body exoskeleton developed by Lockheed Martin for the military field, which provides, as *BLEEX*, a feature of increased strength and endurance. This device features a hydraulic drive system that allows it to have the highest maneuvering speed among other existing portable robots. Besides, it carries loads up to 90 Kg without hindering the user and lowering the user's metabolic cost [6]. For this application, the device's capabilities are high enough to assist the user and carry loads. Furthermore applications also carried the user's weight although they aimed to ease the user to perform specific task.

3.2.1.2 Exoskeletons for Rehabilitation

Rehabilitation implementation includes a clinical insight, adding functionalities needed for the rehabilitation process of the patients [12]. Clinicians' needs are aimed to support physical therapy tasks. These tasks require demanding training to provoke neural plasticity [13]. They are also focused on building muscle strength, balance, and regain healthy patterns. In this application, the principal goal is aimed to reduce the burden of the clinicians. Simultaneously, robotic aid eases repetitive motor tasks ensuring patient recovery and reduced locomotor dysfunctions [14].

This overall goal supports multiple steps along with physical therapy, where each step defines main functionalities to improve a specific activity of the daily living. Focusing on gait, many gait rehabilitation systems have been deployed and designed for a clinical environment, known as treadmill-based lower-limb exoskeletons [15]. These exoskeletons have a standard functionality known as body weight support, widely used in early locomotion training, reducing the user's weight sensation [16, 17]. Besides, the body weight support eases the pelvis aid to mend asymmetric changes among the gait, either passive or active approaches [8, 18].

One of the treadmill-based lower-limb exoskeletons widely deployed in clinical scenarios is the *Lokomat* (Hocoma, Switzerland) and it has been extended around the world with one thousand devices within 646 facilities [19]. Other treadmill-based lower-limb exoskeletons have been developed to research purposes such as *ALEX* (University of Delaware, USA) [20] or *LOPES* (Institute for Biomedical Technology, Netherlands) [21]. Treadmill-based lower-limb exoskeletons are only applicable in clinical scenarios due to their enormous dimensions, either for clinical or for research purposes. However, those devices provide helpful functionalities in people who suffered a neurological injury [22].

Another type of exoskeleton also provides aid to perform specific tasks or daily living activities, which are known as exoskeletons for assistance. They could be used to rehabilitation purposes, although its design process changes.

3.2.1.3 Exoskeletons for Assistance

Other devices have been designed to assist activities of daily living, which are also required by the patient, such as stand to sit, sit to stand, and walking up/downstairs, among others [7]. The execution of activities of daily living could pursue a therapeutic objective, as mentioned before, through gait rehabilitation. Nevertheless, those devices have also been targeted to increase the ability to perform activities of daily living and the user's quality of life [23, 24]. In some cases, these devices are used to cope several activities of daily living due to the user's neurological injury (e.g., spinal cord injury, cerebral palsy) [25]. In this sense, the lower-limb exoskeleton's layout is known as overground exoskeletons, which most of them are designed as a wearable and portable device. Contrary to treadmill-based lower-limb exoskeleton, overground exoskeletons can be used in home-setting scenarios. Therefore, lower-limb exoskeleton's requirements are related to assistive tasks for the user's activities of daily living.

Overground lower-limb exoskeletons have been arranged as portable devices such as *Indego* (Vanderbilt University, USA) [26], *HAL* (University of Tsukuba, Japan) [27], and *MINDWALKER* (University of Twente, The Netherlands) [28]. These portable devices can be deployed for home-setting such as the *ReWalk* exoskeleton [29]. Similarly, overground lower-limb exoskeleton can also be applied for clinical purposes known as *H2* or *Indergo* [26, 30]. In particular, robotic functionalities have been mainly focused on the user's daily independence within these scenarios. These functionalities required a suitable arrangement of sensors and actuators to perform a proper aid, which will be addressed in the following sections.

As mentioned in this section, the exoskeletons are applied for the three principal applications, and each one provides information to the device's design. However, the user's anatomic features are also involved and complement the design parameters.

3.2.2 Anatomic Concepts

The second part of the user-centered features is the anatomic concepts involved in the design process. As shown in Fig. 3.1, this part is also related to the targeted goal delimiting the forward device's features. The targeted population defines the anatomic concepts, usually remarked through gender (e.g., male or female) and age group (e.g., infant, child, adolescent, young adult, adults, and older adults)

[31]. The human joint kinematics are also defined by an underlying range or approximation that follows the gender and age group. Moreover, the human joints are described emphasizing hip, knee, and ankle biological components and their degrees of freedom (DOFs).

Anthropometric measurements are at the forefront for different design features for lower-limb exoskeleton. They can be estimated using three approaches: (1) ergonomic information of the targeted population, (2) cadaver studies, and (3) user's height. The ergonomic approach has been used to determine user measurements. Focusing on lower limbs, these values are highly related to the targeted population's region, and they have been estimated under specific postures (i.e., standing and sit-down) [32]. In addition to these measurements, other user features contribute to the lower-limb exoskeleton's design process. The understanding and definition of the body planes and human joints, which are also the center of attention along the design process, aim for the following subsections.

3.2.2.1 Body Planes and Human Joints

To ease the understanding of joints' movements, the human body is divided into three principal body planes: the sagittal plane, transversal plane, and frontal plane [33], as shown in Fig. 3.2. The body planes' definition is essential for the following joints' definitions, which are established on these planes. Besides, kinematic and kinetic variables will also be described through each body plane.

The anatomical composition of human joints allows different patterns of movements. They provide intricate functionalities, while they collaborate independently. Most human joints' biological behavior is not straightforward due to their intrinsic geometry [34]. Besides, the constituent elements have non-linearities (i.e., configuration, or arrangement, and mechanical properties), increasing the modeling difficulty. These anatomical elements are bones, ligaments, tendons, and muscles. According to the anatomical configuration and bones' geometry, human joints often exhibit several DOFs allowing for complex movements [35].

To correctly interpret the motion generated by human joints, scientists and researchers have been addressed joints' definition through several methods. One of them describes the human joints' as a mathematical approximation where a model resembles their kinematic outcome. Another approach is joint modeling in which its definition is made through 2D and 3D registration techniques such as X-ray fluoroscopy [36, 37]. On the other hand, joint dissection has been aimed at the same problem through a 3D digital electrogoniometer [38]. Although previous approaches could deliver a precise outcome, they require expensive equipment and their results are exclusive to a particular user.

Despite these limitations, joint models are applicable input models as an arrangement of theoretical joints. In this sense, joint's representation could be through n-joints in which each one is a 1-DOF joint (e.g., rotational degree of freedom known as a revolute joint or linear displacement degree of freedom known as prismatic joint), shown in Fig. 3.3a and b, respectively. Depending on their configuration (e.g., series or parallel), it might resemble the joint model [39]. The joints' representation is widely analyzed in several devices through theoretical

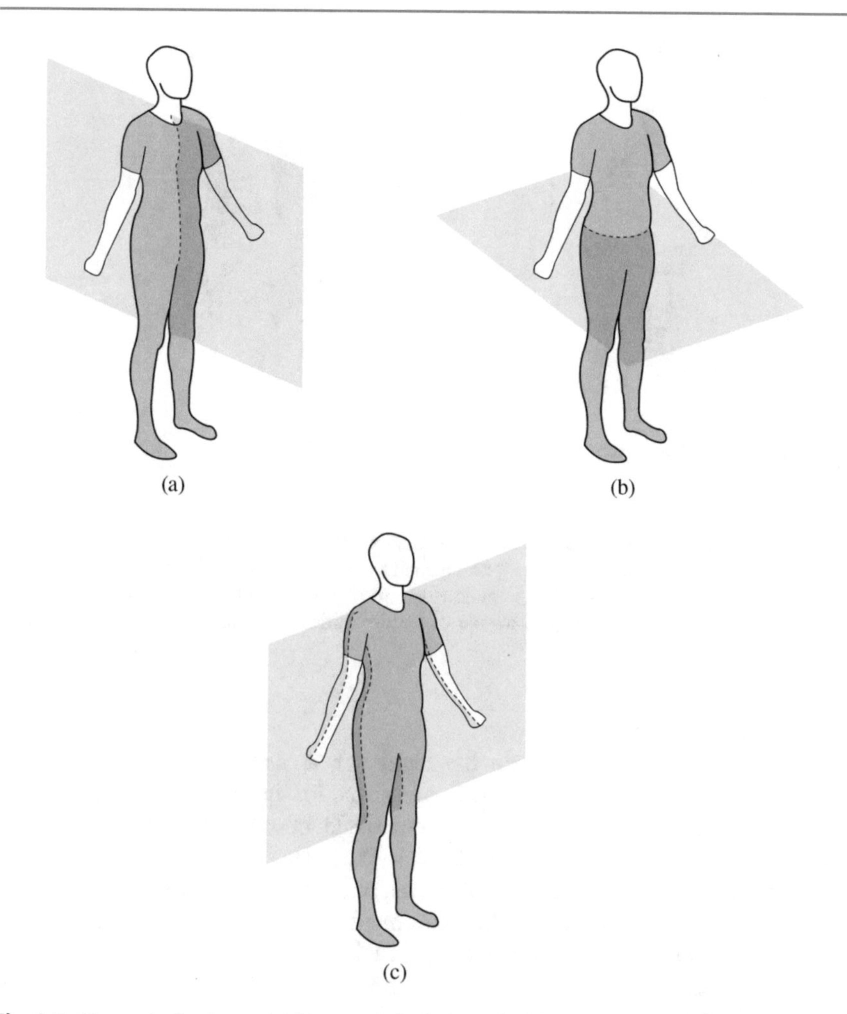

Fig. 3.2 Human body planes. (**a**) Human sagittal plane. (**b**) Human transversal plane. (**c**) Human frontal plane

joints. Besides, the arrangement of joints types and configurations are often referred to as the device's kinematic chain, as shown in Fig. 3.3c.

The human joint and theoretical joint are also delimited by targeted activities addressed in the previous section. These activities establish principal planes of operation, together with the range of motion (ROM), the principal DOF of each joint, and kinematic and kinetic of the user's joints. Each variable establishes a particular waveform compared with gait percentage according to the task. The hip, knee, and ankle joints will be described in this context, highlighting their DOF, muscles and bones involved, and their kinematic outcome during gait.

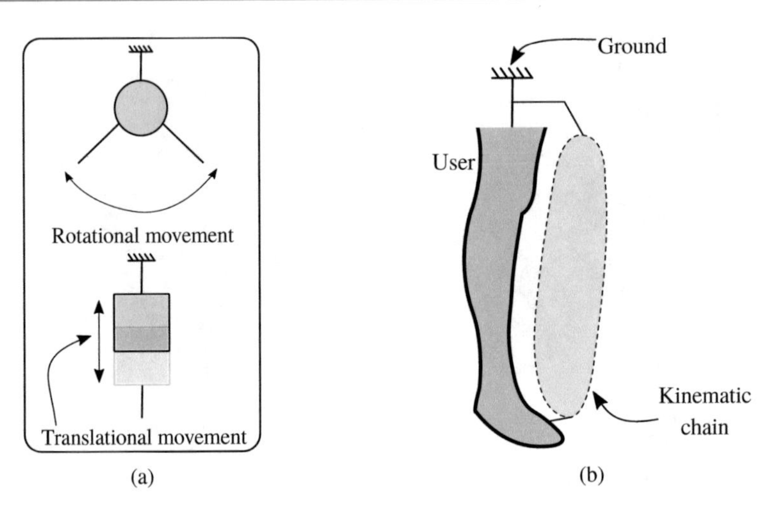

Fig. 3.3 Basic types of theoretical joints. (**a**) A diagram of the kinematic chain represented by the gray zone can be placed in various joint arrangements. Within this zone, the kinematic chain can be built. (**b**) Two types of theoretical joints divided into one degree of freedom revolute joint that allows rotational movement and one degree of freedom prismatic joint that allows translational movement

3.2.2.2 Human Hip Joint

The hip joint articulates between the femoral head and the acetabulum of the pelvis [40]. It has three DOFs, which permit motion in the three planes: sagittal (flexion and extension), frontal (abduction and adduction), and transverse (internal and external rotation) as shown in Fig. 3.4. The bones, as mentioned earlier, are articulated by a set of ligaments (iliofemoral, pubofmeoral, and ischiofemoral) and various muscle groups that ensure hip movements in the different planes and restrict them to a specific range of motion (ROM) [40, 41].

The biomechanics of walking at a natural pace is well established and was presented in the previous chapters. The hip joint angles while walking in all three planes of motion are given in Fig.3.5 [42].

Along the sagittal plane, the hip's maximum flexion and extension angles are, respectively, around 40°and 0°as shown in Fig. 3.5a, for a total ROM of around 40°. Within the coronal plane, the maximal adduction angle is approximately 6°and the maximal abduction angle is nearly 4°as shown in Fig. 3.5b, for a total ROM of 10°. Finally, the hip's maximal internal and external rotation angles are approximately 4°, as presented in Fig. 3.5c, for a total ROM of 8°. A summary of the gait ROMs is presented in Table 3.1.

To achieve the hip angles showed previously, multiple groups of muscles are in charge to perform different motions, as presented in Table 3.1, using between two muscles (e.g., hip extension) and five muscles (e.g., hip external rotation). Through those muscles, the kinetic outcomes are commonly presented through the hip moment (i.e., $\mathrm{Nm} \cdot \mathrm{Kg}^{-1}$) and hip power (i.e., $\mathrm{W} \cdot \mathrm{Kg}^{-1}$). Focusing on gait,

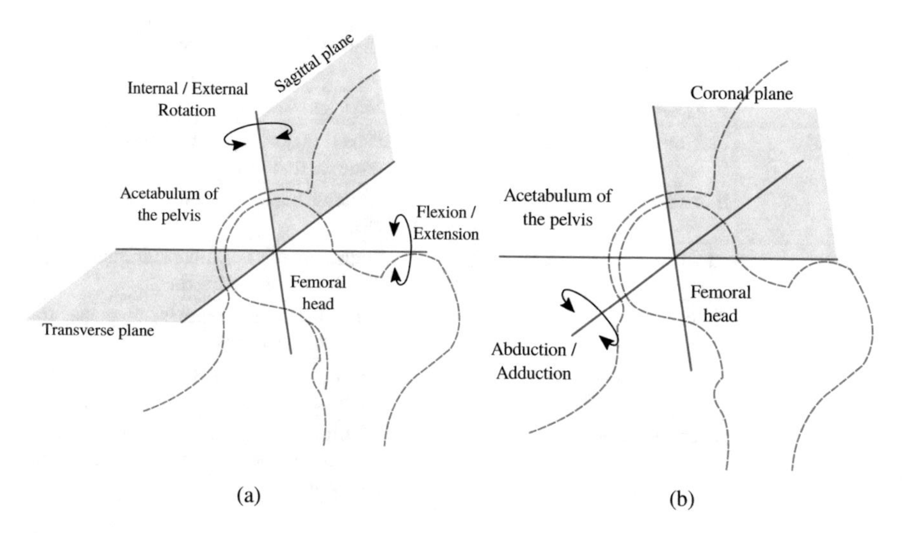

Fig. 3.4 A descriptive scheme of the hip joint degrees of freedom. (**a**) Sagittal and transverse planes. (**b**) Coronal plane

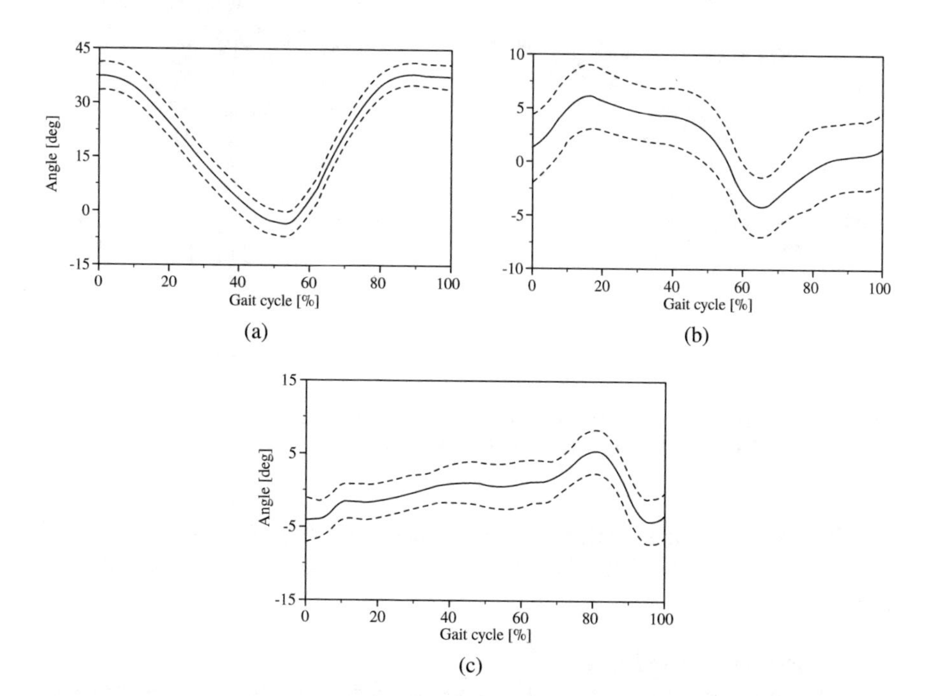

Fig. 3.5 Hip angles behavior along the three principal planes of motion during gait. (**a**) Flexion and extension motion. (**b**) Abd/adduction motion. (**c**) Internal and external rotation motion

Table 3.1 Anatomical features and range of motion of the human hip joint [40,41]. The gait ROM is presented in degrees

Plane	Hip motion	Gait ROM [deg]	Muscles
Sagittal	Flexion	40	Psoas major, iliacus pectineus, rectus femoris, and sartorius.
	Extension		Gluteus maximus and hamstring muscles.
Frontal	Abduction	10	Gluteus medius, gluteus minimus, tensor fascia latae, and sartorius.
	Adduction		Adductor longus, brevis, magnus, gracilis, and pectineus.
Transverse	Internal rotation	8	Tensor fascia, fibers of the gluteus medius, and minimus.
	External rotation		Obturator muscles, quadrutus femoris, gemelli, gluteus maximus, sartorius, and piriformis.

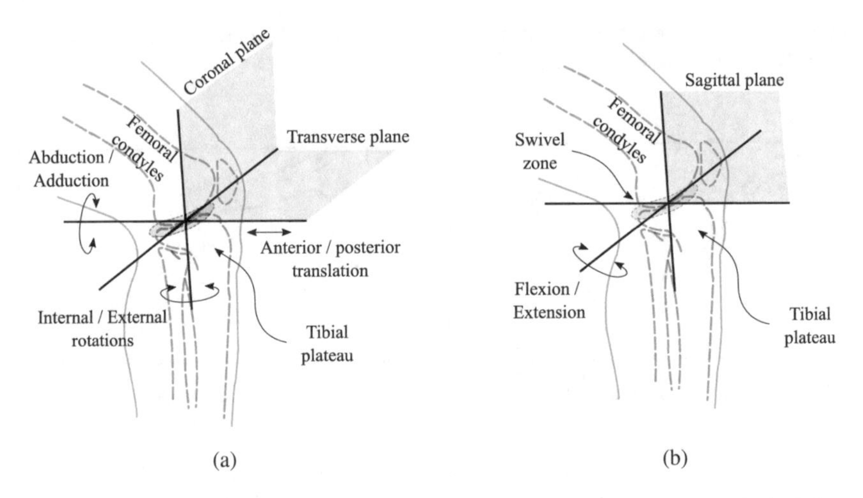

Fig. 3.6 A descriptive scheme of the knee joint's degrees of freedom illustrates its principal planes of motion. (**a**) Coronal and transverse planes. (**b**) Sagittal plane

the hip extensor moment varies between 0.7 and -1.2 Nm $\cdot$ Kg^{-1} for flexion and extension, respectively. Following this, the hip power fluctuates between -0.9 and 1.4 W $\cdot$ Kg^{-1} [43].

3.2.2.3 Human Knee Joint

Several approaches are introduced to understand the knee's kinematic behavior through the methodologies mentioned above. These approaches provided meaningful details of the biological components of the knee. This joint is composed of the patella, the tibial plateau, and femoral condyles, as shown in Fig. 3.6. It has a

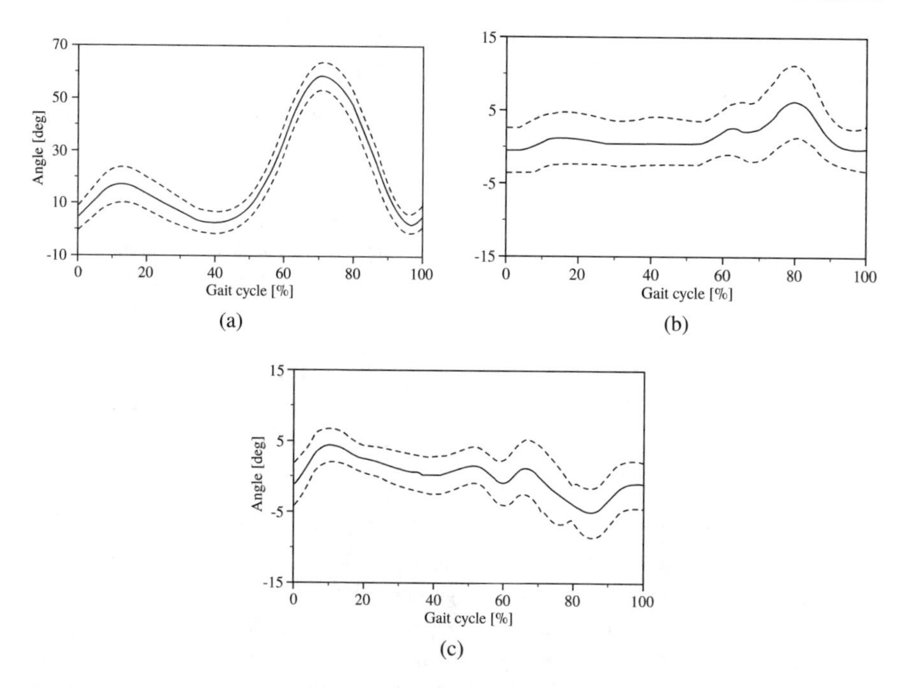

Fig. 3.7 Knee angles behavior along the three principal planes of motion during gait. (**a**) Flexion and extension motion. (**b**) Abd/adduction motion. (**c**) Internal and external rotation motion

joint capsule to provide strength and lubrication, an intricate group of ligaments, and fluids that empower the weight-bearing, such as intracapsular and extracapsular ligaments [44].

Knee kinematics might change according to the task performed by the user. For instance, knee motion during gait has a specific waveform for each plane, as is shown in Fig. 3.7 [42]. The sagittal plane presented in Fig. 3.7a, starts with the knee flexion at the initial contact (i.e., 0% of the gait percentage) to the loading response (i.e., 10% of the gait percentage). During these events, the knee absorbs the energy of the load weight. Then, the knee extends during the mid-stand (i.e., 10% to 30% of the gait percentage) and starts flexing at the terminal stance (i.e., 50% of the gait percentage) to the initial swing. After the initial swing, the knee produces energy to extend the end of the gait cycle [33,45]. Most of the knee motion within the coronal plane relies on the terminal stance and the swing period, as shown in Fig. 3.7b. Finally, the knee motion along the transverse plane varies at the initial contact and the initial swing, as is shown in Fig. 3.7c [42].

Similarly, as presented in the human hip joint, the knee joint's kinetic outcomes are commonly presented through the knee moment (i.e., $\text{Nm} \cdot \text{Kg}^{-1}$) and knee power (i.e., $\text{W} \cdot \text{Kg}^{-1}$). Focusing on gait, the knee extensor moment varies between 0.7 and -0.4 $\text{Nm} \cdot \text{Kg}^{-1}$ for flexion and extension, respectively. Following this, the knee power fluctuates between 1.1 and -1.9 $\text{W} \cdot \text{Kg}^{-1}$ [46]. To achieve this, a predominant group of muscles is involved in each knee motion, as seen in

Table 3.2 Anatomical features and range of motion of the human knee joint [44]

Plane	Knee motion	Gait ROM [deg]	Muscles
Sagittal	Flexion	55–60	Articularis genus, rectus femoris, vastus lateralis, vastus intermedius, and vastus medialis.
	Extension		Biceps femoris, semitendinosus, semimembranosus, gastrocnemius, plantaris, gracilis, and popliteus
Frontal	Abduction	8	Gluteus medius and minimus, obturator externus, gemelli, and sartorius.
	Adduction		Adductor group of muscles.
Transverse	Internal rotation	8	Biceps femoris.
	External rotation		Semimembranosus, semitendinosus, gracilis, sartorius, and popliteus.

Table 3.2. Each group of muscles assists during the user's tasks, and these motions are delimited by four DOF, as is shown in Fig. 3.6. Compared with the hip joint, the knee's muscles and ligaments arrangement absorbs energy and has one less DOF due to the bones and ligaments' coupling. Besides, the gait ROM along the sagittal plane is also greater, contrary to the transverse plane in which the gait ROM is similar for both joints. Additionally, the hip joint has a higher gait ROM along the frontal plane than the knee joint. In kinetic terms, the knee absorbs energy as mentioned before, and as a result of this behavior, the knee moment is lower than the hip moment and ankle moment. Similarly, the knee power is also lower than the hip and ankle power.

3.2.2.4 Human Ankle–Foot Complex

The ankle–foot complex consists of 28 bones divided into the distal tibia and distal fibula, seven tarsals, five metatarsals, and 14 phalangeal bones [47]. These joints, formed by the bones, work synchronously to accomplish distinctive functions in activities of daily living's execution [47]. Like other lower joints, the ankle–foot complex movements involve several motion planes [47, 48]. For instance, supination requires plantarflexion, inversion, and adduction movements, while pronation includes dorsiflexion, eversion, and abduction [48], as is shown in Fig. 3.8.

The foot progression angle (FPA) is defined as the angle between the line from the calcaneus to the second metatarsal and the line of progression averaged from heel strike to toe-off during the stance phase of walking for each step (toe-in angle is positive and toe-out angle is negative).

Regarding the range of motion (ROM), the ankle has variations concerning geographical and cultural differences [48]. Specifically, for the main movements on the ankle, studies estimate standardized ranges between 65 and 75°in the sagittal plane (i.e., 10 to 20°covers the dorsiflexion and 40 to 55°the plantarflexion), −8 to −17°in the transverse plane (i.e., foot progression), and 35°in the frontal plane (i.e., 23°during eversion and 12°in inversion) [33, 48]. However, activities

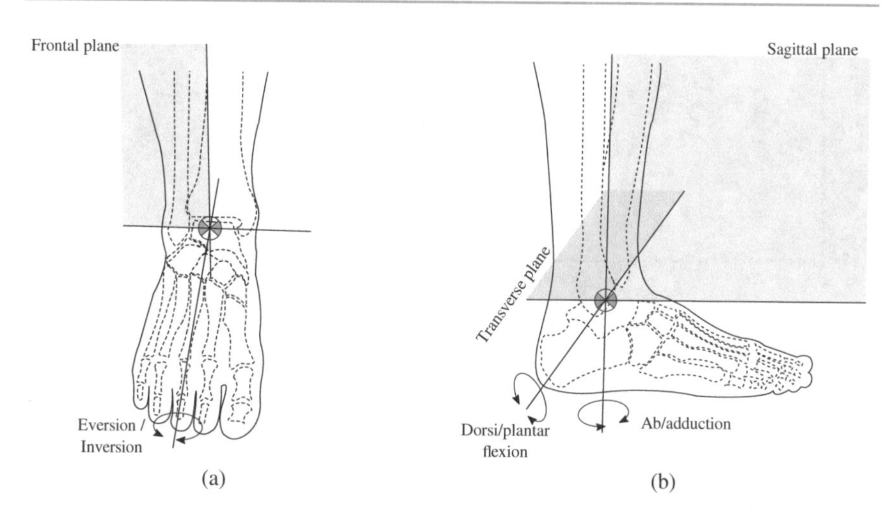

Fig. 3.8 A descriptive scheme of the ankle–foot joint's degrees of freedom. (**a**) Frontal plane. (**b**) Sagittal and transverse planes

Table 3.3 Anatomical features and range of motion of the human ankle joint [48]

Plane	Ankle–Foot	Gait ROM [deg]	Muscles
Sagittal	Dorsiflexion	30	Tibialis anterior, extensor hallucis, peroneus tertius, digitorum longus, peroneus longus, and peroneus brevis.
	Plantarflexion		Gastrocnemius, soleus, plantaris, tibialis posterior, flexor digitorum longus, and flexor hallucis longus
Frontal	Inversion	15	Extensor hallucis longus, tibialis posterior, flexor digitorum longus, and flexor hallucis longus.
	Eversion		Peroneus tertius, peroneus longus, and peroneus brevis.

of daily living's execution require reduced ROM values. For instance, the dorsi–plantarflexion movements need 30°for walking and 37 and 56°for ascending and descending stairs, respectively [47, 48]. Those ankle's movements are achieved through a bigger group of muscles and manage more energy along gait compared to the hip and knee joints. Table 3.3 only presents sagittal and frontal planes due to the dorsi/plantarflexion movements involving multiple movements along sagittal and transverse planes compared to the hip and knee joints. Besides, inversion and eversion movements also require frontal and transverse planes. Hence, anatomical features of the ankle–foot complex can be defined along these two planes [48].

The ankle joint angles while walking in all three planes of motion are given in Fig. 3.9. In the kinetic context, the ankle–foot complex works as a power dissipater, bearing approximately five times body weight during stance in normal walking and up to thirteen times body weight during other activities of daily living (e.g., running)

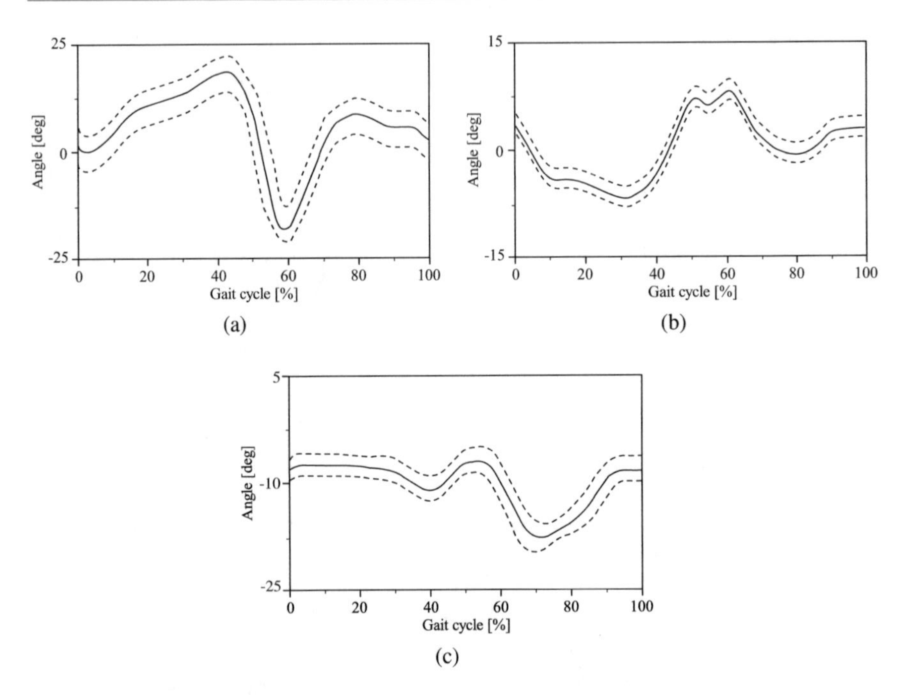

Fig. 3.9 Ankle–foot complex angles behavior along the principal planes of motion during gait. (**a**) Dorsi–plantarflexion motion. (**b**) Inversion–eversion motion. (**c**) Foot progression angle

[48, 49]. Nevertheless, the ankle also can work as a generator, providing enough torque to execute propulsion tasks. This way, to accomplish activities of daily living properly (e.g., walking), this complex joint needs to generate a torque per kilogram of 1.6 $\text{Nm} \cdot \text{Kg}^{-1}$ approximately [33, 48]. According to previous kinetic outcome, the ankle and the hip joints produce energy during the gait because of the amount of torque generated per kilogram. In addition, the ankle also absorbs energy similarly to knee joint.

As mentioned in this section, the user-centered features are forefronts at the beginning of the lower-limb exoskeleton's design process. The targeted goal and anatomic concepts delimited this process, gathering information for the device characteristics and design parameters. This information is defined as device-centered features, explaining multiple features to consider.

3.3 Device-Centered Features

A lower-limb exoskeleton's design process uses user-centered features to establish multiple parameters, limitations, and functionalities to be deployed for the user. Bearing on these features, the device-centered features are organized into four principal characteristic structure's design, joint's design, actuation and control strategies, and physical interfaces. As is shown in Fig. 3.1, each feature is interrelated and co-dependent to ensure proper performance.

3.3.1 Mechanical Design

The understanding of biological processes allows enhancing designs and mechanisms that are deployed for a lower-limb exoskeleton. The bioinspired design implements biological models to provide energetic advantages or maximize a specific function (e.g., weight, torque, shock absorption) [50]. These biological models could come from animals, plants, or humans, and they could be applied along the design process. Following the biological model source, several approaches have been proposed to resemble human lower limbs, aiming to enhance ergonomics and comfort [51]. Most of the solutions deployed for the lower-limb exoskeleton are bioinspired pursuing the user-centered features. In this sense, some design process has an underlying approach or outcome to follow, such as the distribution of energy during gait, the shape of the mechanical structure, or the joints' arrangement.

The human's semblance of the lower-limb exoskeleton is achieved by multiple mechanical features related to the mechanical structure, joints' design, and the physical interfaces, further defined in this section.

3.3.1.1 Structure's Design

As was addressed in the previous sections, the user-centered features are paramount to design the lower-limb exoskeleton, which are used in each device's features. In this context, the structure's design involves the anatomic concepts defined in Sect. 2.2. To do this, the design approach (e.g., bioinspired or biomimetic) allows distinguishing two types of structures that could be used in robotic devices. They have been classified as nonanthropomorphic structures (NS) and anthropomorphic structures (AS) [52]. Even though both types of structures have the same goal, they can be differentiated if they emulate the user's body.

The most common method is the AS, in which the structure's shape follows the assisted limb by placing the actuators near the user's joints. This approach is the most commonly used in active lower-limb exoskeletons, such as *BLEEX*, *HAL-3*, and *ALEX* [20, 53, 54]. In contrast, NS differs from the human shape extending the design possibilities. Although these structures provide several mechanical advantages, the design process's complexity may increase [52].

One of the advantages is the back drivability provided by NS, because of the actuator's location along with the structure. Besides, the load transfer to the ground may be easier than AS, as it is presented by *Van Dijk* et al. [55] through the *Exobuddy* (Intespring BV, Netherlands). This approach allows to overcome misalignment between the user and the exoskeleton and reduces the kinematic constraints. Another advantage is the proper distribution of masses, which reduces the risk of falls and allows a suitable balance for the user. These two design approaches are described as is shown in Fig. 3.10.

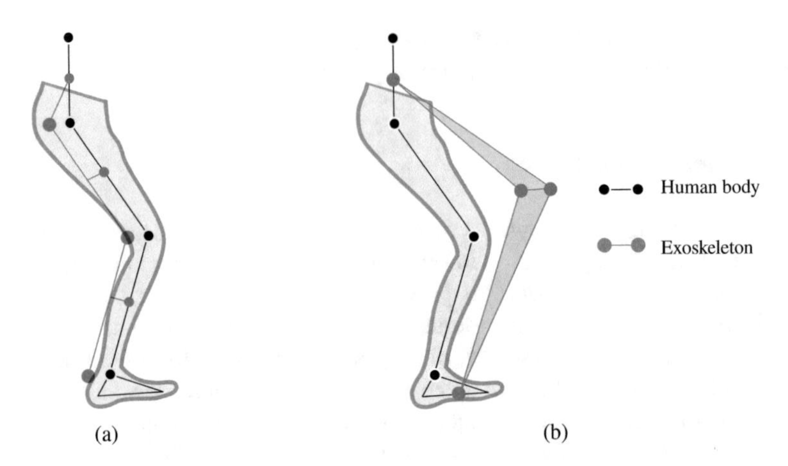

Fig. 3.10 Types of the mechanical structure for lower-limb exoskeletons. (**a**) A descriptive diagram of an anthropomorphic exoskeleton. (**b**) A descriptive diagram that presents one example of a nonanthropomorphic exoskeleton

The definition of the structure's type establishes the foundations of the joints' design. The following section will discuss the kinematic compatibility involved in this feature and the approach to design exoskeleton joints.

3.3.1.2 Joints' Design

The exoskeleton joints have been widely addressed within the structure design. A suitable design can be achieved by mimicking or simplifying the joint structure of interest, according to the type of structure mentioned before. Moreover, the joint structure simplification involves an additional adjustment, causing a misalignment between the user and the device. Focusing on AS, joint proposals are targeted to adjust the alignment through (1) manual adjustment, (2) compliant mechanisms or joints, or (3) kinematic redundancy [10]. The manual adjustment has been deployed in many commercial lower-limb exoskeletons such as *Lokomat* [22], *HAL* [27], *ReWalk* (Argo Medical Technologies Ltd., Israel) [29], and *Indergo* [26]. Nevertheless, their manual adjustment may be exhaustive and requires considerable practice to reduce the time to install the device. Hence, compliant mechanisms and kinematic redundancy approaches are intended to overcome previous limitations.

Two principles are used in compliant joints, which consider (1) the material's mechanical behavior and (2) the represented linkage through the joint's geometry [56]. Regardless of their intricate design, it allows motion in the central plane and multiple differed motions. These joints have a degree of flexibility among other planes, due to the geometry and material's compliance (e.g., low compliance equals rubber, and high compliance equals steel). An example of these types of joints is shown in Fig. 3.11. The geometry allowed multiple displacements. Using these characteristics, a compliant joint allows the misalignment caused between the device and the user, while the device assists or enhances the user's movements.

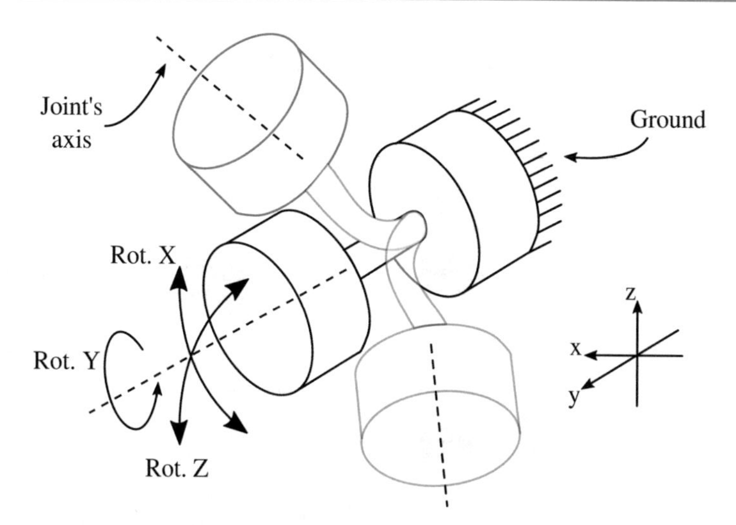

Fig. 3.11 Descriptive example of a spherical compliant joint. The three degrees of freedom of the spherical joint are represented through the rotations X, Y, and Z and their locations illustrated in gray

Another joint proposal approach is the kinematic redundancy by adding DOF for the same joint through revolute (i.e., rotational joint) or prismatic (i.e., linear displacement joint) joints. Both of them have one degree of freedom [57]. Using these joints, the kinematic redundancy can be achieved at different levels. *Näf* et al. [10] proposed multiple joints in series such as three revolute joints (RRR) as presented in Fig. 3.12a, two revolute joints and one prismatic joint (RRP) as seen in Fig. 3.12b, one revolute joint and two prismatic joints (RPP) as shown in Fig. 3.12c, and three prismatic joints (PPP) as represented in Fig. 3.12d. ESA exoskeleton presented by *Schiele* et al. [34] displays RRR and RRP arrangements that ease the interaction of the wrist and elbow, respectively. Following this, the RPP arrangement is deployed for a shoulder exoskeleton. In addition, the joint has to be improved by including passive DOFs [58]. Finally, PPP joints are closely related to a parallelogram setup presented by *Li* et al. [59] to ease the knee alignment.

Each configuration could add advantages and disadvantages to the joint's volume, mass, angle relation, expandability, ability to compensate for misalignment, or transmitting forces. Even though these arrangements enhance the kinematic compatibility [34], the kinematic redundancy will increase the device's volume, mass, and complexity [10].

To this point, the structure and joint design are addressed leaving one last part within the mechanical design of lower-limb exoskeleton. The last part is known as the physical interfaces which will be detailed its functionality, importance, manufacture, material, and layout.

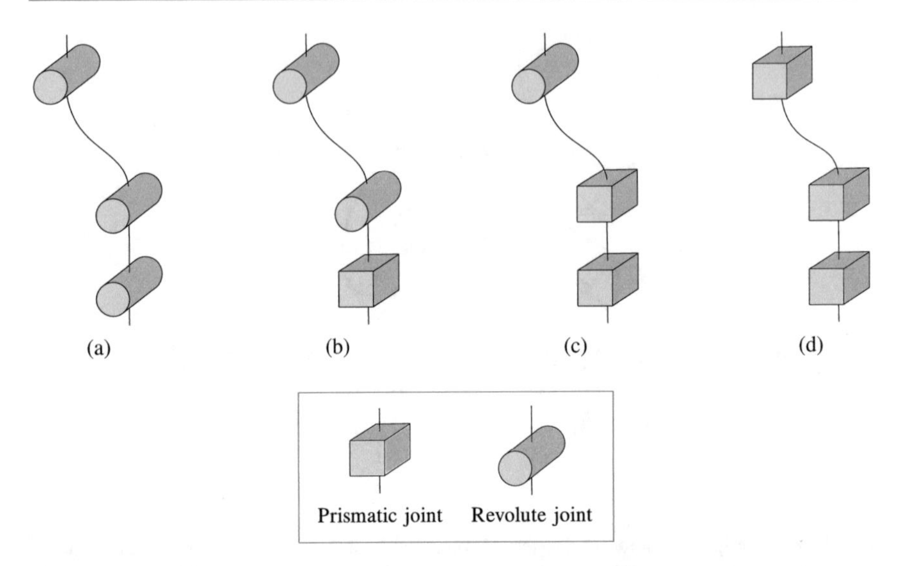

Fig. 3.12 Kinematic redundancy. Proposal kinematic chains using three joints in series to enhance the kinematic compatibility. These kinematic chains could be deployed along with the structure and within the physical interfaces. Adapted from [10]. (**a**) Series of three revolute joints (RRR). (**b**) Series of two revolute joints and one prismatic joint (RRP). (**c**) Series of one revolute joint and two prismatic joints (RPP). (**d**) Series of three prismatic joints (PPP)

3.3.1.3 Physical Interfaces

The last device-centered feature related to the mechanical design of the lower-limb exoskeleton is the physical interface. The interfaces fulfill three functions: first, the transmission of forces between human and robot, second, the correct placement of the exoskeleton relative to the body, and third, a comfortable physical interaction.

These elements ensure the transmission of energy from the device to the user. In other words, they are the attachment between them. The importance of the physical interfaces relies upon the end of the process that transfers energy. The real progress in generating motion begins at the actuators' energy and is then transferred to the mechanical structure. The physical interfaces are attached and carry the energy to human joints [60, 61]. This is challenging as shown in [62], where it was observed that up to 50% of the mechanical power generated by their lower-limb exoskeleton was lost due to soft tissue compression and harness compliance.

Thus, the design and development of physical interfaces must consider the overall features to provide a suitable solution. Moreover, they have to ensure comfort and ergonomics to the user. Within this framework, diverse materials and arrangements have been deployed to secure the lower-limb exoskeleton to the user.

Physical interfaces should promote comfort to the user along with the function related to the targeted goal. From an ergonomic perspective, critically, avoid safety hazards such as skin injuries or pressure points [63]. This is still a major challenge in the field, since soft tissue-related injuries are frequent and repetitive even with commercial devices [64, 65].

Since excessive pressure is known to cause skin injuries and discomfort [66], researchers have developed physical interfaces with integrated pressure sensors [67, 68]. This way clinicians can collect and report sustained wounds related to pressure, which could further help the community to develop safety standards.

In this context, the interaction between the user's skin and the physical interfaces must be compliant and secure its safety against pain and lesions. Moreover, it also has to be strong enough to transfer energy as efficiently as possible. Hence, the physical interfaces' materials are one of the main aspects to enhance the device's performance.

Many approaches include orthopedic components integrating commercial solutions for lower-limb exoskeleton's applications (e.g., Ottobock). These solutions can combine compliant materials (i.e., foams, fabrics, thermoplastics) to leverage ergonomics and comfort [69, 70]. These components embrace a broader population through an adjustable design, and they could reduce the costs of the physical interfaces. They have also been tailored based on user-centered features [51]. However, the design process to achieve a suitable physical interface can be done in different ways.

One of these design approaches is the most common, and the physical interfaces are designed in two sections: (1) rigid interface to transfer the actuators' energy and (2) flexible or soft interface to adapt to user limb geometry and provide comfort. The ratio of each section depends on the material selection. For instance, the proposed exoskeleton by *Asbeck* et al. [71] uses fabrics for the physical interfaces under their Bowden cable actuation. Hence, most of the material used in this case was mainly flexible due to the design's physical interfaces to fit different users. Another option is the development of inflatable interfaces [72, 73]. This type of interface, shown in Fig. 3.13, comprises an inflatable bladder that regulates the pressure applied on the soft tissues. Hereby a truly adaptable design is achieved with the benefits of pressure-monitoring capabilities.

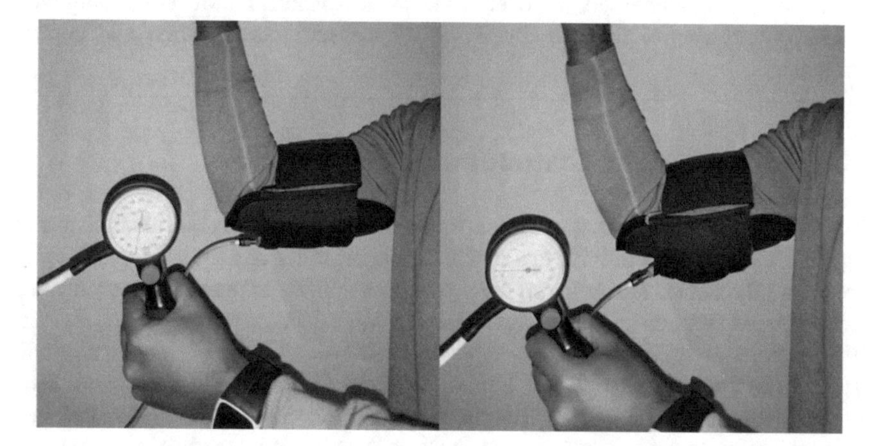

Fig. 3.13 An inflatable interface to adapt to multiple users and to set an exact strapping pressure

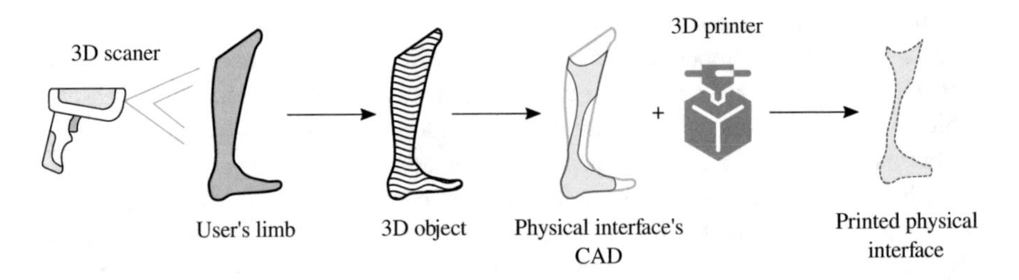

Fig. 3.14 Design and manufacture scheme for customized physical interface for an ankle–foot orthosis

The second approach established a rigid physical interface that can be deployed using a composite material to fix the user's actuator. *Langlois* et al. [74] presented an active ankle–foot exoskeleton (Robotics & Multibody Mechanics Research Group, Vrije Universiteit Brussel, Belgium) that uses this approach. It is made of PLA and carbon fiber reinforcement. The interface was designed through the use of a 3D scanner and was 3D printed using commercially available technologies [51, 74], as is shown in Fig. 3.14. This technique can be applied to develop a customized solution for each individual, allowing clinicians to address user-specific requirements. Furthermore, customization enhance comfort and reduce relative motions while the exoskeleton interact with the user.

Despite the extensive development of lower-limb exoskeleton, physical interfaces have been overlooked compared to other design features. Consequently, several layouts have been defined to the same application (e.g., number of physical interfaces per body segment, best material aimed at a rehabilitation or assistance task) [75]. These layouts involved an ergonomic, kinematic, and kinetic cost, affecting the device's performance. Within the physical interface layout, the user's limb location plays an essential role in ensuring comfort.

As previously addressed, the overall mechanical features allow continuing with actuation and control features, involving the actuator classification and control strategies.

3.3.2 Actuation and Control Strategies

The last device-centered feature is defined by the actuation and control, as shown in Fig. 3.1. The actuators provide energy for the user's joints, and the control strategies command the device according to the user's intention. Over the design process, the user-centered features and previous device's features detailed the main requirements for the actuators and control schemes deployed on the device. This section will address the classification of actuators and the control strategies commonly used for lower-limb exoskeleton, highlighting the main aspects of these topics. Detailed information on each topic is described in Chap. 2, regarding actuation, and Chaps. 5 and 8 expand control strategies.

The energy applied to the lower-limb exoskeleton's joints can be derived from multiple sources to generate motion and assist with the targeted goal. These energy sources delimit lower-limb exoskeleton's response features (i.e., device bandwidth, weight, power) affected by the actuator type and the relation between the supplied torque and the user device's weight [51]. Therefore, more robust or more straightforward actuators may be suitable for different scenarios. These actuators obey different types of control strategies, which ensure proper physical Human–Robot Interaction (pHRI).

3.3.2.1 Actuator Classification

As explained in Chap. 2, the lower-limb exoskeletons have been used in a wide range of actuators. These actuators have been frequently coupled in rehabilitation or assistance lower-limb exoskeletons [76]. The motor drive has been deployed in most commercial exoskeletons because of its ease of implementation contrary to pneumatic actuators, which have a heavyweight energy source, restricting the faculty of deployment. However, as a linear actuator, it has to be outweighed by a control strategy.

Throughout the years, novel actuators emerged in lower-limb exoskeleton applications, and they have been widely used in treadmill-based (e.g., *LOPES* [21], *ALEX* [20,77], *MINDWALKER* [28]) and overground exoskeletons (e.g., *Soft exosuit* (The Wyss Institute for Biologically Inspired Engineering, Harvard University, MA, USA) [71], *Rex* (Rex Bionics Ltd., New Zealand) [78]). These actuators use passive elements coupled to motors (i.e., series or parallel elastic actuators, variable stiffness actuator), giving back drivability to the joint, in which they are capable of uncoupling the shaft [15]. Moreover, their kinematic and kinetic outcome can be specifically designed to joint requirements for mechanical leverage by configuration of springs or dampers [79].

Three main characteristics are summarized in Table 3.4 according to six different kinds of actuators used in lower-limb exoskeleton (detailed information on actuators for lower-limb exoskeleton is provided in Chaps. 2 and 7). The first characteristic presents a general configuration, highlighting the main components according to the type of actuator. These components define the actuator's back drivability, which is related to the characteristic impedance, in other words, the capability of response against a contrary motion [9]. Although this capability depends on their components, as shown in Table 3.4. Another central aspect is the trade-off between power and weight. The actuators showed that electric-based actuators (e.g., series elastic actuator (SEA), variable stiffness actuator (VSA)) weight double than hydraulic or pneumatic actuators (e.g., pneumatic artificial muscle (PAM)), defining the power/weight ratio as low for electric-based actuators and high for hydraulic or pneumatic [80]. However, hydraulic and pneumatic actuators rely on a rich source.

According to the task or activity, these actuators deployed on lower-limb exoskeletons' distributed energy for the user's joints. To do this, the actuators are commanded through control strategies further addressed.

Table 3.4 Type of actuators commonly used in lower-limb exoskeleton. The most common actuators are described according to their generalized configuration, power/weight ratio, and back drivability [80]

Type	Configuration	Power/weight ratio	Back drivability
Electric	Motor—Gearbox—Load	Low	No
SEA	Motor—Spring—Load	Low	Yes
VSA	Motor—Nonlinear spring—Load	Low	Yes
Pneumatic	Air source—Linear or rotatory transfer—Load	High	No
PAM	Air source—Artificial muscle—Load	High	Yes
Hydraulic	Fluid source—Linear or rotatory transfer—Load	High	No

3.3.2.2 Control Strategies

Designers have incorporated mechanical statements or properties (e.g., joints modeling, dynamic constraints, type of actuator) into several control strategies to achieve the targeted main goal based on the features mentioned above. These strategies are defined as high-level control strategies focusing on the user's intention or targeted task that achieves the device [80], in contrast to the low-level control strategies, which handle the response of the actuators directly, either position, velocity, torque, or force.

The control strategies could be based on multiple signals from the user or the device to acquire the user's intention. On the one hand, electromyographic (EMG) signals have been used in a single-joint lower-limb exoskeleton, selecting the principal muscle or muscle group during the task [9]. On the other hand, the device has been used as an input to the control strategies through the actuators or other sensors attached to the structure [80].

Many control strategies have been intended for specific rehabilitation tasks or to improve the interaction regarding the source of the control signals. Moreover, they have been divided into two categories: (1) trajectory tracking control (TTC) and (2) assist-as-needed (AAN) control [81, 82]. The first approach was adapted from industrial robots [82]. Trajectory tracking control is a conventional method used in the initial stages of the rehabilitation process for gait training. This approach has developed several kinds of trajectory tracking control (e.g., reference trajectory, model-based stability, predefined motion, sensitivity magnification control strategies), which predominately guide the user to perform a specific movement [81]. However, these approaches provided a fixed amount of assistance during the tasks.

Current approaches vary the level of assistance according to the user leading to assist-as-needed control strategy, enhancing the patient's neuroplasticity within rehabilitation scenarios. To fully detect the human effort and addition to the control strategies, compensation systems have been defined, which are intrinsical to the device, such as friction and gravity compensation [81]. Considering compensation systems, one of the many approaches is based on bioinspired control that addressed

AAN through impedance controllers [83]. Other control strategies are known as path control algorithms included in compliant virtual walls to amend gait patterns, which could be based on position or force fields [81].

3.4 Design Remarks to Bring Exoskeletons to the Market

Considering that the final goal is to bring novel designs as products to the market, careful consideration has to be paid to the regulatory pathways in place for the planned product. Especially, for start-up companies, obtaining a comprehensive understanding of what is required to legally market a lower-limb exoskeleton as a certified medical device can be a costly challenge, considering the general amount of work and possible late awareness of safety standards and regulations that have to be dealt with. Such late awareness of what is needed often requires implementing technical adaptations to a product (too) late in the development process, which may be costly and lead to a suboptimal result.

To avoid costly technical adaptations at a late stage of product development and avoid discovering that documentation was not generated as required, it is always advisable to be informed on the regulatory pathways and their requirements as soon as the path towards a marketable product is initiated. For a medical device, extensive documentation is typically required for the hardware and software development process, the usability engineering process including user-involved testing, the risk management process, and the clinical reasoning and evidence behind the product. Starting too late with documenting may exponentially increase the work needed to generate it. Work will have to be repeated, or documentation generated retrospectively, which usually involves having to close gaps of information.

The overall work required to achieve a wearable robot or exoskeleton (i.e., implementing safety according to regulations, and documenting a product, a considerable part of the development process in general, the state of the start of safety) is in development and expanding. This is caused by the fact that it is a new emerging technology and that the safety of robots is traditionally regulated by keeping them away from humans, regulated by the ISO. Although it considers close contact with a patient, the safety of medical electrical equipment traditionally does not deal with moving that patient around through an environment and is regulated by the [84]. The overlap between different bodies of standards may confuse or even inconsistent requirements, which is produced by various entities and working groups. Recently a joint working group of ISO and IEC has published the standard IEC 80601-2-78, as a particular standard in the IEC 60601 series of standards, which applies explicitly to medical robots that are used for rehabilitation or associated functions such as the assistance of patients or assessment of function [85]. This could be a starting point to understand what requirements apply to the electrical, mechanical, and software aspects of the safety of an exoskeleton.

3.5 Conclusions

The design process for lower-limb exoskeletons considers multiple features, which are divided into user-centered and device-centered features. The user-centered features provide the primary information and parameters, establishing the device's guidelines through, the anatomic concepts of the population who will use the lower-limb exoskeleton and the targeted goal of the device. Given the user-centered features addressed along with the chapter, the mechanical design embraces three features related to the structure's design, joints' design, and the physical interfaces. Defining these features, the last device-centered feature is the actuation and control strategies that command the lower-limb exoskeleton. Taking into account these features into global guidelines (i.e., ISO) will ease the deployment of lower-limb devices to the market. These overall features defined the fundamental concepts for the design of a lower-limb exoskeleton.

References

1. P.H. King, R.C. Fries, A.T. Johnson, *Design of Biomedical Devices and Systems*, chapter 1, 3rd edn. (CRC Press, Boca Raton, 2014), pp. 1–460
2. S. Sierra, L. Arciniegas, F. Ballen-Moreno, D. Gomez-Vargas, M. Munera, C.A. Cifuentes, Adaptable robotic platform for gait rehabilitation and assistance: design concepts and applications, in *Exoskeleton Robots for Rehabilitation and Healthcare Devices* (Springer, New York, 2020), pp. 67–93
3. C. Majidi, Soft robotics: a perspective—current trends and prospects for the future. Soft Robot. **1**, 5–11 (2014)
4. J.K.S. Nagel, A thesaurus for bioinspired engineering design, in *Biologically Inspired Design* (Springer London, London, 2014), pp. 63–94
5. L.H. Shu, H. Cheong, A natural language approach to biomimetic design, in *Biologically Inspired Design* (Springer London, London, 2014), pp. 29–61
6. S. Yeem, J. Heo, H. Kim, Y. Kwon, Technical analysis of exoskeleton robot. World J. Eng. Technol. **07**, 68–79 (2019)
7. D. Simonetti, N.L. Tagliamonte, L. Zollo, D. Accoto, E. Guglielmelli, Biomechatronic design criteria of systems for robot-mediated rehabilitation therapy. Rehabil. Robot. **2018**, 29–46 (2018, Elsevier Ltd.)
8. D. Shi, W. Zhang, W. Zhang, X. Ding, A review on lower limb rehabilitation exoskeleton robots. Chin. J. Mech. Eng. (English Edition) **32**(1) (2019). Article number: 74
9. B. Kalita, J. Narayan, S.K. Dwivedy, Development of active lower limb robotic-based orthosis and exoskeleton devices: a systematic review. Int. J. Soc. Robot. 0123456789, 1–19 (2020)
10. M.B. Naf, K. Junius, M. Rossini, C. Rodriguez-Guerrero, B. Vanderborght, D. Lefeber, Misalignment compensation for full human-exoskeleton kinematic compatibility: state of the art and evaluation. Appl. Mech. Rev. **70**(5), 1–19 (2018)
11. A.B. Zoss, H. Kazerooni, A. Chu, Biomechanical design of the Berkeley lower extremity exoskeleton (BLEEX). IEEE/ASME Trans. Mech. **11**(2), 128–138 (2006)
12. N. Chia Bejarano, S. Maggioni, L. De Rijcke, C.A. Cifuentes, D.J. Reinkensmeyer, *Robot-Assisted Rehabilitation Therapy: Recovery Mechanisms and Their Implications for Machine Design* (Springer, New York, 2016)
13. M. Pekna, M. Pekny, M. Nilsson, Modulation of neural plasticity as a basis for stroke rehabilitation. Stroke **43**(10), 2819–2828 (2012)

14. N. Koceska, S. Koceski, Review: robot devices for Gait rehabilitation. Int. J. Comput. Appl. **62**(13), 1–8 (2013)
15. G. Chen, C.K. Chan, Z. Guo, H. Yu, A review of lower extremity assistive robotic exoskeletons in rehabilitation therapy. Crit. Rev. Biomed. Eng. **41**(4–5), 343–363 (2013)
16. P.W. Duncan, K.J. Sullivan, A.L. Behrman, S.P. Azen, S.S. Wu, S.E. Nadeau, B.H. Dobkin, D.K. Rose, J.K. Tilson, S. Cen, S.K. Hayden, Body-weight - supported treadmill rehabilitation after stroke. New Engl. J. Med. **364**(21), 2026–2036 (2011)
17. C. Bayón, O. Ramírez, J. Serrano, M.D. Castillo, A. Pérez-Somarriba, J. Belda-Lois, I. Martínez-Caballero, S. Lerma-Lara, C. Cifuentes, A. Frizera, E. Rocon, Development and evaluation of a novel robotic platform for gait rehabilitation in patients with Cerebral Palsy: CPWalker. Robot. Auton. Syst. **91**, 101–114 (2017)
18. S.J. Park, S. Oh, Effect of diagonal pattern training on trunk function, balance, and gait in stroke patients. Appl. Sci. (Switzerland) **10**(13), 4635 (2020)
19. A. Jayaraman, B. Marinov, Y. Singh, S. Burt, W.Z. Rymer, Current evidence for use of robotic exoskeletons in rehabilitation, in *Wearable Robotics: Systems and Applications* (INC, 2019), pp. 301–310
20. S.K. Banala, S.H. Kim, S.K. Agrawal, J.P. Scholz, Robot assisted gait training with active leg exoskeleton (ALEX). IEEE Trans. Neural Syst. Rehabil. Eng. **17**(1), 2–8 (2008)
21. J. Meuleman, E. Van Asseldonk, G. Van Oort, H. Rietman, H. Van Der Kooij, LOPES II - design and evaluation of an admittance controlled gait training robot with shadow-leg approach. IEEE Trans. Neural Syst. Rehabil. Eng. **24**(3), 352–363 (2016)
22. S. Jezernik, G. Colombo, T. Keller, H. Frueh, M. Morari, Robotic orthosis Lokomat: a rehabilitation and research tool. Neuromodulation: Technol. Neural Interface **6**, 108–115 (2003)
23. S. Taki, T. Imura, Y. Iwamoto, N. Imada, R. Tanaka, H. Araki, O. Araki, Effects of exoskeletal lower limb robot training on the activities of daily living in stroke patients: retrospective pre-post comparison using propensity score matched analysis. J. Stroke Cerebrovasc. Dis. **29**, 105176 (2020)
24. R. Mustafaoglu, B. Erhan, I. Yeldan, B. Gunduz, E. Tarakci, Does robot-assisted gait training improve mobility, activities of daily living and quality of life in stroke? A single-blinded, randomized controlled trial. Acta Neurol. Belg. **120**(2), 335–344 (2020)
25. M. Cardona, V.K. Solanki, C.E.G. Cena, Technologies for therapy and assistance of lower limb disabilities: Sit to stand and walking, in *Exoskeleton Robots for Rehabilitation and Healthcare Devices*, pp. 43–66 (Springer, New York, 2020)
26. R.J. Farris, H.A. Quintero, M. Goldfarb, Preliminary evaluation of a powered lower limb orthosis to aid walking in paraplegic individuals. IEEE Trans. Neural Syst. Rehabil. Eng. **19**, 652–659 (2011)
27. H. Kawamoto, Y. Sankai, Comfortable power assist control method for walking aid by HAL-3. Proc. IEEE Int. Conf. Syst. Man Cybernet. **4**, 447–452 (2002)
28. J. Gancet, M. Ilzkovitz, G. Cheron, Y. Ivanenko, H. Van Der Kooij, F. Van Der Helm, F. Zanow, F. Thorsteinsson, MINDWALKER: a brain controlled lower limbs exoskeleton for rehabilitation. Potential applications to space, in *11th Symposium on Advanced Space Technologies in Robotics and Automation*, vol. 1 (2011), pp. 12–14
29. G. Zeilig, H. Weingarden, M. Zwecker, I. Dudkiewicz, A. Bloch, A. Esquenazi, Safety and tolerance of the ReWalk^{TM} exoskeleton suit for ambulation by people with complete spinal cord injury: A pilot study. J. Spinal Cord Med. **35**(2), 96–101 (2012)
30. M. Bortole, A. Venkatakrishnan, F. Zhu, J.C. Moreno, G.E. Francisco, J.L. Pons, J.L. Contreras-vidal, The H2 robotic exoskeleton for gait rehabilitation after stroke: early findings from a clinical study. J. Neuroeng. Rehabil. **12**, 1–14 (2015)
31. N. Geifman, R. Cohen, E. Rubin, Redefining meaningful age groups in the context of disease. Age **35**(6), 2357–2366 (2013)
32. M. De Onis, J.P. Habicht, Anthropometric reference data for international use: recommendations from a World Health Organization expert committee. Am. J. Clin. Nutr. **64**(4), 650–658 (1996)

33. J. Perry, *Gait Analysis* (Slack Incorporated, 1992)
34. A. Schiele, F.C.T. van der Helm, Kinematic design to improve ergonomics in human machine interaction. IEEE Trans. Neural Syst. Rehabil. Eng. **14**, 456–469 (2006)
35. J.C. Perry, J. Rosen, S. Burns, Upper-limb powered exoskeleton design. IEEE-ASME Trans. Mechatron. **12**(4), 408–417 (2007)
36. T. Yamazaki, T. Watanabe, Y. Nakajima, K. Sugamoto, T. Tomita, H. Yoshikawa, S. Tamura, Improvement of depth position in 2-D/3-D registration of knee implants using single-plane fluoroscopy. IEEE Trans. Med. Imag. **23**(5), 602–612 (2004)
37. T. Yamazaki, T. Watanabe, Y. Nakajima, K. Sugamoto, T. Tomita, D. Maeda, W. Sahara, H. Yoshikawa, S. Tamura, Visualization of femorotibial contact in total knee arthroplasty using X-ray fluoroscopy. Eur. J. Radiol. **53**(1), 84–89 (2005)
38. S. Martello, V. Pinskerova, A. Visani, Anatomial investigations on the knee by means of computer-dissection. J. Mech. Med. Biol. **06**, 55–73 (2006)
39. P. Corke, Robotics. Vision and Control (2011)
40. M. Gold, M. Varacallo, *Anatomy, Bony Pelvis and Lower Limb, Hip Joint.* (StatPearls Publishing, Treasure Island (FL), 2019)
41. R. Glenister, S. Sharma, Anatomy, bony pelvis and lower limb, hip. StatPearls [Internet] (2020)
42. M.P. Kadaba, H.K. Ramakrishnan, M.E. Wootten, Measurement of lower extremity kinematics during level walking. J. Orthopaed. Res. **8**, 383–392 (1990)
43. F. Leboeuf, J. Reay, R. Jones, M. Sangeux, The effect on conventional gait model kinematics and kinetics of hip joint centre equations in adult healthy gait. J. Biomech. **87**, 167–171 (2019)
44. A. Ribeiro, J. Rasmussen, P. Flores, L.F. Silva, Modeling of the condyle elements within a biomechanical knee model. Multibody Syst. Dynam. **28**(1–2), 181–197 (2012)
45. J.B. Webster, B.J. Darter, *Principles of Normal and Pathologic Gait*, 5th edn. (Elsevier Inc., Amsterdam, 2019)
46. A. Bonnefoy-Mazure, S. Armand, Normal gait, in *Orthopedic Management of Children with Cerebral Palsy: A Comprehensive Approach* (2015), pp. 199–214
47. S. Angin, İ. Demirbüken, Ankle and foot complex, in *Comparative Kinesiology of the Human Body*, pp. 411–439 (Elsevier, Amsterdam, 2020)
48. C.L. Brockett, G.J. Chapman, Biomechanics of the ankle. Orthopaed. Trauma **30**, 232–238 (2016)
49. R. Burdett, Forces predicted at the ankle during running. Med. Sci. Sports Exerc. **14**(4), 308–316 (1982)
50. J.L. Pons, *Wearable Robots: Biomechatronic Exoskeletons* (Wiley, Hoboken, 2008)
51. M.D.C. Sanchez-Villamañan, J. Gonzalez-Vargas, D. Torricelli, J.C. Moreno, J.L. Pons, Compliant lower limb exoskeletons: a comprehensive review on mechanical design principles. J. NeuroEng. Rehabil. **16**, 55 (2019)
52. D. Accoto, F. Sergi, N.L. Tagliamonte, G. Carpino, A. Sudano, E. Guglielmelli, Robomorphism: a nonanthropomorphic wearable robot. IEEE Robot. Autom. Mag. **21**, 45–55 (2014)
53. H. Kawamoto, S. Lee, S. Kanbe, Y. Sankai, Power assist method for HAL-3 using EMG-based feedback controller, in *SMC'03 Conference Proceedings. 2003 IEEE International Conference on Systems, Man and Cybernetics. Conference Theme-System Security and Assurance (Cat. No. 03CH37483)*, vol. 2 (IEEE, New York, 2003), pp. 1648–1653
54. A. Zoss, H. Kazerooni, A. Chu, On the mechanical design of the Berkeley Lower Extremity Exoskeleton (BLEEX), in *2005 IEEE/RSJ International Conference on Intelligent Robots and Systems, IROS* (2005), pp. 3132–3139
55. W. Van Diik, T. Van De Wijdeven, M.M. Hölscher, R. Barents, R. Könemann, F. Krause, C.L. Koerhuis, Exobuddy - a non-anthropomorphic quasi-passive exoskeleton for load carrying assistance, in *Proceedings of the IEEE RAS and EMBS International Conference on Biomedical Robotics and Biomechatronics*, August, vol. 2018 (2018), pp. 336–341
56. D.F. Machekposhti, N. Tolou, J.L. Herder, A review on compliant joints and rigid-body constant velocity universal joints toward the design of compliant homokinetic couplings. J. Mech. Des. Trans. ASME **137**(3), 1–12 (2015)
57. B. Celebi, M. Yalcin, V. Patoglu, AssistOn-knee: a self-aligning knee exoskeleton, in *2013 IEEE/RSJ International Conference on Intelligent Robots and Systems* (IEEE, 2013), pp. 996–1002

58. T. Nef, M. Mihelj, G. Colombo, R. Riener, ARMin - robot for rehabilitation of the upper extremities, in *Proceedings - IEEE International Conference on Robotics and Automation*, May (2006), pp. 3152–3157
59. J. Li, S. Zuo, C. Xu, L. Zhang, M. Dong, C. Tao, R. Ji, Influence of a compatible design on physical human-robot interaction force: a case study of a self-adapting lower-limb exoskeleton mechanism. J. Intell. Robot. Syst.: Theory Appl. **98**(2), 525–538 (2020)
60. A.T. Asbeck, S.M. De Rossi, K.G. Holt, C.J. Walsh, A biologically inspired soft exosuit for walking assistance. Int. J. Robot. Res. **34**, 744–762 (2015)
61. M.B. Näf, A.S. Koopman, S. Baltrusch, C. Rodriguez-Guerrero, B. Vanderborght, D. Lefeber, Passive back support exoskeleton improves range of motion using flexible beams. Front. Robot. AI **5**, 1–16 (2018)
62. M.S. Cherry, S. Kota, A. Young, D.P. Ferris, Running with an elastic lower limb exoskeleton. J. Appl. Biomech. **32**(3), 269–277 (2016)
63. A. Chiri, M. Cempini, S.M.M. De Rossi, T. Lenzi, F. Giovacchini, N. Vitiello, M.C. Carrozza, On the design of ergonomic wearable robotic devices for motion assistance and rehabilitation, in *2012 Annual International Conference of the IEEE Engineering in Medicine and Biology Society* (IEEE, 2012), pp. 6124–6127
64. J. Bessler, G.B. Prange-Lasonder, R.V. Schulte, L. Schaake, E.C. Prinsen, J.H. Buurke, Occurrence and type of adverse events during the use of stationary gait robots—a systematic literature review. Front. Robot. AI **7**, 158 (2020)
65. Y. He, D. Eguren, T.P. Luu, J.L. Contreras-Vidal, Risk management and regulations for lower limb medical exoskeletons: a review. Med. Dev. (Auckland, NZ) **10**, 89 (2017)
66. T. Kermavnar, V. Power, A. de Eyto, L.W. O'Sullivan, Computerized cuff pressure algometry as guidance for circumferential tissue compression for wearable soft robotic applications: a systematic review. Soft Robot. **5**(1), 1–16 (2018)
67. K. Langlois, E. Roels, G. Van De Velde, C. Espadinha, C. Van Vlerken, T. Verstraten, B. Vanderborght, D. Lefeber, Integration of 3d printed flexible pressure sensors into physical interfaces for wearable robots. Sensors **21**(6), 2157 (2021)
68. J. Tamez-Duque, R. Cobian-Ugalde, A. Kilicarslan, A. Venkatakrishnan, R. Soto, J.L. Contreras-Vidal, Real-time strap pressure sensor system for powered exoskeletons. Sensors **15**(2), 4550–4563 (2015)
69. J. Beil, G. Perner, T. Asfour, Design and control of the lower limb exoskeleton KIT-EXO-1, in *IEEE International Conference on Rehabilitation Robotics*, September, vol. 2015 (2015), pp. 119–124
70. K. Junius, B. Brackx, V. Grosu, H. Cuypers, J. Geeroms, M. Moltedo, B. Vanderborght, D. Lefeber, Mechatronic design of a sit-to-stance exoskeleton, in *5th IEEE RAS/EMBS International Conference on Biomedical Robotics and Biomechatronics* (IEEE, New York, 2014), pp. 945–950
71. A.T. Asbeck, R.J. Dyer, A.F. Larusson, C.J. Walsh, Biologically-inspired soft exosuit, in *IEEE International Conference on Rehabilitation Robotics*, 2013
72. K. Langlois, D. Rodriguez-Cianca, B. Serrien, J. De Winter, T. Verstraten, C. Rodriguez-Guerrero, B. Vanderborght, D. Lefeber, Investigating the effects of strapping pressure on human-robot interface dynamics using a soft robotic cuff, in *IEEE Transactions on Medical Robotics and Bionics*, 2020
73. A. Schiele, F.C. Van der Helm, Influence of attachment pressure and kinematic configuration on PHRI with wearable robots. Appl. Bion. Biomech. **6**(2), 157–173 (2009)
74. K. Langlois, M. Moltedo, T. Bacek, C. Rodriguez-Guerrero, B. Vanderborght, D. Lefeber, Design and development of customized physical interfaces to reduce relative motion between the user and a powered ankle foot exoskeleton, in *2018 7th IEEE International Conference on Biomedical Robotics and Biomechatronics (Biorob)*, August, vol. 2018 (IEEE, 2018), pp. 1083–1088
75. M. Sposito, S. Toxiri, D. Caldwell, J. Ortiz, E. De Momi, Towards design guidelines for physical interfaces on industrial exoskeletons: overview on evaluation metrics, in *International Symposium on Wearable Robotics*, pp. 170–174 (Springer, New York, 2018)

76. X. Zhang, Z. Yue, J. Wang, Robotics in lower-limb rehabilitation after stroke. Behav. Neurol. **2017**, 1–13 (2017)
77. S.K. Banala, S.K. Agrawal, S.H. Kim, J.P. Scholz, Novel gait adaptation and neuromotor training results using an active leg exoskeleton. IEEE/ASME Trans. Mech. **15**(2), 216–225 (2010)
78. R. Bionics, Rex bionics. Web Page (www.rexbionics.com). Accessed October, 2020
79. B. Vanderborght, A. Albu-Schaeffer, A. Bicchi, E. Burdet, D. Caldwell, R. Carloni, M. Catalano, O. Eiberger, W. Friedl, G. Ganesh, M. Garabini, M. Grebenstein, G. Grioli, S. Haddadin, H. Hoppner, A. Jafari, M. Laffranchi, D. Lefeber, F. Petit, S. Stramigioli, N. Tsagarakis, M. Van Damme, R. Van Ham, L. Visser, S. Wolf, Variable impedance actuators: a review. Robot. Auton. Syst. **61**, 1601–1614 (2013)
80. W. Huo, S. Mohammed, J.C. Moreno, Y. Amirat, Lower limb wearable robots for assistance and rehabilitation: a state of the art. IEEE Syst. J. **10**, 1068–1081 (2016)
81. P.K. Jamwal, S. Hussain, M.H. Ghayesh, Robotic orthoses for gait rehabilitation: an overview of mechanical design and control strategies. Proc. Inst. Mech. Eng. Part H: J. Eng. Med. **234**(5), 444–457 (2020)
82. J. Cao, S.Q. Xie, R. Das, G.L. Zhu, Control strategies for effective robot assisted gait rehabilitation: the state of art and future prospects. Med. Eng. Phys. **36**(12), 1555–1566 (2014)
83. J.C. Moreno, J. Figueiredo, J.L. Pons, Exoskeletons for lower-limb rehabilitation, in *Rehabilitation Robotics* (Elsevier, Amsterdam, 2018), pp. 89–99
84. T. Jacobs, J. Veneman, G.S. Virk, T. Haidegger, The flourishing landscape of robot standardization [industrial activities]. IEEE Robot. Autom. Mag. **25**(1), 8–15 (2018)
85. I. Robotics, Medical electrical equipment—Part 2–78: Particular requirements for basic safety and essential performance of medical robots for rehabilitation, assessment, compensation or alleviation. standard, International Organization for Standardization (2019)

4 Fundamentals for the Design of Smart Walkers

Julián Aristizabal-Aristizabal, Rubén Ferro-Rugeles, María Lancheros-Vega, Sergio D. Sierra M., Marcela Múnera, and Carlos A. Cifuentes

4.1 Introduction

Advances in technology, especially in mobile robotics, have grown exponentially in recent years. One of these advances relates to mobile robots for gait rehabilitation and assistance, leading to the emergence of intelligent walkers or robotic walkers. These devices contain electronic components, control systems, and sensory architectures built into their conventional mechanical structure (see Chap. 2) [1]. These improvements have provided a better driving experience for users when controlling the devices, regarding their integrated sensory feedback, physical support, and cognitive support. Another advantage is minimizing the risk for falling and the possibility that users can experience a smoother and more natural gait (see Chap. 1) [2].

Several communication channels took place in walker-assisted gait. As described by *Sierra et al.*, the robotic walkers equip sensors and actuators that allow them to interact with the users and acquire information from them [1]. This concept refers to Human–Robot Interaction (HRI), which comprises strategies that provide cognitive and physical communication between the device and the user [2]. Similarly, the robotic walkers frequently move within complex and dynamic environments, such as clinical and rehabilitation settings. In this sense, the walkers employ sensory architectures and actuation interfaces to acquire information about the environment [2]. This concept refers to Robot–Environment Interaction (REI), which comprises

J. Aristizabal-Aristizabal · R. Ferro-Rugeles · M. Lancheros-Vega
S. D. Sierra M. · M. Múnera · C. A. Cifuentes (✉)
Biomedical Engineering Department of the Colombian School of Engineering Julio Garavito, Bogotá, Colombia
e-mail: julian.aristizabal@mail.escuelaing.edu.co; ruben.ferro@mail.escuelaing.edu.co; maria.lancheros@mail.escuelaing.edu.co; sergio.sierra@escuelaing.edu.co; marcela.munera@escuelaing.edu.co; carlos.cifuentes@escuelaing.edu.co

C. A. Cifuentes, M. Múnera, *Interfacing Humans and Robots for Gait Assistance and Rehabilitation*, https://doi.org/10.1007/978-3-030-79630-3_4

control strategies that provide guidance, autonomous navigation, and safe movement within a particular environment [1].

In addition, recent studies in robotic walkers have coined the term Human–Robot–Environment Interaction (HREI) [1–3]. This concept refers to the communication loop that involves the user, the device, the environment, and the healthcare professional. In this way, the HREI strategies provide natural and compliant user interaction, and effective environment sensing while maintaining safety requirements [1]. As an illustration, Fig. 4.1 shows the communication loops that took place during walker-assisted gait.

The design process of an intelligent walker involves several considerations and parameters. It focuses on the user requirements and analyzes their locomotion impairments and their cognitive and sensory affectations. Moreover, the design process requires the continuous assessment of the user perception to guarantee daily living acceptance [4].

This chapter covers some of the main developments related to intelligent walkers, focusing on: (1) the assistance functionalities, (2) the employed control strategies, and (3) the mechanical structure. This chapter also details fundamentals notions for the design of robotic walkers based on a literature review.

4.2 State of the Art About Smart Walkers

There is a wide variety of robotic walkers reported in the literature, which offer multiple functionalities for gait assistance and rehabilitation. On the one hand, these devices provide natural gait patterns, lateral stability, and weight support. On the other hand, they offer fall prevention, obstacle avoidance, intuitive control

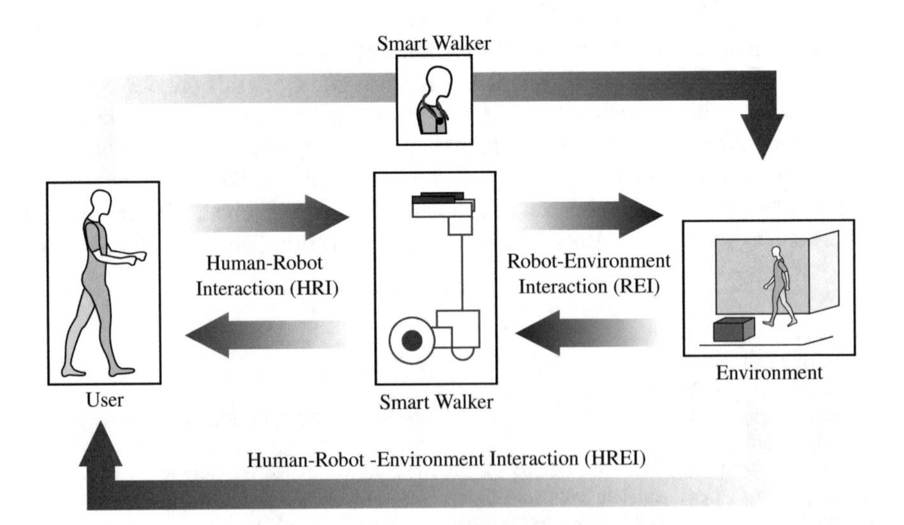

Fig. 4.1 Communication channels during a walker-assisted application

strategies, and comfortable interaction. This section describes the conduction of a literature review to categorize and analyze the most relevant functionalities and control strategies of the robotic walkers.

This literature review followed the PICO search strategy to obtain an evidence-based classification. PICO is an acronym for Patient, Intervention, Comparison, and Outcome. It aims at formulating an appropriate research question to find the most relevant information, using previous experience and reported outcomes [5]. In this chapter, the following constructs yielded the PICO strategy:

- **(P)**: Patients with motor limitation and healthcare professionals who use smart walkers in rehabilitation processes.
- **(I)**: Interventions and validations of a specific control strategy and assistance functionalities, conducted in laboratories or clinical settings.
- **(C)**: Studies comparing the performance of the walker-assisted trials with unassisted gait or passive device such as conventional rollators.
- **(O)**: Research outputs about the operation and performance of the control strategies in smart walkers.

According to the above, the following equation based on Boolean logic was used:

(*"Smart Walker*"* OR *"Robotic Walker*")* AND
(*"Gait Assistance"* OR *"Assistive Gait"* OR
"Walker-assisted Gait" OR *"Gait Rehabilitation"* OR
"Walker functionalities" OR *"Walking Assistance")* AND
(*"Control systems")*

The inclusion criteria were as follows: (1) articles that described the walker functionality and internal systems, (2) articles reporting smart walkers targeted at individuals with gait impairments, (3) articles in English. The exclusion criteria were as follows: (1) articles reporting rehabilitation robots for different conditions, (2) non-intelligent assistive devices, (3) publications before 1998. The search strategy was implemented in the following databases: *IEEE Xplore*, *PUBMED*, and *Google Scholar*.

The PICO search methodology resulted in 407 articles found at Google Scholar, 48 at IEEE Xplore, and one at PUBMED. A total of 65 articles were selected based on the inclusion and exclusion criteria. Table 4.1 describes a summary of the robotic walkers found from the literature review. From these walkers, the following sections explain an analysis of their main characteristics.

Table 4.1 Summary of smart walkers with their purpose and configuration

Year	Name	Main purpose	Configuration
2000	PAMM [6]	Biomechanical monitoring	Active + Caster/Orientable wheels
2003	VA-PAMAID [7]	Safety provision	Active + Caster/Orientable wheels
2003	NOMAD [8]	Guidance and navigation	Active + Fixed/Swedish wheels
2005	MONIMAD [9]	Safety provision	Active + Fixed/Orientable wheels
2008	GUIDO [10]	Safety provision	Active + Caster/Orientable wheels
2008	I-WALKER [11]	Safety provision	Active + Orientable wheels
2010	JARoW [12]	Biomechanical monitoring	Active + Omni-d. wheels
2011	I-GO [13]	Estimation of movement's intention	Active + Fixed/Omni-d. wheels
2011	SIMBIOSIS [14]	Estimation of Movement's Intention	Active + Caster/orientable wheels
2011	ODW [15]	Guidance and navigation	Active + Omni-d. wheels
2012	UFES [16]	Biomechanical monitoring	Active + Fixed/Caster wheels
2012	WCIWAR [17]	Estimation of movement's intention	Active + Omni-d. wheels
2012	CAIROW [18]	Remote control	Active + Fixed/Caster wheels
2013	ASBGo [19]	Guidance and navigation	Active + Fixed/ASOC wheels
2015	EYE WALK [20]	Guidance and navigation	Passive + Orientable/Fixed wheels
2017	MOBOT [21]	Estimation of movement's intention	Active + Fixed/Caster wheels
2017	WACHAJA [22]	Guidance and navigation	Passive + Caster/Orientable wheels
2019	AGoRA [1]	Guidance and navigation	Active + Fixed/Caster wheels

4.3 Physical Structures

This section will provide insights into the structural differences in the construction of robotic walkers and their advantages and disadvantages. Moreover, this section will describe several safety structures that aim at providing additional physical support and balance to the users.

4.3.1 Definition of Physical Structure

At first glance, several factors influence the mechanical structure of a robotic walker. For instance, it involves factors such as the shape, the number and configuration of the wheels, the type of grip, and the locomotion type. The most common structures in robotic walkers involve three or four-wheeled devices. Similarly, depending on the locomotion type, the walkers can be passive, active, or hybrid. More specifically, the active and hybrid configurations employ electric motors such as DC motors, brushless hubs (i.e., motors embedded in wheels), servomotors, and steppers (see Chap. 2).

The physical structure of a robotic walker influences the ability to provide assistance, rehabilitation, stability, and balance during mobility [23, 24]. On the one hand, passive structures introduce mechanical enhancements to improving gait stability without employing traction actuators. Some of these improvements include increasing the base size or distributing weightier elements in the lower part of the structure. Similarly, passive approaches have also reported replacing of conventional handlebars with forearm support platforms [2].

On the other hand, the walkers offer active physical structures by installing traction or steering motors in the wheels and employing electrical braking systems. These actuators aim at compensating the user locomotion requirements, the device kinematics, and the environmental demands (e.g., slopes). The actuators also provide the necessary energy to move the device [2].

4.3.2 Examples of Physical Structures

According to the findings of the search strategy, these are some of the physical structures:

- **Circular shape:** The *JARoW* smart walker (School of Information Science, Japan Advanced Institute of Science and Technology, Japan) is an active device with three structural parts, the base frame, the upper frame, and the connecting rods. This smart walker provides forearm support and offers help during gait by reducing the weight load on the joints of the lower extremities [12]. The *JARoW* has two semi-circular frames that engage with vertical poles, three omnidirectional wheels, and the frame allows forearm support and joint grip [12]. Overall, the *JARoW* smart walker has a circular shape, reducing collisions with obstacles [24].
- **U-Shape and square shape:** The *i-go* smart walker (Chyao Shiunn Electronic Industrial Ltd., Shanghai, China) comprises an *U-shaped* frame and a regulating rod to adjust the height of the handles [24]. Similarly, the *AGoRA* (Department of Biomedical Engineering, Colombian School of Engineering Julio Garavito, Colombia) is an active device with a small square shape, six wheels, and long handles [1]. The *CAIROW* robotic walker (Department of Electrical Engineering, National Taiwan University, Taiwan) is an active device with a medium-sized cubic frame, motorized and steerable wheels, and u-shaped handles that provide support and firmness [18].
- **Adjustable shape:** The *WCIWAR* smart walker (Centre of Neural Interface & Rehabilitation Technology, Huazhong University of Science and Technology, China) is a Width Changeable Intelligent Walking Aid Robot. It integrates a support frame, a mobile base with three servo-driven castor wheels, and two wheels in the middle for weight support. This walker has an array of sensors placed in the armrest to measure the interaction forces, a signal processing system, a rechargeable battery, and a control system for the servo motors to move the base in all directions. The *WCIWAR* also has an electric cylinder aimed at stretching the rods positioned at both sides of the walker. With this system, the walker can vary its width to confer the ability to pass through narrow corridors or give more stability in spacious environments [17].

Another device with an adjustable shape is the *ASBGo* robotic walker (Center for MicroElectroMechanical Systems, University of Minho, Guimarães, Portugal). This device equips electric lifting columns to adjust the device height. Moreover, this walker can also perform lateral adjustments of the handles to meet the patient shoulder width. In addition to this, the *ASBGo* walker includes a handle to assist sit-to-stand transfers [24].

As reported by the *ABSGo* walker, another essential physical feature of the robotic walkers is their ability to assist in sit-to-stand and stand-to-sit transfers. This feature reduces the patient effort and decreases the workload of the healthcare professional [25]. The *MONIMAD* walker (Institut des Systemes Intelligents et de Robotique, Universit e Pierre et Marie Currie, France) also offers this feature. It is an assistive device that combines sit-to-stand transfers with walking assistance for older adults and people with locomotion impairments. In this way, the *MONIMAD* walker allows rehabilitation and mobility assistance and postural stabilization [9]. Particularly, postural stabilization has become a significant characteristic in walkers, as it is an essential step in the rehabilitation process of neurological patients and older adults [13].

4.4 Safety Provisions

Safety is one of the main features of assistive devices such as robotic walkers. In general, walkers can provide safety strategies with physical equipment or through software and sensing architectures. The following sections describe these two approaches.

4.4.1 Safety Physical Provisions

In gait rehabilitation and assistance, the patients commonly exhibit stability and balance impairments. In this sense, walkers often equip safety equipment such as belts and suspension systems. This type of equipment is mainly required at the early stages of rehabilitation or in neurological patients. It provides partial body weight support and increased stability [25]. For instance, the *CPWalker* is a combined platform (i.e., exoskeleton and walker) that provides gait rehabilitation to children with cerebral palsy. This device equips a suspension system that lifts the patients during training sessions [26].

Some solutions help the users during sit-to-stand and stand-to-sit transfers. They include special handlebars attached to the walker structural frame. For instance, the *MOBOT* platform and the walker proposed by *Chugo et al.* are active walkers with this functionality [27, 28]. These implementations are of great relevance, as they reduce the risk for falls during these transfers.

Other safety approaches employ braking and emergency stop systems. These systems aim at stopping the device when the user needs it [2]. For instance, some walkers like the *Wachaja et al.* walker (University of Freiburg, Freiburg, Germany), and the *I-walker* (Technical University of Catalonia, Spain), equip braking systems on the handles [11, 22]. The *I-walker* also uses a vibration alarm system interconnected to the brakes. This system warns the users about nearby obstacles. Thus, they can activate the brakes, avoiding possible falls and injuries [11].

4.4.2 Sensory Provisions for Safety

In addition to the physical approaches, some solutions employ the walker sensory interface. For instance, walkers provide using fall prevention systems and speed limiters. Likewise, other devices also equip modules for detecting obstacles, slopes, and stairs [2, 4]. To this end, the walkers exploit the information retrieved from laser rangefinders, ultrasonic sensors, proximity sensors, force and pressure sensors, infrared sensors, inertial sensors, among others [2, 25].

4.4.2.1 Fall Prevention

Fall prevention systems are essential modules that detect hazardous conditions that could lead to falls. In most cases, these systems employ vision systems and inertial sensors to monitor the users [29].

For instance, the *ASBGo* walker includes a safety strategy that continuously monitors the user center of mass and performs a trajectory and orientation change when an abnormality is detected [30]. Likewise, this walker detects dangerous situations such as excessive separation between the user and the device, abnormal gait patterns, and asymmetrical supporting forces on the handles [30]. These situations are detected to reduce the risk for falls.

Another concept to prevent falls is the Zero Moment Point (ZMP) recognition. This concept is essential for bipedal modeling and consists of detecting the vertical projection of the user center of gravity [31]. A polygon within the walker is defined, and the user center of gravity must lie inside it [31]. Exceeding the limits could mean an imminent fall, and the position of the center of gravity determines the type of fall (e.g., backward, lateral, or frontal) [31]. This strategy combines the information from force sensors and a laser rangefinder. The *SIMBIOSIS* walker (Bioengineering Group, CSIC, Spain) has reported implementation of this strategy [14].

There are several sensors useful to detect and prevent falls. For instance, Fig. 4.2 illustrates a particular sensory architecture for this purpose. Implementations in the literature also report alarm systems to advise when the user is about to fall [32].

4.4.2.2 Obstacle Detection and Avoidance

Detecting surrounding obstacles in robotic walkers is essential to guarantee security in patients with visual, balance, and coordination impairments [2]. These systems help users to avoid obstacles through sound or vibration alerts. Several walkers offer this sensory system employing multiple sensory interfaces as shown in Fig. 4.3 [4, 33].

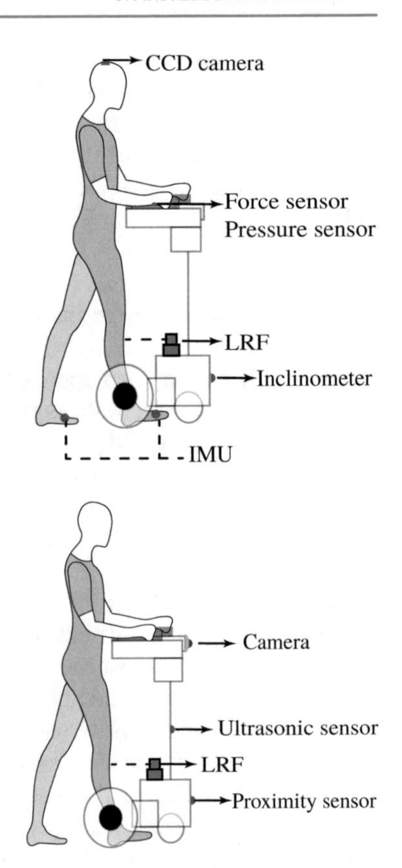

Fig. 4.2 Fall detection systems employ several sensor modalities. Force and pressure sensors measure physical interaction applied to the forearm supports by the user. Laser rangefinders estimate the position of the user's legs. Tilt sensors and IMUs acquire the orientation or inclination of the user. CCD cameras measure the user's head position

Fig. 4.3 Obstacle detection requires laser rangefinders, cameras, and ultrasonic sensors to measure the distance to obstacles

The *PAMM* (Department of Mechanical Engineering Massachusetts Institute of Technology Cambridge, USA) and *CAIROW* walker provide support on unstable terrains because it has a camera and laser to detect obstacles [18, 34]. This walker uses a camera and a sonar to detect possible obstacles [6]. Similarly, the *AGoRA* walker, the *CAIROW*, the *i-Walker* and the *Width Changeable Intelligent Walking Aid Robot (WCIWAR)* are designed for obstacle detection [17]. These robots slow down when the sensors detect an obstacle, ignoring the impulse force of the user. Finally, the platform stops at a short distance no matter how large the force is the force that the user exerts [11].

These systems employ ranging sensors (e.g., laser rangefinders and ultrasonic sensors) to detect the obstacles around the walker [1, 2]. A particular shortcoming of these strategies refers to the inability to sense low-rise obstacles. Likewise, light-based sensors exhibit problems detecting reflecting objects and glass-like surfaces.

Other platforms with obstacle detection systems are the *NOMAD XR4000* walker (School of Computer Science Carnegie Mellon University Pittsburgh, USA) and the *GUIDO* (University of Pittsburgh, USA) walker. Given that visually impaired populations are the target population of these walkers, they employ laser scanners, ultrasonic sensors, and infrared sensors to detect obstacles [8]. Also, the *WACHAJA* walker equips an obstacle detection system with a haptic alarm system. It has a vibration belt and two vibration motors on the handles. These motors send haptic warnings when an obstacle is nearby [22].

4.4.2.3 Stairs and Slopes Detection

The detection of slopes and stairs differs from the detection of obstacles because horizontally placed sensors cannot detect changes in the ground slope [35]. The presence of stairs also requires a different approach based on specific programming, sensors, and functions. Figure 4.4 illustrates a sensory architecture for stairs and slope detection.

As shown in Fig. 4.4, inclination sensors allow sensing significant changes in the ground slope. Cameras can feed video processing algorithms to estimate the presence of stairs. Ranging sensors detect changes in surfaces and detect stairs by sensing distances in the environment [14].

4.4.2.4 Speed Detection

Speed detection systems seek to implement safety measures when the walker moves at high speed. If the users exceed the speed limits, the robot takes action and slows down. Moreover, speed detection systems also aim at adjusting the walker's speed to meet the user walking pattern. To this end, the walker requires odometry sensors to estimate its velocity. Likewise, the speed calculation also requires ranging sensors to locate the lower limbs of the user [12, 15].

For instance, the *JARoW* walker can autonomously adjust the direction and speed of its movement according to the walking behavior of the user. To this end, it equips a rotating infrared system to estimate the current location of the lower limbs of the user. A control algorithm is in charge of sending actuation commands to the motors and matches the user speed [12].

Similarly, the *PAMM* walker automatically adjusts its motion to the user behavior. It monitors the walking direction, gait speed, and grip strength of the walker. This robotic walker has an adaptive shared control system that allows the patient to modulate the speed and direction of movement. If a hazardous condition occurs, the platform takes control of the speed [34]. Figure 4.5 shows sensors related to speed detection on robotic walkers.

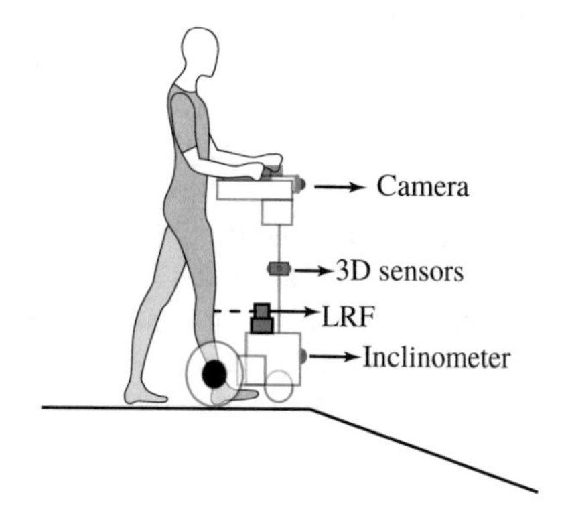

Fig. 4.4 Stairs and slope detection systems require several sensory interfaces. They include cameras to detecting slopes and stairs from images. Laser rangefinders allow to detect stairs. Inclinometers measure displacements in the ground and inclinations of the walkers

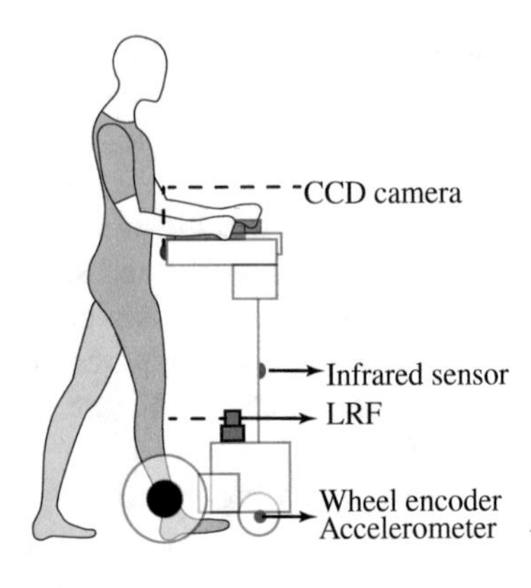

Fig. 4.5 Speed detection require several sensors such as cameras, infrared sensors, laser rangefinders, encoders, and accelerometers. They allow to measuring the speed of the user and the smart walker

4.5 Human–Robot Interaction Strategies

There is evidence in several fields of society about the need for human–robot interaction to reduce human workload, increase productivity, reduce costs, and reduce fatigue-associated factors [36]. In walker-assisted gait also occurs a human–robot interaction loop between the user and the device (see Fig. 4.1). This interaction can be whether cognitive or physical, and the device sensors, actuators, and physical structure facilitate it [23]. Multiple robotic walkers employ human–robot interfaces to provide compliant, safe, and natural interaction [1]. The following sections will describe the main functionalities of such interfaces and the used sensors.

4.5.1 Estimation of Movement's Intention

Making the robot understand the motion intention of the user is a critical problem in human–robot interaction. It allows the robot to actuate compliantly. There are several ways to achieve this interaction. On the one hand, admittance controllers employ information about the force exerted by the user to generate linear and angular velocities from the interaction force and torque (see Chap. 9). These strategies should guarantee that the device is easy to maneuver and does not add additional load to the user. On the other hand, other approaches employ cognitive or non-physical interaction between the user and the robotic walker. These strategies commonly apply touch-less follow-in-front controllers that use the orientation and distance between the user and the device.

In general, the estimation of motion intention facilitates the control of the walker by the user. These controllers are present in multiple robotic walkers reported in the literature. For instance, the *UFES* walker (Electrical Engineering Department at the Universidade Federal do Espírito Santo, Brazil) employs a sensor fusion strategy to extract the force and torque signals. An adaptive admittance controller uses these signals to generate motion on the device, while a modulation strategy sets the gains to make the user follow a specific path [23]. Other developments with the *UFES* walker have proposed follow-in-front controllers based on an inertial sensor and a laser rangefinder [23]. Likewise, the *ODW* (Department of Intelligent Mechanical Systems Engineering, Kochi University of Technology, Japan) robotic walker and the *AGoRA* walker also equip force and torque sensors that measure the intention and intensity of the user support [1, 3].

On the one hand, to identify the directional motion intention, the walkers use force sensors and pressure sensors [1, 2, 23]. On the other hand, cognitive strategies employ cameras and sensors to detect user intention [37]. In some cases, these sensors can be expensive. Therefore, recent studies proposed variants based on optical fiber, dynamometers, or instruments to transfer the force pattern [38].

The estimation of motion intention ensures the safety of the walker and offers an intuitive walker-user interaction. This functionality requires several sensor modalities, as shown in Fig. 4.6.

4.5.2 Biomechanical and Health Monitoring

Biomechanical monitoring aims at obtaining physiological measurements of the user through multiple sensory modalities. For this, the robotic walkers employ heart rate sensors, inertial sensors, SpO2 sensors, thermometers, laser rangefinders, infrared sensors, among others [2]. Non-wearable technologies such as motion analysis laboratories (e.g., VICON and BTS) are also helpful to extract kinematic information of the human-walker interaction [39].

On the one hand, these sensors monitor the users and make them feel more secure. Monitoring also has a direct impact on the walker acceptability and helps the users get used to it. On the other hand, these sensors provide the physiotherapists with data from the user gait pattern and rehabilitation progress [1, 25].

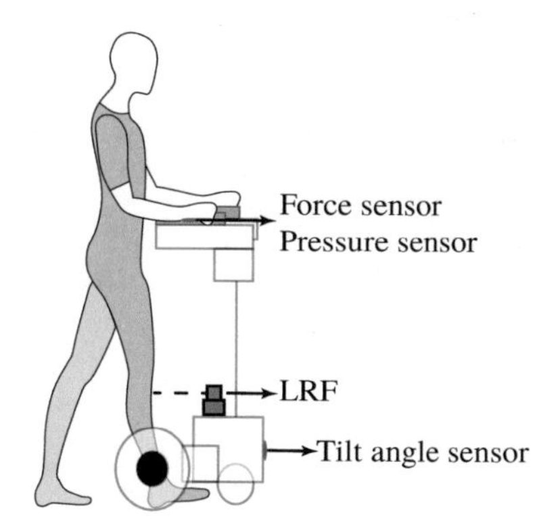

Fig. 4.6 The estimation of motion intention employs force sensors, pressure sensors, laser rangefinders, and tilt sensors to detect the user's motivational demands

4.5.2.1 Gait Parameters Monitoring

Gait monitoring system detects and records the movements, characteristics, and events of human walking. These parameters are of great importance to recognize pathological gait patterns in patients with neuromusculoskeletal diseases [4, 40]. These elements provide safety and assistance to the user because it allows the assessment of alignment and plantar support imbalances that can trigger the appearance of overloads and possible injuries [41,42].

As previously explained in Chap. 5, these systems detect gait phases such as heel strike, toe-off, heel-rise, and toe contact events [4]. To this end, the walkers require several sensors, as shown in Fig. 4.7. For instance, the *PAMM* walker and the *AGoRA* walker can track the user speed, calculate the step-by-step variability, the gait symmetry, the step length, and estimate the user's frequency stride [6, 34].

To this end, these systems require the implementation of ultrasonic sensors, laser rangefinders, inertial sensors, force sensors, among others [34]. In some cases, the detection of human gait requires additional devices worn on the human body, e.g., belts and inertial sensors. However, this can cause difficulties in outdoor environments.

For instance, the *JARoW* walker uses rotating infrared sensors to detect the user's lower limbs [12]. The *SIMBIOSIS* walker includes a monitoring system that tracks the gait trajectory [43]. The *CAIROW* walker utilizes a laser to monitor the gait pattern, especially for the Parkinsonian gait [18]. The device employs an advanced gait analysis system, two integrated lasers, and an arrangement of sensors on the handles.

4.5.2.2 Health Monitoring

This concept refers to the detection and analysis of physiological parameters related to the user-health. Similarly, health monitoring also seeks to support the therapist in monitoring the user's motor skills and supervising daily exercises [4]. The user-related information is often processed and analyzed by medical partners for rehabilitation purposes [11].

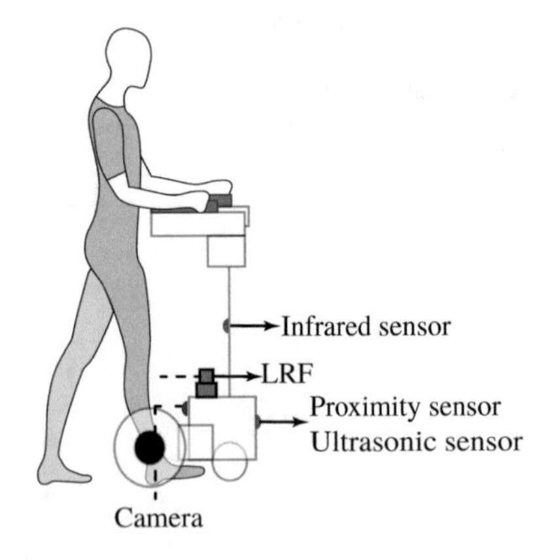

Fig. 4.7 Gait monitoring requires laser rangefinders, cameras, ultrasonic sensors, proximity sensors, inertial sensors, and infrared sensors to measure the lower-limb's kinematics

Continuous monitoring of patients is a challenging task in physical rehabilitation. Moreover, patient self-assessments are often unreliable, either because of poor memory or to avoid therapeutic interventions. Therefore, the walkers can help therapists obtain a complete and valid assessment of the user-health condition. To do this, the walker must have the ability to collect and recognize the user activity [2, 25]. These types of systems include several sensors as shown in Fig. 4.8.

For instance, the *VA-PAMAID* walker (Human Engineering Research Laboratories, University of Pittsburgh, USA) includes an application that collects relevant data taken through physiological sensors. It employs a heart rate sensor, a SpO2 sensor, a blood pressure sensor, and thermometers that display vital signs. The system is also externally supervised via Wi-Fi, allowing therapists to obtain real-time information on the user-health [7]. In addition, research with the UFES walker shows that this technology can store emergency reports and patient medical history [7]. In the case of the *PAMM* walker, it incorporates an ECG-based monitor intended to detect short-term changes and long-term health trends [34].

4.5.3 Guidance and Navigation

Guidance and navigation require odometry sensors and ranging sensors. For instance, encoders, GPS, compasses, and inertial sensors provide the position and orientation information of the walker. Similarly, laser rangefinders, cameras, and ultrasonic sensors provide environmental data. These navigation systems use software that allows a shared control between the user and the walker. Studies with the *AGoRA* walker report that when walkers provide this type of control, users feel more comfortable and natural when interacting with the environment [1]. This shared control is helpful in crowded navigation environments because the device has a general map of the environment, while the user controls the decisions on the local navigation [22].

The navigation system commonly requires path monitoring modules that handle discrete and continuous planning, providing unobstructed routes. This feature also

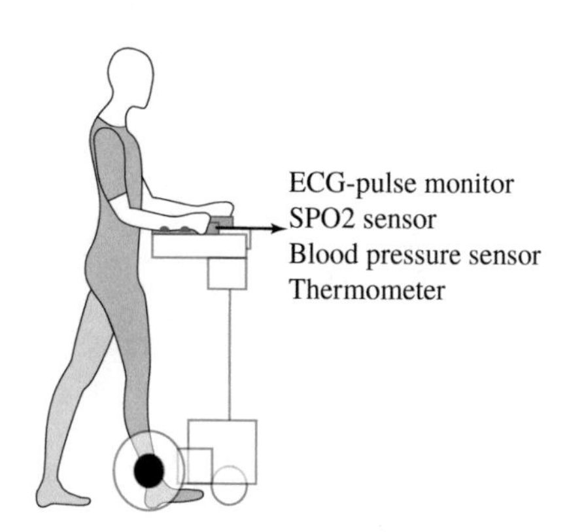

Fig. 4.8 User-health monitoring require sensors such as heart rate sensors, Sp02 sensors to measure the oxygen saturation, blood pressure sensor, and thermometers

allows the dynamic detection of obstacles and the safe locomotion of the robot while guiding the human. These systems help users walking more naturally as they reduce their cognitive load [1, 44]. It requires several sensors, as shown in Fig. 4.9.

For instance, the *GUIDO* robotic walker is a healthcare robot that serves as a support and navigation aid for the fragile and visually impaired population. Many technologies are implemented in *GUIDO* to let it achieve its tasks: simultaneous localization and map building, pose tracking, path planning, and human–robot interaction [10].

Likewise, the *Nomad XR4000* walker is for people with cognitive impairments. This walker provides navigation and global orientation through robot localization and navigation software combined with a shared control interface [45]. This robot has an omnidirectional drive. It provides physical support to users and is ideal for navigating in corridors, similar to the *WCIWAR* walker. It uses a motion model of the user, combining force data with navigation. The force sensor records the reading and logs it to the trajectory commands [17].

4.5.3.1 Autopilot System

Autopilot systems intend to allow a robot that navigates by itself following the desired route while avoiding obstacles. They also comprise systems like GPS, infrared sensors, cameras, magnetometer, and onboard microcontrollers to help route following processes [46].

The *C-walker* walker developed by Siemens has this system, and the target population is people with cognitive impairments. The walker equips a Kinect sensor (see Fig. 4.10) that enables the system to monitor its spatial surroundings in real-time [47].

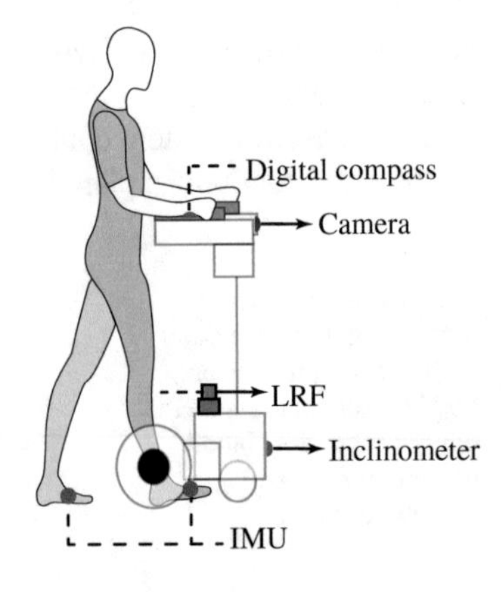

Fig. 4.9 Path monitoring requires sensors such as digital compasses, cameras, laser rangefinders, inclinometers, and inertial sensors (IMU)

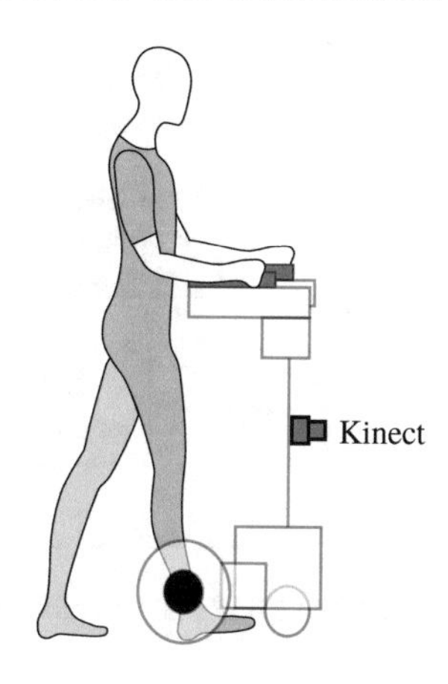

Fig. 4.10 For the autopilot system the smart walker comprises a Kinect sensor that enables the device to determine obstacles, the direction in which people are moving and warning signals on buildings [47]

4.6 Control Strategies

A control strategy is in charge of executing the corresponding actuation commands to achieve a desired functionality or objective. The robotic walkers employ different control strategies for multiple purposes. This chapter describes several controllers such as (1) fuzzy logic controllers, (2) kinematic controllers, (3) admittance controllers, (4) follow-in-front controllers, among others. These control strategies aim at providing motion intention detection, safe and natural interaction, obstacle detection, and gait monitoring.

4.6.1 Fuzzy Logic Controller

Fuzzy logic is a concept that uses expressions that are neither true nor false. It applies to statements that can take any value within a set of values that oscillate between absolute truth and total falsehood. This term allows treating imprecise information in terms of fuzzy sets to define actions [48].

Many walkers use fuzzy logic controllers since they can infer environmental data even under motion uncertainties. These receive as inputs distances obtained from sensors, and their output is differential velocities for the walker [14]. The idea of these drivers is that they employ information obtained from experimental situations (see Fig. 4.11). New ways of developing fuzzy controlled systems include neuro-fuzzy systems that allow the programmer to obtain more data from prediction [49].

The *ODW* and the *SIMBIOSIS* walkers employ fuzzy logic to develop their control system. These devices establish relationships between forearm pressure and directional intention extracted from fuzzy logic [15]. The measurements used to carry out fuzzy reasoning were forearm pressure while turning right/left while moving forward/backward and going right/left from the start. Finally, the knowledge from fuzzy reasoning has its basis on the compatibility grade between the *fact* and

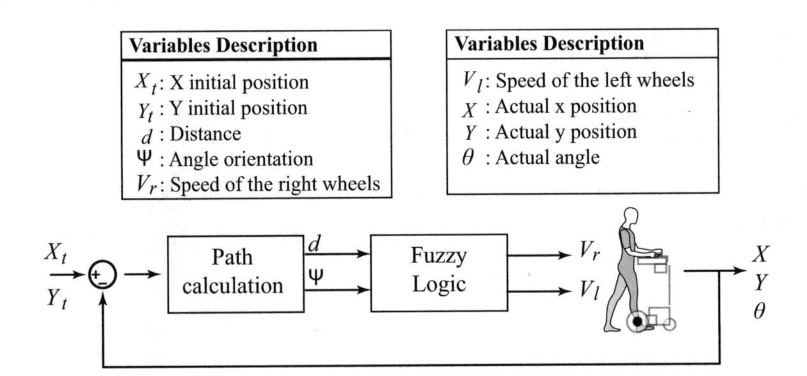

Fig. 4.11 Block diagram of an application from a fuzzy controller in a smart walker

the antecedent [50]. In this case, *the fact* is the force exerted by users with walking impairments on the sensors. *The antecedent* is the force exerted by healthy users on the sensors [15].

4.6.2 Admittance Controller

The admittance control transforms the forces and torques to the desired velocities for the walker. When the admittance control is in the task space needs the Jacobian matrix, while the joint space requires inverse kinematics [51]. In robotic walkers, these controllers generate reference speeds from movement intention [1, 3].

The admittance controller allows driving the walker from the forces and torsions exerted on the handles. The controller gains can be constant or periodically re-configured to give users the feeling of ease and naturalness during physical interaction with the walker [1]. Admittance controllers model robotic walkers as first-order mass damping systems. The inputs are force (F) and torque (t) applied to the device by the user, and the outputs of these controllers are linear (v) and angular (w) velocities [1, 3].

Robotic walkers such as the *UFES* walker, *AGoRA* walker, and the *PAMM* walker extract force and torque signals and feed admittance controllers for motion control [1, 3, 34]. Figure 4.12 shows the block diagram of an admittance controller.

4.6.3 Kinematic Controller

The kinematic models of a mobile robot are used within the design of controllers when the robot performs tasks or missions at low speed and with little load about its structure. Path tracking is possible to achieve using a control law, in which the mobile robot reaches and follows with zero error desired states that vary with time. This trajectory control uses two subsystems in cascade: (1) kinematic control is in

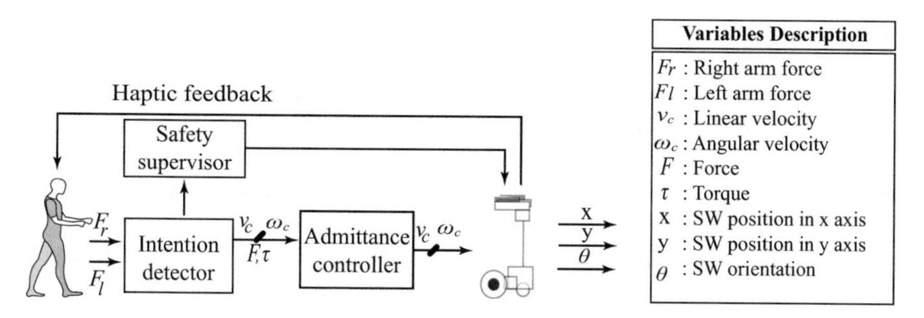

Fig. 4.12 Block diagram of an application from an admittance controller in a smart walker

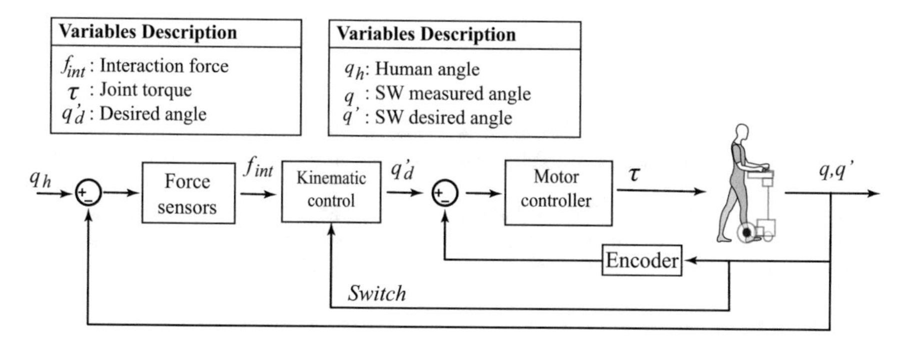

Fig. 4.13 Block diagram of an application from a Kinematic controller in a smart walker

charge of fulfilling the objective of the task; (2) dynamic compensation control is the one in charge of compensating the robot dynamics [52].

The diagram shown in Fig. 4.13 shows the general scheme of a kinematic controller used in an intelligent walker application. The system broadly indicates the interaction between the sensors and the movement angles of the user's knees. In this interaction, the controller ensures correct and harmonic movement [53].

A device developed from the frontal tracking of the user is the *Rollator*, which proposes a virtual push approach through a kinematic controller. An equilibrium distance is defined when the system is at rest. If the user passes the balance point and approaches the robot, the robot starts moving depending on the human–robot distance [54].

The *iReGo* walker is a rehabilitation platform designed to facilitate lower extremity rehabilitation training based on movement intention recognition. The walker first identifies the user's intention to move from the interaction forces on the left and right sides of the pelvis. Then it uses the kinematic model to generate the appropriate riding speeds to support the weight of the body and improve mobility. For this device, the workspace, dexterity, and force field are analyzed based on a Jacobian system [55].

4.7 Conclusions

This chapter analyzed a total of 18 relevant robotic walkers from the literature. At first, it described several physical structures according to some of the found walkers. Then, it presented an overall description of the sensors, functionalities, and interfaces of these walkers. Similarly, this chapter described some of the most common interaction strategies for robotic walkers. The information described in this chapter provides fundamental concepts in the design process of new smart walkers. Chapter 9 delves into the mathematical formulation and implementation of multiple control strategies for human–robot interaction, robot–environment interaction, and human–robot–environment interaction.

References

1. S.D. Sierra M., M. Garzón, M. Múnera, C.A. Cifuentes, Human–Robot–environment interaction interface for smart walker assisted gait: AGoRA walker. Sensors (Switzerland) **19**(13), 1–29 (2019)
2. S. Sierra, L. Arciniegas, F. Ballen-Moreno, D. Gomez-Vargas, M. Munera, C.A. Cifuentes, Adaptable robotic platform for gait rehabilitation and assistance: design concepts and applications, in *Exoskeleton Robots for Rehabilitation and Healthcare Devices* (Springer, Singapore, 2020), pp. 67–93
3. M.F. Jiménez, M. Monllor, A. Frizera, T. Bastos, F. Roberti, R. Carelli, Admittance controller with spatial modulation for assisted locomotion using a smart walker. J. Intell. Robot. Syst. Theory Appl. **94**(3–4), 621–637 (2019)
4. M.M. Martins, C.P. Santos, A. Frizera-neto, R. Ceres, Assistive mobility devices focusing on Smart walkers: classification and review. Robot. Auton. Syst. **60**(4), 548–562 (2012)
5. C.M.D.C. Santos, C.A.D.M. Pimenta, M.R.C. Nobre, The PICO strategy for the research question construction and evidence search. Revista Latino-Americana de Enfermagem **15**, 508–511 (2007)
6. J.S. Henry, Gait monitoring for the elderly using a robotic walking aid, in *2010 IEEE 26-th Convention of Electrical and Electronics Engineers in Israel* (2010), pp. 392–394
7. A.J. Rentschler, R.A. Cooper, B. Blasch, M.L. Boninger, Intelligent walkers for the elderly: performance and safety testing of VA-PAMAID robotic walker. J. Rehabil. Res. Develop. **40**(5), 423–432 (2003)
8. A. Morris, R. Donamukkala, A. Kapuria, A. Steinfeld, J. Matthews, J. Dunbar-Jacob, S. Thrun, A robotic walker that provides guidance, in *2003 IEEE International Conference on Robotics and Automation* (IEEE, New York, 2003)
9. V. Pasqui, L. Saint-Bauzel, O. Sigaud, Characterization of a least effort user-centered trajectory for sit-to-stand assistance. IUTAM Bookseries **30**, 197–204 (2011)
10. A.J. Rentschler, R. Simpson, R.A. Cooper, M.L. Boninger, Clinical evaluation of Guido robotic walker. J. Rehabil. Res. Dev. **45**(9), 1281–1293 (2008)
11. U. Cort, A. Mart, C. Barru, E.X. Mart, R. Annicchiarico, C. Caltagirone, Towards an intelligent service to elders mobility using the i-Walker, in *Proceedings of AAAI Fall Symposia AI in Eldercare*
12. G. Lee, T. Ohnuma, N.Y. Chong, Design and control of JAIST active robotic walker. Intell. Serv. Robot. **3**(3), 125–135 (2010)
13. C.-K. Lu, Y.-C. Huang, C.-J. Lee, Adaptive guidance system design for the assistive robotic walker. Neurocomputing **170**, 12 (2015)

14. R. Ceres, Empowering and assisting natural empowering and assisting natural human mobility: The simbiosis mobility: The simbiosis walker. Int. J. Adv. Robot. Syst. **8**(3), 34–50. ISSN 1729-8806
15. Y. Jiang, S. Wang, Adapting directional intention identification in running control of a walker to individual difference with fuzzy learning, in *2010 IEEE International Conference on Mechatronics and Automation, ICMA 2010* (2010), pp. 693–698
16. C.A. Cifuentes, C. Rodriguez, A. Frizera, T. Bastos, Sensor fusion to control a robotic walker based on upper-limbs reaction forces and gait kinematics, in *Proceedings of the IEEE RAS and EMBS International Conference on Biomedical Robotics and Biomechatronics*, 2016 (2014), pp. 1098–1103
17. J. Ye, J. Huang, J. He, C. Tao, X. Wang, Development of a width-changeable intelligent walking-aid robot, in *2012 International Symposium on Micro-NanoMechatronics and Human Science (MHS)* (2012), pp. 358–363
18. M.F. Chang, W.H. Mou, C.K. Liao, L.C. Fu, Design and implementation of an active robotic walker for Parkinson's patients, in *Proceedings of the SICE Annual Conference*, June 2015 (2012), pp. 2068–2073
19. I. Caetano, J. Alves, J. Goncalves, M. Martins, C.P. Santos, Development of a biofeedback approach using body tracking with active depth sensor in ASBGo smart walker, in *2016 International Conference on Autonomous Robot Systems and Competitions (ICARSC)* (IEEE, New York, 2016), pp. 2068–2073
20. V. Weiss, A. Korolev, G. Bologna, S. Cloix, T. Pun, An embedded ground change detector for a "smart walker", 2015. Artificial Computation in Biology and Medicine, Elche, June 1–5, 2015 (Springer, New York, 2015), pp. 533–542
21. E. Efthimiou, S.-E. Fotinea, T. Goulas, A.-L. Dimou, M. Koutsombogera, V. Pitsikalis, P. Maragos, C. Tzafestas, The MOBOT platform—showcasing multimodality in human-assistive robot interaction, in *Proceedings of the 20th Pan-Hellenic Conference on Informatics* (2016)
22. A. Wachaja, P. Agarwal, M. Zink, M.R. Adame, K. Möller, W. Burgard, Navigating blind people with walking impairments using a smart walker. Auton. Robots **41**(3), 555–573 (2017)
23. C.A. Cifuentes, A. Frizera, Human–Robot Interaction Strategies for Walker-Assisted Locomotion. Springer Tracts in Advanced Robotics, vol. 115 (Springer International Publishing, Cham, 2016)
24. J. Alves, I. Caetano, E. Seabra, C. Santos, Design considerations of ASBGo++ (Plus Plus) smart Walker. Robótica: Automação, Controlo, Instrumentação **3**(108), 4–8 (2017)
25. M. Martins, C. Santos, A. Frizera, R. Ceres, A review of the functionalities of smart walkers. Med. Eng. Phys. **37**(10), 917–928 (2015)
26. C. Bayon, O. Ramirez, M. Del Castillo, J. Serrano, R. Raya, J. Belda-Lois, R. Poveda, F. Molla, T. Martin, I. Martinez, S. Lerma Lara, E. Rocon, CPWalker: robotic platform for gait rehabilitation in patients with Cerebral Palsy, in *2016 IEEE International Conference on Robotics and Automation (ICRA)*, May 2016 (IEEE, New York, 2016), pp. 3736–3741
27. D. Chugo, T. Asawa, T. Kitamura, S. Jia, K. Takase, A moving control of a robotic walker for standing, walking and seating assistance, in *2008 IEEE International Conference on Robotics and Biomimetics*, February (IEEE, New York, 2009), pp. 692–697
28. C. Werner, M. Geravand, P.Z. Korondi, A. Peer, J.M. Bauer, K. Hauer, Evaluating the sit-to-stand transfer assistance from a smart walker in older adults with motor impairments. Geriatr. Gerontol. Int. **20**, 312–316 (2020)
29. J. Huang, P. Di, K. Wakita, T. Fukuda, K. Sekiyama, Study of fall detection using intelligent cane based on sensor fusion, in *2008 International Symposium on Micro-NanoMechatronics and Human Science* (IEEE, New York, 2008)
30. A. Pereira, N.F. Ribeiro, C.P. Santos, A preliminary strategy for fall prevention in the ASBGo smart walker, in *2019 IEEE 6th Portuguese Meeting on Bioengineering (ENBENG)* (IEEE, New York, 2019), pp. 1–4
31. S. Taghvaei, K. Kosuge, Image-based fall detection and classification of a user with a walking support system. Front. Mech. Eng. **13**(3), 427–441 (2018)

32. S.T. Londei, J. Rousseau, F. Ducharme, A. St-Arnaud, J. Meunier, J. Saint-Arnaud, F. Giroux, An intelligent videomonitoring system for fall detection at home: perceptions of elderly people. J. Telemed. Telecare **15**, 12 (2009)
33. I.-S. Weon, S.-G. Lee, Intelligent robotic walker with actively controlled human interaction. ETRI J. **40**, 8 (2018)
34. M. Spenko, H. Yu, S. Dubowsky, Robotic personal aids for mobility and monitoring for the elderly. IEEE Trans. Neural Syst. Rehabil. Eng. **14**(3), 344–351 (2006)
35. S. Cloix, G. Bologna, V. Weiss, T. Pun, D. Hasler, Low-power depth-based descending stair detection for smart assistive devices. EURASIP J. Image Video Process. **2016**(1), 1–15 (2016)
36. S. Haddadin, E. Croft, Physical human–robot interaction, in *Springer Handbook of Robotics* (Springer International Publishing, Cham, 2016), pp. 1835–1874
37. W.M. Scheidegger, R.C. de Mello, S.D. Sierra M., M.F. Jimenez, M.C. Munera, C.A. Cifuentes, A. Frizera-Neto, A novel multimodal cognitive interaction for Walker-assisted rehabilitation therapies, in *2019 IEEE 16th International Conference on Rehabilitation Robotics (ICORR), June 2019* (IEEE, New York, 2019)
38. S. Poeggel, D. Tosi, D. Duraibabu, G. Leen, D. McGrath, E. Lewis, Optical fibre pressure sensors in medical applications. Sensors **15**(7), 17115–17148 (2015)
39. S. Sierra, M. Munera, T. Provot, M. Bourgain, C.A. Cifuentes, Evaluation of physical interaction during Walker-assisted gait with the AGoRA Walker: strategies based on Virtual mechanical stiffness. Sensors **21**(9), 3242 (2021)
40. P. Müller, A.J. Del Ama, J.C. Moreno, T. Schauer, Adaptive multichannel FES neuroprosthesis with learning control and automatic gait assessment. J. Neuroeng. Rehabil. **17**(1), 1–20 (2020)
41. K.-R. Mun, B.B.S. Yeo, Z. Guo, S.C. Chung, H. Yu, Resistance training using a novel robotic walker for over-ground gait rehabilitation: a preliminary study on healthy subjects. Med. Biol. Eng. Comput. **55**(10), 1873–1881 (2017)
42. N.D.P. Silva Suárez, Interacción en la marcha asistida con caminador robótico: Evaluación con pacientes en actividades de la vida diaria y la integración de clínicos en el ciclo de control (2016)
43. A. Frizera, R. Raya, J. Pons, A. Abellanas, R. Ceres, The smart walkers as geriatric assistive device, in *6th International Conference of the International Society for Gerontechnology* (2008)
44. G.P. Moustris, C.S. Tzafestas, Assistive front-following control of an intelligent robotic rollator based on a modified dynamic window planner, in *2016 6th IEEE International Conference on Biomedical Robotics and Biomechatronics (BioRob)* (IEEE, New York, 2016), pp. 588–593
45. M. Morris, R. Iansek, T. Matyas, J. Summers, Abnormalities in the stride length-cadence relation in Parkinsonian gait. Movem. Disorders: Offic. J. Movem. Disorder Soci. **13**(1), 61–69 (1998)
46. B. Seçkin, T. Ayan, E. Germen, Autopilot project with unmanned robot. Proc. Eng. **41**, 958–964 (2012)
47. Siemens, A smart walker that guides its users (2013)
48. F. Sabahi, M.-R. Akbarzadeh-T, A qualified description of extended fuzzy logic. Inf. Sci. **244**, 60–74 (2013)
49. P. Rusu, E.M. Petriu, T.E. Whalen, A. Cornell, H.J. Spoelder, Behavior-based neuro-fuzzy controller for mobile robot navigation. IEEE Trans. Instr. Measure. **52**(4), 1335–1340 (2003)
50. S.X. Yang, H. Li, M. Meng, Fuzzy control of a behavior-based mobile robot, in *The 12th IEEE International Conference on Fuzzy Systems, 2003. FUZZ'03*, vol. 1 (IEEE, 2003), pp. 319–324
51. W. Yu, A. Perrusquía, Simplified stable admittance control using end-effector orientations. Int. J. Soc. Robot. **12**(5), 1061–1073 (2020)
52. G. Andaluz, V. Andaluz, A. Rosales, Modelación, Identificación y Control de Robots Móviles, in *Escuela Politécnica Nacional* (2013), p. 9
53. A. Taherifar, G. Vossoughi, A. Ghafari, M. Jokar, A fast kinematic-based control method for lower-limb power augmentation exoskeleton, in *2014 Second RSI/ISM International Conference on Robotics and Mechatronics (ICRoM)* (IEEE, 2014), pp. 678–683

54. G.P. Moustris, C.S. Tzafestas, Intention-based front-following control for an intelligent robotic rollator in indoor environments, in *2016 IEEE Symposium Series on Computational Intelligence (SSCI)* (IEEE, New York, 2016)
55. J.C. Ji, S. Guo, F.J. Xi, L. Zhang, Design and analysis of a smart rehabilitation walker with passive pelvic mechanism. J. Mech. Robot. **12**, 6 (2020)

Sensing Methodologies for Gait Parameters Estimation and Control

5

Maria J. Pinto-Bernal, Sergio D. Sierra M., Marcela Múnera, and Carlos A. Cifuentes

5.1 Introduction

Mobility is one of the essential faculties and can be defined as the ability of an individual to freely move through multiple environments and perform activities of daily living with ease [1, 2]. Following a neurological dysfunction, such as stroke, mobility may be affected and only a short period might remain to take advantage of the inherent adaptability and plasticity of the central nervous system [3]. Reestablishing adequate mobility for individuals with lower-limb impairments is often a complex challenge and frequently involves the interdisciplinary efforts of many medical, surgical, and rehabilitative specialists [4]. Thus, robotics-based training is considered a potential aid, not only for patients but also for healthcare professionals.

Although these diseases that compromise mobility are well identified and studied, just a small group of individuals can be entirely reversed by surgical or rehabilitation procedures [5]. In other words, most of the patients who suffer disorders of gait are left with consequences. In this context, it is paramount to mitigate disability and the deterioration of the quality of life of these individuals. It is necessary to develop techniques that enhance the rehabilitation processes to improve patient mobility safely and efficiently [6]. Therefore, gait analysis has been used to help therapists who wish to monitor the recovery of patients going through rehabilitation processes [7]. Within clinical settings, gait classification can be implemented as part of the control parameters for functional electrical stimulation [8, 9], estimation of the risk of older adults fall [10], the detection of

M. J. Pinto-Bernal · S. D. Sierra M. · M. Múnera · C. A. Cifuentes (✉)
Biomedical Engineering Department of the Colombian School of Engineering Julio Garavito, Bogotá, Colombia
e-mail: maria.pinto@mail.escuelaing.edu.co; sergio.sierra@escuelaing.edu.co; marcela.munera@escuelaing.edu.co; carlos.cifuentes@escuelaing.edu.co

C. A. Cifuentes, M. Múnera, *Interfacing Humans and Robots for Gait Assistance and Rehabilitation*, https://doi.org/10.1007/978-3-030-79630-3_5

abnormal gait pattern in patients with paretic limbs, and their classification based on known pathologies [11]. Besides, an atypical gait pattern can be an indicator of the progression of neurological disorders. For instance, atypical gait patterns have been proven to predict if seniors will develop dementia or cognitive decline [12].

Regarding the field of robotics, researchers have managed to program humanoid robots to use human-based gait trajectories generated via gait classification [13], as well as consistently control wearable assistive devices such as robotic prostheses [14] and orthoses [15] for the recovery of lower-limb mobility. In particular, gait phase detection methods have been used in robotic lower-limb orthoses to command force-field behaviors according to the detected gait sub-phase. Due to the recent rise in lower-limb exoskeletons as an alternative for gait rehabilitation, gait phase detection has become an increasingly important feature in controlling these devices.

This chapter aims to present strategies for the automatic identification of gait phases and their applications. To this end, firstly, it is essential to identify the importance of gait parameters to have a successful gait analysis in rehabilitation scenarios; and secondly, it is necessary to recognize the most commonly used portable devices for gait analysis with their advantages and disadvantages. The main content of this chapter is organized into five thematic sections, addressing relevant aspects regarding gait phase estimation and essential aspects covering their applications in rehabilitation settings. Section 5.2 begins with the definition of the spatiotemporal parameters that describe the gait pattern. Section 5.3 presents the most commonly used wearable gait analysis devices. Section 5.4 describes two methodologies to automatically classify and detect the gait phases for assistive and rehabilitation applications. Section 5.5 illustrates a walker-assisted gait case study, where an online methodology is presented to estimate gait parameters. Finally, Sect. 5.6 presents the conclusions and recommendations for future works in this field and the challenges of gait phases estimation in the rehabilitation context.

5.2 Spatiotemporal Gait Parameters

Spatial and temporal parameters or indicators characterize the gait cycle (presented in Chap. 1). These indicators commonly refer to the step time (seconds), stride time (seconds), step length (meters), stride length (meters), cadence (steps per minute), walking speed (meters per second), foot angle (grades), single limb support time (seconds), double limb support time (seconds), and stance-to-swing ratio. These time and distance parameters provide an index of an individual's walking patterns. It is essential to highlight that these parameters are dependent on an individual's walking speed. Therefore, it is recommended that individuals walk and their freely selected cadence during a gait analysis exam. On the other hand, although temporal gait parameters are often helpful when diagnosing pathological conditions and evaluating treatment efficacy, these parameters rarely provide sufficient insight into the origin of gait abnormalities [16].

Step length is the longitudinal distance from heel strike (HS) of one foot to contralateral HS. Step time is the elapsed time associated with the step length.

Stride length is the longitudinal distance between the occurrences of the same event (e.g., HS) with the same foot. Normal gait is symmetrical; hence, stride length is equal to twice the step length. Stride time is the elapsed time associated with the stride length. Cadence is defined as the rate at which an individual ambulates and is measured in steps per minute. The rate at which an individual ambulates at a self-selected comfortable speed is termed natural cadence. Walking speed represents the overall performance of walking. It is the rate of displacement change along the predefined direction of progression per unit time. Walking speed is also the product of step length and cadence. Foot angle is the angle between the line of progression and the foot axis. Foot angle is positive when the axis points lateral to the line of progression. Foot angle is zero when the foot axis is parallel to the line of progression. Foot angle is negative when the foot axis points medially to the line of progression. Single limb support is the elapsed time of the gait cycle during which one foot contacts the ground. Double limb support is the elapsed time of the gait cycle during which both feet are in contact with the ground. Single and double limb support may also be expressed as a percentage of the overall gait cycle. The stance-to-swing ratio is the stance interval divided by the swing interval [16–19].

5.3 Wearable Gait Analysis Devices

Gait analysis has become an essential task in clinical and rehabilitation programs, as it provides powerful insights into the individual's gait quality, the behavior of the gait pattern, and other dynamic factors [20]. Moreover, the output from a gait analysis process can offer information that is characteristic of a particular gait pathology or impairment. Thus, individualized treatments can be proposed.

Nowadays, the applications of gait analysis are divided into two main categories: clinical gait assessment and gait research purposes. Despite both seek to improve the human quality of life, clinical gait assessment (addressed in Chap. 10) has the purpose of helping individual patients directly, whereas gait research aims to improve medical diagnosis or treatment by improving the understanding of gait [21]. For instance, the gait spatiotemporal parameters are widely used in control algorithms for robotic applications and several rehabilitation programs [2, 22]. The smart walkers use gait information to provide natural and safe control strategies [2,23]. Similarly, estimating users' gait speed is useful to implement follow-in-front controllers or intention-based strategies in smart walkers [24–26]. Furthermore, it has been demonstrated that the tracking of gait parameters during rehabilitation processes may offer an overall indicator of patients' gait health [27].

Several gait analysis methods have been used and employed in these applications according to: (1) the nature of the clinical condition, (2) the individual's skills, (3) the available facilities in the clinic or laboratory, and (4) the purpose for which the analysis is being performed [21]. In general terms, the analysis methodology strongly relies on the type of sensor used. Among the most common wearable sensors are: inertial sensors, ultrasonic sensors, laser rangefinder systems, and force sensors.

5.3.1 Inertial Sensors

Inertial sensors are gaining increasing popularity in human motion analysis, as they are commonly worn by the user, provide motion data directly, and do not require external sources or devices. Using inertial sensors can typically achieve high accuracy at moderate to high walking speeds or in self-paced walking. However, their performance noticeably degrades at lower speeds, usually the pace for individuals with walking difficulties [28, 29].

The typical inertial sensor is the Inertial Measurement Unit (IMU), a combination of three components: accelerometers, gyroscopes, and magnetometers. This device can measure gravitational force, speed, and orientation. With these parameters it is possible to make estimations of the gait phases, as well as spatiotemporal parameters [30]. The implementation of different sources of information in IMUs makes them a very robust sensor, and they often require complex fusion algorithms to get improved estimations [27].

Accelerometers are the most widely used option if an outpatient gait analysis is required; these have certain advantages such as reduced size, highly mobile, low cost, and power consumption [30]. Accelerometers are transducers used to measure linear and angular accelerations. They can be arranged in either uni- or multi-axial configurations. These devices are designed according to Newton's second law of motion and Hooke's law [31]. Displacement and velocity sensors can be used in combination with derivative circuits to measure acceleration. Direct measurement of acceleration can also be obtained with the use of compact accelerometers [31]. However, its use carries several factors such as: (1) the need for gravity compensations, (2) increased computational load in the post-processing stage, (3) the occurrence of drift error in position data, and (4) the need for system's calibration to properly locate the sensors in the required application [30].

Gyroscopes are angular velocity sensors. This velocity is a factor whose signal is not influenced by the vibrations that occur when hitting the heel, and additionally, this variable is not affected by the force of gravity. In gyroscope, the output is the obtained periodic results whose patterns are repeated during the gait cycle [30]. To get the references to the framework to these sensors as it is the orientation of the axis are commonly used the Direction Cosine Matrix (DCM) and Kalman filter [32]. Today, commercially available inertial sensors measure both linear and angular accelerations with six degrees of freedom.

Finally, magnetometers provide information related to changes in magnetic fields. By definition, these devices measure the air's magnetic flux density and detect fluctuations in Earth's magnetic field. With this information, the magnetometers offer the possibility to find the vector towards Earth's magnetic North. This is often used to improve the accuracy of the measurement system through the use of data from the magnetometer, accelerometer, and gyroscope. Commercially available inertial sensors with these three types of sensors are considered as nine degrees of freedom IMUs. They provide a more robust estimation of orientation angles (i.e., yaw, pitch, roll), as well as linear and angular accelerations, and employ better drift correction strategies [33].

5.3.2 Ultrasonic Sensors

An ultrasonic sensor is an electronic device that measures the distance of a target object using ultrasonic sound waves and converts the reflected sound into an electrical signal [34]. An ultrasonic sensor uses a transducer to send and receive ultrasonic pulses that relay back information about an object's proximity. These sensors, for instance, estimate the gait parameters by measuring the distance between the user's feet and the floor [27]. In general, these measure kinematic variables such as the stride length, step length, the distance separating the two feet, and the distance separating the swinging foot from the ground.

5.3.3 Laser Rangefinders (LRFs)

Laser-based systems or laser rangefinders are optical sensors that use infrared laser beams for distance measurement in two dimensions. In general, these systems consist of a transmitter of light pulses arranged on a rotation system that allows distance measurements at different angles. Most common LRFs are based on the time-of-flight principle. Under this method, the time it takes for the light beam to travel to a target and return is measured [35, 36].

The data delivered by the laser sensors can be organized as an ordered sequence of points in polar coordinates (S), as shown in Eq. 5.1, where ρ corresponds to the measured distance and θ to the angle.

$$S = [s_1, s_2, \ldots, s_n], \quad s_i = (\rho, \theta). \tag{5.1}$$

In some applications it is helpful to express the points acquired by the laser in Cartesian coordinates. Considering that the plane of laser readings corresponds to the XY plane, Eq. 5.2 illustrates this conversion.

$$P = [(x_1, y_1), (x_2, y_2), \ldots, (x_n, y_n)], \ : \ x_i = \rho \sin(\theta), \ \rho \cos(\theta). \tag{5.2}$$

For instance, in walker-assisted gait applications, these sensors track the user's legs position and are placed in front of the user at approximately the knee height [2, 22–24, 37]. Clustering algorithms then process the information retrieved from the LRFs to estimate the average position of each leg.

5.3.4 Foot Pressure Sensors

The ground reaction force is an external force acting on the sole during standing, walking, or running activities. In this sense, the ground reaction force has also been of interest in human motion analysis [38]. To date, numerous measurement techniques have been utilized in the study of this type of force. For instance, the ground reaction force can be measured by sensors placed on the floor [27, 39] or foot pressure sensors that measure the foot contact with the ground [40]. Other techniques include floor-mounted transducer matrices, pressure mats, instrumented shoes, force plates, insole-based pressure systems, and glass plates using the critical light reflection technique [27].

Studies utilizing floor-mounted force sensing resistors (FSR) or transducers illustrate barefoot, isolated steps, and insole systems that allow investigation of step-to-step alterations in gait. Their output requires a straightforward processing, but they do not provide any information regarding the swing phase of the gait [41]. Specifically, an FSR is a sensor whose electrical resistance changes in proportion to an applied pressure; as applied to gait phase detection, these sensors are located in shoe soles so that changes in the plantar pressure can be directly correlated to the gait phase, since each gait phase can be related to a specific pressure pattern [42–45]. Flexible pressure sensors experience changing resistances as a function of pressure. These sensors are inexpensive and have a convenient input composition [42]. Nevertheless, their use in everyday activities is not recommended as they need to be placed at optimal locations to accurately detect gait phases, thus requiring an experienced professional to determine their optimal placement [46]. Additionally, pressure insoles must be tailored for each subject's foot, which incurs higher research costs, and are continuously exposed to tear and friction, which results in a shorter lifetime [47].

Overall, although these sensors provide essential real-time information correlated to locomotion, they are low-cost and small size enabling applications in both clinical settings and home environment applications [41]. Besides, in plantar pressure studies consideration should be given to possible sources of error. These include sensor hysteresis, non-linearity, bending, humidity and temperature changes, and stress shielding secondary to sensor–tissue or sensor–insole interface mechanics [38].

5.4 Classification of Gait Phases: Exoskeletons' Case Study

The evolution of technology allows using more sophisticated tools such as pattern analysis and artificial intelligence to analyze and interpret the motion and gait analysis. These models allow accurate biomechanical analysis and precise analysis of the biomechanical effects of orthotics, prosthetics, and assistive devices. This section will show the theoretical approach, implementation methodology, and applications of two classification strategies that have been implemented in this book

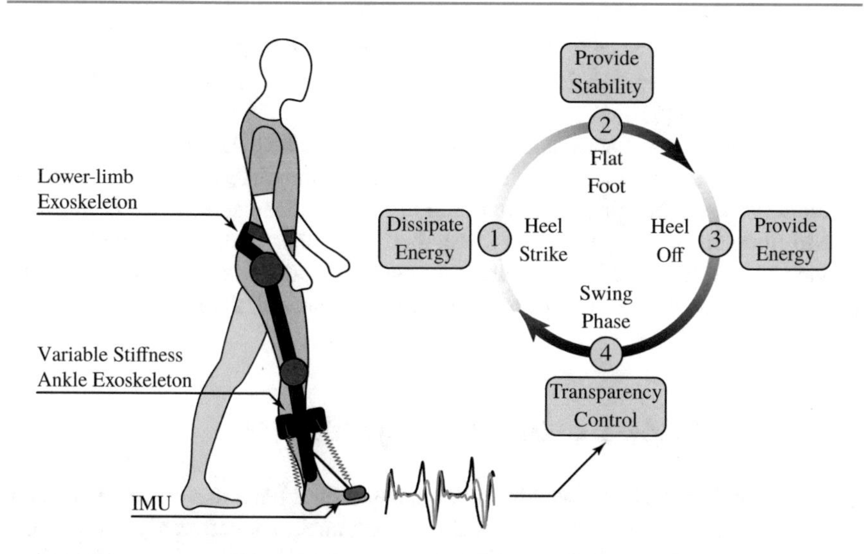

Fig. 5.1 Illustration of an assisted gait application using a lower-limb exoskeleton and a variable stiffness ankle exoskeleton. Gait phases' information is used to determine the behavior of the assistive devices

for the automatic identification of gait cycle.[1] As an illustration, Fig. 5.1 describes an example of a lower-limb exoskeleton and a variable stiffness ankle exoskeleton in a gait assisted application. The assistive devices employ the information extracted from an IMU to determine the most appropriate control strategy at each gait phase.

First, it is essential to note that several computational methods have been previously proposed for the automatic segmentation of the gait cycle, which fall into two main categories. The first category is comprised of algorithms, which divide the gait phases based on the threshold selection of either raw or processed data [44,48]. Second, some machine learning approaches have emerged in recent years to substitute the techniques mentioned above that rely on hand-crafted feature extraction. These adaptive methods extract patterns based on Support Vector Machines (SVM) [11], Artificial Neural Network (ANN) [43,49,50], hybrid algorithms [51], hidden Markov model (HMM) [52–58], among others.

This chapter implements two classification strategies for the automatic identification of four gait phases, drawn from inertial data coming from a single Inertial Measurement Unit (IMU) located at the foot instep. This sensor is used as means of gait phase detection thanks to its cost-effectiveness [59] and the fact that inertial quantities present typical waveform features during a gait cycle [60]. Studies have been conducted positioning IMUs on the waist [61], thigh [62], shank [63], and foot instep [8,64]. The IMU placed at the foot instead was because scalar classifiers have

[1]The implementation of the two classification strategies and the dataset are available at: https://github.com/midasama3124/hmm_gait_phase_classifier.

shown better performance with the sensor placed at this location, even compared to other vector classifiers involving more inertial sensors placed at different locations on the lower limb [53].

Regarding the two classification strategies implemented, the first and most easily implemented strategy is a threshold-based algorithm that determines the gait phases of interest by establishing specific decision rules and thresholds, which must be met to jump from one gait phase to another. The other partitioning method may be viewed as a machine learning algorithm since it requires a training stage and a posterior testing stage [52]. Specifically, the implemented algorithm is based on a continuous HMM.

5.4.1 Threshold-Based Detection Algorithm (TB)

A Threshold-Based Detection Algorithm is based on a finite state machine (FSM), which consists of a set of states s_i and a set of transitions between pairs of states s_i, s_j. A transition is labeled condition/action: a *condition* that causes the transition to be taken and *action* that is performed when the transition is taken [65]. State machines are a method of modeling systems whose output depends on the entire history of their inputs and not just on the most recent input, compared to purely functional systems, in which the input purely determines the output. State machines have a performance determined by their history [66] and provide means to control decisions.

The developed Threshold-Based Detection (TB) Algorithm is based on the mediolateral axis rotation component of the foot accelerometer (A_y) and gyroscope (G_y) signals as input, since the lower-limb joints movement occurs mainly along the sagittal plane. Timestamps are also used as algorithm inputs, since this detection algorithm uses spatial thresholds and temporal limits. This means that each gait phase can be associated with a sequence of wave-related features without any complex processing that would result in a high computational load [67]. For more information on the TB's feature extraction process, the author recommends to read the publication associated with this chapter [68]. Figure 5.2 illustrates the feature-based conditions for the transitions between gait phases. Similar strategies based on curve characteristics could be carried out in different inertial signals drawn from different locations in the human body (i.e., waist, shank, thigh).

To be more precise, the flowchart in Fig. 5.3 summarizes the main detection features for the transition between gait phases and their extraction process. First, feature extraction from linear acceleration and angular velocity signals begins with creating a feature list, since several features must be found before any gait phase is claimed as detected. Each time a new gait phase is updated, this list is emptied. The input data ($D(i)$ in Fig. 5.3: G_y or A_y as appropriate) are updated at a sampling rate of 100 Hz, which matches the inertial sensor sampling rate. Each feature should meet certain conditions to be included in the list. These conditions are evaluated sequentially as follows:

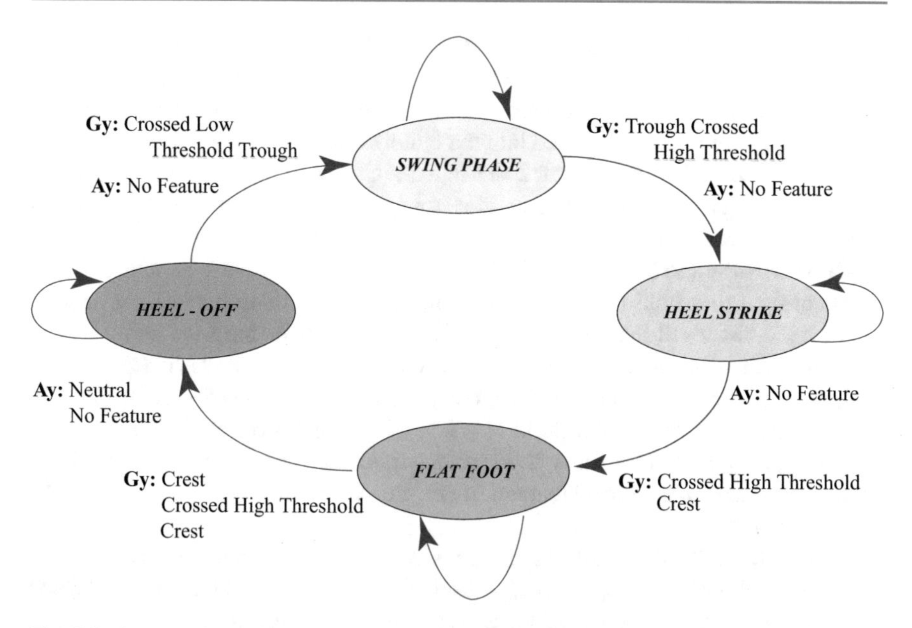

Fig. 5.2 *State machine of the threshold-based algorithm.* The transition conditions are based on features found in the angular velocity (G_y) and linear acceleration (A_y) signals

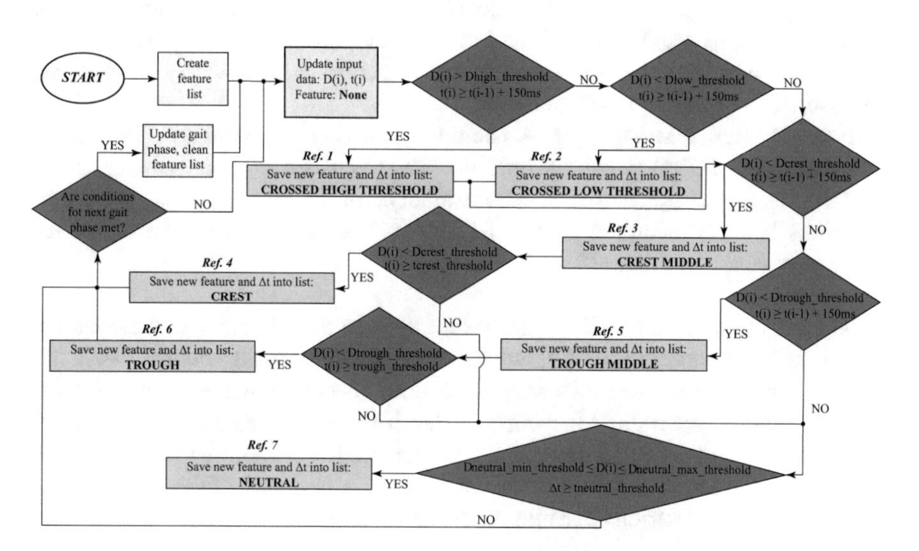

Fig. 5.3 *Flowchart of feature extraction of the threshold-based algorithm from inertial motion data.* This chart shows how a feature list is updated based on the fulfillment of certain conditions. The occurrence of each feature and their corresponding conditions are sequentially assessed in the following manner: crossed high threshold (Ref. 1), crossed low threshold (Ref. 2), Crest Middle (Ref. 3), crest (Ref. 4), Trough Middle (Ref. 5), Trough (Ref. 6), and neutral (Ref. 7). When a new gait phase is detected, the feature list is emptied for further searches. Adapted from [68]

- **Crossed High Threshold (Ref. 1 in Fig. 5.3):** Current data ($D(i)$) must be above a pre-established threshold ($D_{\text{high\_threshold}}$) and at least 150 ms should have elapsed since the last saved feature. The time difference between features (Δt) and the spotted feature is saved into the feature list.
- **Crossed Low Threshold (Ref. 2 in Fig. 5.3):** Current data ($D(i)$) must be below a pre-specified threshold ($D_{\text{low\_threshold}}$) and at least 150 ms should have elapsed since the last saved feature. Also Δt is saved into the feature list.
- **Crest Middle (Ref. 3 in Fig. 5.3):** Current data ($D(i)$) must be above a pre-established threshold ($D_{\text{crest\_threshold}}$), and at least 150 ms should have elapsed since the last saved feature. Also, Δt is saved into the feature list.
- **Crest (Ref. 4 in Fig. 5.3):** This feature is only assessed if *Crest Middle* has been saved into the list. Hence, the current data ($D(i)$) must have crossed the already exceeded pre-specified threshold ($D_{\text{crest\_threshold}}$), and a certain amount of time ($t_{\text{crest\_threshold}}$), which differs between acquired signals (A_y, G_y), should have elapsed since the last saved feature. Therefore, a crest may be reported. Also, Δt is saved into the feature list.
- **Trough Middle (Ref. 5 in Fig. 5.3):** Current data ($D(i)$) must be below a pre-specified threshold ($D_{\text{trough\_threshold}}$) and at least 150 ms should have elapsed since the last saved feature. Also, Δt is saved into the feature list.
- **Trough (Ref. 6 in Fig. 5.3):** This feature is only assessed if *Trough Middle* has been saved into the list. Hence, the current data ($D(i)$) should again be above the already crossed pre-specified threshold ($D_{\text{trough\_threshold}}$), and a certain amount of time ($t_{\text{trough\_threshold}}$), which differs between acquired signals (A_y, G_y), should have elapsed since the last saved feature. Therefore, a crest may be reported. Also, Δt is saved into the feature list.
- **Neutral (Ref. 7 in Fig. 5.3):** A neutral region is only reported as long as the current data ($D(i)$) remains within the range between $D_{\text{neutral\_min\_threshold}}$ and $D_{\text{neutral\_max\_threshold}}$, and if a certain amount of time ($t_{\text{neutral\_threshold}}$), which differs between acquired signals (A_y, G_y), has elapsed since the last saved feature. Also, Δt is saved into the feature list.

The selection of correct threshold values used in this classifier was carried out as reported in *Kotiadis et al.* [69], whose research validated all possible thresholds within a range, and whose limits were determined visually from the signals captured in a preliminary analysis. After checking each condition included in the feature extraction process (as evidenced above), the feature list is reviewed to determine if a new gait phase has been found. The following is a summary of the feature-based rules governing the various transitions between gait phases.

- **Heel Strike (HS) → Flat Foot (FF):** To detect the FF onset, the current angular velocity signal must have exhibited a crest, while the linear acceleration data must be right in the middle of a trough, after entering the crest corresponding to the HS in the feature list (see Fig. 5.4).

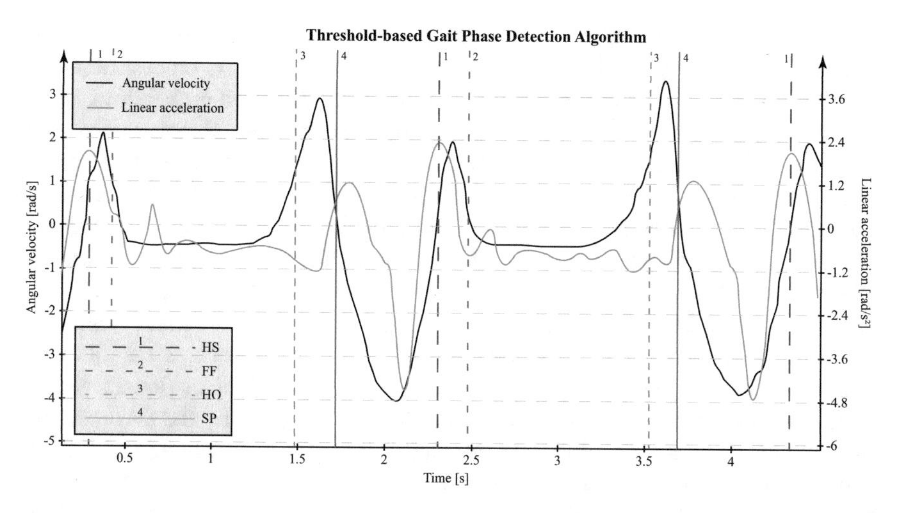

Fig. 5.4 Threshold-based gait phase detection using an inertial detection system over two gait cycles. Feature-based decisions are made to identify the onset of each gait phase: heel strike (first dashed line), Flat Foot (second dashed line), heel off (third dashed line), and swing phase (fourth complete line). Adapted from [68]

- **Flat Foot (FF) → Heel Off (HO):** To detect the HO onset, the current angular velocity signal has a neutral region, since linear acceleration data also must remain within a neutral region, followed by a high cross threshold (see Fig. 5.4).
- **Heel Off (HO) → Swing Phase (SP):** To detect the SP onset, the current angular velocity signal must have shown a crest, while the linear acceleration data must have crossed a predefined threshold (see Fig. 5.4).
- **Swing Phase (SP) → Heel Strike (HS):** To detect the HS onset, the current velocity signal should have shown a trough and a crossed high threshold, while the linear acceleration data should be in the middle of a crest, and after another crest, a trough and a crossed high threshold have been sequentially entered in the feature list (see Fig. 5.4).

5.4.2 Classification Using a Hidden Markov Model (HMM)

A Markov process is a stochastic extension of a finite state automaton. It provides a way to model the dependencies of current information with previous information. In a Markov process, state transitions are probabilistic and there is in contrast to a finite state automaton no input to the system: besides, it is composed of states, transitions, scheme between states, and emission of outputs (discrete or continuous) [70, 71]. The following can be achieved with Markov's models: Learning sequential data statistics, making predictions or estimates, and recognizing patterns.

An HMM is a double stochastic process in which the existence of a set of discrete states is assumed for a given system. The first stochastic process describes how the system may jump from one state to another (transition probability), under the hypothesis that the next state depends only on the state at present (Markov property). It means that this process has N underlying discrete states that are not observable, i.e., its state sequence is hidden to the observer who only has access to the emissions of each state [72]. In this case, this first stochastic process refers to the gait cycle, which was divided into four phases (this division is the most used as illustrated in Chap. 1). On the other hand, the second stochastic process yields the statistical description governing the emissions of each observed variable (emission probability). It means that the second embedded stochastic process describes the emissions from Y observations, i.e., either the sensor readout or feature vectors extracted from them (in this case they were the signals given by the IMU sensor), in terms of discrete probabilities or Probability Density Functions [54]. HMM is a statistical model widely used to estimate a sequence of hidden states in a time series [52], which for the case of gait phase detection corresponds to the gait events ($N = 4$), i.e., Flat Foot, Toe Off, Swing, and heel strike.

The HMM can be expressed as a function, as presented in Eq. 5.3 as a set λ characterized of three parameters A, B, and π.

$$\lambda = (A, B, \pi), \tag{5.3}$$

which includes the probability distribution matrix of state transition A, the probability distribution matrix of observation symbols B, and the initial state distribution vector π.

The typical gait pattern repeats itself indefinitely with a known sequence of gait events, which in terms of probability means that it can either remain in the current state or eventually transition to the consecutive state. This behavior has recently been modeled using a left–right model [53, 54], whose main feature is to limit transitions to consecutive states of the Markov chain. Since transitions represent a narrow fraction of the gait cycle, their associated probabilities assume lower values than those related to permanence in the same state. Thus, diagonal elements assume a higher value than the others. Therefore, the transition matrix A may be implemented as shown in Eq. 5.4 [53].

$$A = \{a_{ij}\} = \begin{bmatrix} 0.9 & 0.1 & 0 & 0 \\ 0 & 0.9 & 0.1 & 0 \\ 0 & 0 & 0.9 & 0.1 \\ 0.1 & 0 & 0 & 0.9 \end{bmatrix}, \tag{5.4}$$

where a_{ij} denotes the transition probability from state S_i to state S_j. The possible transitions among gait phases are reported in Fig. 5.5. The state $S1$ was paired to the gait phase delimited by swing phase and heel strike events. Further pairings were:

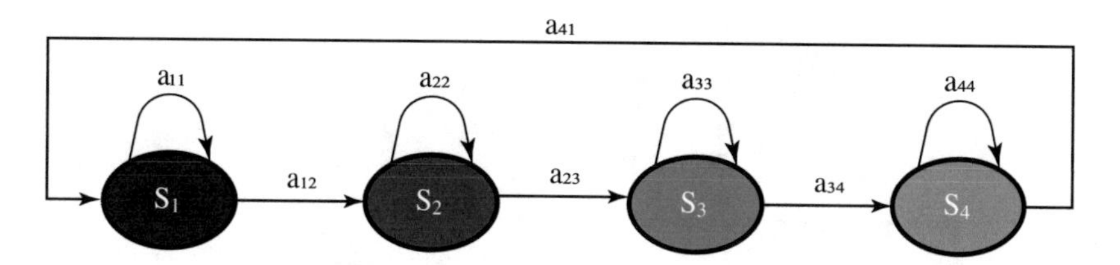

Fig. 5.5 Possible transitions (a_{ij}) among four states (S_i) of a continuous HMM according to a left–right model. Each model state is paired to a gait phase, whose borders are identified by corresponding gait events and whose emissions are modeled using a Gaussian Mixture Model with three components: a_{ij} denotes the transition probability from state S_i to state S_j

S2: Heel Strike—Stance Phase; *S3*: Stance Phase—Toe Off; *S4*: Toe Off—Swing of the next stride.

Because the initial state t_0 of the model is unknown, an initial state distribution vector π allocates the same probability to all states (see Eq. 5.5), i.e., each state has the same probability of being the first in a chosen state sequence.

$$\pi = \{t_0\}_{Qx1} = \begin{pmatrix} \frac{1}{N} \\ \frac{1}{N} \\ \frac{1}{N} \\ \frac{1}{N} \end{pmatrix} = \begin{pmatrix} 0.25 \\ 0.25 \\ 0.25 \\ 0.25 \end{pmatrix}. \tag{5.5}$$

Finally, a Bivariate Gaussian Mixture Model with three components was utilized to describe the emissions from each state. These emissions allude to feature vectors that include the angular velocity measured at any sampling time, and its time derivative computed employing a first-order finite difference approximation, i.e., the angular acceleration [55]. This particular stochastic model yields the best trade-off between complexity and accuracy for gyroscope signals [54, 62].

It is essential to highlight that to develop an HMM is necessary to consider the three problems learning/interference and three algorithms to help to solve these problems: the Baum–Welch (BW) algorithm computes the Maximum Likelihood (ML) estimates of model parameters; the Viterbi algorithm estimates the most likely sequence of hidden states; the forward–backward algorithm [72].

The continuous HMM development involves two main procedures: a training phase and a test phase. The first stage concerns the adjustment of model parameters λ to optimally fit them to an observed training dataset [72]. In the present classifier, the Baum–Welch algorithm, which is the most common solution to this problem, is implemented. This training procedure starts with a set of initial parameters (*first-phase training* in Fig. 5.6), based on which it extracts probabilistically weighted state sequences. The initial model is repeatedly updated with these new transition and emission probabilities until a desired level of convergence is reached [62].

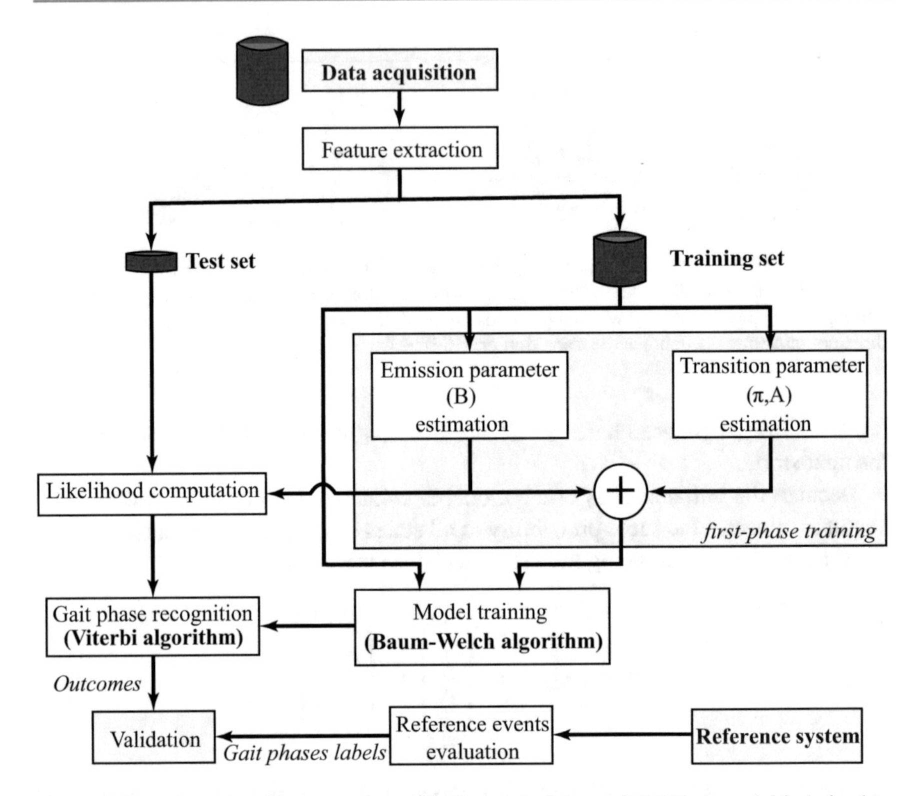

Fig. 5.6 *Flowchart that illustrates the validation methodology of HMM.* A model is trained to employ the Baum–Welch algorithm after applying feature extraction to the acquired dataset. The optimal state sequence is then computed through the Viterbi algorithm by using feature vectors from the test dataset, and the performance evaluation is conducted concerning gait phase labels drawn from the reference system. Adapted from [68]

Afterward, the testing stage allows the classification of features based on the trained model reached in the training phase, i.e., the search for the optimal state sequence is carried out. The Viterbi algorithm uses a common optimality criterion to find the most likely/probable state sequence [72]. Despite its computational efficiency, this algorithm is not suitable for real-time application, since the indicators it uses are computed based on a whole observation dataset. Therefore, the validation of the classifier outputs is compared offline concerning an FSR-based reference system that provides the actual gait phase labels. The reference system should be matched with each subject's shoe size and equipped with four force-sensitive resistors (FSRs) on the foot sole. The first sensor is located at the hallux (toe), two more sensors are located at the first and fifth metatarsophalangeal articulation, and one more is located at the heel.

5.5 Estimation of Gait Parameters: Smart Walkers' Case Study

As previously described, gait characterization is often accomplished by calculating spatiotemporal parameters, such as speed, cadence, stride length, step length, among others. Some of these features are referred to as the general gait parameters (GGPs) and have been widely used as standard indicators for gait assessment [73]. Among the GGPs, three spatial and temporal parameters are found: (1) the stride cadence (SC), (2) the stride length (STL), which is composed of two step lengths (SPL); and (3) the gait speed (GS) [73, 74].

To estimate the GGPs, many studies have proposed and assessed sensing technologies comprising wearable and non-wearable sensors [37]. However, most of them have been developed within laboratory conditions, with non-wearable constraints, and do not allow online estimations [75]. Therefore, an accurate online estimation of these parameters with ambulatory technologies for practical scenarios is described in the following sections [37].

5.5.1 Gait Data Acquisition

The first step in estimating gait parameters is acquiring data from the patient or user of an assistive device. For this description, a walker-assisted gait application was proposed, in which a laser rangefinder (LRF) sensor is used to detect and track the user's legs relative position. Figure 5.7a shows an example of the LRF's location and orientation in a passive rollator application.

In ambulatory applications employing LRF sensors and mobile assistive devices, such as smart walkers, it is essential to place the sensor at an appropriate height guaranteeing that the laser's field of view is not occluded and the user's legs are always visible. Given the variability in gait patterns from one individual to another, a proper recommendation on the LRFs location is to place them approximately at the user's knees height [37]. This location provides a clear field of view, even in self occlusions during walking (i.e., the legs are too close or one in front of the other). Figure 5.7b and c illustrates examples of LRF's readings during walker-assisted gait.

In addition to the above, to avoid noisy readings with objects not related to users' legs, the measurement area is often constrained to a narrow polygon. Notably, the LRF's field of view is constrained between 45° and 135° and between 1 m and 1.5 m.

5.5.2 Clustering of Legs' Data

Once data are obtained from the LRF sensor, it is essential to label the laser readings, to identify which leg they are from. To this end, different machine learning algorithms are often used for classification purposes. One of the most common techniques relies on an unsupervised learning classifier that returns clusters of laser

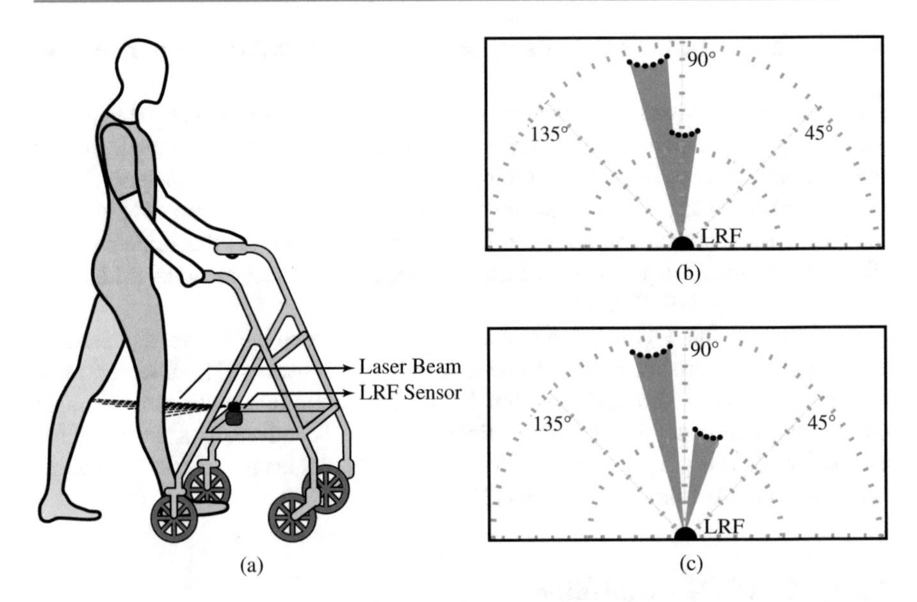

Fig. 5.7 (**a**) Illustration of an assisted gait application with a passive rollator, equipped with a laser rangefinder (LRF) to detect and track the user's legs. (**b**) Example of LRF scan where one leg is covering part of the other leg. (**c**) Example of LRF scan where legs are enough separated

points [24, 37]. For instance, density-based spatial clustering of applications with noise (DBSCAN) is one of the most used data clustering algorithms that, in this case, allows the classification of laser points either as legs or as noise [24].

These clustering algorithms require two parameters to be executed. The first one refers to the minimum distance to group two consecutive points. The second one defines the minimum number of grouped points to save a cluster [24]. After that, the position of each leg and relative distance to the walker can be defined by calculating the center of each cluster and the mean distance between them [24].

5.5.3 Legs' Distance Difference (LDD) Signal

During gait, the human trunk exhibits oscillatory behavior, and so does the movement of the legs. In particular, the distance between the legs is characterized by a sinusoidal behavior. Therefore, to estimate the GGPs, this signal can be used. The frequency and amplitude of the principal component of the signal obtained from a distance between each leg correspond to the stride cadence (SC) and the step length (SPL), respectively [2]. Moreover, it is possible to estimate the gait speed (GS), by multiplying the SC and the SPL [76]. Using the LRF readings, the LDD signal can be calculated as described in Eq. 5.6.

$$LDD = d_R * \sin(\theta_R) - d_L * \sin(\theta_L), \tag{5.6}$$

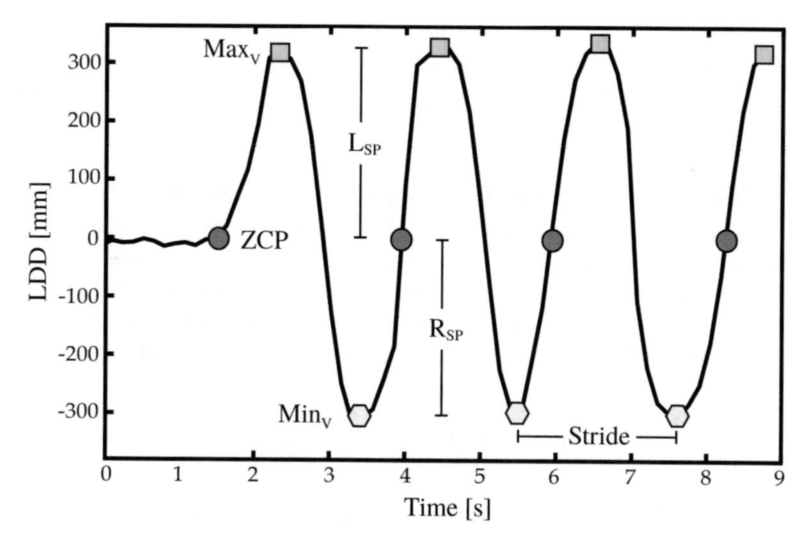

Fig. 5.8 Illustration of the legs' distance difference (LDD) signal, obtained from the readings of the LRF sensor

where d_R and d_L are the distances to the right and left legs, respectively; θ_R and θ_L are the angles of the right and left legs. Each cycle of the LDD signal illustrates the behavior of a stride cycle composed of two steps. The right step (R_{SP}) can be seen as a decrease in the LDD signal and the left step (L_{SP}) can be seen as an increase in the LDD signal. Moreover, the length of each step can be calculated by the maximum value (Max_{LDD}) and minimum values (Min_{LDD}) of each cycle. Thus, the stride length (STL) can be estimated as the sum of both step lengths.

To extract every cycle of the LDD signal, the zero crossings with positive slopes (ZCP) provide the starting and ending points of each stride. With this information, the SC can be calculated as the inverted value of the period of each cycle [2]. Figure 5.8 illustrates the LDD signal, describing the left and right steps (R_{SP}, L_{SP}), the zero crossings with positive slopes (ZCP), and the maximum values (Max_LDD) and the minimum values (Min_LDD) of each cycle.

5.5.4 Adaptive Filters for LDD Processing

A simple approach to estimate the GGPs would use the Max_{LDD}, Min_{LDD}, and ZCP values to calculate the frequency and amplitude of the LDD signal. However, this method might be affected by sudden objects sensed by the LRF or noisy readings. Therefore, a robust methodology based on two adaptive filters has been proposed to estimate the GGPs [37]. In particular, the Weighted Frequency Fourier Linear Combiner (WFLC) and the Fourier Linear Combiner (FLC) are filters that allow a smooth online estimation of the frequency and amplitude of the principal Fourier component of the LDD signal [37]. In this sense, the WFLC takes the LDD signal as input and estimates the stride cadence, while the FLC takes the LDD signal and the stride cadence to estimate the step length. According to the literature evidence, these filters have proven to be valuable and efficient in several real-time applications [2, 22, 77, 78].

5.5.4.1 Weighted Frequency Fourier Linear Combiner (WFLC)

In general terms, the WFLC filter is a powerful tool capable of calculating the frequency, amplitude, and phase of the Fourier components from a real-time signal [79]. The WFLC filter uses the least mean square algorithm to reduce the error between the actual signal and the estimated signal conformed by the Fourier components. The process of the WFLC filter can be described as follows in Eqs. 5.7, 5.8, 5.9, and 5.10 [37]:

$$x_{rk} = \begin{cases} \sin(r \sum_{t=1}^{M} \omega_{0_t}), & 1 \leq r \leq M \\ \cos(r \sum_{t=1}^{M} \omega_{0_t}), & M+1 \leq r \leq 2M \end{cases} \tag{5.7}$$

$$\varepsilon_k = S_k - \mu_b - \mathbf{W_k^T X_k} \tag{5.8}$$

$$\omega_{0_{k+1}} = \omega_{0_k} + 2\mu_0 \varepsilon_k \sum_{r=1}^{M} r(W_{rk} X_{m+rk} - W_{m+rk} X_{rk}) \tag{5.9}$$

$$\mathbf{W_{k+1}} = 2\mu_1 \varepsilon_k \mathbf{X_k} + \mathbf{W_k}. \tag{5.10}$$

Equation 5.7 describes the estimation of the Fourier components with an initial guess of the frequency ω_{0_t}. Equation 5.8 illustrates the calculation of the error between the input signal (S_k) and the estimated signal conformed by the Fourier components ($\mathbf{W_k}$ represents a matrix containing the weights of each Fourier component, and $\mathbf{X_k}$ represents a matrix with each Fourier component value). Equations 5.9 and 5.10 show the implementation of the least mean square algorithm to update the frequency ($\omega_{0_{k+1}}$) and the amplitudes ($\mathbf{W_{k+1}}$) [37].

It is worth mentioning that the WFLC formulation requires four parameters to be set or tuned: (1) M, the number of required harmonics to estimate the input signal (set to 1); (2) μ_0, the gain used to adapt the frequency estimation (set to 0.14); (3) μ_1, the gain used to adapt the amplitude estimation (set to 0.4); and (4) μ_b, the gain used to compensate low-frequency errors (set to 0) [80]. A normalization value (NV) of 1000 was used to set the signal between -1 and 1 [37, 79].

As an illustration of a walker-assisted gait application, Fig. 5.9 shows a comparison between the stride cadence obtained by the WFLC (SC_{WFLC}) and the estimated

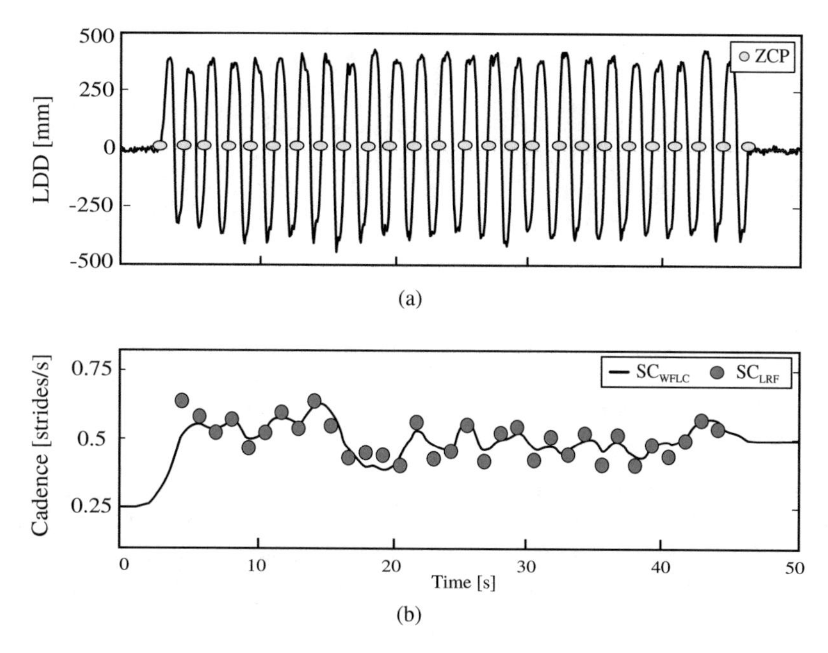

Fig. 5.9 Example of the Weighted Fourier Linear Combiner (WFLC) behavior in a walker-assisted gait application at a speed of 1.8 km/h. (**a**) Illustration of the legs' distance difference signal (LDD) and the zero crossings with positive values (ZCP). Illustration of the stride cadence obtained with the WFLC filter and the ZCP values

one with the period of each cycle (SC_{LRF}). The WFLC filter generates the SC_{WFLC} by taking the LDD signal as input, whereas the SC_{LRF} is obtained with the ZCP values of the LDD signal. The WFLC returns the SC value as soon as a sample of the input signal appears. For this example, the Hokuyo URG-04LX-UG01 was used, which works at a sample rate of 10 Hz. In this sense, the SC_{WFLC} was updated every 0.1 s. Similarly, the SC_{LRF} was only updated every time a new ZCP was detected [37]. Figure 5.9a shows the LDD signal and ZCP values obtained with the LRF at a walking speed of 1.8 km/h. Moreover, Fig. 5.9b shows the cadence obtained with the WFLC and with the ZCP values.

5.5.4.2 Fourier Linear Combiner (FLC)

The FLC is an adaptive algorithm used to achieve a continuous estimation of quasi-periodical signals based on the M harmonics dynamic Fourier mode [81]. Using a frequency and number of harmonics as inputs, the FLC can estimate the amplitudes and phases of the Fourier components. In this application, the algorithm requires the frequency (ω_0) of the signal produced by the WFLC as an input parameter [37].

Even though the WFLC also estimates the Fourier components' amplitudes, these calculations can be affected by the frequency estimation. Therefore, it is better to estimate such amplitudes employing the FLC [2]. The formulation of this filter uses

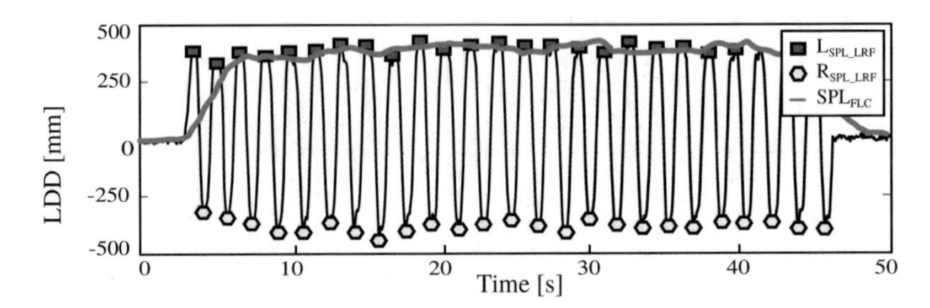

Fig. 5.10 Example of the Fourier Linear Combiner (FLC) behavior in a walker-assisted gait application at a speed of 1.8 km/h. The step length ($SP_F LC$, in black) is estimated from the legs' distance difference (LDD, in gray) signal. The left and right step lengths (L_{SPL_LRF}, R_{SPL_LRF}) are also shown for comparison purposes

two parameters: (1) M, the number of harmonics to estimate the input signal (set to 1) and (2) μ, the gain used to calculate the harmonics weights (set to 0.2). Similar to WFLC, the FLC also employs the least mean square recursion to update the estimations of the amplitude and phase. This algorithm is described as follows by Eqs. 5.11, 5.12, and 5.13 [37, 81].

$$x_{rk} = \begin{cases} sin(r\omega_{0_k}), & 1 \le r \le M \\ cos((r-M)\omega_{0_k}), & M+1 \le r \le 2M \end{cases} \tag{5.11}$$

$$\varepsilon_k = Y_k - \mathbf{W_k^T X_k} \tag{5.12}$$

$$\mathbf{W_{k+1}} = 2\mu\varepsilon_k \mathbf{X_k} + \mathbf{W_k}, \tag{5.13}$$

where Y_k is the input signal, **W** and **X** matrices are the weights and values of the Fourier components, and ε_k is the error between the input signal and the estimated one. The FLC also requires a normalization value for setting the input values, between -1 and 1.

As previously described, with the amplitude estimated by the FLC, the step length values can be directly computed. Following the walker-assisted gait example used for the WFLC illustration, the step length was estimated using the FLC output (SPL_{FLC}). This estimation was compared with the left step length, (L_{SPL_LRF}) and the right step length (R_{SPL_LRF}) calculated with maximum value Max_{LDD} and minimum value Min_{LDD} of each LDD signal's cycle. Figure 5.10 illustrates these estimations.

Finally, to obtain the gait speed (GS), the stride cadence (SC) and the step length (SPL) calculated with the WFLC and FLC are multiplied as shown in Eq. 5.14 [37].

Table 5.1 Comparison of estimation errors between multiple technologies reported in the literature and the online methodology using an LRF sensor and adaptive filters

	Ultrasonic	Accelerometers	Gyroscopes	IMUs	LRF
SC	4.1 [82] %	5% [59]	10% [39]	2.3% [83]	4.9%
SPL	–	7% [59]	–	7% [83]	4.1%

$$GS = 2(SC)(SPL). \tag{5.14}$$

This process is not a heavy computational task and it is also carried out with every scan reading. Thus all the GGPs are computed every 0.1 s for this scenario.

5.5.5 Online Estimation

This methodology can be easily extended to an online version with simple mathematical correction. In *Aguirre et al.*, the authors found that the stride cadence does not exhibit significant differences at any speed, when comparing this methodology with an automated motion tracking system (e.g., VICON, BTS) [37]. In contrast, the step length presented significant differences with a reference system (i.e., motion tracking system). To fix this discrepancy, the authors proposed a linear model was to adjust the step length measurements.

In particular, the error between the estimation of this methodology and a reference system increases when the stride cadence does. By comparing the GGPs measurement in ten volunteers at four different speeds, the authors found the linear model presented in Eq. 5.15.

$$K = 0.157(SC) + 1.069. \tag{5.15}$$

In this way, the online estimation process goes as follows: (1) the user's legs are detected and tracked with the LRF sensor, (2) a clustering algorithm is applied to estimate the distance difference between legs (i.e., the LDD signal), (3) the WFLC estimates the frequency of the LDD's Fourier principal component (i.e., the SC), (4) the FLC uses the SC as input to estimate the amplitude of the LDD's Fourier principal component (i.e., the SPL), (5) the linear model described in Eq. 5.15 is used to adjust the SPL, and finally (vi) the GS is obtained using Eq. 5.14.

As reported by the authors in *Aguirre et al.* [37], this methodology was able to attain very accurate estimations compared to the literature evidence. In particular, Table 5.1 summarizes the estimation errors of real-time technologies, and the estimation errors for the SC and the SPL, using this online methodology. Considering that LRF sensors are commonly used to control robotic devices, such as smart walkers, rather than to do gait assessment, an average error of 5% can be considered as a good accuracy [2, 23, 24, 37].

According to the collected data in [37], it was reported that a scanning frequency of at least ten times the stride cadence is required to ensure the proper performance of the WFLC and FLC filters. This means that to obtain an online GGPs estimation system at walking speeds of 5 km/h to 6 km/h (i.e., the average speed of healthy individuals), an LRF working at frequencies from 10 Hz to 20 Hz would be required.

5.6 Conclusions

The analysis of gait patterns, indicators, and phases is a fundamental aspect when evaluating assistive technologies, both conventional and robotic. In general, gait evaluation techniques allow to: (1) track and report a patient's rehabilitation progress, (2) detect anomalies in the gait pattern, and (3) obtain reference inputs for high-level controllers in mobile and wearable robotic devices.

According to evidence in the literature, multiple sensing technologies have been developed to acquire information about an individual's gait. In this sense, this chapter presented a brief description of the most relevant spatial and temporal indicators used to characterize human gait, and some wearable sensors that allow their acquisition for rehabilitation and daily living scenarios. Finally, multiple methodologies that allow the concise description of the different gait characteristics, including their classification by phases and calculation of the most relevant indicators, were also presented.

References

1. D.A. Winter, *Biomechanics and Motor Control of Human Movement*, 4th edn. (Wiley, Hoboken, NJ, 2009)
2. C.A. Cifuentes, A. Frizera, *Human-Robot Interaction Strategies for Walker-Assisted Locomotion*. Springer Tracts in Advanced Robotics, vol. 115 (Springer International Publishing, Cham, 2016)
3. D.M. Wolpert, J. Diedrichsen, J.R. Flanagan, Principles of sensorimotor learning. Nat. Rev. Neurosci. **12**, 739–751 (2011)
4. P.F. Pasquina, C.G. Emba, M. Corcoran, *Lower Limb Disability: Present Military and Civilian Needs* (Springer, New York, NY, 2017)
5. Z.O. Abu-Faraj, G.F. Harris, P.A. Smith, S. Hassani, Human gait and Clinical Movement Analysis, in *Wiley Encyclopedia of Electrical and Electronics Engineering* (Wiley, Hoboken, NJ, 2015), pp. 1–34
6. C. Vaughan, B. Davis, J. O'Connor, The three-dimensional and cyclic nature of gait, in *Dynamics of Human Gait*, vol. 168 (1999), pp. 16–17
7. J.F. Veneman, R. Kruidhof, E.E. Hekman, R. Ekkelenkamp, E.H. Van Asseldonk, H. Van Der Kooij, Design and evaluation of the lopes exoskeleton robot for interactive gait rehabilitation. IEEE Trans. Neural Syst. Rehabil. Eng. **15**(3), 379–386 (2007)
8. J. Rueterbories, E.G. Spaich, O.K. Andersen, Gait event detection for use in FES rehabilitation by radial and tangential foot accelerations. Med. Eng. Phys. **36**, 502–508 (2014)
9. J.-U. Chu, K.-I. Song, S. Han, S.H. Lee, J.Y. Kang, D. Hwang, J.-K.F. Suh, K. Choi, I. Youn, Gait phase detection from sciatic nerve recordings in functional electrical stimulation systems for foot drop correction. Physiol. Measure. **34**, 541–565 (2013)
10. W.-C. Wen-Chang Cheng, D.-M. Ding-Mao Jhan, Triaxial accelerometer based fall detection method using a self-constructing cascade-AdaBoost-SVM classifier. IEEE J. Biomed. Health Inf. **17**, 411–419 (2013)

11. A. Mannini, D. Trojaniello, A. Cereatti, A. Sabatini, A machine learning framework for gait classification using inertial sensors: application to elderly, post-stroke and Huntington's disease patients. Sensors **16**, 134 (2016)
12. O. Beauchet, G. Allali, G. Berrut, C. Hommet, V. Dubost, F. Assal, Gait analysis in demented subjects: interests and perspectives. Neuropsychiatr. Dis. Treat. **4**, 155–160 (2008)
13. J. Figueiredo, C. Ferreira, C.P. Santos, J.C. Moreno, L.P. Reis, Real-time gait events detection during walking of biped model and humanoid robot through adaptive thresholds, in *2016 International Conference on Autonomous Robot Systems and Competitions (ICARSC)*, May 2016 (IEEE, New York, 2016), pp. 66–71
14. H.T.T. Vu, F. Gomez, P. Cherelle, D. Lefeber, A. Nowé, B. Vanderborght, ED-FNN: a new deep learning algorithm to detect percentage of the gait cycle for powered prostheses. Sensors (Basel, Switzerland) **18**(7), 2389 (2018)
15. S. Murray, M. Goldfarb, Towards the use of a lower limb exoskeleton for locomotion assistance in individuals with neuromuscular locomotor deficits, in *2012 Annual International Conference of the IEEE Engineering in Medicine and Biology Society*, vol. 2012 (IEEE, New York, 2012), pp. 1912–1915
16. R.B. Davis, Reflections on clinical gait analysis. J. Electromyogr. Kinesiol. **7**(4), 251–257 (1997)
17. M.W. Whittle, Clinical gait analysis: a review. Human Movement Sci. **15**(3), 369–387 (1996)
18. D. Sutherland, R. Olshen, E. Biden, *The Development of Mature Walking* (Cambridge University Press, Cambridge, 1988)
19. J. Perry, J.M. Burnfield, *Gait Analysis. Normal and Pathological Function*, 2nd edn. (Slack, Thorofare, 2010)
20. J. Marín, T. Blanco, J.J. Marín, A. Moreno, E. Martitegui, J.C. Aragüés, Integrating a gait analysis test in hospital rehabilitation: a service design approach. PLoS One **14**, e0224409 (2019)
21. M.W. Whittle, *Gait Analysis: An Introduction* (Butterworth-Heinemann, Oxford, 2014)
22. J. Casas, N. Cespedes, M. Múnera, C.A. Cifuentes, Human–robot interaction for rehabilitation scenarios, in *Control Systems Design of Bio-Robotics and Bio-mechatronics with Advanced Applications* (Elsevier, 2020), pp. 1–31
23. S. Sierra, M. Garzon, M. Munera, C.A. Cifuentes, Human–Robot–Environment interaction interface for smart walker assisted gait: AGoRA walker. Sensors **19**(13), 2897 (2019)
24. W.M. Scheidegger, R.C. de Mello, S.D. Sierra M., M.F. Jimenez, M.C. Munera, C.A. Cifuentes, A. Frizera-Neto, A novel multimodal cognitive interaction for walker-assisted rehabilitation therapies, in *2019 IEEE 16th International Conference on Rehabilitation Robotics (ICORR)*, June (IEEE, New York, 2019), pp. 905–910
25. J. Ballesteros, C. Urdiales, A.B. Martinez, M. Tirado, Online estimation of rollator user condition using spatiotemporal gait parameters, in *International Conference on Intelligent Robots and Systems (IROS)* (2016), pp. 3180–3185
26. C.A. Cifuentes, C. Rodriguez, A. Frizera, T. Bastos, Sensor fusion to control a robotic walker based on upper-limbs reaction forces and gait kinematics, in *5th IEEE RAS & EMBS International Conference on Biomedical Robotics and Biomechatronics* (2014), pp. 1098–1103
27. A. Muro-de-la Herran, B. Garcia-Zapirain, A. Mendez-Zorrilla, Gait analysis methods: an overview of wearable and non-wearable systems, highlighting clinical applications. Sensors **14**, 3362–3394 (2014)
28. A. Mannini, A.M. Sabatini, Gait phase detection and discrimination between walking–jogging activities using hidden Markov models applied to foot motion data from a gyroscope. Gait Posture **36**(4), 657–661 (2012)
29. S. Crea, S.M.M. De Rossi, M. Donati, P. Reberšek, D. Novak, N. Vitiello, T. Lenzi, J. Podobnik, M. Munih, M.C. Carrozza, Development of gait segmentation methods for wearable foot pressure sensors, in *2012 Annual International Conference of the IEEE Engineering in Medicine and Biology Society* (IEEE, New York, 2012), pp. 5018–5021
30. J. Taborri, E. Palermo, S. Rossi, P. Cappa, Gait partitioning methods: a systematic review. Sensors **16**(1), 66 (2016)

31. J.D. Bronzino, *Biomedical Engineering Handbook 2*, vol. 2 (Springer Science & Business Media, Berlin/Heidelberg, 2000)
32. G. Welch, G. Bishop, Others, An introduction to the Kalman filter. Technical report, University of North Carolina at Chapel Hill (1995)
33. F. Wittmann, O. Lambercy, R. Gassert, Magnetometer-based drift correction during rest in IMU arm motion tracking. Sensors **19**, 1312 (2019)
34. A. Carullo, M. Parvis, An ultrasonic sensor for distance measurement in automotive applications. IEEE Sensors J. **1**(2), 143 (2001)
35. Acuity, *Principles of Measurement Used by Laser Sensors and Scanners* (Acuity, Toronto, 2016)
36. SICK Sensor Intelligence, *Operating Instructions S300 Safety Laser Scanner* (SICK AG, Waldkirch, 2019)
37. A. Aguirre, S.D. Sierra M., M. Munera, C.A. Cifuentes, Online system for gait parameters estimation using a LRF sensor for assistive devices. IEEE Sens. J. 1–1 (2020). https://doi.org/10.1109/JSEN.2020.3028279
38. A. Ancillao, S. Tedesco, J. Barton, B. O'Flynn, Indirect measurement of ground reaction forces and moments by means of wearable inertial sensors: a systematic review. Sensors **18**(8), 2564 (2018)
39. J. Taborri, E. Palermo, S. Rossi, P. Cappa, Gait partitioning methods: a systematic review. Sensors 2016 **16**, 66 (2016)
40. A.M. Howell, T. Kobayashi, H.A. Hayes, K.B. Foreman, S.J.M. Bamberg, Kinetic gait analysis using a low-cost insole. IEEE Trans. Biomed. Eng. **60**, 3284–3290 (2013)
41. N.C. Bejarano, E. Ambrosini, A. Pedrocchi, G. Ferrigno, M. Monticone, S. Ferrante, A novel adaptive, real-time algorithm to detect gait events from wearable sensors. IEEE Trans. Neural Syst. Rehabil. Eng. **23**(3), 413–422 (2014)
42. D.-H. Lim, W.-S. Kim, H.-J. Kim, C.-S. Han, Development of real-time gait phase detection system for a lower extremity exoskeleton robot. Int. J. Prec. Eng. Manuf. **18**, 681–687 (2017)
43. M. Islam, E.T. Hsiao-Wecksler, Detection of gait modes using an artificial neural network during walking with a powered ankle-foot orthosis. J. Biophys. (Hindawi Publishing Corporation: Online) 2016, 7984157 (2016)
44. S. Ding, X. Ouyang, Z. Li, H. Yang, Proportion-based fuzzy gait phase detection using the smart insole. Sensors Actuat. A: Phys. **284**, 96–102 (2018)
45. X. Jiang, K.H.T. Chu, M. Khoshnam, C. Menon, A wearable gait phase detection system based on force myography techniques. Sensors (Basel, Switzerland) **18** (2018)
46. B. Smith, D. Coiro, R. Finson, R. Betz, J. McCarthy, Evaluation of force-sensing resistors for gait event detection to trigger electrical stimulation to improve walking in the child with cerebral palsy. IEEE Trans. Neural Syst. Rehabil. Eng. **10**, 22–29 (2002)
47. D. Gouwanda, A.A. Gopalai, A robust real-time gait event detection using wireless gyroscope and its application on normal and altered gaits. Med. Eng. Phys. **37**, 219–225 (2015)
48. L. Yu, J. Zheng, Y. Wang, Z. Song, E. Zhan, Adaptive method for real-time gait phase detection based on ground contact forces. Gait Posture **41**, 269–275 (2015)
49. W. Kong, M.H. Saad, M.A. Hannan, A. Hussain, Human gait state classification using artificial neural network, in *2014 IEEE Symposium on Computational Intelligence for Multimedia, Signal and Vision Processing (CIMSIVP)*, December 2014 (IEEE, New York, 2014), pp. 1–5
50. J.-Y. Jung, W. Heo, H. Yang, H. Park, A neural network-based gait phase classification method using sensors equipped on lower limb exoskeleton robots. Sensors **15**, 27738–27759 (2015)
51. R. Evans, D. Arvind, Detection of gait phases using Orient specks for mobile clinical gait analysis, in *2014 11th International Conference on Wearable and Implantable Body Sensor Networks*, June 2014 (IEEE, New York, 2014), pp. 149–154
52. J. Taborri, E. Scalona, S. Rossi, E. Palermo, F. Patane, P. Cappa, Real-time gait detection based on Hidden Markov Model: is it possible to avoid training procedure?, in *2015 IEEE International Symposium on Medical Measurements and Applications (MeMeA) Proceedings*, May 2015 (IEEE, New York, 2015), pp. 141–145

53. J. Taborri, S. Rossi, E. Palermo, F. Patanè, P. Cappa, A novel HMM distributed classifier for the detection of gait phases by means of a wearable inertial sensor network. Sensors **14**, 16212–16234 (2014)
54. A. Mannini, A.M. Sabatini, Gait phase detection and discrimination between walking–jogging activities using hidden Markov models applied to foot motion data from a gyroscope. Gait Posture **36**, 657–661 (2012)
55. A. Mannini, A.M. Sabatini, A hidden Markov model-based technique for gait segmentation using a foot-mounted gyroscope, in *2011 Annual International Conference of the IEEE Engineering in Medicine and Biology Society*, August 2011, vol. 2011 (IEEE, New York, 2011), pp. 4369–4373
56. N. Abaid, P. Cappa, E. Palermo, M. Petrarca, M. Porfiri, Gait detection in children with and without hemiplegia using single-axis wearable gyroscopes. PloS One **8**(9), e73152 (2013)
57. A. Mannini, V. Genovese, A.M. Sabatin, Online decoding of hidden Markov models for gait event detection using foot-mounted gyroscopes. IEEE J. Biomed. Health Inf. **18**, 1122–1130 (2014)
58. J. Taborri, E. Scalona, E. Palermo, S. Rossi, P. Cappa, Validation of inter-subject training for hidden Markov models applied to gait phase detection in children with cerebral palsy. Sensors **15**, 24514–24529 (2015)
59. R. Caldas, M. Mundt, W. Potthast, F. Buarque de Lima Neto, B. Markert, A systematic review of gait analysis methods based on inertial sensors and adaptive algorithms. Gait Posture **57**, 204–210 (2017)
60. J. Taborri, E. Palermo, S. Rossi, P. Cappa, Gait partitioning methods: a systematic review. Sensors **16**, 66 (2016)
61. M. Yuwono, S.W. Su, Y. Guo, B.D. Moulton, H.T. Nguyen, Unsupervised nonparametric method for gait analysis using a waist-worn inertial sensor. Appl. Soft Comput. **14**, 72–80 (2014)
62. E. Guenterberg, A. Yang, H. Ghasemzadeh, R. Jafari, R. Bajcsy, S. Sastry, A method for extracting temporal parameters based on hidden Markov models in body sensor networks with inertial sensors. IEEE Trans. Inf. Technol. Biomed. **13**, 1019–1030 (2009)
63. P. Catalfamo, S. Ghoussayni, D. Ewins, Gait event detection on level ground and incline walking using a rate gyroscope. Sensors (Basel, Switzerland) **10**(6), 5683–702 (2010)
64. A. Sabatini, C. Martelloni, S. Scapellato, F. Cavallo, Assessment of walking features from foot inertial sensing. IEEE Trans. Biomed. Eng. **52**, 486–494 (2005)
65. Y. Gurevich, Sequential abstract-state machines capture sequential algorithms. ACM Trans. Comput. Log. **1**(1), 77–111 (2000)
66. E. Börger, The abstract state machines method for high-level system design and analysis, in *Formal Methods: State of the Art and New Directions* (Springer, New York, 2010), pp. 79–116
67. D. Kotiadis, H. Hermens, P. Veltink, Inertial gait phase detection for control of a drop foot stimulator. Med. Eng. Phys. **32**, 287–297 (2010)
68. M.D. Sánchez Manchola, M.J. Pinto Bernal, M. Munera, C.A. Cifuentes, Gait phase detection for lower-limb exoskeletons using foot motion data from a single inertial measurement unit in hemiparetic individuals. Sensors **19**, 2988 (2019)
69. D. Kotiadis, H.J. Hermens, P.H. Veltink, Inertial gait phase detection for control of a drop foot stimulator: inertial sensing for gait phase detection. Med. Eng. Phys. **32**(4), 287–297 (2010)
70. N. Limnios, G. Oprisan, *Semi-Markov Processes and Reliability*. (Springer Science & Business Media, Heidelberg, 2012)
71. P. Blunsom, Hidden Markov models. Lecture Notes **15**(18–19), 48 (2004)
72. L. Rabiner, A tutorial on hidden Markov models and selected applications in speech recognition. Proc. IEEE **77**(2), 257–286 (1989)
73. M.W. Whittle, Clinical gait analysis: a review. Hum. Movement Sci. **15**, 369–387 (1996)
74. H. Stolze, J.P. Kuhtz-buschbeck, C. Mondwurf, A. Boczek-funcke, K. Jo, Gait analysis during treadmill and overground locomotion in children and adults. Electroencephalogr. Clin. Neurophysiol. **105**, 490–497 (1997)

75. J. Rueterbories, E.G. Spaich, B. Larsen, O.K. Andersen, Methods for gait event detection and analysis in ambulatory systems. Med. Eng. Phys. **32**(6), 545–552 (2010)
76. N. Sekiya, H. Nagasaki, H. Ito, F. Taketo, The invariant relationship between step length and step rate during free walking. J. Hum. Movement Stud. **30**(6), 241–257 (1996)
77. W.T. Latt, U.-X. Tan, K.C. Veluvolu, C.Y. Shee, W.T. Ang, Real-time estimation and prediction of periodic signals from attenuated and phase-shifted sensed signals, in *2009 IEEE/ASME International Conference on Advanced Intelligent Mechatronics*, July (IEEE, New York, 2009), pp. 1643–1648
78. V. Bonnet, C. Mazzà, J. McCamley, A. Cappozzo, Use of weighted Fourier linear combiner filters to estimate lower trunk 3D orientation from gyroscope sensors data. J. NeuroEng. Rehabil. **10**(1), 29 (2013)
79. V. Bonnet, C. Mazzà, J. McCamley, A. Cappozzo, Use of weighted Fourier linear combiner filters to gyroscope sensors data. J. Neuroeng. Rehabil. **10**, 29 (2013)
80. G. Data, Real-time estimation of pathological tremor parameters from gyroscope data. Sensors **10**, 2129–2149 (2010)
81. A.F. Neto, J.A. Gallego, E. Rocon, J.L. Pons, R. Ceres, Extraction of user's navigation commands from upper body force interaction in walker assisted gait. BioMed. Eng. OnLine **1–16**, 37 (2010)
82. Y. Qi, C.B. Soh, E. Gunawan, K.-S. Low, R. Thomas, Assessment of foot trajectory for human gait phase detection using wireless ultrasonic sensor network. IEEE Trans. Neural Syst. Rehabil. Eng. **24**, 88–97 (2016)
83. N.-H. Ho, P. Truong, G.-M. Jeong, Step-detection and adaptive step-length estimation for pedestrian dead-reckoning at various walking speeds using a smartphone. Sensors **16**, 1423 (2016)

Experimental Characterization of Flexible and Soft Actuators for Rehabilitation and Assistive Devices

6

Daniel Gomez-Vargas, Felipe Ballen-Moreno, Orion Ramos, Marcela Múnera, and Carlos A. Cifuentes

6.1 Introduction

Several actuators have been developed for robotic devices, focused on assisting or enhancing the human limbs or joints. In this sense, considering both (1) the human limb or joint to assist and (2) the goal task, the actuators' energy should be efficiently provided to the user during the assistance process [1]. Therefore, to guarantee a proper human–robot interaction, multiple tests aimed at measuring and delimiting the device's functional capabilities should be carried out. Currently, this assessment has been commonly accomplished through experimental studies applied directly on subjects, either healthy or pathological. Notwithstanding, the test benches' inclusion in the characterization process could provide a general understanding related to (1) how the device interacts with the user and (2) what are its maximum capabilities.

Regarding the actuator type implemented on the device, the characterization process can evidence different techniques and the assessed variables. However, the goal of those processes remains focused on measuring the device's performance in the assisted activity. This way, characteristics such as the system response, stability, and device's limitations, among others can be measured to improve the robot behavior. Likewise, this characterization allows estimating accurate experimental

D. Gomez-Vargas
Biomedical Engineering, Department of the Colombian School of Engineering Julio Garavito, Bogotá D.C., Colombia

Institute of Automatics, National University of San Juan, San Juan, Argentina
e-mail: daniel.gomez-v@mail.escuelaing.edu.co

F. Ballen-Moreno · O. Ramos · M. Múnera · C. A. Cifuentes (✉)
Biomedical Engineering Department of the Colombian School of Engineering Julio Garavito, Bogotá, Colombia
e-mail: felipe.ballen@mail.escuelaing.edu.co; orion.ramos@mail.escuelaing.edu.co; marcela.munera@escuelaing.edu.co; carlos.cifuentes@escuelaing.edu.co

C. A. Cifuentes, M. Múnera, *Interfacing Humans and Robots for Gait Assistance and Rehabilitation*, https://doi.org/10.1007/978-3-030-79630-3_6

models, starting from the actuator or the device coupled in the application, which can be used to enhance the control strategies and consequently the human–robot interaction.

As mentioned before, the methodologies to measure the device's capabilities commonly have been divided into experimental tests (1) within the goal application (i.e., involving users) and (2) through test bench setups. Furthermore, depending on the stage of the actuator's development, this assessment could include different previous tests. For instance, in *Fang et al.*, the initial phases of a pneumatic bending actuator were focused on determining an analytical model and building an actuator's prototype [2]. Subsequently, the actuator was characterized using a test bench structure based on the device's application, and then it was assessed in the goal users. In the same line, another study presented the characterization of an ankle exoskeleton based on a variable stiffness actuator, focusing on measuring the device capabilities in terms of bandwidth, system response, assisted torque, and saturation non-linearities for different stiffness values [3].

On the other hand, the characterization process can also include complex variables aimed at the device's specific application. For instance, *Yandell* et al. [4] defined an analytical kinetic model to estimate the energy losses of a powered ankle–foot orthosis based on cable-driven actuators. For this case, the proposed assessment included experiments applied directly to subjects.

In this context, this chapter presents the characterization of two types of actuators focused on rehabilitation scenarios, showing the trends and essential variables measured in the experiments. The first actuator consists of an ankle exoskeleton based on variable stiffness' concepts detailed in Chap. 7, and the second actuator describes a soft hand exoskeleton based on pneumatic actuation concepts.

6.2 Characterization of Actuators

The importance of understanding the actuators' capabilities arises from the motivation to enhance the device's performance and improve the human–robot interaction. However, considering the goal user and the desired scenario, the assessed variables change depending on the actuator type. In this sense, the following sections present the trends and essential variables according to the actuator principle, addressing an experimental characterization of a variable stiffness actuator and a soft actuator based on pneumatic principles.

6.2.1 Characterization of a Variable Stiffness Actuator in Gait Rehabilitation

Different actuators for assistive scenarios (e.g., pneumatic, hydraulic, electric actuators) were presented in Chap. 2. This way, aspects such as the actuator type and the assisted joint involve determining (1) what should be the proper amount of energy provided to the user during the task and (2) how this energy should be provided [3]. Concerning the variable stiffness actuators, devices based on this actuation type allow changing the system behavior to be adapted to the user's physical conditions, as Chap. 7 shows. Therefore, those devices could exhibit multiple responses affecting the control performance.

In this sense, this actuation type requires a complete dynamic and static model that describes the system behavior under the goal application. Nevertheless, some devices could have complex mechanical designs, resulting in intricate models. Hence, a characterization process could simplify the model being delimited within the conditions where the device should operate. Moreover, this process could also provide information about the device performance and physical interaction in those scenarios.

6.2.1.1 Trends and Essential Variables

In general terms, robotics aimed at gait rehabilitation integrate principles applied in passive orthotic structures, although incorporating benefits such as the capacity of providing torque, feedback during the assistive process, modification of the device performance, consistency during the exercise, and support for the therapists, among others [5]. From the motivation to enhance the physical interaction during the assistance, compliant actuators are emerging as a solution to preserve the device's actuation system and improve the transparency effect during an interaction forces scenario. Particularly, series elastic actuators and variable stiffness actuators are being exhibited as potential principles.

Robotic devices based on variable stiffness actuators are widely recommended in applications where the robot interacts intensively with the human [6, 7] because of advantages evidenced by this principle under rehabilitation scenarios, as Chap. 7 presents. Specifically, a variable stiffness effect enables the device to modify its behavior considering the desired physical interaction. Moreover, from the bioinspired concepts applied in assistive robots, variable stiffness resembles the motor human functions. Those devices aim at assisting the lower limb joints' sagittal plane (i.e., hip, knee, and ankle). However, each one of these joints has challenges related to the required torque, movements on the other planes (e.g., add–abduction and internal–external rotation), reaction forces, and torque transmission, among others.

In the ankle rehabilitation context, powered ankle–foot orthoses have applied this actuation type, exploiting the spring's inclusion characteristics, i.e., shock loads and backdrivability. This way, kinematic and kinetic models for control purposes could be complex, and additionally, they could not provide information related to the environmental constraints and users' requirements. Thus, an experimental characterization, using a test bench either static or dynamic, or through human trials, is exhibited as an alternative to determine these aspects [1, 8]. For a static setup, the structure fixes the actuators' output (see Fig. 6.1), restricting their motions and coupling sensors to measure the interaction forces and the device performance [8]. The actuators execute the set-point values resembling the device's application and inducing a physical effect in the output (i.e., torque, force, pressure, angular or linear position). This way, it can be understood the actuator capabilities (i.e., apparent bandwidth, peak, and continuous torque), adjusting different controllers to reduce

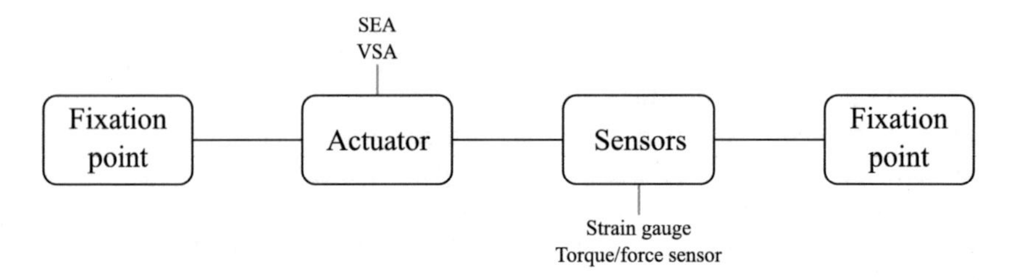

Fig. 6.1 Descriptive scheme of the principal components for a static bench

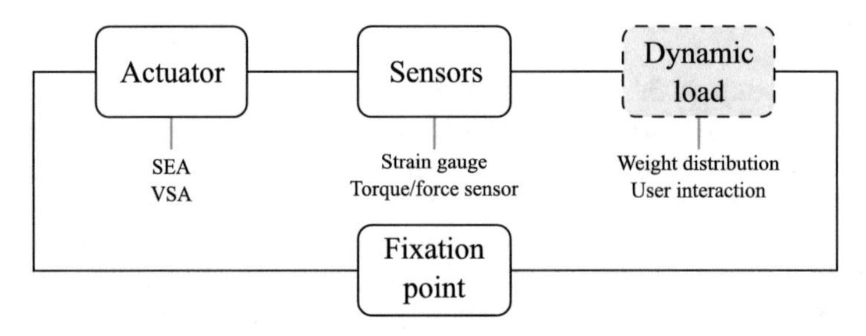

Fig. 6.2 Descriptive scheme of the principal components for a dynamic bench

position or torque errors, increase the time of response. Nevertheless, it does not represent its actual performance during operation [9].

On the other hand, a dynamic test bench intends to closely assess the actuator in the application, including the user's external forces. In contrast to a static test bench, the actuators' output is coupled to a dynamic load [10] (see Fig. 6.2). This way, the trials could recreate a realistic scenario, including the user's dynamic that affects the device's performance (i.e., bandwidth and temporal response).

6.2.1.2 T-FLEX Design and Test Bench Structure

T-FLEX is a wearable and portable ankle exoskeleton based on variable stiffness actuators (see Chap. 7) to assist the dorsi–plantarflexion movements without restricting the other ankle motions [11]. This ankle exoskeleton comprises two servomotors attached to bioinspired tendons (see Fig. 6.3) whose mechanical behavior is similar to the human Achilles tendon [12]. T-FLEX integrates a bidirectional system of stiff filaments to (1) involve both actuators during the assistance process and (2) correct the foot pathological postures. Additionally, this device is manually adjustable and usable for both limbs.

For the variable stiffness effect, passive elements resemble the stiffness of a human tendon as a spring-like component. Thus, T-FLEX integrates a braided material formed by (1) elastic filament (Filaflex, 2.85 mm, Recreus, Spain) and (2) fishing rod (eight filaments, Sufix 832, USA). These filaments were intertwined following a volumetric fraction of 14% to accomplish a variable stiffness performance regarding the elongation.

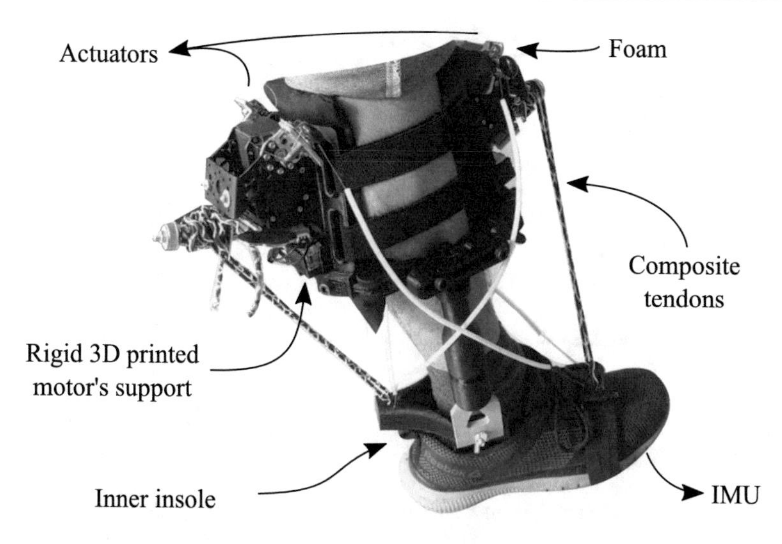

Fig. 6.3 Wearable and portable ankle exoskeleton T-FLEX. The remarked elements refer to the main parts of the device

In terms of functionality, T-FLEX includes two operational modes, i.e., (1) stationary therapy and (2) gait assistance, which employ a calibration stage customized for each user. For the first modality, the exoskeleton executes dorsi–plantarflexion repetitions, integrating an inertial sensor placed on the foot tip to detect the user movement intention. The implemented therapy model allows varying parameters such as the repetition number, repetition frequency, and movement speed. This modality has exhibited promising outcomes in a rehabilitation context with a stroke patient [13].

For the gait assistance, T-FLEX assists the user's gait phases (i.e., mid-stance, heel-off swing, and heel strike), providing dorsi–plantarflexion movements and increasing the system stiffness, following the actuators' combination shown in Fig. 6.4. In this sense, the device incorporates a gait phase detector based on a hidden Markov model and machine learning [14], detailed in Chap. 5. A preliminary study focused on assessing the T-FLEX's actuation system in a walking application evidenced significant potential for the lower-limb kinematics of patients who suffered a stroke [15].

From the T-FLEX's design and goal applications presented above, the test bench structure employed to characterize the device should consider different conditions that affect the device response and performance. Thus, the user's anthropometric measurements play a significant role in the torque provided by the exoskeleton. Specifically, the distance between the ankle and the fixation points of the tendons (i.e., plastic part placed on the foot tip and the structure adapted to the heel), the shank's length, and the user's body composition modify the torque provided to this joint (see Fig. 6.5). Additionally, the ankle torque, tendon force, and the electrical

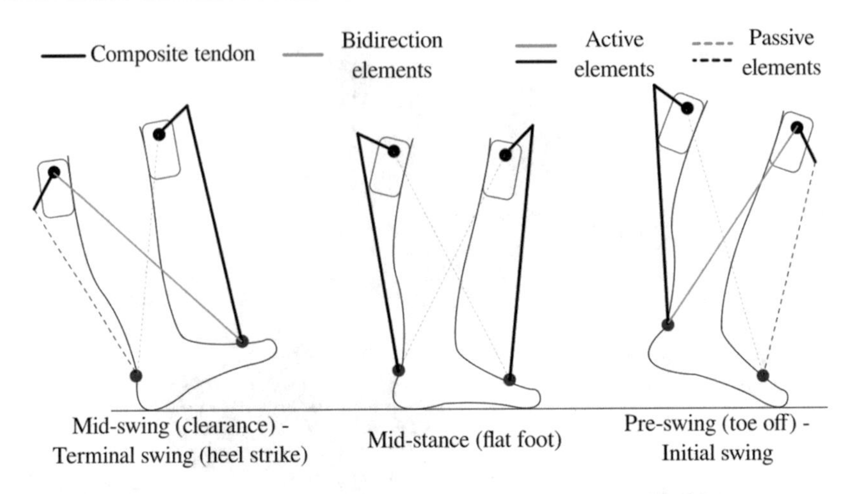

Fig. 6.4 Descriptive scheme of the mechanical configuration of the T-FLEX's functionalities

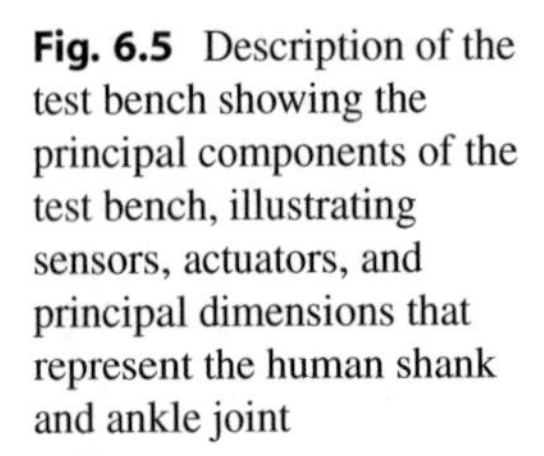
Fig. 6.5 Description of the test bench showing the principal components of the test bench, illustrating sensors, actuators, and principal dimensions that represent the human shank and ankle joint

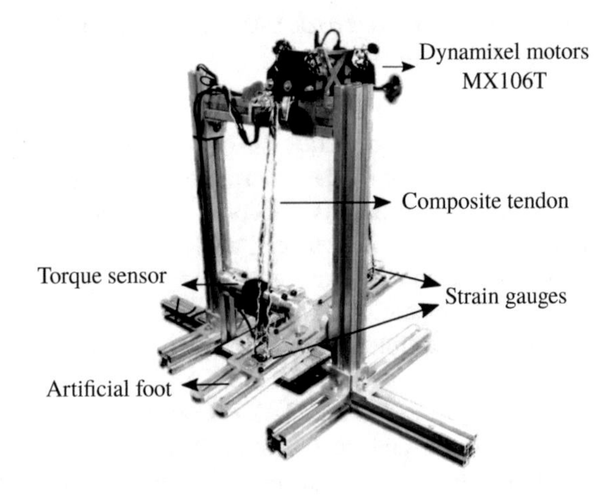

and physical characteristics of the T-FLEX's actuators are the main variables in an experimental process that could assess the device's performance.

In this context, a mechanical structure composed of aluminum frames was developed to adjust the variable distances mentioned above, applying the T-FLEX's operation concept in an actual application (see Fig. 6.5). This test bench structure included a torque sensor FY01 (Forsentek, China) coupled on the artificial ankle and a set of strain gauges (RS PRO, UK) placed on the tendons' fixation points. For the actuators, 3D printed pieces fixed the smart servomotors to the mechanical structure, resembling the shape of a human shank.

6.2.1.3 Experimental Procedure

Firstly, the experimental procedure intended to analyze the tendon behavior resembling the T-FLEX operation principle. For this purpose, a tensile test was carried

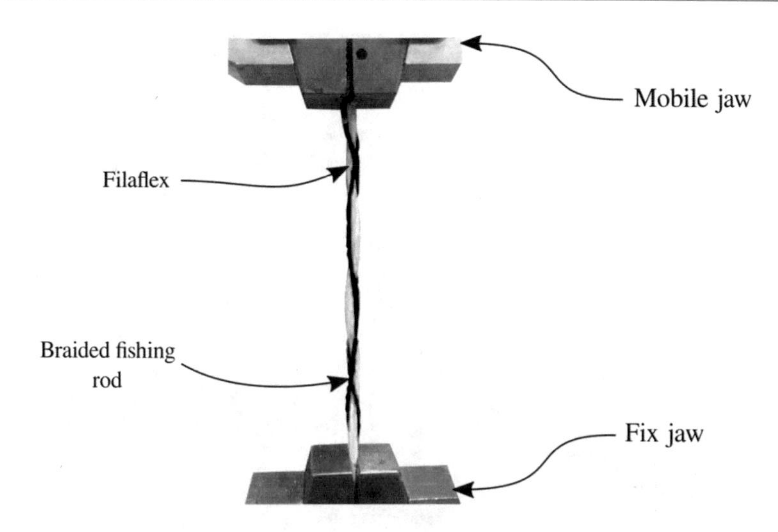

Fig. 6.6 Tensile test experimental setup to measure the T-FLEX's tendons behavior

out through a universal machine, fixing a specimen between two jaws, as shown in Fig. 6.6. Besides, the tensile tests followed the ASTM C1557-14 [16].

On the other hand, considering the T-FLEX variability, the second part of this characterization analyzed the tendon effect under a pretension level of 10 N in an assistance process. The selected force level corresponds to the medium force concerning the maximum value that induces actuator saturation. From this value, the study included two signals to measure the device response and estimate the device capabilities (i.e., step and chirp). These signals were sent to the actuators as position commands with an amplitude between 3 and 15 degrees, which is a common value applied in an actual application. Likewise, the set-points resembled the T-FLEX operation in a gait assistance application, following the movements presented in Fig. 6.4.

6.2.1.4 Results

From the tendon trials, stress–strain tests estimated two elastic zones and their Young's modulus, as Fig. 6.7 shows. A range of strain defines each zone: zone A between 0 and 0.10 mm/mm and zone B between 0.1 and 0.15 mm/mm. Nevertheless, zone C presented inconsistent stress values and rupture point. However, this last zone is not taken into account in the analysis because the bioinspired tendons will be loaded with forces smaller than the required for the rupture point.

On the other hand, considering the assistance application in the proposed test bench structure, the step function measured both the response of the T-FLEX's actuation system and the behavior of the composite tendons under tension. Figure 6.8 shows the curve obtained during the dorsi–plantarflexion movements in terms of the assisted torque and tendons force. The first set-point (i.e., segmented black line) describes the dorsiflexion command. In this movement, the anterior motor turns

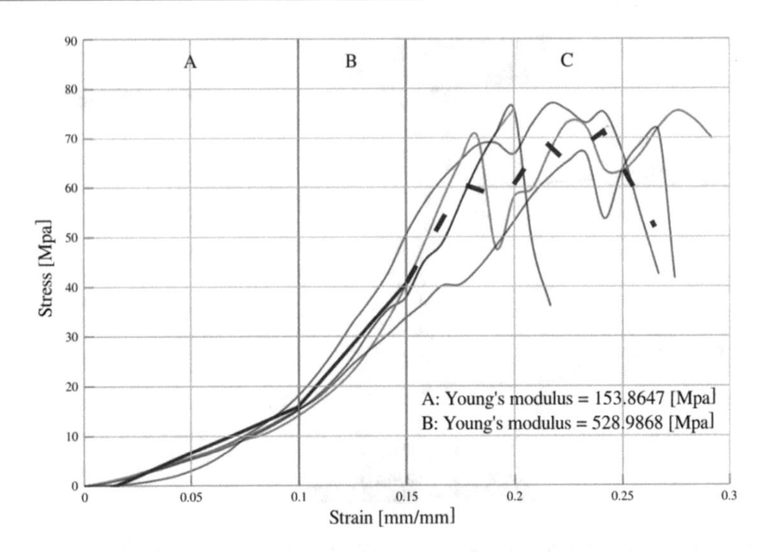

Fig. 6.7 Tensile results of the bioinspired tendons. The stress–strain curve presents the Young's modulus for the A and B zones

to pull the foot, and the posterior actuator works reversely. This way, the system transmits positive torque to assist the ankle. For the second set-point, the actuators operate in opposite directions concerning the movement mentioned above. Thus T-FLEX assists the plantarflexion, providing negative torque on the ankle joint.

The chirp signal measured the system response to frequency changes and the maximum values of torque on the ankle for the different amplitude values. Figure 6.9 shows the system behavior in terms of torque on the ankle, the tendons force, and T-FLEX's actuators. The responses illustrated in Fig. 6.9 occurred in the tendons-alone configuration for the dorsi–plantarflexion movements with a force level of 10 N.

Considering the obtained responses, the trials reported that T-FLEX exhibits a bandwidth close to 6.8 Hz for the measured amplitude range. This value was estimated using the system identification toolbox of MATLAB (MathWorks, US). Likewise, the maximum provided torque measured on the ankle was 12 Nm for propulsion and 20 Nm for the dorsiflexion movement. In terms of the system response, the trials evidenced (1) a delay related to the tendon elasticity (i.e., close to 45 ms) and (2) a stabilization time less than 284 ms. In general terms, T-FLEX can assist the human gait under this configuration. However, the control architecture should include an adaptive stage that accelerates the device response concerning the set-point value and the pretension level. This way, the device could anticipate the stabilization time and the tendon delay, ensuring the maximum torque transmission in the specific gait phase.

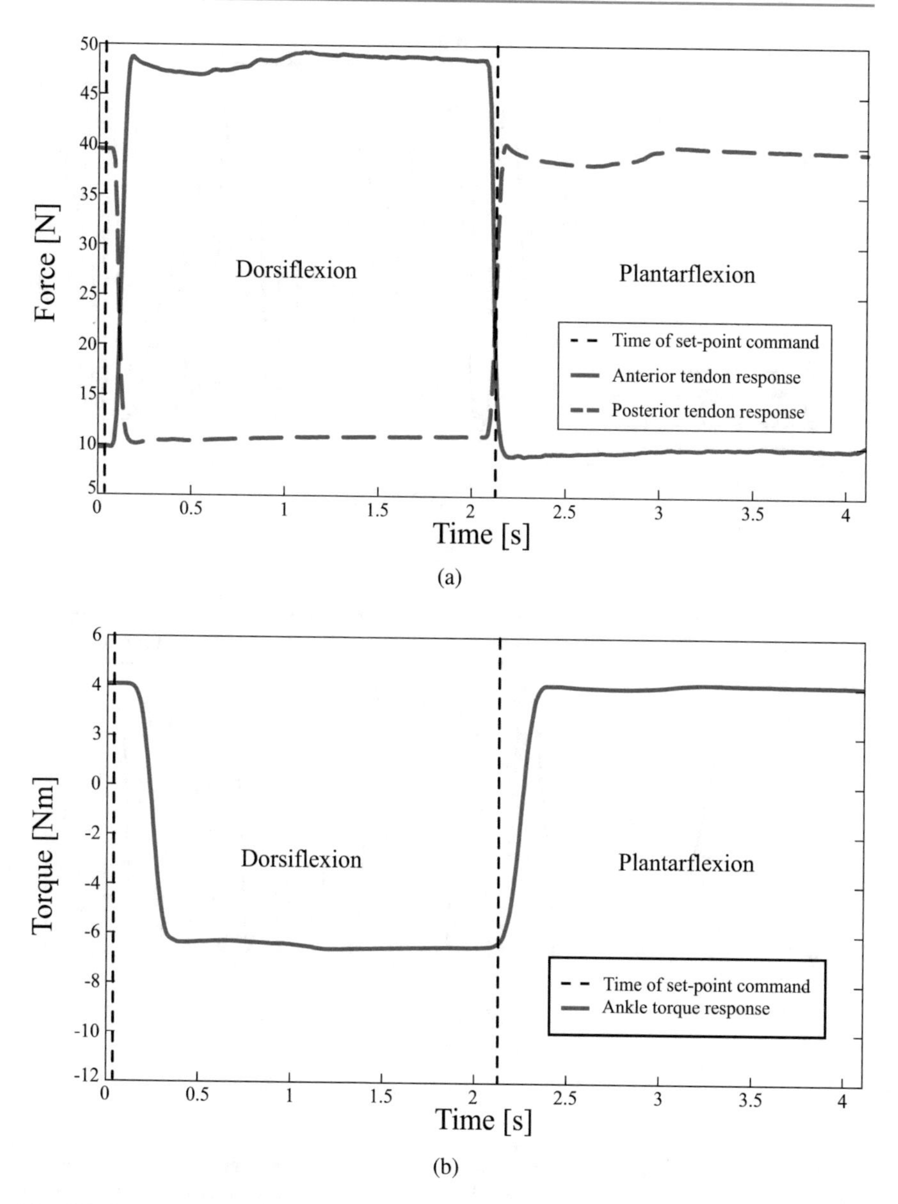

Fig. 6.8 Step response of the system for the dorsi–plantarflexion movement. (**a**) Force outcomes. (**b**) Torque outcomes

[ht!]

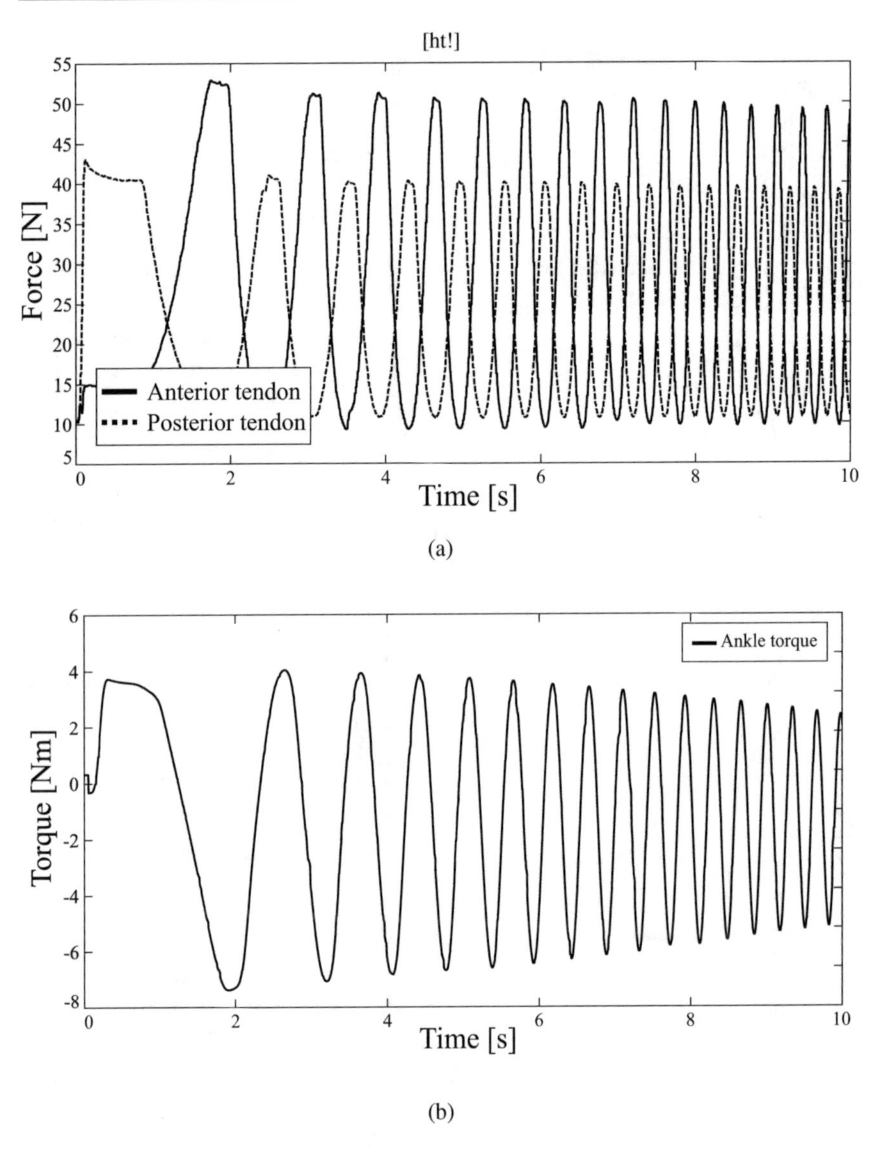

Fig. 6.9 System responses for the chirp signal in the tendons-alone configuration executing dorsi–plantarflexion movements under a force level of 10 N. (**a**) Force outcomes. (**b**) Torque outcomes

6.2.2 Characterization of a Soft Actuator Based on Pneumatic Actuation in Hand Rehabilitation

New technologies in the engineering field have raised new design paradigms such as soft robotics strategies. These technologies are differentiated by using soft materials such as elastomers or flexible fabrics such as lycra. These techniques facilitate assistive or rehabilitative devices by eliminating mechanisms that require exact alignment of the limb joints with the degrees of freedom of the device [17]. This is possible because these types of soft actuators do not have defined degrees of freedom; on the contrary, they are considered to have infinite degrees of freedom [18].

Another advantage proposed by this new technology is the separation of the actuation source from the device attached to the person. This is to reduce the weight that the limb to be assisted or rehabilitated must support. Usually, systems are built to be stored in a backpack or control box apart from the device. These properties of systems built using soft robotics make it possible to reduce the device's weight concerning rigid technologies considerably. However, modeling and mathematically characterizing actuators based on pneumatic actuation are considerably more complex than performing this procedure on electric actuators or DC motors due to the non-linearities that pneumatic systems can generate and their infinite degrees of freedom.

6.2.2.1 Trends and Essential Variables

As seen in Chap. 2, the amount and form of motion of each degree of freedom of a robot allow finding the kinematic model to the robot. However, in actuators based on soft robotics techniques, it is impossible to perform these calculations due to their high deformation properties. In these cases, it is usual to characterize the relevant variables of the problem to be solved with the actuator through experimental techniques [19, 20]. In other cases, computational models based on finite element analysis [21, 22] are used. In this section, characterization methodology of soft actuators to be used in an upper-limb exoskeleton will be explained. Therefore, the relevant variables must be related to this problem.

Since a hand exoskeleton developed for assistance or rehabilitation is a wearable system that must be comfortable for the user, the most relevant variables in this application will be those that achieve the assistance of the human hand. These variables are the range of motion and the forces needed to intervene in the movements of the fingers without affecting comfort and efficiency. Specifically, the requirements for the design of a hand rehabilitation or assistive device according to the state-of-the-art review are divided into different categories. One of these is practical considerations, such as the number of fingers to be assisted, the device's weight, and its dimensions.

The other category focuses on the kinematic requirements necessary for the device to be considered functional in clinical rehabilitation or assistive applications. This category defines the ranges of motion and forces required for each actuator of the exoskeleton, as mentioned above. Finally, the control category specifies actuator response speed, system bandwidth, and device sensing. For each of the categories, values and requirements are defined based on engineering studies and clinical recommendations.

In the category of general considerations, the wearable device's total weight should not exceed 3 kg. The weight supported in hand should be around 0.5 kg [23], and the actuator dimensions should be in the range of human fingers' size. In terms of kinematic requirements, each actuator must bend at least 250° to be considered capable of assisting in human finger flexion [23]. The force to be exerted by the actuators establishes that forces higher than 7 N allow any assistance in daily life tasks [23]. However, only 3 N is enough to generate practical assistance for rehabilitation cases [17]. In the control section, the sampling rate should be at least 20 times faster than the response speed of the actuators [23]. The actuators' speed should be around the human hand's average values, which is 0.3 m/s [24]. Finally, the system must be powered by a small power supply and air supply while avoiding exceeding the total allowable weight.

Since the general considerations and control category's characterization are usual in engineering fields, they will not be explained in detail. Measuring the device's weight and actuators does not involve any complexity, although it gives relevant information. Similarly, calculating the system's sample time is not part of the actuator characterization and becomes a requirement of the complete device; in this case, only the actuators will be studied. The kinematic requirements category is the one that will be detailed in this section. In this category, variables such as the pressure required to generate the maximum bending and the force that the soft actuators can generate are mentioned.

To evaluate the necessary pressure required to generate actuator bending, a test is performed that relates air pressure measured with air pressure sensors to the maximum actuator tip bending angle measured by video processing. A comparative test of actuator tip pressure versus actuator tip angle reported that it is possible to achieve mean bending at pressures in the range of 42 kPa to 52 kPa depending on actuator length. In that study, silicon actuators built using the PneuNets technique were compared. The values were rectified with finite element analysis simulations specializing in this type of actuator [25]. Another comparison of silicone actuators was performed. It was found that the fiber reinforced-type actuator of Elastosil M4601 material with a length of 16 cm achieves full bending at 243 kPa [26].

Moreover, a study was conducted to identify how construction parameters, such as actuator length, internal air chamber inner radius, and actuator wall thickness, affect the pressure required to achieve full bending. It was concluded that the smaller the actuator length and internal radius, the higher the pressure must be to achieve full bending. Also the chamber wall is directly proportional to the required pressure, so the thicker the actuator, the higher the pressure must be to achieve full bending. The comparison actuator in the study has a length of 16 cm, a wall thickness of 2 mm, and an inner radius of 8 mm. With this actuator, full bending is achieved at 200 kPa (approximately 30 psi) [27]. Finally, silicone actuators achieved medium bending (close to 90 degrees bending) at a pressure of 110 kPa. It was also shown that the actuator reduces this angle if attached to a finger simulating an exoskeleton [28]. As seen in these studies, varying the actuator dimensions, the material of construction, and the type of reinforcement changes the actuator behavior for the required input pressure that generates the maximum bending.

On the other hand, several ways to find the force generated by soft actuators have been studied. The most common ones are based on measuring the actuator tip force with a load cell. Two configurations are regularly presented to measure the force exerted on the actuator tip. One measures the bending force. The other configuration measures the blocked force, which is greater than the bending force. Some studies report bending average force values close to 4.5 N at 407 kPa in silicone actuators with a length of 80 mm. In that study, different silicone actuators of different lengths are compared. It is evident that shorter actuators generate more force, e.g., a 60 mm long silicone actuator reaches the maximum force (5.58 N) at 450 kPa [29]. Another study using hydraulically actuated fiber reinforced silicone actuators can generate bending forces of 9N at the tip of the actuator [30]. Silicone actuators were compared for use in rehabilitation or assistive hand devices. In that study, bending force values were obtained for actuators of different elastomer references. Thus actuators constructed with Dragon-skin 10 achieved a force of 3.19 N at 180 kPa and actuators constructed with Dragon-Skin 20 reached 3.5 N at 380 kPa [19].

Moreover, the blocked force test has been used more frequently to characterize this actuator, so more information is available for comparisons. The force recorded in this test is usually higher than that recorded in the bending force test. For example, a 13 cm long PneuNets silicone actuator generates a blocked force of 1.2 N at only 43 kPa [25]. In fiber reinforced actuators, blocked force values of 1 N at 200 kPa were reported for Dragon-Skin 10 silicone. Furthermore, for references such as Elastosil M4601, forces of 5 N were obtained at pressures of 400 kPa. These actuators were constructed with a length of 17 cm [26]. Likewise, for these same types of fiber reinforced actuators, force values close to 8.8 N at 180 kPa were achieved for materials such as Dragon-Skin 10 and 9.96 N at 380 kPa for Dragon-skin 20 [19]. Finally, in 2017, forces of 9.12 N at only 120 kPa were achieved in a hybrid silicone and textile actuator [20].

As presented above, performing experiments where peak forces, efficiency, and kinematic variables are determined is the easier way to obtain the device's models. Fiber reinforced- and textile-type actuators with pleats have been the most developed and used in hand assistance and rehabilitation applications to create new exoskeletons based on soft robotics [31, 32]. Therefore, comparing them experimentally is relevant to select the most efficient one. Starting from the premise of designing a wearable device for hand assistance and rehabilitation, test benches were designed for fiber reinforced- and textile-type actuators with pleats characterizing the efficiency of the two types of actuators and another test bench to find the bending force and blocked force that these actuators can generate.

6.2.2.2 ExHand Design and Test Bench Structure

Based on how the behavior of soft actuators is evaluated according to state of the art, three types of tests were performed for a textile and a silicone actuator. According

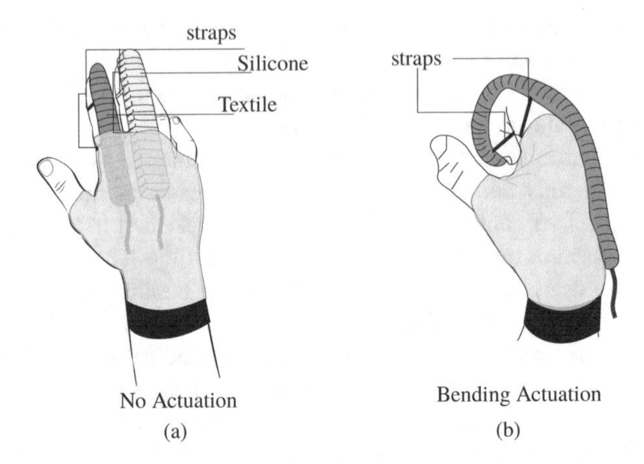

Fig. 6.10 Diagram of actuators' locations. (**a**) Possible placement of the soft actuators to create the assistive or rehabilitative hand device. (**b**) Demonstration of how the bending actuator generates the assisting motion on the human finger

to the designs that have been made, the ExHand actuators' location on the hand can be seen in Fig. 6.10a. The figure also shows how the actuator's bending motion allows the bending of the fingers without concern for the degrees of freedom of the actuator Fig. 6.10b. These actuators are very different from each other in terms of dimensions, materials, and construction methods, so each actuator's characteristics will be explained in general terms.

The silicone fiber reinforced actuator is built by pouring elastomeric materials such as silicone into 3D printed molds. Depending on the type of motion generated, reinforcements are made with rigid elements such as layers of carbon fiber layers and inelastic thread [27]. The type of movement that was evaluated is elemental bending. This motion works to simulate human fingers in assistive or rehabilitative applications in a hand exoskeleton. The fiber reinforced actuator built to explain this section was made with an elastomer (Dragon-Skin 00–30 from Smooth on). The entire construction process can take about two days, as the silicone must cure in the molds. One of the features that these actuators can provide is based on the fact that a single actuator can be configured to generate different movements, such as bending, extension, and torsion. However, it should be noted that this later construction cannot be modified, and the actuator will always have the same behavior. An example of this property is a thumb actuator's design, which integrates different motions into a single actuator [33].

On the other hand, the textile actuator uses elastic and inelastic fabric materials for the creation of the bending motion and an elastic–plastic element (Stretchlon 200, FibreGlast) to contain the air internally. The construction process consists of sewing a layer of rigid fabric with a layer of elastic fabric, creating a finger-sized pocket. This layer must be sewn together, generating pleats that facilitate the bending motion [34]. The construction time of a textile actuator with pleats for the

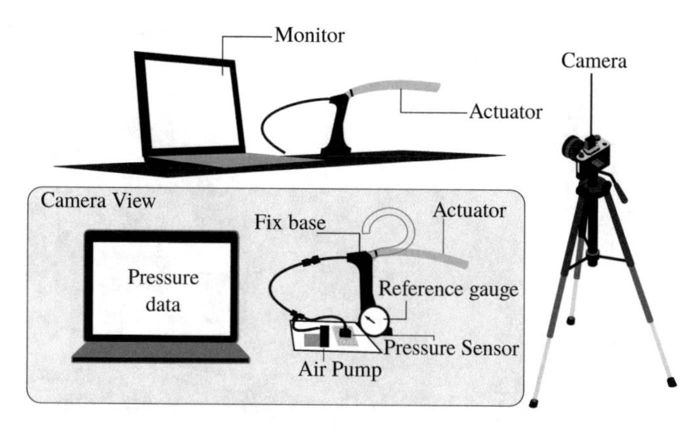

Fig. 6.11 General setup and parameters displayed through the camera for the test bench where the bending angle measurement concerning the input pressure is performed

bending motion can take about 6 h. The advantage of this type of actuator is that it can generate different motions. However, in contrast to silicone actuators, these can be actuated independently, generating actuators with bending and extension movements with independent control [34]. From the ExHand design presented previously, Fig. 6.11 shows the test bench setup developed for the experimental characterization.

6.2.2.3 Experimental Procedure

An experiment was designed to measure the pressure required by the actuators to generate the maximum bending angle to characterize the actuators' efficiency. This variable is the angle required when the actuator is used to assist in closing the human hand. An ASDXACX100PAAA5 analog pressure sensor (Honeywell, USA) and a D2028 air pump (Karlsson Robotics, USA) were used for this purpose. To define when the actuator reaches the maximum bending angle, the open-source video post-processing software Kinovea was used. For this purpose, the entire experiment was recorded at 30 FPS ensuring that the camera scene takes the actuator's motion and the data provided by the microcontroller display. The critical data to synchronize the video and microprocessor values were the time, the number of samples, and the sensor's pressure. The processing speed of the microprocessor was synchronized with the 30 FPS of the video. For each actuator, five repetitions were performed, which consisted of moving the actuator from steady state (no air pressure) to maximum deflection employing the air pump.

Other variables that could be obtained to study the actuators' behavior could include the actuation speed, acceleration, and even bending shape or path trajectory. All this information can be obtained in the same test bench of Fig. 6.11 using the Kinovea software (Kinovea org, France). This software uses artificial vision methods to process some marker changes in each video frame and estimate the variables listed above.

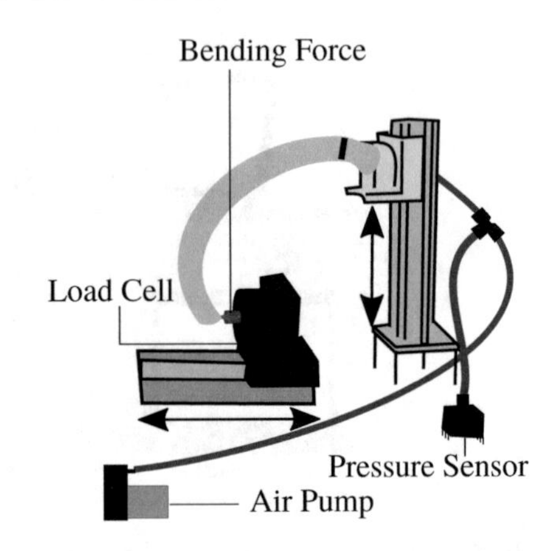

Fig. 6.12 Bending force test bench assembly where the necessary elements can be seen. Here the distance from the sensor and the height of the actuator base are variable to redirect the actuator tip and achieve that the force is applied correctly on the sensor

The two types of forces that the actuators can generate at the same pressure were measured. The bending force, which consists of measuring the force generated by the tip of the actuator when it reaches the maximum bending [29], was the first experiment. Moreover, the second one was the blocked force. This test consists of restricting the actuator's bending motion using a rigid sheet and measuring the tip force [19]. In both cases, the air pressure used was that which in the previous experiment achieved maximum bending. The instrument used to measure the actuator force was an FC2211-0000-0010-L load cell (TE Connectivity, Switzerland). Fig. 6.12 shows the test bench for performing the maximum bending force measurement. As in the first experiment, five repetitions were performed for each actuator, which was averaged to estimate the applied force. In this experiment, the two crucial variables (input pressure and force) are acquired by the microcontroller, so it is unnecessary to synchronize video data with sensor data. Since the two types of actuators have different behaviors in their bending path trajectory, the test bench must change the height and the distance at which the force sensor is located.

The experiment to measure the blocked force uses the same elements. It is based on the same five repetitions of pressurizing the actuator with the pressure that generates full bending and recording the force values at that point for each actuator. The set-up in Fig. 6.13 is the one that represents the test bench for this experiment. In this case, neither the height nor the distance needs to be modified. In this case, the two actuators are constructed of the same length.

6.2.2.4 Results

The pressure required for the actuator to achieve full bending was measured. The five values were summarized by the mean and their standard deviation to be plotted and compared in Fig. 6.14. In this graph, the comparison of the pneumatic energy required to perform the same task for the two types of actuators can be seen. Note that the textile actuators reach the maximum bending angle at a value of 11.04 psi (76 kPa). And the silicone actuators at a higher value of 28.49 psi (196.4 kPa). It is important to note that both systems reach the full bending angle at air pressure below

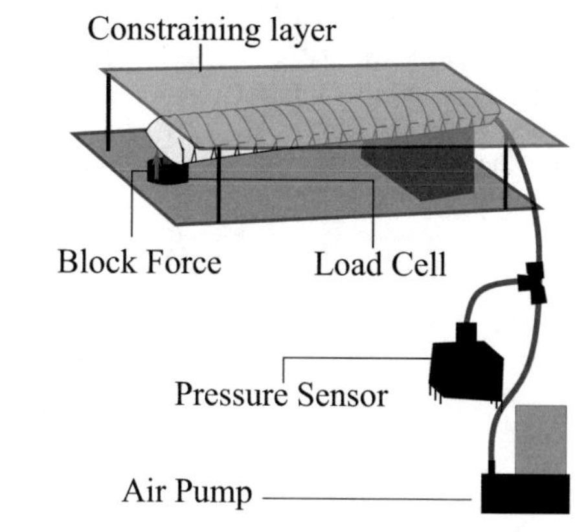

Fig. 6.13 Test bench setup to perform the blocked force test. The figure shows how the top layer restricts the bending motion and allows redirecting the force to the actuator tip

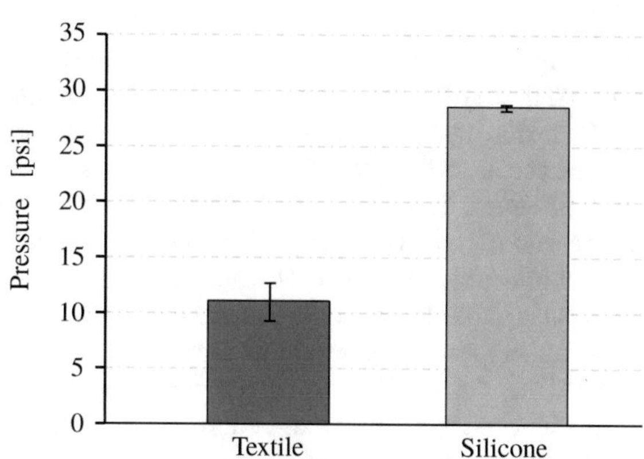

Fig. 6.14 Air pressure needed to achieve full bending performance for both types of actuators

200 kPa. This indicates that both types of actuators can be used with small size and low-power air sources. Nevertheless, the difference between the results of textile actuators and silicone actuators is high. It can be affirmed that textile actuators are more energy efficient in achieving the maximum bending angle. In other words, the silicone actuators, while meeting the test objective and the stated requirements, are less efficient than the textile actuators. This means that silicone actuators require more air pressure to perform the same task as textile actuators.

It is vital to identify that the deviation of the textile actuators' data is higher than in the silicone actuators. In this sense, silicone actuators are more stable and accurate of the input pressure required to achieve full bending. However, considering the purpose of those devices, the found deviation is not relevant. Suppose the actuator is pressurized with the highest value of the five values found. In that case, it is assumed that the actuator will be in the position of full bending.

The experiments that found the actuator forces are shown in Fig. 6.15. As with the previous results, the graphs represent the means and deviations of the five values found in the test. It is possible to observe how the bending force for the two actuators is very similar. In comparison, in the blocked force test, there are

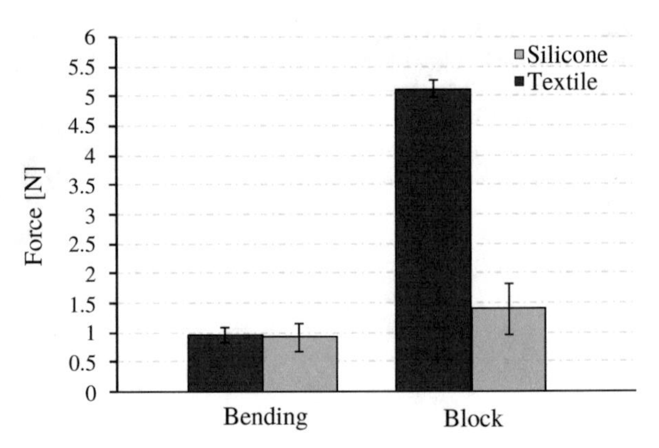

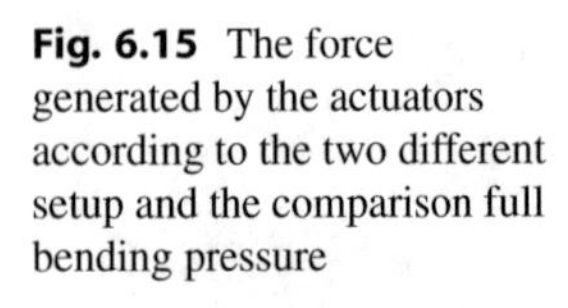
Fig. 6.15 The force generated by the actuators according to the two different setup and the comparison full bending pressure

pretty high differences between the two actuators. In detail in the bending force test performed, very similar results were obtained regarding the force exerted.

The textile actuators generated an average peak force of 0.95 N at 11 psi, and the silicone actuators 0.92 N at 28 psi. These values are very close to each other. There is no difference between the two actuators in this type of experiment. Therefore, it can be said that the two types of actuators generate the same bending force at the maximum bending position. The difference is in the pressure required to generate this value. If the force exerted at the actuator's tip at the same input pressure is compared, a lower value would be seen in the silicone actuators. With this test, it is rectified that the textile actuators are more efficient than the silicone actuators.

The blocked force test results show that the textile actuators generate more force than the silicone actuators. In this case, the textile actuators generate 5.1 N of locked force, much higher than the 1.4 N of the silicone actuator. Another critical point seen in Fig. 6.15 is the deviation of the fiber reinforced-type silicone actuator's data. In both the bending force test and the blocked force test, the data are more scattered in this actuator type. This may be related to the fact that silicone actuators' behavior in terms of actuator tip forces is less repeatable and more complex to control than in textile actuators.

Knowing the application for which the two types of actuators are intended to be used and the design requirements stated in state of the art is easy to establish which are the most feasible results according to each test. For example, in the first experiment, the actuator needs less pressure to achieve full bending and, therefore, reduce the power source's size, and the air source is desired. All this reduces the device's weight, which is an essential factor in wearable devices for assistance or rehabilitation [33]. According to the results obtained, textile actuators are better than silicone actuators for the pressure required to achieve the entire bending movement. So in the design of a wearable assistive or rehabilitation device, they should be considered over silicone actuators.

Similarly, the results of the force tests are analyzed in these cases. For this type of application, the desired results are the higher force values. The more force the actuator can generate, the more possibilities it will have to assist human fingers' movements. It will have a greater capacity to propose rehabilitation tools for different pathologies and at different rehabilitation stages. The force tests shown in Fig. 6.15 show that the force generated by the two types of actuators flexion is very similar. Regardless of the efficiency, this force value is achieved with the same pressure value used in the previous experiment.

Similarly, the results of the force tests are analyzed in these cases. For this type of application, the desired results are the higher force values. The reason for this is that the more force the actuator can generate, the more possibilities it will have to assist human fingers' movements. It will have a greater capacity to create therapies for different types of pathologies and at different rehabilitation stages. The force tests shown in Fig. 6.15 show that the force generated by the two types of actuators in flexion is very similar, so there would not be a key factor to decide which one is better.

The only notable difference in these results is the deviation of the results. In this case, the textile actuators generate the most stable values. Analyzing the blocked force test results shows that the type of actuator that generates more force is the textile actuator (30% more than the silicone actuators). This result is essential for selecting the actuator type if an assistive device is to be built. Based on these experiments' results, it is suggested to select textile actuators with pleats to develop assistive or rehabilitative hand devices. As they meet the basic requirements in terms of force, they can be actuated without requiring much pneumatic energy, and the construction time is shorter than silicone actuators.

6.3 Actuators for Assistive Applications

Proper selection of actuation for an application in biomedical fields is a critical task in engineering. The design involves understanding the fundamental variables to be considered and how to characterize them on a test bench. The next step after measuring the device's performance consists of applying the technology in an actual scenario considering the trials executed during the characterization process. In this sense, it is possible to determine the device's capabilities and guarantee the safe human–robot interaction. This section presents assistive devices based on pneumatic actuation used and evaluated in clinical fields.

Based on the tests performed to characterize the types of soft actuators, the textile-type actuator was selected to construct rehabilitation and assistive device. The device was built using bending and extension actuators in the thumb and index finger. With these two actuators, it is possible to assist in the type of grip called the pincer grip, which was evaluated through the modified Jebsen Taylor test [35]. Figure 6.16 shows the device built with the textile actuators previously characterized and compared with the silicone actuators.

The Jebsen Taylor test consists of a series of 7 subtests to assess various hand functions related to activities of daily living. The subtests include tasks such as writing, turning letters, and picking up small objects. In the study conducted to evaluate the device's functionality, an adaptation of the test based on measuring the

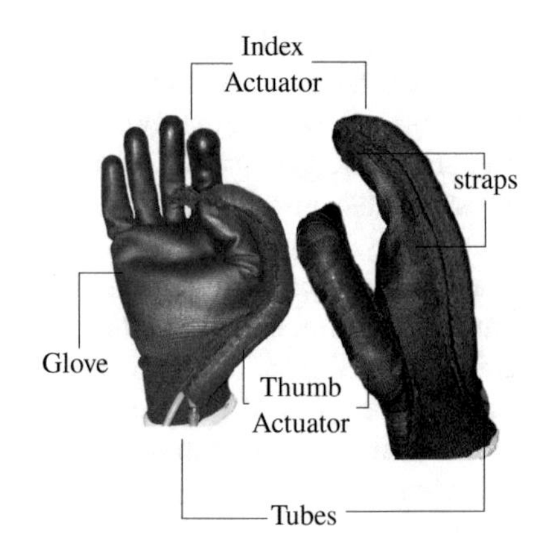

Fig. 6.16 Front and side view of the implementation of the textile-type actuator in a glove for the creation of the assistive and rehabilitative hand device

times required to hold different objects was used [36]. Five objects were selected for this evaluation; a coin, an eraser, a sphere, a plastic cup, and a book. The test compared the time required to hold each object without the device in a healthy subject. The study results show that the device can assist in the grasping of objects for activities of daily living. However, it is less efficient with small objects than with large objects. In the healthy subject, the time to grasp everyday objects is higher with the use of the device than without its use.

In the literature review, different hand rehabilitation devices based on wearable robotics and soft robotics techniques were found. The technique based on pneumatic actuators has been the most varied over the years. Initially, this style's devices began using silicone actuators of the PneuNets type due to their ease of construction and modeling. For example, a flexing glove with PneuNets actuators was built with the ability to control the velocity and position of each actuator by pressure input to the system [37]. In parallel, exoskeleton designs with other types of silicon actuators were realized. For example, the development of an exoskeleton with fiber reinforced actuators was evaluated in rehabilitation therapy [23]. Specifically, fiber reinforced-type silicone actuators are most commonly used to construct these assistive and rehabilitative hand devices. An exoskeleton for rehabilitation and task-specific training help to perform tasks faster and more accurately in a subject with stroke [23]. In this work, each actuator has different stretches of movement, which allows the actuator to generate assistance in finger flexion and twist these. Another exoskeleton built with fiber reinforced actuators was presented in [38]. The actuators assisted finger flexion and were tested by inflating and deflating the actuators at 100 kPa to see each actuator's life cycle. Specifically, these actuators functioned properly for 62.2 cycles of the previous test. Following the previous works, with the same fiber reinforced silicone actuator, an exoskeleton was made for hand rehabilitation that assists in flexion and extension movements. This movement is achieved with a brace that generates a torque on the actuators that keeps them extended until they are pressed for flexion. It is important to note that each actuator in this device weighs

37 g, which is a factor for improvement in future versions. This device was tested in a study with stroke patients.

On the side of exoskeletons built with textile-type actuators, the devices usually assist in both directions through layers of fabrics of different properties. For example, an exoskeleton for hand rehabilitation and assistance was built with corrugated textile actuators with the possibility to assist in both flexion and extension movements. This device's manufacture is particular because it is not made with elastic elements such as lycra-type fabrics. In this case, flexible materials such as TPU-coated fabric are used, and employing the geometry in some actuator layers, the flexion and extension movements are achieved. Also, in the study, a control interface is designed to switch between different operation modes such as specific tasks, bilateral rehabilitation, and grip types, among others [39]. With some external improvements in the device, a functionality study was performed based on the box and block test. It was found that the device assists in finger flexion, but the force generated in extension is not enough for patients with high muscle spasticity [20].

A fabric exoskeleton was built using the previous case's geometrical principles but with elastic materials such as lycra. That study uses textile actuators with pleats that facilitate the bending motion and reduce the input pressure. The designed device can assist in opening and closing the hand and was validated utilizing muscle activity in the forearm when performing the movements with and without the exoskeleton. Muscle activity in these tasks with the device is lower than without the exoskeleton [34]. Another following work [40] added force and deflexion sensors; bending actuators segmented by phalanges also were modified to apply force on only the parts needed. In this version of the device, the system requires 25 psi of power and can generate three movements; power closure, gripper closure, and finger extension. For evaluating the device, grip strength tests were performed, which obtained an 87% increase using the device. The Jebsen Hand Function Test and Box and Blocks tests evidenced improvements in patients with high-level injuries and reduced functionality in low-level injuries.

6.4 Conclusions

The characterization of actuators has several steps to fully understanding the capacities and limitations required for the aimed application. The test bench is the forefront tool to assess the actuator's performance through essential variables, depending on the implemented mechanical principles. This chapter presented the characterization process for two actuator types (i.e., variable stiffness actuator and pneumatic actuator) regarding the goal application and interaction scenario. Thus, the overall process defined the interesting variables, signals applied, and setup proposed to assess the actuator coupled in an assistive robot in terms of system response and device capabilities.

References

1. M.D.C. Sanchez-Villamañan, J. Gonzalez-Vargas, D. Torricelli, J.C. Moreno, J.L. Pons, Compliant lower limb exoskeletons: a comprehensive review on mechanical design principles. J. NeuroEng. Rehabil. **16**(1), 1–16 (2019)
2. J. Fang, J. Yuan, M. Wang, L. Xiao, J. Yang, Z. Lin, P. Xu, L. Hou, Novel accordion-inspired foldable pneumatic actuators for knee assistive devices. Soft Robot. **7**(1) 95–108 (2020)
3. M. Moltedo, G. Cavallo, T. Baček, J. Lataire, B. Vanderborght, D. Lefeber, C. Rodriguez-Guerrero, Variable stiffness ankle actuator for use in robotic-assisted walking: control strategy and experimental characterization. Mech. Mach. Theory **134**, 604–624 (2019)
4. M.B. Yandell, B.T. Quinlivan, D. Popov, C. Walsh, K.E. Zelik, Physical interface dynamics alter how robotic exosuits augment human movement: implications for optimizing wearable assistive devices. J. NeuroEng. Rehabil. **14**(1), 1–11 (2017)
5. J.-M. Belda-Lois, S. Mena-del Horno, I. Bermejo-Bosch, J.C. Moreno, J.L. Pons, D. Farina, M. Iosa, M. Molinari, F. Tamburella, A. Ramos, et al., Rehabilitation of gait after stroke: a review towards a top-down approach. J. Neuroeng. Rehabil. **8**(1), 66 (2011)
6. P.K. Jamwal, S. Hussain, S.Q. Xie, Review on design and control aspects of ankle rehabilitation robots. Disabil. Rehabil. Assist. Technol. **10**, 93–101 (2013)
7. M. Moltedo, T. Baček, B. Serrien, K. Langlois, B. Vanderborght, D. Lefeber, C. Rodriguez-Guerrero, Walking with a powered ankle-foot orthosis: the effects of actuation timing and stiffness level on healthy users. J. Neuroeng. Rehabil. **17**(1), 1–15 (2020)
8. Y. Li, K.A. Shorter, E.T. Hsiao-Wecksler, T. Bretl, Simulation and experimental analysis of a portable powered ankle-foot orthosis control, in *ASME 2011 Dynamic Systems and Control Conference and Bath/ASME Symposium on Fluid Power and Motion Control, DSCC 2011*, vol. 1(1), 77–84 (2011)
9. M. Moltedo, T. Bacek, K. Langlois, K. Junius, B. Vanderborght, D. Lefeber, Design and experimental evaluation of a lightweight, high-torque and compliant actuator for an active ankle foot orthosis, in *2017 International Conference on Rehabilitation Robotics (ICORR)* (IEEE, New York, 2017), pp. 283–288
10. Y.-L. Park, B.-R. Chen, D. Young, L. Stirling, R.J. Wood, E. Goldfield, R. Nagpal, Bio-inspired active soft orthotic device for ankle foot pathologies, in *2011 IEEE/RSJ International Conference on Intelligent Robots and Systems* (IEEE, New York, 2011), pp. 4488–4495
11. M. Manchola, D. Serrano, D. Gómez, F. Ballen, D. Casas, M. Munera, C.A. Cifuentes, T-FLEX: variable stiffness ankle-foot orthosis for gait assistance, in *Wearable Robotics: Challenges and Trends*, ed. by J. González-Vargas, J. Ibáñez, J.L. Contreras-Vidal, H. van der Kooij, J.L. Pons. Biosystems & Biorobotics (Springer International Publishing, New York, 2017), pp. 160–164
12. J. Casas, A. Leal Junior, C. Díaz, A. Frizera, M. Munera, C.A. Cifuentes, Large-range polymer optical-fiber strain-gauge sensor for elastic tendons in wearable assistive robots. Materials **12**, 1443 (2019)
13. D. Gomez-Vargas, M.J. Pinto-Bernal, F. Ballen-Moreno, M. Munera, C.A. Cifuentes, Therapy with t-flex ankle-exoskeleton for motor recovery: a case study with a stroke survivor, in *8th IEEE RAS & EMBS International Conference on Biomedical Robotics and Biomechatronics (BioRob)* (2020)
14. M.D. Manchola, M.J. Bernal, M. Munera, C.A. Cifuentes, Gait phase detection for lower-limb exoskeletons using foot motion data from a single inertial measurement unit in hemiparetic individuals. Sensors (Switzerland) **19**(13), 2988 (2019)
15. D. Gomez-Vargas, F. Ballen-Moreno, P. Barria, R. Aguilar, J. M. Azorín, M. Munera, C.A. Cifuentes, The actuation system of the ankle exoskeleton t-flex: first use experimental validation in people with stroke. Brain Sci. **11**(4), 412 (2021)
16. ASTM C1557-14, *Standard Test Method for Tensile Strength and Young's Modulus of Fibers* (ASTM International, West Conshohocken, PA, 2014)
17. G. Agarwal, N. Besuchet, B. Audergon, J. Paik, Stretchable materials for robust soft actuators towards assistive wearable devices. Sci. Rep. **6**(1), 1–8 (2016)

18. D. Bruder, B. Gillespie, C.D. Remy, R. Vasudevan, Modeling and control of soft robots using the Koopman operator and model predictive control (2019). Preprint. arXiv:1902.02827
19. H.K. Yap, J.H. Lim, F. Nasrallah, J. Cho Hong Goh, C.-H. Yeow, Characterisation and evaluation of soft elastomeric actuators for hand assistive and rehabilitation applications. J. Med. Eng. Technol. **40**(4), 199–209 (2016)
20. H.K. Yap, F. Sebastian, C. Wiedeman, C.-H. Yeow, Design and characterization of low-cost fabric-based flat pneumatic actuators for soft assistive glove application, in *2017 International Conference on Rehabilitation Robotics (ICORR)* (IEEE, 2017), pp. 1465–1470
21. M. Pozzi, E. Miguel, R. Deimel, M. Malvezzi, B. Bickel, O. Brock, D. Prattichizzo, Efficient fem-based simulation of soft robots modeled as kinematic chains, in *2018 IEEE international conference on robotics and automation (ICRA)* (IEEE, New York, 2018), pp. 4206–4213
22. E. Coevoet, A. Escande, C. Duriez, Soft robots locomotion and manipulation control using FEM simulation and quadratic programming, in *2019 2nd IEEE International Conference on Soft Robotics (RoboSoft)* (IEEE, 2019), pp. 739–745
23. P. Polygerinos, K.C. Galloway, E. Savage, M. Herman, K. O'Donnell, C.J. Walsh, Soft robotic glove for hand rehabilitation and task specific training, in *2015 IEEE International Conference on Robotics and Automation (ICRA)* (IEEE, New York, 2015), pp. 2913–2919
24. H.A. Varol, S.A. Dalley, T.E. Wiste, M. Goldfarb, Biomimicry and the design of multigrasp transradial prostheses, in *The Human Hand as an Inspiration for Robot Hand Development* (Springer, New York, 2014), pp. 431–451
25. P. Polygerinos, S. Lyne, Z. Wang, L.F. Nicolini, B. Mosadegh, G.M. Whitesides, C.J. Walsh, Towards a soft pneumatic glove for hand rehabilitation, in *2013 IEEE/RSJ International Conference on Intelligent Robots and Systems* (IEEE, New York, 2013), pp. 1512–1517
26. K.C. Galloway, P. Polygerinos, C.J. Walsh, R.J. Wood, Mechanically programmable bend radius for fiber-reinforced soft actuators, in *2013 16th International Conference on Advanced Robotics (ICAR)* (IEEE, New York, 2013), pp. 1–6
27. P. Polygerinos, Z. Wang, J.T. Overvelde, K.C. Galloway, R.J. Wood, K. Bertoldi, C.J. Walsh, Modeling of soft fiber-reinforced bending actuators. IEEE Trans. Robot. **31**(3), 778–789 (2015)
28. M. Ariyanto, J.D. Setiawan, R. Ismail, I. Haryanto, T. Febrina, D.R. Saksono, Design and characterization of low-cost soft pneumatic bending actuator for hand rehabilitation, in *2018 5th International Conference on Information Technology, Computer, and Electrical Engineering (ICITACEE)* (IEEE, 2018), pp. 45–50
29. Y. Sun, X. Liang, H.K. Yap, J. Cao, M.H. Ang, R.C.H. Yeow, Force measurement toward the instability theory of soft pneumatic actuators, IEEE Robot. Autom. Lett. **2**(2), 985–992 (2017)
30. J. Peters, E. Nolan, M. Wiese, M. Miodownik, M. Spurgeon, A. Arezzo, A. Raatz, H. Wurdemann, Actuation and stiffening in fluid-driven soft robots using low-melting-point material, in *Proceedings of the 2019 IEEE/RSJ International Conference on Intelligent Robots and Systems (IROS 2019)*, vol. 2019 (IEEE, New York, 2019)
31. C. Correia, K. Nuckols, D. Wagner, Y.M. Zhou, M. Clarke, D. Orzel, R. Solinsky, S. Paganoni, C.J. Walsh, Improving grasp function after spinal cord injury with a soft robotic glove. IEEE Trans. Neural Syst. Rehabil. Eng. **28**(6), 1407–1415 (2020)
32. K.H. Heung, R.K. Tong, A.T. Lau, Z. Li, Robotic glove with soft-elastic composite actuators for assisting activities of daily living. Soft Robot. **6**(2), 289–304 (2019)
33. P. Maeder-York, T. Clites, E. Boggs, R. Neff, P. Polygerinos, D. Holland, L. Stirling, K. Galloway, C. Wee, C. Walsh, Biologically inspired soft robot for thumb rehabilitation. J. Med. Dev. **8**(2) (2014)
34. L. Cappello, K.C. Galloway, S. Sanan, D.A. Wagner, R. Granberry, S. Engelhardt, F.L. Haufe, J.D. Peisner, C.J. Walsh, Exploiting textile mechanical anisotropy for fabric-based pneumatic actuators. Soft Robot. **5**(5), 662–674 (2018)
35. A. Peñas, J. Maldonado, O. Ramos, M. Munera, P. Barria, M. Moazen, H.A. Wurdemann, C.A. Cifuentes, Towards a fabric-based soft hand exoskeleton for various grasp taxonomies, in *The International Symposium on Wearable Robotics (WeRob2020) and WearRAcon Europe* (2020)
36. P. Tran, S. Jeong, S.L. Wolf, J.P. Desai, Patient-specific, voice-controlled, robotic FLEXotendon Glove-II system for spinal cord injury. IEEE Robot. Autom. Lett. **5**(2), 898–905 (2020)

37. M. Haghshenas-Jaryani, R.M. Patterson, N. Bugnariu, M.B. Wijesundara, A pilot study on the design and validation of a hybrid exoskeleton robotic device for hand rehabilitation. J. Hand Therapy **33**(2), 198–208 (2020)
38. Y. Jiang, D. Chen, J. Que, Z. Liu, Z. Wang, Y. Xu, Soft robotic glove for hand rehabilitation based on a novel fabrication method, in *2017 IEEE International Conference on Robotics and Biomimetics (ROBIO)*, pp. 817–822 (IEEE, New York, 2017)
39. H.K. Yap, P.M. Khin, T.H. Koh, Y. Sun, X. Liang, J.H. Lim, C.-H. Yeow, A fully fabric-based bidirectional soft robotic glove for assistance and rehabilitation of hand impaired patients. IEEE Robot. Autom. Lett. **2**(3), 1383–1390 (2017)
40. Y.M. Zhou, D. Wagner, K. Nuckols, R. Heimgartner, C. Correia, M. Clarke, D. Orzel, C. O'Neill, R. Solinsky, S. Paganoni, et al., Soft robotic glove with integrated sensing for intuitive grasping assistance post spinal cord injury, in *2019 International conference on robotics and automation (ICRA)* (IEEE, New York, 2019), pp. 9059–9065

Variable Stiffness Actuators for Wearable Applications in Gait Rehabilitation

7

Daniel Gomez-Vargas, Diego Casas-Bocanegra, Marcela Múnera, Flavio Roberti, Ricardo Carelli, and Carlos A. Cifuentes

7.1 Introduction

Currently, there are mainly two types of robots for rehabilitation and assistance: (1) platform-based robots, intended solely for the improvement of joints function, and (2) wearable devices, which can contribute in the rehabilitation of joints in stationary scenarios, as platform-based robots, but can also improve joints performance during gait in daily activities outside controlled environments. Therefore, this second type of robot exhibits considerable advantages concerning stationary platforms in aspects such as multimodality and applicability [1, 2].

In general terms, devices applied to gait rehabilitation integrate principles implemented in passive orthotic structures, although incorporating the robotics' benefits (i.e., energy supply using actuators, user monitoring through sensors, programmed functionality profiles, among others) [3]. In this sense, those devices mainly aim at improving the patients' gait pattern or decrease the metabolic effort during walking [4]. However, considering the complexity of developing robotic

D. Gomez-Vargas
Biomedical Engineering, Department of the Colombian School of Engineering Julio Garavito, Bogotá D.C., Colombia

Institute of Automatics, National University of San Juan, San Juan, Argentina
e-mail: daniel.gomez-v@mail.escuelaing.edu.co

D. Casas-Bocanegra · M. Múnera · C. A. Cifuentes (✉)
Biomedical Engineering, Department of the Colombian School of Engineering Julio Garavito, Bogotá D.C., Colombia
e-mail: diego.casas-b@mail.escuelaing.edu.co; marcela.munera@escuelaing.edu.co; carlos.cifuentes@escuelaing.edu.co

F. Roberti · R. Carelli
Institute of Automatics, National University of San Juan, San Juan, Argentina
e-mail: froberti@inaut.unsj.edu.ar; rcarelli@inaut.unsj.edu.ar

C. A. Cifuentes, M. Múnera, *Interfacing Humans and Robots for Gait Assistance and Rehabilitation*, https://doi.org/10.1007/978-3-030-79630-3_7

devices aimed at physical interaction scenarios, wearable devices challenge in aspects such as portability, adaptability to the human body, and compliance [2, 5].

On the other hand, from the capacity of assisting the human body's movements, robotic devices require providing high torque levels during assistive scenarios [6]. Therefore, different actuation systems have been applied in those systems, intending to improve human–robot interaction and assist pathological motor functions. Specifically, current developments integrate principles based on (1) Stiff Actuators, (2) Serial Elastic Actuators (SEAs), and (3) Pneumatic Actuators [4]. Some devices also include other mechanisms such as (4) Hydraulic Actuators and (5) Magnetorheological Actuators [7, 8]. Furthermore, other actuation systems widely implemented nowadays, particularly in wearable systems, applies concepts of (6) Variable Stiffness Actuators (VSAs) and (7) Cable-Driven Actuators [9].

Devices based on pneumatic actuators have potential in aspects such as compliance and physical interaction with the user. However, this actuation type exhibits disadvantages related to the overweight power supply required to assist human movements, as well as hydraulic actuators [9, 10]. On the other hand, wearable robots based on magnetorheological actuators include drawbacks associated with (1) the complex and heavyweight equipment implemented and (2) the high energy consumption to achieve this principle [8].

From the other actuation systems' drawbacks presented above, electrical power supplies could be appropriate for portable devices applied in rehabilitation and assistance scenarios because of their lightweight and autonomy [11]. However, to ensure this portability, those machines need to have reduced sizes and low weights, resulting in a limited torque capacity provided by the system [9]. Consequently, actuators generally include gear mechanisms to enhance the torque capabilities and assist the human body's movements, although reducing the actuator speed response [11]. Nonetheless, the gears' inclusion also leads to non-backdrivable mechanisms, which affect the human–robot interaction [12].

Within the mechanical principles that use electrical power supplies, stiff actuators appear to be an efficient solution to assistive devices. Specifically, this actuation system exhibits relevant characteristics such as high provided torque and wide bandwidth, which are beneficial in assistance applications [13]. Notwithstanding, stiff actuators remain the non-passive backdrivability due to the gears system, resulting in a hard physical interaction [11–13]. Likewise, for human limbs that involve movements in multiple planes, designs with stiff actuators generally restrict several motions, inducing abnormal compensatory movements. Moreover, in terms of interaction, these actuators can present damages derivated from external forces (e.g., impacts or unexpected motions) during real applications [13].

In this context, wearable devices based on cable-driven mechanisms, series elastic actuators, and VSA are emerging to overcome the stiff systems' limitations and preserve the actuators in interaction scenarios. These mechanical principles include elements or mechanisms in the actuator's output to decouple the load, improving the human–robot interaction although reducing the system capacities [9]. This chapter is focused on the VSAs and their potential applications in gait rehabilitation scenarios. The first part explains the variable stiffness principle and several configurations and techniques to accomplish this behavior. The second part shows the T-FLEX exoskeleton's design based on VSA, and finally, the third part presents two experimental validations in gait assistance and stationary therapy.

7.2 Variable Stiffness Actuators

VSA's concept has arisen from the theory on (1) impedance and (2) Series Elastic Actuators (SEAs) [14], which were published by Hogan [15] and Pratt [13], respectively. Specifically, SEAs involve the spring elements' inclusion between the actuator and the load (see Fig. 7.1a). In this sense, series elastic elements can give back the actuators' lost qualities when it includes a gears system. Geared motors intend to increase the provided torque, reducing the motor speed. However, the gear system introduces characteristics such as (1) high friction, (2) backlash, (3) torque ripple, and (4) noise, affecting the device performance [11, 13].

SEAs work as a low-pass filter for shock loads, reducing peak gear forces [11]. This way, the interaction forces in assistive applications are dissipated mainly by the elastic element, preserving the actuator's mechanical structure. Likewise, this characteristic also affects the torque supplied by the actuator. However, the proper amount of elasticity can solve this drawback without limiting the absorption capability. In terms of control, SEAs turn the force in the impedance concept into a position control problem. Thus, the output force becomes proportional to the position difference across the series elasticity multiplied by its spring constant. Moreover, increased spring constants provide higher control stability, like in a stiff system, even though including the elastic elements' benefits [13].

Actuators based on variable stiffness follow the concept applied in SEAs, where the elastic element is included between the actuator and the load (see Fig. 7.1a). However, the difference lies in the variable impedance's inclusion for the actuator design [14]. This variable impedance allows deviating the equilibrium position (i.e., where the actuator generates zero force or torque) concerning the external forces and the actuators' mechanical properties [16, 17]. This way, VSAs include elastic elements whose spring magnitude takes different values conditioned by an active device (see Fig. 7.1b). Specifically, the adaptive stiffness can be achieved by (1) changing the spring preload, (2) varying the transmission ratio between the output link and the elastic elements, and (3) altering the spring's physical structure [17]. Therefore, this functionality allows adjusting the actuator's stiffness and adapting the device to a specific task [18].

Two setups have mainly been applied in devices that exhibit a variable stiffness behavior: (1) the agonist–antagonist, and (2) the independent motor setup [14]. In the agonist–antagonist principle, two motors modify the stiffness output (see Fig. 7.2b). This way, when actuators turn in the same direction, the output results in movement transmission. On the other hand, when the actuators turn opposite directions, the system exhibits the springs' co-contraction, changing the output stiffness [14, 18]. In terms of design, each motor-spring set is opposite to the other, and usually, the spring size is the same (see Fig. 7.2b).

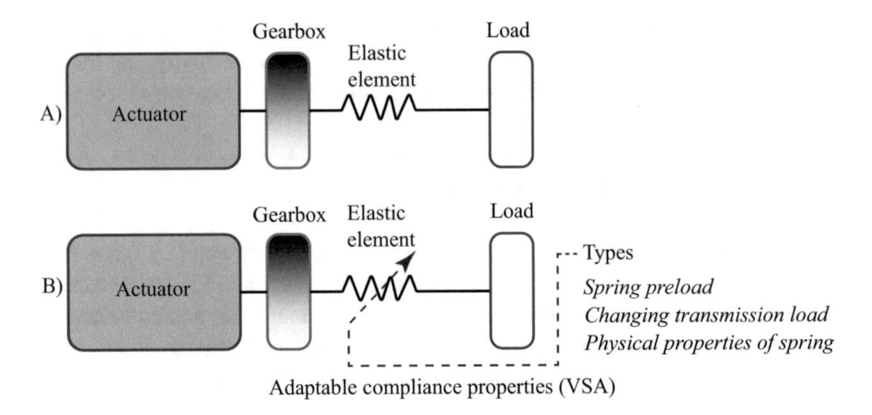

Fig. 7.1 Schematic representation of the actuators studied in this chapter (**a**) Series Elastic Actuators (SEAs) and (**b**) Variable Stiffness Actuators (VSAs) in a geared motor

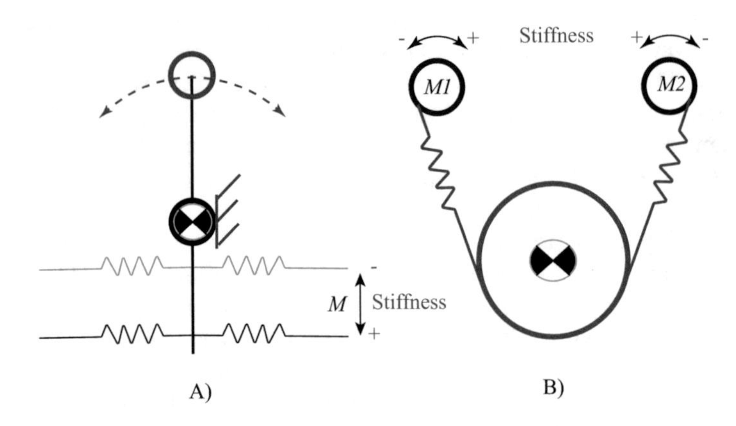

Fig. 7.2 Actuators configurations to generate a variable stiffness behavior. The left part (**a**) shows the independent motor configuration, and the right part (**b**) indicates the actuators' configuration for the agonist–antagonist principle

For a setup with independent motors, a motor varies the output position, resulting in a variable stiffness behavior (see Fig. 7.2a). In this sense, only one actuator is required to change the system's stiffness concerning the agonist–antagonist configuration. Likewise, that actuator is selected to achieve the needed power for this purpose, which is usually smaller than the primary system's actuator [14].

The positive effects for the agonist–antagonist configuration effects are related to both motors' contribution to the stiffness generation [18, 19]. Moreover, for systems that couple tendon-driven mechanisms, this configuration can quickly compensate the stiffness change when the length between the actuator and the joint varies. On the other hand, the independent motor setup ensures smaller and lighter devices because the actuator used to change the stiffness is selected only for this purpose

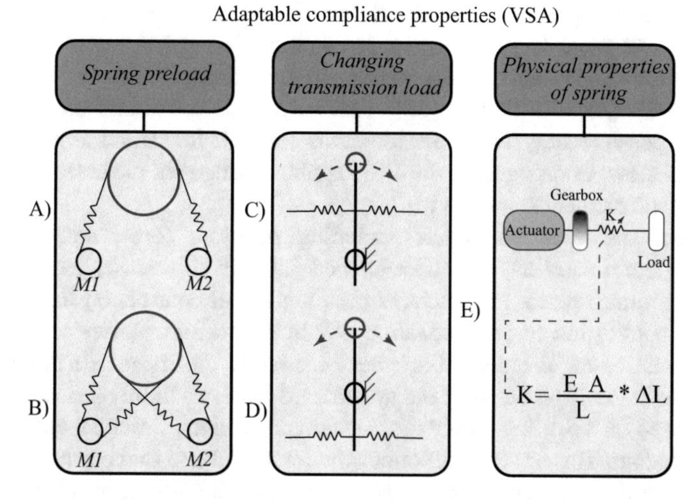

Fig. 7.3 Strategies implemented in a system to change the stiffness. The left part shows the spring preload strategy under agonist–antagonist configuration with two actuators (i.e., M1 and M2). The (**c**) and (**d**) parts illustrate the variable stiffness generated by a change in the transmission load. The right part shows the variation in stiffness through the spring's physical properties (i.e., K parameter shown in Eq. 7.1)

[19]. However, this configuration only uses one motor to move the joint, and consequently, generate the system output torque.

Following these setups, different strategies can be applied to VSAs, looking for changing the system's stiffness: (1) spring preload, (2) transmission load variation, and (3) spring's physical properties [14] (see Fig. 7.3).

In the spring preload (see Fig. 7.3a and b), the system response changes concerning the spring pretension. This way, the spring force is directly proportional to the spring displacement, accomplished by the actuator coupled in the system. However, to generate a variable stiffness behavior, energy should be stored in the springs and may not be retrievable. Therefore, an agonist–antagonists setup explained above is implemented, resulting in a large passive angular deflection [14]

For the transmission load variation, the distance between the output link and the spring element (i.e., the transmission ratio) varies continuously, leading to a variable stiffness effect (see Fig. 7.3c and d). Thus, when the springs are close to the pivot, the stiffness is less than when they are far. This strategy does not require energy to change the stiffness because the spring force is orthogonal to the spring displacement [14].

Finally, the variable stiffness strategy is achieved through the spring's physical properties (see Fig. 7.3e), which is derivated from the basic elasticity law as shown in Eq. 7.1.

$$F = \frac{EA}{L_0}\Delta L = K\Delta L, \qquad (7.1)$$

where F is the spring force, E the material modulus, A the cross-sectional area, L the effective beam length, and ΔL is the spring's displacement. Likewise, the stiffness K is defined by EA/L_0, to control the structural stiffness, any of these parameters should be modified [14]. This way, E is a material property whose change is possible only for some materials through temperature, although this variation is slow. Consequently, the VSA applies changes in the cross-section area and the elastic element's length [14].

Different low-level controllers including position, force, admittance, and impedance controllers have been developed [20]. In this sense, those strategies support elaborate control architectures (i.e., high-level controllers) focused on the actuator's application in real scenarios [9]. In general terms, devices applied to human scenarios require control architectures based on impedance controllers to improve the interaction between the system and the user. In this context, wearable devices based on VSA commonly use position controllers, which inherently results in torque output. This strategy simplifies the force problem to the actuator position problem as in SEAs, being more simple to implement and control [13]. However, considering the non-linearities exhibited by different VSA designs, this strategy could be challenging [16].

7.3 VSA in Rehabilitation Scenarios

Given the possibility of changing the system output stiffness in several interaction cases with the environment (e.g., constant load and constant position) [14], different device's performances can be achieved according to a particular topic. This characteristic is advantageous for VSAs in physical interaction scenarios, something that was a limitation for devices based on SEA [18].

Specifically, in dynamic scenarios, the human body changes its stiffness' properties to accomplish different tasks such as (1) limb movements, (2) shock absorptions, and (3) weight support. Hence, the wearable devices' designs focus on replicating human functions (i.e., bioinspired concepts), intending to improve the physical interaction [21]. Furthermore, spring's inclusion in a system, i.e., based on both SEAs and VSAs, preserves the actuators' mechanisms during scenarios with complex interaction forces (e.g., impacts, mechanical locking, and unexpected events against the actuators' movements) [22, 23]. Likewise, considering the changing stiffness' capability in VSA, the system response could be adapted to modify the user–device interaction during the same task.

In compliance terms, robotic devices applied to rehabilitation scenarios should guarantee safer and more natural interaction with the user [16]. This way, for gait rehabilitation, neurological patients often exhibit sporadic spasms and spastic events. Hence, when robotic devices implement stiff actuators, the response controller tries to correct the position errors generated by those movements. Thus, this correction could cause larger forces, injuring the user's limbs.

Considering the mentioned advantages, this actuation type is widely recommended for robotic applications where the robot interacts intensively with humans

[19,24]. However, despite the benefits of the human–machine interaction, the elastic materials' inclusion limits the actuator's features in terms of bandwidth, supplied torque, response time, among others [25].

Notwithstanding, devices based on VSA allow mitigating the effects caused by the spring's inclusion compared to SEAs. In this sense, higher pretensions lead to a system response similar to a stiff actuator without losing the benefits of spring's inclusion. Hence, system characteristics improve (i.e., increasing the bandwidth and supplied torque and decreasing the response time), and the compliance remains.

In the first part, this section presents different devices that integrate variable stiffness concepts to assist human movements in rehabilitation scenarios. On the other hand, the second part is focused on the T-FLEX ankle exoskeleton, which is a novel bioinspired device based on an agonist–antagonist configuration to achieve a variable stiffness behavior.

7.3.1 VSA in Wearable Robotics

In general terms, robotic devices applied to rehabilitation scenarios arisen from the promising results of including robotics in these applications: (1) neuroplasticity induction, (2) improvement of motor recovery, and (3) regaining functional independence [26–28]. The current developments have been focused on designing soft structures and compliant actuators to guarantee a proper and safer user–device interaction [22]. Specifically, as mentioned previously, wearable devices exhibit benefits regarding platform-based systems regarding a multi-functionality characteristic. Those benefits are mainly related to potential applications of wearable robots in both gait assistance and stationary therapy.

On the other hand, considering the advantages presented in the previous section, wearable devices based on VSA have also shown high potential in rehabilitation scenarios, from the bioinspired operation principle and the system response variation regarding its configuration. In this context, one of the most representative actuators based on VSA concepts is MACCEPA. This actuator consists of 3 bodies pivoting around a common rotation axis whose working principle is a torsion spring able to control its equilibrium position and joint stiffness, independently [23, 29]. Thus, this actuator's design has evidenced high assistance torque levels (i.e., from 50 Nm), comprehensive frequency response (i.e., close to 30 Hz), and consequently, potential use in gait rehabilitation scenarios [23, 30]. Several assistive devices have applied this principle to interact with patients in those scenarios [23, 31–34].

Another actuator focused on gait assistance based on VSA is ARES. This design exhibits relevant characteristics such as lightweight compared to other VSAs, faster response to change the device's stiffness, and considerable assistance torque levels (i.e., providing up to 70 Nm) [30]. ARES includes a stiff set coupled with a compliant mechanism able to change the stiffness using a DC motor. The device aims at controlling the joints' equilibrium position, as occurs in the MACCEPA actuator [35]. ARES has been implemented in the ATLAS exoskeleton, a pediatric device with actuation in the hip, knee, and ankle joints [35, 36].

As well as the previous designs, actuators based on a changing transmission load have also been applied in wearable devices [30, 37]. Specifically, AwAS-II and CompAct-VSA were developed and implemented in lower-limb exoskeletons, including two springs antagonistically attached to the lever that can move toward or away from the pivot [37–39]. In the capacity terms, AwAS-II provides a torque range of 80 Nm with a faster response to adjust the stiffness (i.e., 0.8 s), and CompAct-VSA registers a high torque capacity (i.e., up to 117 Nm) with a faster change stiffness response [30].

In this sense, all of those presented actuators remark the capabilities of devices based on VSA in human interaction scenarios. Notwithstanding, mechanical structures to fix the actuator to the human joints are usually rigid structures that block the joints' movements on the different assisted planes. The following section presents an agonist–antagonist configuration of a VSA ankle exoskeleton with a fully compliant structure.

7.3.2 T-FLEX Ankle Exoskeleton

T-FLEX is a wearable and portable ankle exoskeleton that is part of the AGoRa lower-limb exoskeleton [40]. This device can operate independently (i.e., supporting the ankle movements) or cooperatively with the AGoRA exoskeleton (i.e., assisting the hip, knee, and ankle joints). T-FLEX integrates VSA concepts in its mechanical principle to support the dorsi-plantarflexion movements without restricting the other ankle motions (i.e., foot rotations, inversion–eversion, and pronation–supination) [41]. This exoskeleton integrates two servomotors attached to elastic elements whose mechanical behavior is similar to the human Achilles tendon [42]. It uses an agonist–antagonist configuration with a bidirectional movement mechanism to assist the ankle motions on the sagittal plane (see Fig. 7.4). The principal torque transmission is generated by a composite elastic element that attaches the motor and the user's foot. Moreover, crossed stiff filaments involve both actuators in the torque output, as Fig. 7.4 shows.

The variable stiffness system intends to change the user–device interaction according to the application (i.e., dorsi-plantarflexion repetitions in stationary therapy and gait phases during walking assistance). Likewise, the spring's inclusion also allows modifying the interaction with the patients from their motor capabilities (i.e., increasing initial pretension for a weak ankle or decreasing this value for a spastic ankle). For this purpose, the device includes a composite tendon whose mechanical behavior, tested in stress trials, is similar to the human Achilles tendon (i.e., Young's modulus between 500–1800 Mpa) [42]. The tendon braids flexible materials (i.e., thermoplastic elastomer and fibers of polyethylene) and stiff filaments (i.e., polytetrafluoroethylene) to achieve an exponential stress–strain curve.

The exoskeleton integrates a soft structure, where two actuators are placed on the user's shank (i.e., anterior and posterior sides), as Fig. 7.4 shows. Thus, 3D-printed pieces of polylactic acid (PLA) support the T-FLEX's actuators. Moreover, flexibles interfaces of polyurethane-coated, coupled to the support system, improve the

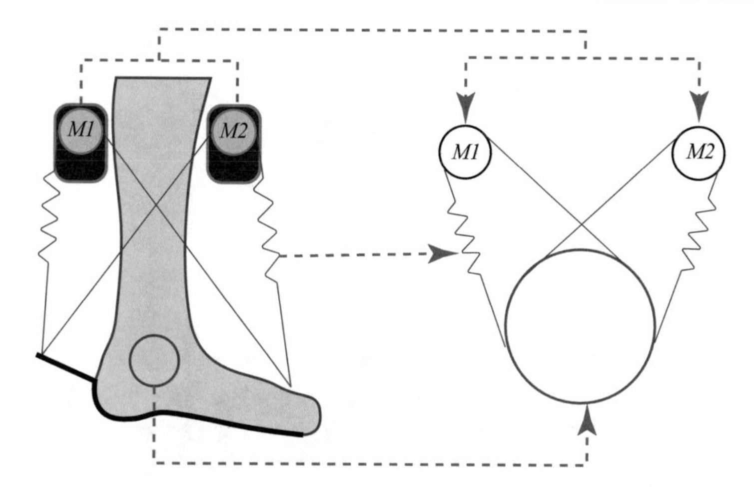

Fig. 7.4 Variable stiffness configuration applied in the T-FLEX exoskeleton. The right part shows the agonist–antagonist setup, including spring elements and stiff filaments. The left part shows this concept implemented on a user's limb

device-user physical interaction. This way, the device allows portable applications, avoiding slipping and reducing pressure points on the limb related to reaction forces when the device actuates. On the other hand, T-FLEX includes elastic elements and stiff filaments to transmit torque from actuators to the user's foot. Hence, the exoskeleton uses an insole adapted with 3D-printed pieces to attach the composite tendons to the foot (i.e., heel for plantarflexion and metatarsals for dorsiflexion), as Fig. 7.5. This way, the device comprehends a four-bar mechanism by each actuator, where one of them is a spring with variable stiffness (see Fig. 7.5).

In the electronic system context, the device has two smart servomotors, an inertial sensor, and a processing unit coupled within the open-source robotic meta-operating system (ROS). The actuators are smart servomotors Dynamixel MX106-T (Robotis, Korea) placed on the user's affected shank (see Fig. 7.6). Each actuator has a stall torque of 10 Nm with a maximum no-load speed of 55 rpm for a power supply of 14.8 V.

The sensing system integrates an Inertial Measurement Unit (IMU) BNO055 (Bosch, Germany) placed on the foot tip. This sensor runs to 60 Hz, using the angular velocity and acceleration to trigger the device. Specifically, for gait assistance, an algorithm based on machine learning estimates the user's gait phases in real-time [43]. Notwithstanding, for the stationary therapy, a statistical algorithm determines the user's movement intention.

For the processing, the exoskeleton uses a Raspberry Pi 3 Board under a Debian operating system. This computer acquires sensor information, runs the control algorithms, and sends the control commands to the actuators. Finally, in the power supply context, the device has a LiPo battery of 4000 mah 4S 14.8V 30C, which allows an autonomy close to 4 h in non-extreme conditions (i.e., high level of spasticity and excessive strain on the tendons or stiff filaments).

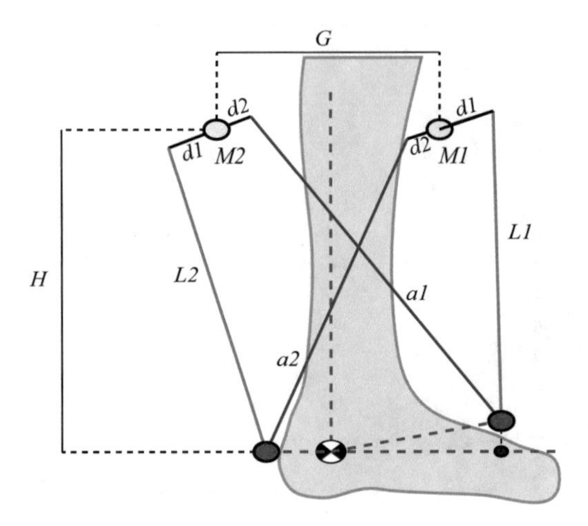

Fig. 7.5 Mechanical diagram of the T-FLEX exoskeleton implemented in the user's limb conditioned to the anthropometric measurements. The device includes two actuators, i.e., anterior (M1) and posterior (M2), placed on the shank and distanced by the user's body compositions (G). The exhibited parameters are elastic elements' length (i.e., L1 for the anterior tendon and L2 for the posterior), stiff element' lengths (i.e., a1 for the filament attached from M2 to the foot tip, and a2 for the filament from the M1 to the heel), the distance between the actuators and the ankle joint (H), and finally, the attachment systems' dimensions for the elastic and stiff elements, i.e., d1 and d2, respectively

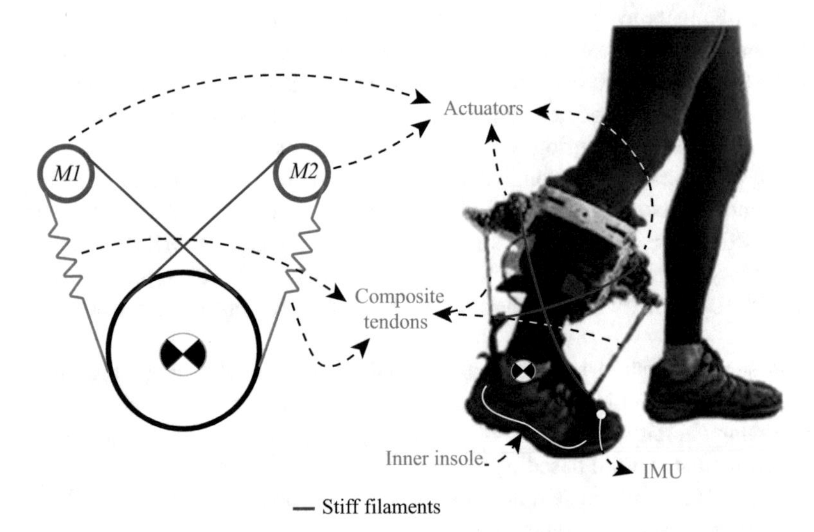

Fig. 7.6 Real prototype of the T-FLEX ankle exoskeleton based on a variable stiffness principle

Finally, T-FLEX has a mechatronic model divided into two main parts: (1) actuators controller and (2) mechanical design's effect. The first part refers to the internal PID controller implemented on each motor. This controller has as input a goal motor position converted to profiles of velocity and acceleration. The PID controller calculates the PWM output based on those profiles. Finally, an inverter supplies the PWM value to the actuator and an encoder closes the control loop.

The second part covers effects due to the elastic elements' inclusion. Considering the VSA's characteristics, the system response has a dependency on the spring behavior in terms of provided torque and the system's bandwidth. Thus, the composite tendon included in T-FLEX leads to limited bandwidth and actuators' torque reduction. Likewise, these effects depend on each tendon's pretension, during the initial configuration. On the other hand, from the T-FLEX's mechanical design shown in Fig. 7.5, the provided torque on the ankle increases proportionally concerning the user's foot length. Moreover, regarding the assisted movements, the torque in dorsiflexion is more significant than in the plantarflexion movement. This characteristic is because the distance between the ankle and the foot part, where the torque is transmitted is different (see Fig. 7.5). The torque for each movement can be simplified as shown in Eqs. 7.2 and 7.3.

$$\tau_{dorsiflexion} = (F_{M1} - F_{tendon}) \cdot \cos\alpha_1 \cdot L_m + F_{M2} \cdot \cos\alpha_2 \cdot L_m \tag{7.2}$$

$$\tau_{plantarflexion} = (F_{M2} - F_{tendon}) \cdot \cos\alpha_1 \cdot L_h + F_{M1} \cdot \cos\alpha_2 \cdot L_h, \tag{7.3}$$

where L_m and L_h are the distances between the attached tendon position and the ankle joint, $alpha_1$ and $alpha_2$ are the angles between the attached tendon position and the motor lever arm, F_{tendon} is the loss force related to the elastic element, and F_{M1} and F_{M2} are the provided forces by the motors expressed as

$$F_{1,2} = \frac{\tau_{motor(1,2)}}{d_{1,2}}. \tag{7.4}$$

Considering the mechanical and electronic design of T-FLEX presented previously, the following sections show two preliminary validations of the device in real scenarios. This way, from the wearable robotics' multi-functionality, these validations include experiments in both applications: (1) stationary therapy and (2) gait assistance with healthy people.

7.4 Experimental Validations of the T-FLEX

Wearable devices based on VSA exhibit advantages for rehabilitation and assistive scenarios, as previous sections stated. Specifically, those devices show benefits in aspects such as (1) multimodality, (2) variable physical interaction adjustable to the user performance, and (3) actuation systems based on the human body. Considering these advantages, VSAs are widely recommended for robotic applications where the device interacts intensively with humans [19, 24]. Specifically, studies have evidenced metabolic cost' reductions related to the changing stiffness level of the actuator [24].

On the other hand, preliminary studies have shown the significant potential of the T-FLEX ankle exoskeleton in rehabilitation scenarios for (1) stationary therapy and (2) gait assistance. For the therapy, a stroke survivor evidenced improvements in the ankle kinematics and spatiotemporal parameters after a rehabilitation process during 18 sessions [44]. Likewise, in terms of gait assistance, the study exhibited relevant outcomes related to the lower-limb kinematics when stroke patients wore the T-FLEX's actuation system [45].

In this context, this section presents two validations of the T-FLEX ankle exoskeleton in rehabilitation scenarios (i.e., stationary therapy and gait assistance). This device is a wearable and portable powered ankle–foot orthosis that applies bioinspiration concepts based on a variable stiffness actuators principle under an agonist–antagonist configuration (see Fig. 7.6), as extensively presented in the previous section.

7.4.1 T-FLEX in Gait Assistance

Considering the T-FLEX's applications, this section presents the experimental validation of this ankle exoskeleton for assistive applications in gait. Thus, the proposed protocol aimed at assessing the first-use condition with T-FLEX to analyze the effect on the user's kinematics during walking over a treadmill.

The experimental validation consisted of assessing the T-FLEX's first-use effect in a healthy subject. This way, the protocol included three modes: (1) no device, (2) unpowered, and (3) powered. For the three modalities, an Electromyography (EMG) sensor (Shimmer, Ireland) measured the muscular activity on the participant's gastrocnemius and tibialis anterior muscles (see Fig. 7.7). Likewise, a G-Walk sensor (BTS Bioengineering, Italy), placed on L5, estimated the spatiotemporal parameters during the trials. Finally, an inertial sensor (Shimmer, Ireland) measured the ankle's kinematics in the actuated limb side. This sensor was located on the participant's foot tip, as Fig. 7.7 shows.

The participant performed three trials of 6 minutes over a treadmill: (1) no device, (2) unpowered, and (3) powered. Moreover, a previous stage, where the volunteer accomplished three 10-meter walk tests to estimate the average walking speed, was included. Thus, the treadmill was configured to this speed value for the different executed modalities. In the first trial, the participant walked over the treadmill without wearing the device. The data acquired in this modality was used as the reference for the other modalities. For the second trial, the user wore the T-FLEX exoskeleton, although the actuators were deactivated. Finally, in the third trial, the participant walked with the device powered and assisting the user gait phases.

This experiment also included an additional calibration stage to adjust the device's mechanical structure to the user's anthropometric measurements for the assisted gait mode. Likewise, this stage allowed configuring the T-FLEX's movements concerning the user's range of motion (ROM).

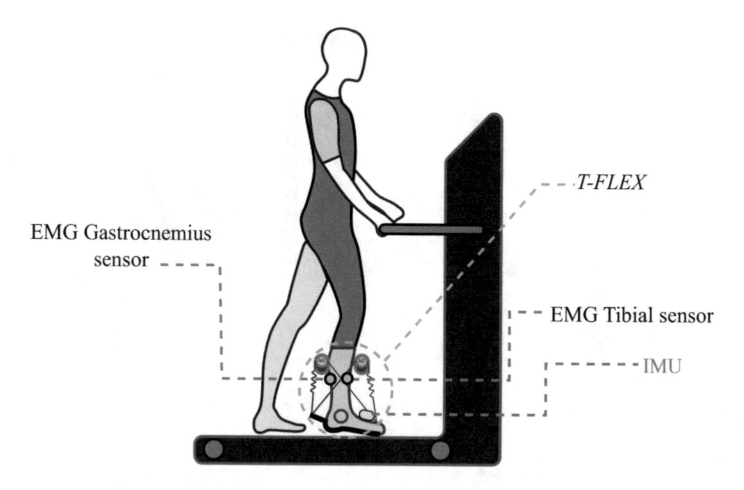

Fig. 7.7 Experimental setup for the validation in a gait assistance application on the treadmill

Data processing was performed offline using the MATLAB software (MathWorks, 2018b, USA) and information acquired through the rosbag package within the ROS operating system framework. Thus, for the acquisition and processing, the HP Pavilion Gaming laptop (IntelCore i5-8300H, CPU@2.30 GHz, Taiwan) was used, running Windows 10 Home. On the one hand, in terms of kinetic parameters, the G-Studio software (BTS Bioengineering, USA) estimated the user performance during trials. On the other hand, for the EMG information, a band-pass filter was applied to remove noise.

The most relevant kinematic parameters are summarized in Fig. 7.8. The device's inclusion in the passive mode shows a decrease in the ankle's ROM. Specifically, the dorsiflexion and plantarflexion exhibited reductions greater than 10% concerning the no-device condition.

This behavior responds to the spring's pretension in the variable stiffness system implemented in T-FLEX. This way, although the mechanical structure does not restrict the user's movements completely, the interaction force between the tendons and the foot is enough to reduce the dorsi-plantarflexion during gait. Consequently, the cadence and the gait cycle exhibited slight changes, and the step length showed increases to compensate for the mentioned reductions. In contrast, when the T-FLEX exoskeleton assisted the user's gait, the dorsi-plantarflexion movements increased concerning the baseline state. Specifically, these movements exhibited increases from 10% to 16%, which could be related to suitable foot-ground contact on the heel strike phase. Likewise, the dorsiflexion's variation for this mode indicates improvements in foot clearance during the swing phase, resulting in a fall risk reduction [44].

On the other hand, this modality also showed reductions in cadence and step length, increasing the user's gait cycle duration. These variations could be associated with the training's lack in the experimental procedure. Thus, multiple sessions with

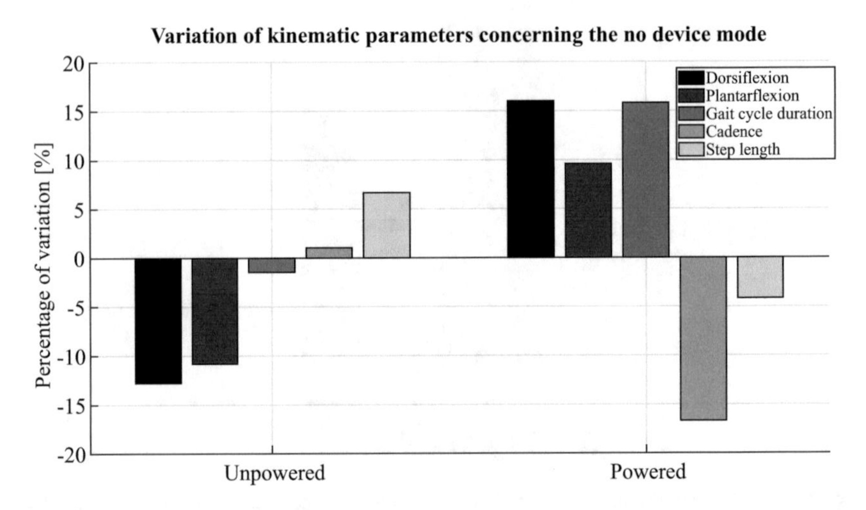

Fig. 7.8 Percentage variation of the kinematic parameters registered during the assistance with T-FLEX. The values were estimated concerning the baseline (i.e., no-device condition). Positive values indicate increases in the parameter, and negative values evidence decreases

the exoskeleton could improve the user–device adaptability until achieving at least values determined in the baseline state.

In the muscular activity context, Fig. 7.9 shows the variations in the gastrocnemius and tibialis anterior muscles. In general terms, the unpowered condition exhibited no significant changes (i.e., less than 10%) in the measured electrical activity concerning the no-device condition. Thus, it could be inferred that the T-FLEX exoskeleton does not cause an additional effort in the patient related to the device's weight and the mechanical structure.

On the other hand, the powered mode (i.e., T-FLEX assisting user gait) showed (1) an increase in the tibialis anterior and (2) a reduction in the gastrocnemius muscle (see Fig. 7.9). This way, these variations relate to the variable stiffness system's effects and the device's assistance capacity. Specifically, the change in the dorsiflexion movement could increase the tibialis anterior's electrical activity. Moreover, the EMG's reduction in the gastrocnemius could indicate that T-FLEX assisted the leg propulsion during the toe-off phase.

7.4.2 T-FLEX in a Stationary Scenario

This section shows a preliminary validation of the device under a stationary condition from the T-FLEX's applicability in rehabilitation scenarios. This way, the proposed experimental protocol was intended to measure the muscular activity response in a healthy subject using the exoskeleton.

The experimental protocol's goal intended to assess the device's effect during a stationary therapy scenario, measuring the user's EMG and ROM (see Fig. 7.10). To

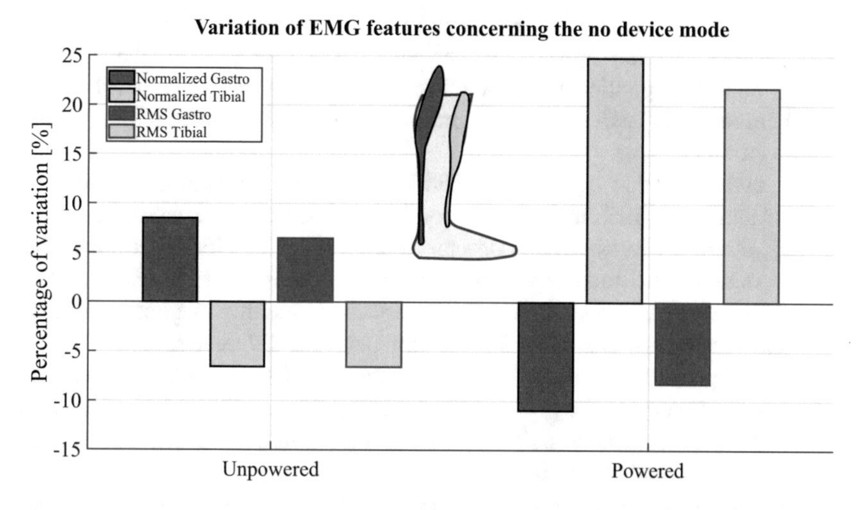

Fig. 7.9 Percentage variation of the electrical activity measured on the gastrocnemius and tibialis anterior muscle during the gait assistance. The values were estimated concerning the no-device condition. Positive values indicate an increase in the EMG, and negative values a decrease

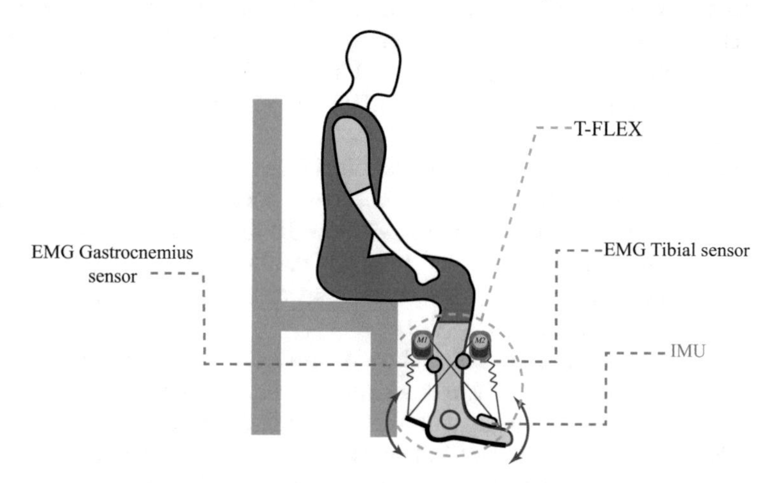

Fig. 7.10 Experimental setup proposed to assess the T-FLEX exoskeleton in a stationary application

this end, this study enrolled a healthy participant who had no exhibited orthopedic, metabolic, or neurological impairment that could modify his muscular activity. The participant was equipped with electrodes located on the tibialis anterior and gastrocnemius muscles on their dominant side. Likewise, two inertial sensors (Shimmer, Ireland) were placed on their foot instep and at the shank (i.e., 10 cm on the tibia proximal to the ankle joint). Subsequently, the user wore the T-FLEX exoskeleton as Fig. 7.10 shows.

For this study, T-FLEX was configured in therapy mode, where the device assists the dorsi-plantarflexion movements concerning the user's ROM, the actuators' velocity, and the repetition frequency. Moreover, the participant was asked to sit in a 90-degree knee flexion with his dominant lower limb elevated without contact with the ground.

This study included one session divided into two modalities: (1) no-device condition and (2) T-FLEX assisting the dorsi-plantarflexion movements. For the first modality, the user accomplished continuous dorsi-plantarflexion repetitions with a self-determined speed and a repetition frequency of 0.8 Hz. This trial was used as a baseline for the other assessed modality. The second modality integrated three tests where the device assisted the ankle movements with different speeds. Thus, the continuous repetitions had an actuator's velocity of 30% (low), 50% (medium), and 100% (high) concerning the maximum device's speed (i.e., 55 rpm for the no-load condition). Likewise, the set-point commands were sent to the actuators to achieve a repetition frequency of 0.8 Hz.

Data processing was performed offline using MATLAB software (MathWorks, 2018b, USA) and an Asus VivoBook S15 S510UA (IntelCore i5-8250U, CPU@1.80 GHz, Taiwan) running Windows 10 Home. For the processing, a band-pass filter removed the atypical values and noise in the EMG signals. Subsequently, the signals were rectified (i.e., from the absolute values), and data smoothing was performed using a 100 ms motion average window. Finally, to provide information about the EMG signal's amplitude during the trials, the root-mean-square (RMS) was calculated.

The muscular activity measured on the tibialis anterior evidenced a significant decrease for all assisted trials (i.e., low, medium, and high), as Fig. 7.11 shows. Specifically, the exhibited reduction was 93% for the low velocity, being the maximum value obtained in the experiment. However, the muscular activity measured on the gastrocnemius muscle presented its maximum variation in this trial (i.e., 270% concerning the baseline state).

On the other hand, medium and high velocities led to a shorter increase in the gastrocnemius' electrical activity, although the tibialis anterior evidenced a slight increase compared to the low speed. This way, the device's speed evidenced a relevant impact on the user's electrical activity. In general, when T-FLEX assisted the movements with a low velocity, the participant registered significant increases in gastrocnemius' activity. However, as the actuators' speed is increased, the electrical activity reached similar values obtained in the baseline state. In contrast, the tibialis anterior showed a significant decrease for all trials, which could be related to the user's posture and the T-FLEX's assistance capacity.

In the user's ROM context, Fig. 7.12 shows the values obtained in the sagittal plane for the baseline (passive) and assisted motion (active) modalities. The user's ROM exhibited a decrease of 11.9% when T-FLEX supported the dorsi-plantarflexion in the different velocities.

This result can be explained by the elongation of the spring while the device assists the movement (see Fig. 7.12), resulting in a position set-point loss. Therefore, a high level of pretension could improve the reduced ROM registered by the device.

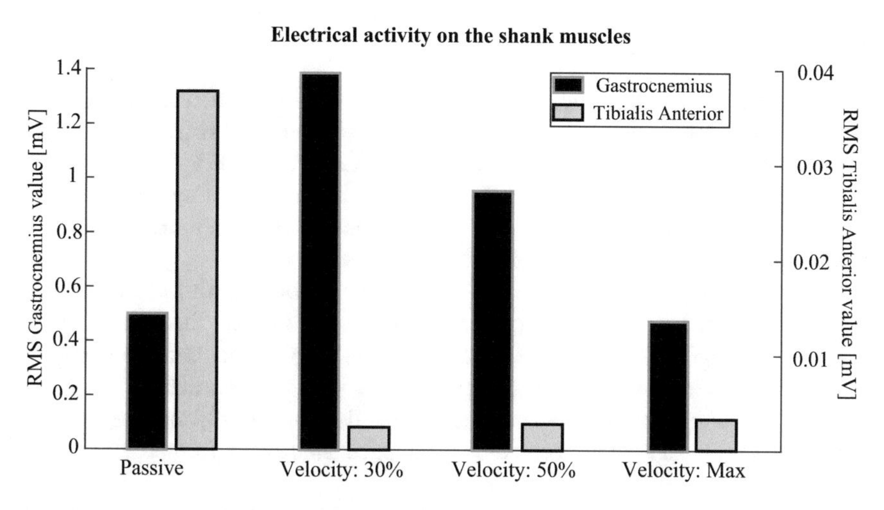

Fig. 7.11 RMS values of the EMG signals acquired for tibialis anterior and gastrocnemius muscles in the proposed trials. The velocity's percentage was configured concerning the actuators' maximum speed (i.e., 55 rpm in a no-load state)

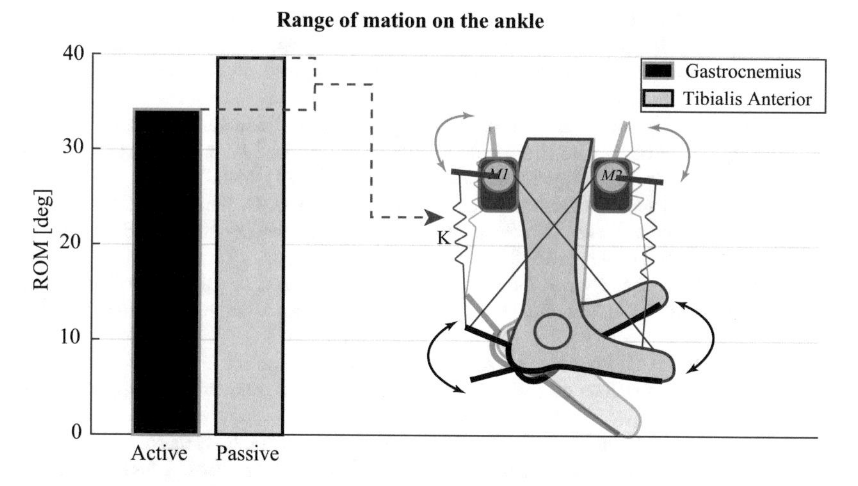

Fig. 7.12 Range of Motion on the ankle joint for the sagittal plane during the assessed modalities, i.e., active and passive. The left part shows the values for the dorsi-plantarflexion movements, and the right part shows the effect on these movements related to the element's elasticity (K)

Additionally, this reduction can also be related to the calibration methodology used for this experiment, i.e., storing the user's ROM and employing these values in the exercise. Hence, an automated calibration, including a sensor to measure the user's current state, could improve these values.

7.5 Conclusions

This chapter presented an overview of the variable stiffness actuators (VSAs) in terms of principles, setups, and characteristics applied to assistive applications. In this sense, it also presented an ankle exoskeleton T-FLEX based on VSA, focusing on its mechanical design and operating principles during gait assistance and stationary therapy. Likewise, this chapter showed two preliminary studies of healthy participants using T-FLEX in these scenarios.

In conclusion, devices based on VSA evidence advantages in assistive applications compared to other actuation mechanisms. Moreover, in a rehabilitation context, the application of these systems, integrating bioinspired concepts and modifying its performance concerning the user's capabilities, enables to improve the physical interaction. However, aspects as the design, control, and integration of these devices in real scenarios could be challenged and require further research.

References

1. M. Zhang, T.C. Davies, S. Xie, Effectiveness of robot-assisted therapy on ankle rehabilitation—a systematic review. J. NeuroEng. Rehab. **10**(1), 30 (2013)
2. G. Onose, V. Cârdei, Ş.T. Crăciunoiu, V. Avramescu, I. Opriş, M.A. Lebedev, M.V. Constantinescu, Mechatronic wearable exoskeletons for bionic bipedal standing and walking: a new synthetic approach. Front. Neurosci. **10**, 343 (2016)
3. J.-M. Belda-Lois, S. Mena-del Horno, I. Bermejo-Bosch, J.C. Moreno, J.L. Pons, D. Farina, M. Iosa, M. Molinari, F. Tamburella, A. Ramos. et al., Rehabilitation of gait after stroke: a review towards a top-down approach. J. NeuroEng. Rehab. **8**(1), 66 (2011)
4. M. Moltedo, T. Baček, T. Verstraten, C. Rodriguez-Guerrero, B. Vanderborght, D. Lefeber, Powered ankle-foot orthoses: the effects of the assistance on healthy and impaired users while walking. J. NeuroEng. Rehab. **15**, 86 (2018)
5. B. Chen, H. Ma, L.-Y. Qin, F. Gao, K.-M. Chan, S.-W. Law, L. Qin, W.-H. Liao, Recent developments and challenges of lower extremity exoskeletons. J. Orthopaedic Transl. **5**, 26–37 (2016)
6. H.P. Crowell III, A.C. Boynton, M. Mungiole, Exoskeleton power and torque requirements based on human biomechanics, tech. rep., Army Research Lab Aberdeen Proving Ground Md, 2002
7. A. Weerasingha, W. Withanage, A. Pragnathilaka, R. Ranaweera, R. Gopura, Powered Ankle exoskeletons: Existent designs and control systems. Proc. Int. Conf. Artif. Life Robot. **23**, 76–83 (2018)
8. M. Alam, I.A. Choudhury, A.B. Mamat, Mechanism and design analysis of articulated ankle foot orthoses for drop-foot. Sci. World J. **2014**, 1–14 (2014)
9. S. Sierra, L. Arciniegas, F. Ballen-Moreno, D. Gomez-Vargas, M. Munera, C.A. Cifuentes, Adaptable robotic platform for gait rehabilitation and assistance: Design concepts and applications. Exoskeleton Robot. Rehab. Healthcare Dev. 67–93 (2020)
10. A. Petcu, M. Georgescu, D. Tarniţă, Actuation systems of active orthoses used for gait rehabilitation. Appl. Mech. Mater. **880**, 118–123 (2018)
11. K. Kong, J. Bae, M. Tomizuka, A compact rotary series elastic actuator for human assistive systems. IEEE/ASME Trans. Mechatron. **17**(2), 288–297 (2011)
12. S. Toxiri, A. Calanca, T. Poliero, D.G. Caldwell, J. Ortiz, Actuation requirements for assistive exoskeletons: Exploiting knowledge of task dynamics, in *International Symposium on Wearable Robotics* (Springer, 2018), pp. 381–385

13. G.A. Pratt, M.M. Williamson, Series elastic actuators, in *Proceedings 1995 IEEE/RSJ International Conference on Intelligent Robots and Systems. Human Robot Interaction and Cooperative Robots*, vol. 1 (IEEE, 1995), pp. 399–406
14. S. Wolf, G. Grioli, O. Eiberger, W. Friedl, M. Grebenstein, H. Höppner, E. Burdet, D.G. Caldwell, R. Carloni, M.G. Catalano, et al., Variable stiffness actuators: Review on design and components. IEEE/ASME Trans. Mechatron. **21**(5), 2418–2430 (2015)
15. N. Hogan, Impedance control: An approach to manipulation, in *1984 American Control Conference* (IEEE, 1984), pp. 304–313
16. R. Van Ham, T.G. Sugar, B. Vanderborght, K.W. Hollander, D. Lefeber, Compliant actuator designs. IEEE Robot. Autom. Mag. **16**(3), 81–94 (2009)
17. B. Vanderborght, A. Albu-Schäffer, A. Bicchi, E. Burdet, D. G. Caldwell, R. Carloni, M. Catalano, O. Eiberger, W. Friedl, G. Ganesh. et al., Variable impedance actuators: A review. Robot. Autonom. Syst. **61**(12), 1601–1614 (2013)
18. F. Petit, M. Chalon, W. Friedl, M. Grebenstein, A. Albu-Schäffer, G. Hirzinger, Bidirectional antagonistic variable stiffness actuation: Analysis, design & implementation, in *2010 IEEE International Conference on Robotics and Automation* (IEEE, 2010), pp. 4189–4196
19. P.K. Jamwal, S. Hussain, S.Q. Xie, Review on design and control aspects of ankle rehabilitation robots. Disability Rehab. Assist. Tech. **10**, 93–101 (2013)
20. L. Marchal-Crespo, D.J. Reinkensmeyer, Review of control strategies for robotic movement training after neurologic injury. J. NeuroEng. Rehab. **6**, 20 (2009)
21. E. Rocon, J.L. Pons, *Exoskeletons in Rehabilitation Robotics: Tremor Suppression*, vol. 69 (Springer, 2011)
22. M.D.C. Sanchez-Villamañan, J. Gonzalez-Vargas, D. Torricelli, J.C. Moreno, J.L. Pons, Compliant lower limb exoskeletons: A comprehensive review on mechanical design principles. J. NeuroEng. Rehab. **16**(1), 1–16 (2019)
23. M. Moltedo, G. Cavallo, T. Baček, J. Lataire, B. Vanderborght, D. Lefeber, C. Rodriguez-Guerrero, Variable stiffness ankle actuator for use in robotic-assisted walking: Control strategy and experimental characterization. Mech. Mach. Theory **134**, 604–624 (2019)
24. M. Moltedo, T. Baček, B. Serrien, K. Langlois, B. Vanderborght, D. Lefeber, C. Rodriguez-Guerrero, Walking with a powered ankle-foot orthosis: the effects of actuation timing and stiffness level on healthy users. J. NeuroEng. Rehab. **17**(1), 1–15 (2020)
25. G. Carpino, D. Accoto, F. Sergi, N. Luigi Tagliamonte, E. Guglielmelli, A novel compact torsional spring for series elastic actuators for assistive wearable robots. J. Mech. Des. **134**, 1–10 (2012)
26. M.A. Dimyan, L.G. Cohen, Neuroplasticity in the context of motor rehabilitation after stroke. Nature Rev. Neurol. **7**(2), 76–85 (2011)
27. L.R. Sheffler, J. Chae, Technological advances in interventions to enhance poststroke gait. Phys. Med. Rehab. Clin. North America **24**, 305–323 (2013)
28. T. Mikolajczyk, I. Ciobanu, D.I. Badea, A. Iliescu, S. Pizzamiglio, T. Schauer, T. Seel, P.L. Seiciu, D.L. Turner, M. Berteanu, Advanced technology for gait rehabilitation: An overview. Adv. Mech. Eng. **10**(7), 1–19 (2018)
29. R. Van Ham, B. Vanderborght, M. Van Damme, B. Verrelst, D. Lefeber, MACCEPA, the mechanically adjustable compliance and controllable equilibrium position actuator: Design and implementation in a biped robot. Robot. Autonom. Syst. **55**(10), 761–768 (2007)
30. M. Cestari, D. Sanz-Merodio, J.C. Arevalo, E. Garcia, An adjustable compliant joint for lower-limb exoskeletons. IEEE/ASME Trans. Mechatron. **20**(2), 889–898 (2014)
31. T. Bacek, M. Moltedo, K. Langlois, G.A. Prieto, M.C. Sanchez-Villamañan, J. Gonzalez-Vargas, B. Vanderborght, D. Lefeber, J.C. Moreno, Biomot exoskeleton—towards a smart wearable robot for symbiotic human-robot interaction, in *2017 International Conference on Rehabilitation Robotics (ICORR)* (IEEE, 2017), pp. 1666–1671
32. P. Cherelle, V. Grosu, P. Beyl, A. Mathys, R. Van Ham, M. Van Damme, B. Vanderborght, D. Lefeber, The MACCEPA actuation system as torque actuator in the gait rehabilitation robot ALTACRO, in *2010 3rd IEEE RAS & EMBS International Conference on Biomedical Robotics and Biomechatronics* (IEEE, 2010), pp. 27–32

33. B. Brackx, J. Geeroms, J. Vantilt, V. Grosu, K. Junius, H. Cuypers, B. Vanderborght, D. Lefeber, Design of a modular add-on compliant actuator to convert an orthosis into an assistive exoskeleton, in *5th IEEE RAS/EMBS International Conference on Biomedical Robotics and Biomechatronics* (IEEE, 2014), pp. 485–490
34. K. Junius, B. Brackx, V. Grosu, H. Cuypers, J. Geeroms, M. Moltedo, B. Vanderborght, D. Lefeber, Mechatronic design of a sit-to-stance exoskeleton, in *5th IEEE RAS/EMBS International Conference on Biomedical Robotics and Biomechatronics* (IEEE, 2014), pp. 945–950
35. M. Cestari, D. Sanz-Merodio, J.C. Arevalo, E. Garcia, Ares, a variable stiffness actuator with embedded force sensor for the atlas exoskeleton. Ind. Robot Int. J. **41**(6), 518–526 (2014)
36. J. Sancho-Perez, M. Perez, E. Garcia, D. Sanz-Merodio, A. Plaza, M. Cestari, Mechanical description of atlas 2020, a 10-dof paediatric exoskeleton, in *Advances in Cooperative Robotics* (World Scientific, 2017), pp. 814–822
37. N.G. Tsagarakis, I. Sardellitti, D.G. Caldwell, A new variable stiffness actuator (compAct-VSA): Design and modelling, in *2011 IEEE/RSJ International Conference on Intelligent Robots and Systems* (IEEE, 2011), pp. 378–383
38. A. Jafari, N.G. Tsagarakis, D.G. Caldwell, AwAS-II: A new actuator with adjustable stiffness based on the novel principle of adaptable pivot point and variable lever ratio, in *2011 IEEE International Conference on Robotics and Automation* (IEEE, 2011), pp. 4638–4643
39. A. Jafari, N.G. Tsagarakis, I. Sardellitti, D.G. Caldwell, A new actuator with adjustable stiffness based on a variable ratio lever mechanism. IEEE/ASME Trans. Mechatron. **19**(1), 55–63 (2012)
40. M. Sánchez-Manchola, D. Gómez-Vargas, D. Casas-Bocanegra, M. Múnera, C. A. Cifuentes, Development of a robotic lower-limb exoskeleton for gait rehabilitation: Agora exoskeleton, in *2018 IEEE ANDESCON* (IEEE, 2018), pp. 1–6
41. M. Manchola, D. Serrano, D. Gómez, F. Ballen, D. Casas, M. Munera, C.A. Cifuentes, T-flex: Variable stiffness ankle-foot orthosis for gait assistance, in *International Symposium on Wearable Robotics* (Springer, 2018), pp. 160–164
42. J. Casas, A. Leal Junior, C. Díaz, A. Frizera, M. Munera, C. Cifuentes, Large-range polymer optical-fiber strain-gauge sensor for elastic tendons in wearable assistive robots. Materials **12**, 1443 (2019)
43. M.D. Sánchez Manchola, M.J.P. Bernal, M. Munera, C.A. Cifuentes, Gait phase detection for lower-limb exoskeletons using foot motion data from a single inertial measurement unit in hemiparetic individuals. Sensors **19**(13), 2988 (2019)
44. D. Gomez-Vargas, M.J. Pinto-Betnal, F. Ballén-Moreno, M. Múnera, C.A. Cifuentes, Therapy with t-flex ankle-exoskeleton for motor recovery: A case study with a stroke survivor, in *2020 8th IEEE RAS/EMBS International Conference for Biomedical Robotics and Biomechatronics (BioRob)* (IEEE, 2020), pp. 491–496
45. D. Gomez-Vargas, F. Ballen-Moreno, P. Barria, R. Aguilar, J.M. Azorín, M. Munera, C.A. Cifuentes, The actuation system of the ankle exoskeleton t-flex: First use experimental validation in people with stroke. Brain Sciences **11**(4), 412 (2021)

Impedance Control Strategies for Lower-Limb Exoskeletons 8

Luis Arciniegas-Mayag, Carlos Rodriguez-Guerrero, Eduardo Rocon, Marcela Múnera, and Carlos A. Cifuentes

8.1 Introduction

A user who has suffered a stroke usually has several limitations in the motor control of his or her limbs. In general terms, the quality of life of users and other dependency-related factors are affected. For example, daily living activities comprise some tasks that the user's efficiency has decreased. These activities are defined as the fundamental processes that allow humans to have a high independence level in daily life. Examples of them are feeding, dressing, and actions related to personal hygiene [1, 2]. The activities of daily living are classified according to their complexity into basic and instrumental. Basic activities are related to self-care and personal mobility [3] and include cleaning, eating, and physical exercise. The instrumental activities of daily living require a higher cognitive level development [2–4]. Examples are buying, preparing food, cleaning, maintaining the house, among others. However, both types depend on mobility, which is deeply affected in patients who suffer a stroke. Some activities such as sitting/standing, ascending, and descending stairs are related to mobility and the user change of position [5]. In a particular case, the walking activity has been included in these tasks that provide the

L. Arciniegas-Mayag · M. Múnera · C. A. Cifuentes (✉)
Biomedical Engineering, Department of the Colombian School of Engineering Julio Garavito, Bogotá D.C., Colombia
e-mail: luis.arciniegas@mail.escuelaing.edu.co; marcela.munera@escuelaing.edu.co; carlos.cifuentes@escuelaing.edu.co

C. Rodriguez-Guerrero
Robotics & Multibody Mechanics Research Group, Department of Mechanical Engineering, Vrije Universiteit Brussel, Elsene, Belgium
e-mail: carlos.rodriguez.guerrero@vub.be

E. Rocon
Centro de Automática y Robótica, CSIC-UPM, Arganda del Rey, Madrid, Spain
e-mail: e.rocon@csic.es

C. A. Cifuentes, M. Múnera, *Interfacing Humans and Robots for Gait Assistance and Rehabilitation*, https://doi.org/10.1007/978-3-030-79630-3_8

user's mobility and improve quality of life (QoL). For this reason, different session therapies have been designed for the recovery of this activity in users that suffered a stroke [6]. Subsequently, the development of these therapy sessions represents a high workload for the post-stroke user and the physiotherapist [7, 8]. Additionally, it represents a heavy burden of the disease on the user [9, 10]. In this sense, in the last decades, various alternatives have been proposed to improve the effectiveness of therapy sessions focused on improving the gait pattern in post-stroke users. One of the options has been the use of a wearable robots, in this case, the development and implementation of lower-limb exoskeletons [11] used to recover the primary movements of the user's lower limbs retraining the user in the walking activity.

As mentioned in Chap. 1, the lower-limb exoskeletons have been implemented in different workspaces. This wearable robot has been focused on three objectives: Power augmentation, assistance, and rehabilitation, where each one is targeted to a different population. Each one of these objectives is related to some aspects that define a lower-limb exoskeleton. Therefore, it is defined some control strategies according to the activities of daily living that the lower-limb exoskeletons complement the human body's primary movements.

This chapter presents the development and implementation of two control strategies based on the principle of impedance. The control is explained through the Human–Robot interface of the *AGoRA* exoskeleton (Colombian school of engineering Julio Garavito, Colombia) and a case study is presented for the controllers.

8.2 Human–Robot Interaction

In the definition of (HRI), it is mandatory to determine some parameters of the human used as an input to a wearable robot. In this concept, an initial stage called cognitive process is developed, defined some phases that a human executes for the development of an activity. The user's movements are divided into three steps: a reasoning phase, a planning phase, and the executing phase [12]. As a result, the lower-limb's movements are generated to execute activities of daily living. Currently, the cognitive process development is estimated using a cognitive human–robot interaction (cHRI). As a result, a connection is created between the user and the wearable robot for the acquisition and of these cognitive processes. For this purpose, brain-machine interfaces (BCI) or human–machine interfaces (HMI) are used. The development of these interfaces uses various alternatives to acquire signals from the human body. As an example, interfaces are used for the acquisition of Electroencephalographic (EEG) signals [13–15] and electromyography (EMG) signals [16–18] adapted to the sensory interface of a wearable robot. Thus, using these signals an alternative for estimating the user's cognitive processes was implemented. the process of pHRI generation is shown in Fig. 8.1

EEG and EMG signals were implemented to estimate cHRI in the control strategies development based on user-generated parameters. On the other hand, these signals require the user's instrumentation on various parts of the body. For

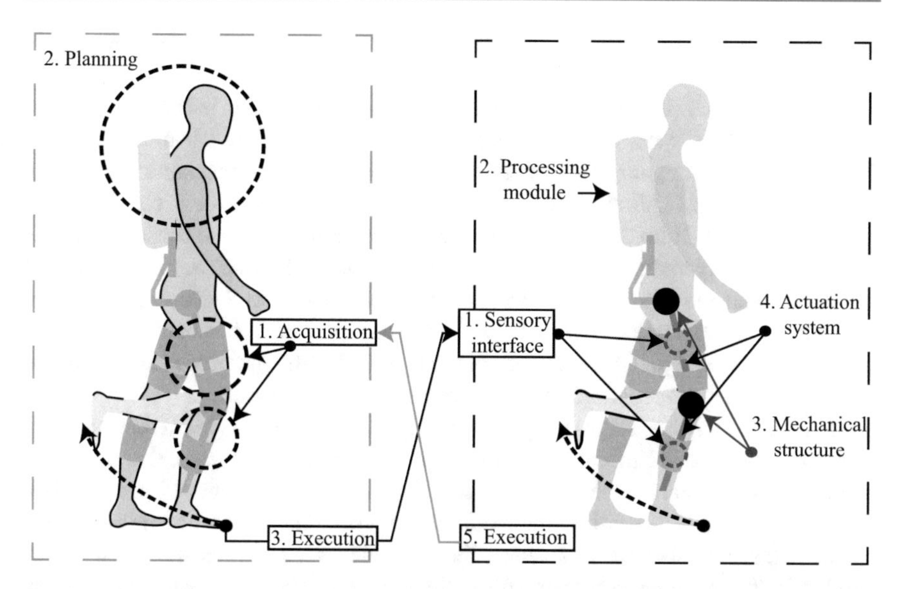

Fig. 8.1 pHRI definition using lower-limb exoskeleton in users that suffer a neurological pathology that affects the lower-limb's motor control; (**a**) shows the lower-limb movement generation through the Acquisition, planning, and lower-limb movement execution phases, respectively; (**b**) presents the lower-limb exoskeleton modules that capture the force/torque generated by the user implementing a sensory interface and a processing module, the actuation system and the mechanical structure generate the calculated torque and transmit the torque to the user's lower-limbs

example, the interface for acquiring EMG is placed on the user's muscle groups and for EEG acquisition. For this reason, the literature shows other alternatives where communication between the user and the wearable robot is generated. In this sense, the estimation of (pHRI) has been proposed [19–21]. In contrast to the cHRI, the pHRI evaluates some parameters that estimate the force/torques generated between the user and the wearable robot [22]. These parameters do not estimate the cognitive process in which the motion is generated as presented in the cHRI. The estimation of pHRI is usually implemented in lower-limb exoskeletons to assist or complement the user's movements in therapy sessions. This information is used as input to various control strategies to generate force/torque or angular velocity applied by the wearable robot employing a mechanical structure or actuation system.

The pHRI recently mentioned is applied in various workspaces. The literature has defined different categories of pHRI: the supportive category, where the robot device does not execute the task, this primary function is providing the tools to the user that develop the task [23]; collaborative category, the user and the robotic device develop the activity; and the cooperative category includes the interaction forces between the user and the robotic device, is to say, the robot and the user work in direct physical contact [21, 23]. In this sense, the lower-limb exoskeletons and the user are classified into a cooperative category. Using these interaction forces and control strategies generates natural user primitive movements using lower-limb exoskeletons [24, 25].

Currently, multiple tools are implemented to estimate the user's movements, where some of these methods are equipped in the lower-limb mechanical structure. For example, Strain gauges are located in the robotic device mechanical structure to calculate the lower-limb movement perceived by the exoskeleton [12, 26]. As a result, each lower-limb exoskeleton joint is used as a force/torque sensor [27]. The force sensor implementation depends on the mechanical structure morphology of the device and the anatomical plane of the human body from which it is desired to estimate these forces. The use of force sensors in the mechanical structure simplifies the use of pHRI-based control strategies. Besides, no user instrumentation is required and system calibration is performed in a short period.

In conclusion, this section explains the generation of the pHRI between the user and the lower-limb exoskeleton that implements the user's movements (force/torque), which categorized the pHRI in the user and the wearable robot. Additionally, Fig. 8.1 describes the method where pRHI and the outcome calculated by the wearable robot. In this sense, the following section presents the methods applied in the lower-limb movements estimation and the force/torque acquisition, establishing the inputs in implementing some control strategies based on the pHRI.

8.3 Sensors in the HRI of the AGoRA Lower-Limb Exoskeleton

The pHRI has focused on bidirectional communication between the exoskeleton and the patient. This development requires various sensors located in the exoskeleton mechanical structure and the user's lower limbs. As an example, the use of force sensors, incremental and absolute encoders, and IMU sensors can allow the estimation of patient force, angular position of the joints, and the estimation of space-time parameters of some sections of the lower limbs for the recognition of various tasks. This section comprises the sensory interface configuration used in the *AGoRA* lower-limb exoskeleton in the estimation of these parameters. To learn the modules that involve a lower-limb exoskeleton, this section presents the information applied as an input for the acquisition of the lower-limbs movements and the force/torque estimation to apply the pHRI in the design of some control strategies.

8.3.1 Force Sensing

The interaction forces estimation between users and rehabilitation devices provides the implementation of stationary therapy [28], and the execution of several activities where an exoskeleton complements the user's movements [29]. Generally, resistive sensors have been used in robotic devices such as lower-limb prostheses [30] and lower-limb exoskeletons [12, 26, 30, 31] to acquire the forces generated by the user using the rehabilitation or assistance wearable robot. The *AGoRA* exoskeleton use strain gauges (632–180, RS Pro, UK) located in the link of the robotic device. As a result, the strain gauge measures the link deflection caused by the user's lower-limb movements. In this way, the user movements generate deflections in different sections of the mechanical structure. Hence, the force/torque sensor is comprised of each link of the *AGoRA* exoskeleton instrumented by the Strain gauge. Consequently, the user's lower-limb force/torque is estimated in terms of (Nm) and implemented as an input of the pHRI control strategies. The force-torque sensor development is divided into sensor location and value acquisition, and signal processing and characterization.

The sensor location and value acquisition comprise the mechatronic integration for the force sensor reading and signal processing. Generally, the force sensor measurement is acquired using either a half Wheatstone bridge configuration or a full Wheatstone bridge configuration. These configurations do not affect the measured value of the force sensor by temperature fluctuations. Figure 8.2 shows an example of the strain gauge location based on the motion intention acquisition in the sagittal plane of the user's thigh and shank. Subsequently, an analog-to-digital conversion (ADC) and a signal amplification step are performed. Finally, the acquired signal is filtered applying a mean filter where the torques data is stored in a vector (mean_v(n)) and is calculated the average vector value. The next vector sample delete the value stored in the vector zero position (mean_v(0)), move each vector data to the left one position, store a new data in the vector n position (mean_v(n)), and recalculate the average value of this vector. As a result, the signal is smoothed.

The signal processing and characterization involve the characterization of the filtered signal, where the acquired value will be expressed in terms of torque (Nm). The filtering and characterization process requires embedded systems adapted for implementing digital filters and ensuring the real-time acquisition of these parameters. For example, the *AGoRA* exoskeleton uses devices such as a RasberryPi to apply the digital filters and implement the force sensor characterization function. The main objective of the sensor characterization process is to calculate the rate of change of the force generated in the sensor vs the voltage variation generated in the Wheatstone half-bridge. Generally, the behavior of these sensors shows a linear behavior expressed by the function $\tau = mx + b$ showed in Fig. 8.2. Where τ represents the force/torque value, m is the change ratio of the force/torque vs voltage value, and x is the value obtained by the force sensor. The value of b equals the value acquired by the force sensor at a force/torque of 0 [N]/ 0 $[Nm]$.

8.3.2 Position and Motion Sensing

Lower-limb exoskeletons implement various control strategies that use parameters such as angular position and angular velocity, which are acquired from the hip, knee, and ankle joints in the three anatomical planes of the human body [32, 33]. The acquisition of these parameters is aimed to monitor the effectiveness of the lower-limb exoskeleton. Additionally, the kinematic parameters acquisitions are

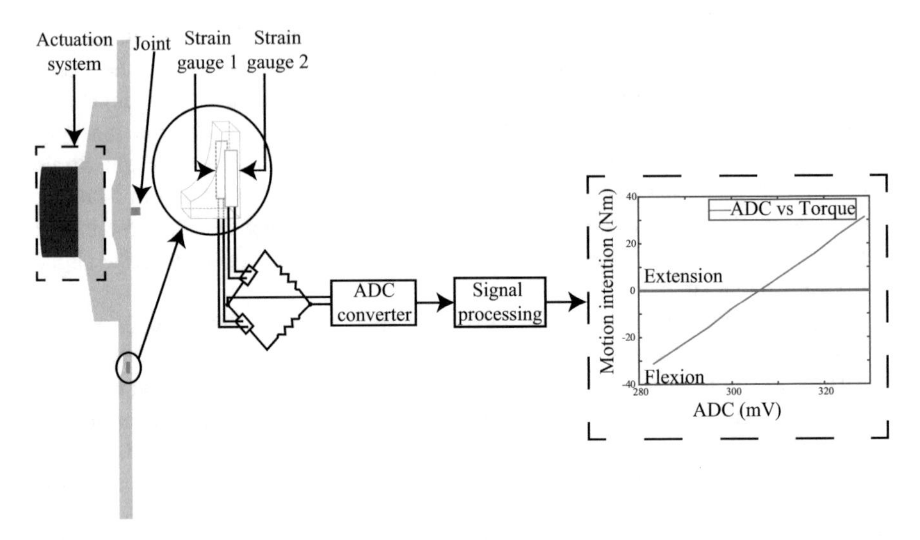

Fig. 8.2 Strain gauge's location and the data processing to obtain the user's generated torque and transmitted to the exoskeleton's mechanical structure

implemented as an input of the control strategies generating calculated torques or gait trajectories to rehabilitation and assistance.

The kinematic parameters acquisition is performed through encoders. These sensors allow estimating the angular position and the angular velocity of the exoskeleton joints. Two types of encoders are used for these applications. The first is the incremental encoder, which counts the motor shaft turns using the encoder position when powered. This feature may represent a disadvantage for exoskeletons adapted to develop more than one activity of daily living. This sensor is used in exoskeletons for stationary therapy [34–36] and the development of various actuation systems focused on the pHRI [37, 38]. The second is the absolute encoder that provides a reference axis to the joints of lower-limb exoskeletons. Generally, the absolute encoder is used in lower-limb exoskeletons focused on assistance and rehabilitation. Figure 8.3 shows the magnetic encoder location for each joint of the *AGoRA* exoskeleton.

The implementation of this sensor in lower-limb exoskeletons can be done in several ways depending on the sensor composition offered by the provider. For example, the AGoRA exoskeleton uses absolute magnetic encoders (AS5600, ams AG, Austria). The sensor instrumentation requires performing sensor location on the lower-limb exoskeleton structure, filtering, and sensor characterization. For this case, the encoder location is performed directly on the center of rotation (CR) of each joint of the exoskeleton as seen from the sagittal plane. The ADC is performed by the magnetic encoder mentioned above. The behavior of the magnetic encoder is expressed by the function $\theta = mx + b$. Where m is equivalent to 0.01371 and indicates the change ratio of the angular position with the voltage value. The value of x is equivalent to the value obtained by the magnetic encoder. Finally, b equal to

Fig. 8.3 Presents the magnetic encoder located in the Center of Rotation (CR) in the lower-limb exoskeleton joint. The figure shows the sensor location in the sagittal plane (**a**) and the frontal plane (**b**)

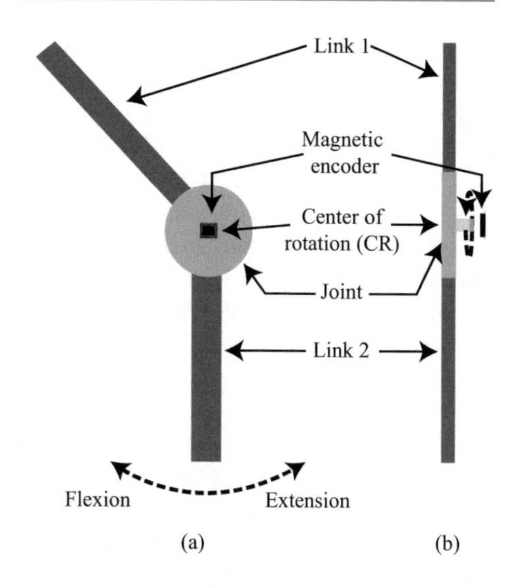

the magnetic encoder value where the joint angular position equal to 0° for the hip and knee joints.

IMU sensors provide data related to the angular position, angular velocity, and angular acceleration in various coordinate axes. These tools are applied to monitor patient movements, classify activities of daily living using machine learning methodology [39–41], or apply the acquired parameters as an input to various control systems. These applications allow increasing user's participation in therapy sessions [42]. In this way, it is possible to define an assistance level in various robotic assistive and rehabilitation devices [43]. For instance, A gait phase detection algorithm mentioned in Chap. 5 is implemented in the AGoRA exoskeleton Control strategies to support the user's gait pattern. This classifier uses an IMU sensor to acquire the inputs in control systems. For example, Chap. 5 presents an online gait phase detection module applied in the *AGoRA* exoskeleton.

8.4 Actuation in the HRI of the AGoRA Lower-Limb Exoskeleton

As mentioned in Sect. 8.3, obtaining the kinematic and kinetic parameters allowed defining the inputs for the various control strategies. Subsequently, this information is processed to estimate the system response. This response expressed in terms of torque, angular velocity, or angular position is generated as mechanical energy transmitted to the user's joints. For this purpose, the lower-limb exoskeletons are composed of a mechanical structure that provides the coupling between the wearable robot and the user. Additionally, an actuation system generates this mechanical energy to complement the user's movements in the execution of walking activity.

The *AGoRA* lower-limb exoskeleton consists of a rigid structure composed of 2 active joints on the right limb to assist flexion/extension movements for the hip and knee joints. Additionally, a passive joint enables abduction and adduction movements in the patient's hip [27]. The rigid structure is made up of duraluminium, lightweight, resistant, and low-corrosion material [27]. Additionally, the links coupled with the thigh and shank are telescopic bars, which are adjusted according to the user's anthropometric measurements. Thus, the device can be used by people between 1.70 and 1.83 m tall with an approximate weight of 90 kg.

The *AGoRA* lower-limb exoskeleton is equipped with stiff actuators to generate the required torques in the user's lower limbs. This actuation system comprises brushless DC electric motors (EC-60 flat 408057, *Maxon Motor AG*, Switzerland), which can provide a nominal torque of 0.228 Nm, reaching a maximum of 6000 rpm or 628.318 rad/s [44]. Taking as reference the torque values generated to assist using 50–80 Nm exoskeletons [45,46], the actuation system is complemented with a speed reduction gearbox (CSD-20-160-2AGR, *Harmonic Drive LLC*, USA) with a gear ratio of 160:1. As a result, torques of 35 Nm nominal and 180 Nm peak torque are generated, with an angular velocity of 37.5 rpm or 3.92 rad/s, torques and angular velocity similar to [47].

8.5 Impedance Control of Human–Robot Interaction

Literature shows the modeling of a lower-limb exoskeleton through the concepts of an n degree of freedom manipulator robot. Robot dynamics is generated, defined as a compensation system to forces/torques that affect the robot's movement. Subsequently, the wearable robots present several behavioral models of active actuation systems based on the mass-spring-damper system implementation and a mass-damper system. As a result, several control strategies are implemented that consider the forces/torques generated in the pHRI. In this sense, an impedance control applied in wearable robot proposed two concepts considering the generation of calculated force/torque or change the joint's stiffness according to the lower-limb exoskeleton workspaces. This section aims at defining these concepts from the viewpoint of the *AGoRA* lower-limb exoskeleton. This section will be covered as follows: (1) problem statement, where the *AGoRA* exoskeleton dynamics and the definition of a mass-spring-damper system will be explained, and two types of impedance control (2) impedance controllers; and (3) admittance controllers.

8.5.1 Problem Statement

Control strategies applied to lower-limb exoskeletons involve two key concepts. The first defines the forces/torques that are present in the wearable robot. As an example, moments of inertia, gravity, friction, among others are defined. The definition of these parameters is named the Feedforward or the robot's dynamic system. The second is the implementation of a feedback system to the robotic device. Therefore, a mass-spring-damper system or a mass-damper system is used in the exoskeleton joints. In this sense, this section shows the key concepts for implementing the control system for a lower-limb exoskeleton.

8.5.1.1 Robot's Dynamics

The development of control strategies for a lower-limb exoskeleton involves the estimation of two relevant factors. One of these focuses on the forces/torques identification to generate the lower-limb exoskeleton joints motion. The second implements the pHRI and the lower-limb movements to estimate the motion generated by the actuation systems. Each of these factors is involved as system feedback and feedforward, respectively. In this sense, this section shows the feedforward estimation of a lower-limb exoskeleton. Note that exoskeletons present various kinematic models. In this section, the kinematic model for the AGoRA exoskeleton will be presented as an example.

The exoskeleton feedforward estimation is performed using applied concepts to manipulator robots. One of these involves the total energy calculation of the system by obtaining the kinetic energy and the potential energy related to the exoskeleton links. This is expressed in Eq. 8.1:

$$\varepsilon = \kappa(q, \dot{q}) + U(q). \tag{8.1}$$

The estimation of these energies involves the conservative and non-conservative forces of the system. Using the Lagrange equation of motion (Eq. 8.2), the compensation system value is obtained, expressed in torque (Nm).

$$[L] = \kappa(q, \dot{q}) - U(q, \dot{q}). \tag{8.2}$$

The differential kinematics value for each link must be estimated to obtain this value for a device such as the *AGoRA* exoskeleton. This will provide information on the angular velocity of the robot link at an instant of time. For this purpose, Eqs. 8.3 and 8.4 are used by applying the geometric Jacobian expressed in Eq. 8.5:

$$Link_{1(x,y)} = \begin{bmatrix} l_{c1} \sin q_1 \\ -l_{c1} \cos q_1 \end{bmatrix} \tag{8.3}$$

$$Link_{2(x,y)} = \begin{bmatrix} l_1 \sin q_1 & l_{c2} sen(q_1 + q_2) \\ -l_1 \cos q_1 & -l_{c2} \cos(q_1 + q_2) \end{bmatrix} \tag{8.4}$$

$$\begin{bmatrix} \Delta x \\ \Delta y \end{bmatrix} = \begin{bmatrix} \frac{\delta F_x(q)}{dq_1} & \frac{\delta F_x(q)}{dq_2} \\ \frac{\delta F_y(q)}{dq_1} & \frac{\delta F_y(q)}{dq_2} \end{bmatrix} \begin{pmatrix} \Delta q_1 \\ \delta q_2 \end{pmatrix}. \tag{8.5}$$

In this sense, the differential kinematic for the lower-limb joints is presented in Eqs. 8.6 and 8.7:

$$\upsilon_1 = \begin{bmatrix} l_{c1} \cos q_1 \\ -l_{c2} \sin 1_1 \end{bmatrix} \dot{q_1} \tag{8.6}$$

$$\upsilon_2 = \begin{bmatrix} l_1 \cos q_1 + l_{c2} \cos q_1 + q_2 & l_{c2} \cos q_1 + q_2 \\ l_1 \sin q_1 + l_{c2} \sin q_1 + q_2 & l_{c2} \sin q_1 + q_2 \end{bmatrix} \begin{bmatrix} \dot{q_1} \\ \dot{q_2} \end{bmatrix}, \tag{8.7}$$

where l_1 is the link 1 length, l_{c2} is the center of mass for the link 2 joint length, q_1 and q_2 are the joint's angular position. The next step to calculate the total system energy is the estimation of the exoskeleton kinetic energy. This parameter is expressed in Eq. 8.8:

$$\kappa = \sum_{i=1}^{n} \frac{1}{2} m_i {\upsilon_i}^T \upsilon_i + \frac{1}{2} I_i^2 \dot{q_i}^2. \tag{8.8}$$

Equation 8.8 presents the kinetic energy. In the lower-limb exoskeleton, where m_n is the link mass, υ_n is the differential kinematic, I_n the link inertia moment. Finally, $\dot{q_n}$ is the angular velocity for each joint. In this order, the kinematic energy of the *AGoRA* lower-limb exoskeleton is expressed in Eq. 8.9:

$$\begin{aligned} \kappa = &\frac{1}{2} \left[m_2 l_1^2 + I_1^2 + I_2^2 + m_2 l_1^2 + m_2 l_{c2} + 2 m_2 l_1 l_{c2} \cos q2 \right] \dot{q_1} \\ &+ \frac{1}{2} \left[I_2 + m_2 l_{c2}^2 \right] \dot{q_2}^2 + [m_2 l_1 l_{c2} \cos q_2 + m_2 l_{c2} + I_2] \dot{q_1} \dot{q_2}. \end{aligned} \tag{8.9}$$

The second step in calculating the Lagrange equation of motion is estimating the potential energy that involves the conservative forces. This parameter is expressed in Eq. 8.10:

$$U(q) = mgh, \tag{8.10}$$

where m is the link mass, g is the gravity acceleration value. This function is applied for each *AGoRA* exoskeleton link. The *AGoRA* exoskeleton potential energy is expressed in Eq. 8.11:

$$U(q) = mgl_{c1}[1 - \cos q_1] - m_2 g[(l_1)]. \tag{8.11}$$

8.5.1.2 The Mass-Spring-Damper System

The mass-spring-damper system is the concept applied in the joints of robotic devices, whose main function is focused on applying pHRI [48]. The implementation of this system in the knee and hip joints approximates the human muscle biomechanical model. Currently, the literature presents the human muscle as a component composed of a spring-like elastic element in parallel with a viscous element [48–50]. Implementing this concept in the knee and hip actuators assists the actuated limb through calculated torques [50]. Additionally, the patient participates in developing the motion to complement the movements generated by the lower-limb exoskeleton [50]. This section comprises the definition of the mass-spring-damper system applied in various mid-level control strategies for lower-limb exoskeletons. Some examples of the implementation of this concept for the *AGoRA* exoskeleton will be shown.

The mass-spring-damper system is mainly composed of a spring element and a damping element. The spring element is referred to as a force restoring component. The response of this element is expressed in terms of force, where the elongation of the element is multiplied with the spring elasticity constant [51, 52]. On the other hand, a delaying element is required to decrease the system oscillations generated by the spring element. For this purpose, a damping element is implemented to provide stiffness to the system. The system's response is expressed in terms of force, which depends on the linear velocity of the system, multiplied with a viscosity constant characteristic of a damper [51,52], resulting in a retarding force to the force generated by the spring element.

The implementation of this system has been called an impedance controller proposed in 1984 [48]. This controller presented an interaction method between a manipulator robot and the workspace environment using calculated forces. Subsequently, it was implemented to apply the pHRI concept to lower-limb exoskeletons for various purposes [36, 53]. Lower-limb exoskeletons such as the AGoRA exoskeleton implement this system at the hip and knee joints for sagittal plane assistance as seen in Fig. 8.4. Although this exoskeleton does not have an actuation mechanism based on a physical mass-spring-damper system at each joint, it applies the system virtually using the Eq. 8.12:

$$F = -\overbrace{k(x_d - x)}^{spring} - \overbrace{\beta(\dot{x_d} - \dot{x})}^{damping}, \tag{8.12}$$

where k equals the elasticity constant, x_d represents the desired position, and x is the joint's actual position, the operation of these constants represents the elastic element's effect on the joint. Additionally, β is the viscosity constant of the system, $\dot{x_d}$ equals the desired velocity of the system, and $\dot{x}$ equals the system's actual velocity. These parameters and the operation of these constants represent the effect of the damping element on the joint. For a mass-spring-damper system to be in equilibrium, the sum of these forces would equal zero (0), so Eq. 8.12 is expressed in Eq. 8.13:

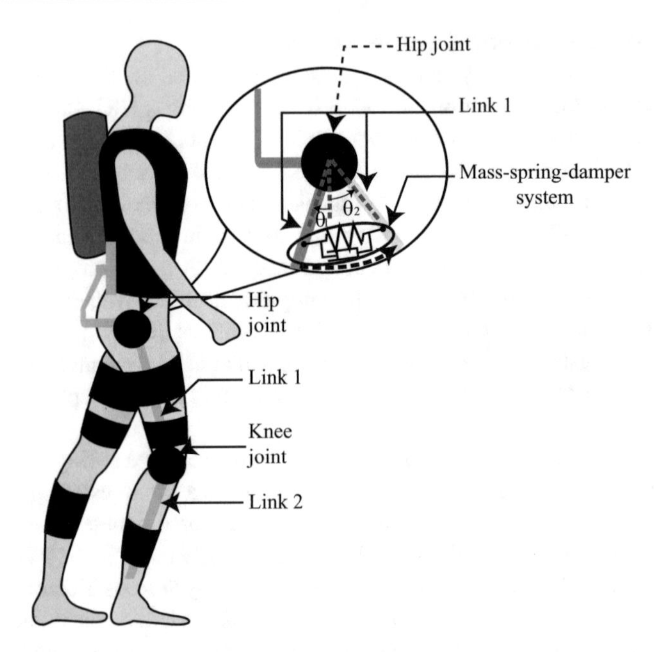

Fig. 8.4 Lower-limb exoskeleton schematic implementing the mass-spring-damper system in the knee and hip joints

$$
\begin{aligned}
0 &= m\ddot{x} + \beta(\dot{x_d} - \dot{x}) + k(x_d - x) \\
&= \ddot{x} + \frac{\beta}{m}(\dot{x_d} - \dot{x}) + \frac{k}{m}(x_d - x).
\end{aligned} \tag{8.13}
$$

If $\frac{\beta}{m}$ equal to 2λ and $\frac{k}{m}$ is ω^2, Eq. 8.13 represents

$$
\ddot{x} + 2\lambda(\dot{x_d} - \dot{x}) + \omega^2(x_d - x) = 0 \tag{8.14}
$$

$$
r^2 + 2\lambda r + \omega^2 = 0. \tag{8.15}
$$

As a result, Eq. 8.15 is equal to a second-order system where $\ddot{x}$ is the acceleration, $\dot{x}$ velocity, and x equal to the system's position. This equation provides the k and β system values.

The application of this concept and its a variation in the various actuation systems comprise the mid-level control strategies. These are based on the acquisition or generation of torque profiles that complement the lower-limb movements of a stroke patient. As has been shown, the lower-limb movements are a significant factor that will allow the patient to be included in the control strategy of the device. The following sections will present two control methods based on this concept to develop mid-level control strategies.

8.5.2 Impedance Controller

The impedance controller presented by Hogan et al. in [48] has been fundamental in the development of control strategies applying the lower-limb movements for rehabilitation devices. Its performance mainly focuses on varying the assistance level provided by the device, by increasing or decreasing the maximum torque profile transmitted by the actuation system [53–55]. The development of the impedance controller is implemented employing a mass-spring-damper system as shown in Fig. 8.5a.

This system considers inputs to the system kinematic parameters such as the angular position and angular velocity of the lower-limb exoskeleton joints to obtain a calculated torque profile applied to the actuation systems (Fig. 8.5b). In this sense, the impedance controller is considered a restoring element of forces that varies according to the value of the spring element elasticity constant(k) shown in Eq. 8.16:

$$\tau = \overbrace{k(q_d - q)}^{spring} + \overbrace{\beta(\dot{q_d} - \dot{q})}^{damper}. \tag{8.16}$$

The k and β are the elasticity constant and the damping constant, respectively, q_d equal to the desired joint angular position; q is the current joint position, $\dot{q_d}$ equal to the desired angular velocity. In this case, when the joint angular position is equal to

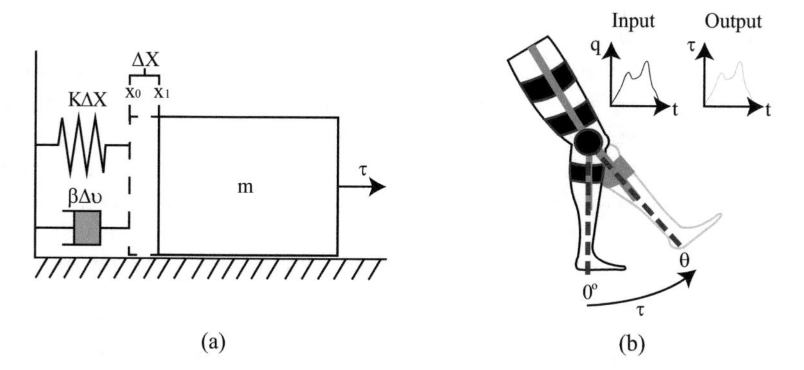

Fig. 8.5 Impedance controller presentation; (**a**) impedance controller schematic using a mass-damper system where the $\beta \times \Delta v$ equal to the damper element torque response, $K \times \Delta X$ equal to the spring element torque response, and τ is the impedance controller output; (**b**) shows the impedance controller applied in the lower-limb exoskeleton's joint, where the input signal is the joint angular position, the impedance controller response is a calculated torque applied in the joint

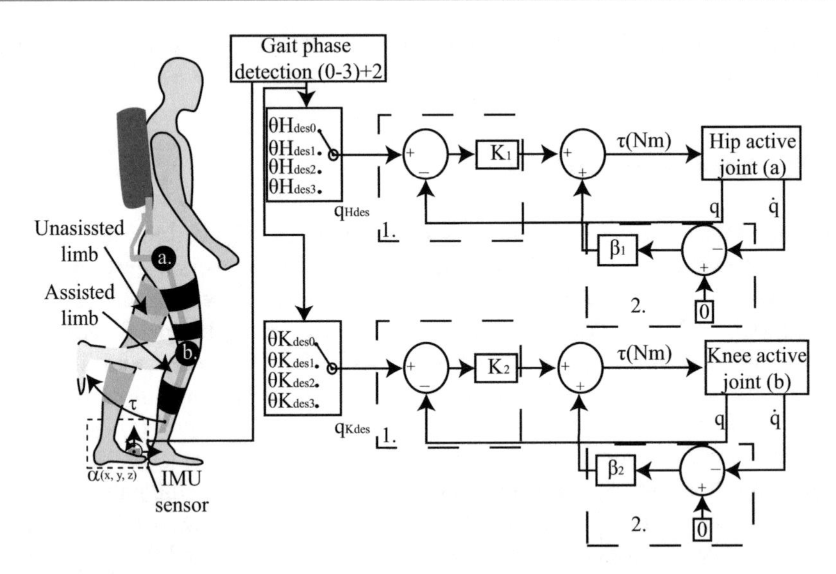

Fig. 8.6 Assistance mode schematic; the IMU sensor is located in the foot tip to the non-assisted limb to estimate the angular acceleration ($\alpha_{(x,y,z)}$); this parameter is applied input signal of the gait phase detection module. As a result, the gait phase detection module provides a number in the range from zero to three according to the phase gait detected. The unassisted limb gait phase equals to the assisted limb gait phase ahead of two phases. This detected phase as an input for a desired angular position selector for the hip (θH_{des0}, θH_{des1}, θH_{des2} and θH_{des3}) and the knee (θK_{des0}, θK_{des1}, θk_{des2} and θk_{des3}) joints. This value is operated for the mass-spring(1.)-damper(2.) system to calculate the actuation system's torque

the joint desired angular position, the desired angular velocity equal to zero. Finally, $\dot{q}$ represents the joint current angular velocity.

The assistance mode of the *AGoRA* exoskeleton, showed in Fig. 8.6, implements an impedance controller for the hip and knee joints that requires the estimation of the parameters of the spring element and the damping element. This control strategy guides each joint to the desired position, providing corrections employing torque profiles. In this way, the patient is involved in the task, making movements in the same direction in which the device applies the assisting torque. Once the gait phase is identified (by employing the sensors IMU [43]), the desired angular position values are established for each gait phase used as input to the impedance controller.

The assistance mode is developed for people with right-sided hemiparesis, where the robotic device considers the movements of the unaffected limb. Therefore, the left limb is instrumented with a 9 degree of freedom IMU sensor for the user's gait sub-phase detection used as input to the control strategy. The values of this input parameter are given in a value from 0 to 3, where Heel Strike equals zero, Flat Foot equals 1, Heel Off equals 2, and the Toe Off phase equals 3. Each sub-phases a desired angular position is assigned to the joint assisted by the *AGoRA* exoskeleton. Taking into account what was presented in Villa Parra et al. [26], El Zahraa Wehbi et al. [34], and Webster et al. [56], the gait characteristic as a symmetric activity, it is identified that the gait phase that the assisted limb should execute corresponds to 2 sub-phases in advance of the four detected gait sub-phases of the unaffected limb.

It is necessary to keep into account that this control is designed to provide an assistance level. Therefore, the adjustment of these variables depends directly on the assistance level required in the walking activity. An approximation of the estimated values for these variables is made using an approximation of the Eq. 8.15 in Eq. 8.17:

$$\ddot{q} + 2\lambda\Delta\dot{q} + \omega^2\Delta\dot{q} = 0 \tag{8.17}$$

$$2\lambda = \frac{\beta}{ml^2} \tag{8.18}$$

$$f_n = \frac{\omega_n}{2\pi} \tag{8.19}$$

$$\omega_n = \sqrt{\frac{k}{ml^2}}. \tag{8.20}$$

To apply Eq. 8.17, λ is defined in Eq. 8.18, where β is the damping constant, m is the lower-limb mass that involves the thigh mass and the shank mass, l is the thigh and shank length. In Eq. 8.19 the bandwidth is defined as equals to $f_n = 0.5936$ Hz. Finally, in Eq. 8.20 k is the elasticity constant. These equations are taken into account to estimate the k and β varying the assistance level and the system stiffness [57].

8.5.3 Admittance Controller

Admittance control is focused on acquiring human–robot interaction forces in the implementation of haptic applications [58,59]. This characteristic allows to simulate the stiffness of a system virtually, or the system inertia to be reduced employing this strategy [22]. This same principle is applied to robotics for rehabilitation and assistance of people with some neurological disease. Exoskeletons such as *ALLOR* (Federal University of Espirito Santo, Brazil) developed for walking, feature an admittance controller that changes the system stiffness at different gait phases [26]. *BioMot* (Future and Emerging Technologies (FET), Spain) applies this control to decrease the inertia of the exoskeleton to provide movement freedom for the identification of the user's gait pattern [60]. Likewise, the knee orthoses used in stationary therapy use an admittance controller to record the therapy trajectory to be performed, complemented by an impedance controller executed in the trajectory reproduction. The implementation of an admittance control in rehabilitation and assistance robotics becomes a significant contribution. It allows the patient to be involved in the control strategy applied to the device, without requiring the patient's instrumentation using invasive sensors.

The literature shows that the implementation of this kind of control using a mass-spring-damper system in each joint of the exoskeleton or robotic orthosis [16]. However, implementing a spring element may affect the system response because the force generated by a person does not necessarily correspond to a desired angular position. Therefore, in some cases an admittance control is not developed for the generation of restoring forces. For this reason, some lower-limb exoskeletons implemented a mass-damper system for the hip and knee joints [26]. This system provides for a viscous coefficient that varies the inertia of the exoskeleton joints, obtaining different levels of stiffness in the system. In rehabilitation and assistance of lower limbs, applied methodologies are focused on providing assistance when needed (AAN) [61]. This concept uses this type of control by taking advantage of the force generated by the patient and used for therapy sessions in gait rehabilitation. As a result, an admittance control allows a device to be called back drivable.

Currently, several ways of implementing this theory can be identified which depend on the actuation system used in the lower-limb exoskeleton. Some actuation systems use a physical mass-damper system by using a damping element in the joint. Other devices use a rigid actuator which generates an angular velocity response at the joint. These actuation systems complement a sensory interface that will enable the device the force/torque user's estimation.

The control strategy based on an *AGoRA* exoskeleton admittance control considers a rigid actuation system. This system is coupled to the mechanical structure of the exoskeleton which is instrumented with force sensors mentioned in Sect. 8.3.1, for the estimation of the force/torque estimation. In this sense, admittance control in the *AGoRA* exoskeleton is implemented, where the system input is the pHRI expressed in terms of torque, and the response to the system is implemented by the actuation system in terms of angular velocity, as shown in Fig. 8.7b. As a result, the device generates movements on the joints according to the patient's voluntary movements. The *AGoRA* exoskeleton follows the patient's movements simulating a back drivable device, or it can operate as a rigid device according to the value of the damping constant configured for the system.

The admittance controller is defined in Eq. 8.21:

$$\tau = \beta(\dot{q_d} - \dot{q_c}), \tag{8.21}$$

where τ equals the torque to be generated by the damping element, $\dot{q_d}$ is the desired velocity, and $\dot{q_c}$ is equal to the current angular velocity of the joint, the admittance controller applied in the *AGoRA* exoskeleton is showed in Fig. 8.7.

The transparency mode presented in Fig. 8.8 is implemented using the admittance controller on the hip and knee joints. This control strategy can change the inertia of each lower-limb joint, produced by the implementation of gearboxes coupled with the exoskeleton actuators.

As is showed in Fig. 8.8, the transparency mode comprises the torque acquisition. Hence, the thigh torque is estimated using the calculated torque sensed by the Strain

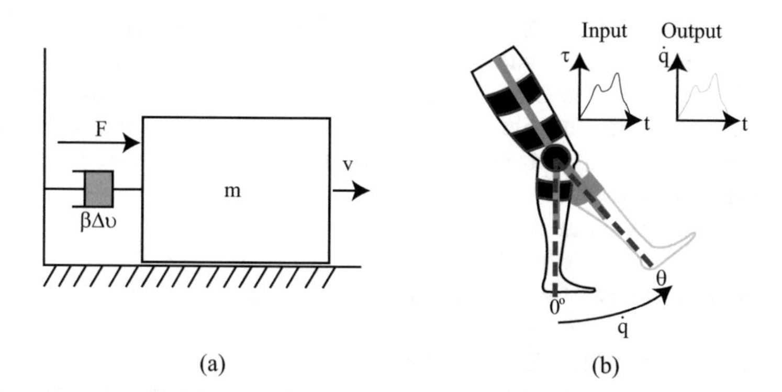

Fig. 8.7 Admittance controller presentation; (**a**) admittance controller schematic using a mass-damper system where the $\beta \times \Delta v$ equal to the damper element torque response, F is the applied force in the system, and v is the admittance controller outcome; (**b**) shows the admittance controller applied in the lower-limb exoskeleton's joint, where the input signal is the torque generated transmitted to the lower-limb mechanical structure, the admittance controller response is an angular velocity applied in the joint

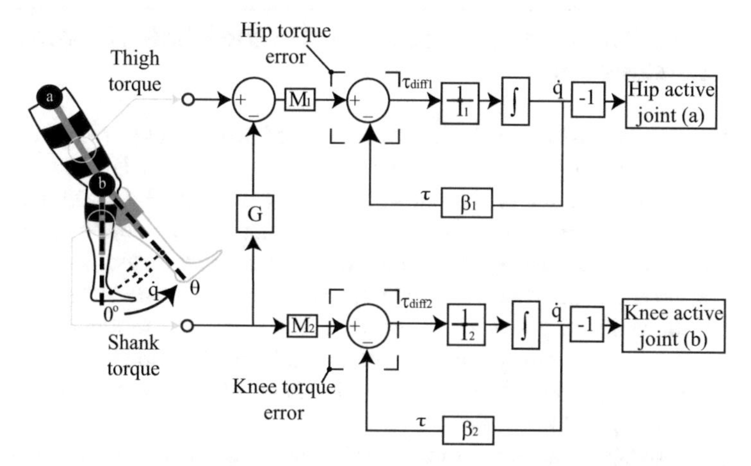

Fig. 8.8 Transparency mode schematic; the thigh and shank torque are estimated in terms of torque. The thigh torque value, and the shank torque value are operated with M_1 M_2 to attenuate the torque value. Subsequently, these parameters are applied into ad admittance controller to generate angular velocity in each actuation system's joint

gauge located in the thigh. Subsequently, this value is operated with the shank torque multiplied by a G gain. As a result, the real thigh torque value generated by the user's hip is calculated. The thigh and shank torque generated by the hip and the knee user's joint, are used to estimate the hip and knee torque error to obtain the difference between the joint torque and the torque obtained from the damper element. The next step in the estimation of the admittance controller response uses the torque error (τ_{diff}) divided into inertia moment (I) for each joint ($I = l \times m$)

where m equal to the joint's mass and l equal to the length for each link. This value is operated using an integral by obtaining the joint's angular velocity ($\dot{q}$). Finally, the rotational orientation of the joint generated by the gearbox is corrected by multiplying the angular velocity value by a factor of -1. As a result, the system produces angular velocity profiles applied to the actuation system according to a user's lower-limb movements estimated by the force sensors.

The transparency mode is organized as follows:

- **Torque acquisition:** Corresponds to the torque acquisition for the thigh and shank sections.
- **Calculate the angular velocity:** Estimating the τ_{diff} through the torque for each joint and the torque generated by the damper element. Subsequently, the division of the inertia moment and the integral of the obtained value.
- **Generate the torques through the actuation system:** The angular velocities are sent for each actuation system to generate movements in the hip and the knee user's joints.

8.6 Case Study: Impedance Control in the AGoRA Lower-Limb Exoskeleton

The test using the assistance mode was performed with a subject who does not suffer any gait pathology. The subject was a male, 1.72 m in height and weighed 73 Kg . The test environment is a flat surface of 20 m walked by the user in a flat surface in a straight line using the *AGoRA* exoskeleton. This device assisted the right limb in the hip and knee joints in the sagittal plane. In this case, the mode uses the joint angular position to correct the trajectory using calculated torque profiles. Figure 8.9 shows the response of the assistance mode in terms of angular position, and torque generated by the controller. Additionally, the system performance uses the gait phase detection module mentioned in Chap. 5 to assign a desired angular position for the hip and knee joints.

The gait assistance mode outcomes are showed at the hip and knee joint during three gait cycles. Figure 8.9a shows the angular knee position vs. the joint desired angular position. The knee joint showed different maximum knee flexion values in three gait cycles, considering the gait phases exhibited different periods. However, the desired angular position was adjusted according to the gait phase detection during the gait phase. As a result, the assistance mode output (Fig. 8.9b) generated several peak torques that were adjusted according to the desired angular position. Additionally, the implementation of impedance control in assistance mode fulfills the objective of generating calculated torques that complement the user's movement. Which is to say, there is not a 100% assistance level assistance. For this reason, an error between the desired angular position and the angular knee position is observed in the test. The system can be adjusted to provide a higher assistance level than the test by increasing the value of the system's elasticity constant (k). As a result, this would decrease the error shown in Fig. 8.9.

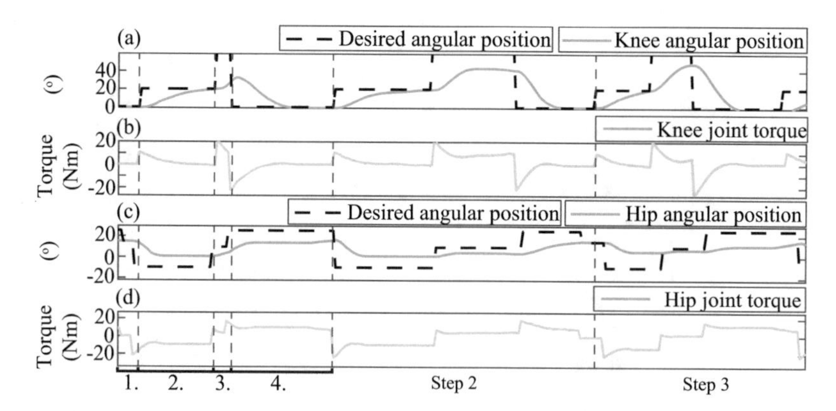

Fig. 8.9 Assistance mode outcome implemented for the walking activity in a healthy user; (**a**) shows the angular knee position in comparison to the desired angular position; (**b**) shows the assistance mode response for the knee joint in terms of torque (Nm); (**c**) presents the hip angular position in comparison to the desired angular position; (**d**) shows the assistance mode response for the hip joint in terms of torque (Nm); 1.-"*Heel Strike*," 2.-"*Flat Foot*," 3.-"*Heel Off*," 4.-"*Toe Off*"

In the hip joint actuation, similar behavior is observed in Fig. 8.9.c shows an error between the angular position of the hip and the angular position as the system complements the motion performed by the user. The desired angular positions at the hip are adjusted according to the gait phase detection module. As a result, the system response generated peak torques to guide the joint to the desired position. Each calculated torque profile is adjusted according to the system's desired position; increasing the elasticity constant of the controller will decrease the angular position error presented at the hip.

8.7 Case Study: Admittance Control in the AGoRA Lower-Limb Exoskeleton

The test performed using the transparency mode was executed with a subject who does not suffer any gait-associated pathology. The subject is a male that is 1.72 m high and weighs 73 Kg. The test was developed on a flat floor surface of 7 m in length without obstacles. The subject wears the *AGoRA* lower-limb exoskeleton that actuates the right lower limb at the hip and knee joints in the sagittal plane during the test. As mentioned in Sect. 8.4, the stiff actuator joints of the *AGoRA* exoskeleton can simulate the back drivable response using an admittance controller applied in this transparency mode. Therefore, this test presented the controller response by showing the system input (thigh/shank estimation torque), the controller output expressed in angular velocity, and the knee joint response in terms of angular position. Figure 8.10 is presented the transparency mode outcome in the knee joint during the walking activity.

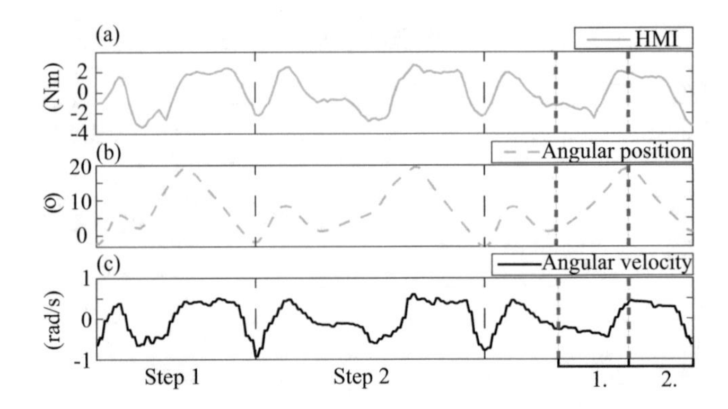

Fig. 8.10 Transparency mode response in the knee joint developing the walking activity. The response is shown in three knee gait cycles. Section 8.1. represents the knee flexion, Sect. 8.2. represents the knee extension; (**a**) presents the knee torque (Nm); (**b**) is the angular knee position during the walking activity; (**c**) equals to the transparency mode output in terms of angular velocity (rad/s)

The results of the pilot test in the walking activity are expressed in 3 gait cycles. The maximum torque generated was 4 Nm in knee extension and 2 Nm in joint flexion. As seen in Fig. 8.10a, the signal trajectory is not uniform to the angular knee position signal. However, the signal response is expressed into angular velocity profiles to obtain a smoothed response in the angular knee position. Likewise, the controller output signal is given since the admittance controller does not consider the joint's angular position. However, the signal is translated into angular velocity profiles and thus a smoothed trajectory of the angular knee position is obtained. Likewise, the behavior of the output signal of the admittance controller does not take into account the angular position of the joint. Finally, the ranges of motion observed in the knee joint are reduced compared to the ranges presented by a healthy person in a walking activity. The gain β is configured for the knee admittance controller that generates a stiffness in the joint that does not allow the free movement of the joint. On the other hand, the system stiffness is minor than generating joint motion with the rigid actuation system. The knee joint movements can be improved by increasing the value of the damper constant β of the admittance control.

8.8 Chapter Conclusions

This chapter presents the design of control strategies for a lower-limb exoskeleton. In this process, several tools that contribute to the development of these control systems were explained. The tools and concepts used in developing the *AGoRA* lower-limb exoskeleton are mentioned as an example. First, the pHRI was defined divided into different phases: (1) the generation of lower-limb movements by performing a process of acquisition, planning, and the user's lower-limb movement generation; (2) the pHRI estimation by the wearable robot, which processes the information and provides a response as mechanical energy, generating the user's joint movements. Second, various inputs used in the pHRI estimation were presented, which mentioned the acquisition of kinetic and kinematic parameters and the signal characterization phase. In this process, the user's lower-limb movement acquisition options were non-invasive implementing a wearable robot in the walking activity. Third, the function of the actuation system that generates the mechanical energy for the movement of each joint is defined. As a complement, the mechanical structure that composes the *AGoRA* exoskeleton is mentioned in the aspects of the material that composes it and DoF, where it fulfills the primary function of being the coupling between the wearable robot and the user's joints.

In the analysis of the processing phase, the critical concepts for implementing the control strategies of the *AGoRA* exoskeleton are mentioned. In this way, the design of the robot dynamics and critical concepts of implementing a mass-spring-damper system and a mass-damper system were presented. As a result, the applied control concepts for the use of force/torque within the control strategies are introduced. As a result, the implementation of two control strategies, involving the integration of each module mentioned, is presented. As an example, the control strategies outcome was shown in a pilot study during the walking activity.

References

1. M.E. Mlinac, M.C. Feng, Assessment of activities of daily living, self-care, and independence. Arch. Clin. Neuropsychol. **31**(6), 506–516 (2016)
2. E. Peter, B. Deb, S. Sukesh, L. Shoshana, Activities of daily living (ADLs) (2021). https://www.ncbi.nlm.nih.gov/books/NBK470404/
3. A. Balaguer, Actividades de la vida diaria (2016)
4. C. Blomgren, K. Jood, C. Jern, L. Holmegaard, P. Redfors, C. Blomstrand, L. Claesson, Long-term performance of instrumental activities of daily living (IADL) in young and middle-aged stroke survivors: results from SAHLSIS outcome. Scand. J. Occup. Ther. **25**(2), 119–126 (2018)
5. K. Han, J. Lee, W.K. Song, Application scenarios for assistive robots based on in-depth focus group interviews and clinical expert meetings, in *2013 44th International Symposium on Robotics, ISR 2013* (2013)
6. J.-M. Belda-Lois, S.M.-d. Horno, I. Bermejo-bosch, J.C. Moreno, J.L. Pons, D. Farina, M. Iosa, M. Molinari, F. Tamburella, A. Ramos, A. Caria, T. Solis-escalante, C. Brunner, M. Rea, Rehabilitation of gait after stroke: a top down approach. J. NeuroEng. Rehabil. **66**, 66 (2011)
7. B. Koopman, E.H.F.V. Asseldonk, H.V.D. Kooij, Selective control of gait subtasks in robotic gait training: foot clearance support in stroke survivors with a powered exoskeleton. J. NeuroEng. Rehabil. **10**, 1–21 (2013)
8. L.D. da Silva, T.F. Pereira, V.R. Leithardt, L.O. Seman, C.A. Zeferino, Hybrid impedance-admittance control for upper limb exoskeleton using electromyography. Appl. Sci. **10**(20), 1–19 (2020)
9. S.C. Ouriques Martins, C. Sacks, W. Hacke, M. Brainin, F. de Assis Figueiredo, O. Marques Pontes-Neto, P.M. Lavados Germain, M.F. Marinho, A. Hoppe Wiegering, D. Vaca McGhie, S. Cruz-Flores, S.F. Ameriso, W.M. Camargo Villareal, J.C. Durán, J.E. Fogolin Passos, R. Gomes Nogueira, J.J. Freitas de Carvalho, G. Sampaio Silva, C.H. Cabral Moro, J. Oliveira-Filho, R. Gagliardi, E.D. Gomes de Sousa, F. Fagundes Soares, K. de Pinho Campos, P.F. Piza Teixeira, I.P. Gonçalves, I.R. Santos Carquin, M. Muñoz Collazos, G.E. Pérez Romero, J.I.

Maldonado Figueredo, M.A. Barboza, M. Celis López, F. Góngora-Rivera, C. Cantú-Brito, N. Novarro-Escudero, M. Velázquez Blanco, C.A. Arbo Oze de Morvil, A.B. Olmedo Bareiro, G. Meza Rojas, A. Flores, J.A. Hancco-Saavedra, V. Pérez Jimenez, C. Abanto Argomedo, L. Rodriguez Kadota, R. Crosa, D.L. Mora Cuervo, A.C. de Souza, L.A. Carbonera, T.F. Álvarez Guzmán, N. Maldonado, N.L. Cabral, C. Anderson, P. Lindsay, A. Hennis, V.L. Feigin, Priorities to reduce the burden of stroke in Latin American countries. Lancet Neurol. **18**(7), 674–683 (2019)
10. G.D. Whitiana, V. Vitriana, A. Cahyani, Level of activity daily living in post stroke patients. Althea Med. J. **4**(2), 261–266 (2017)
11. G. Colombo, M. Joerg, R. Schreier, V. Dietz, Treadmill training of paraplegic patients using a robotic orthosis. J. Rehabil. Res. Develop. **37**(6), 693–700 (2000)
12. J.L. Pons, Human-robot cognitive interaction, in *Wearable Robots: Biomechatronic Exoskeletons*, chap. 4 (Wiley, Hoboken, 2008)
13. A. Kilicarslan, S. Prasad, R.G. Grossman, J.L. Contreras-Vidal, High accuracy decoding of user intentions using EEG to control a lower-body exoskeleton, in *Proceedings of the Annual International Conference of the IEEE Engineering in Medicine and Biology Society, EMBS* (2013), pp. 5606–5609
14. A. Costa, R. Salazar-Varas, E. Ianez, A. Ubeda, E. Hortal, J.M. Azorin, Studying cognitive attention mechanisms during walking from EEG signals, in *Proceedings - 2015 IEEE International Conference on Systems, Man, and Cybernetics, SMC 2015* (2016), pp. 882–886
15. L.I. Minchala, F. Astudillo-Salinas, K. Palacio-Baus, A. Vazquez-Rodas, Mechatronic design of a lower limb exoskeleton, in *Design, Control and Applications of Mechatronic Systems in Engineering* (2017)
16. F.L. Haufe, A.M. Kober, K. Schmidt, A. Sancho-puchades, J.E. Duarte, P. Wolf, R. Riener, User-driven walking assistance: first experimental results using the MyoSuit *, in *2019 IEEE 16th International Conference on Rehabilitation Robotics (ICORR)* (2019), pp. 944–949
17. H. Kawamoto, S. Lee, S. Kanbe, Y. Sankai, Power assist method for HAL-3 using EMG-based feedback controller, in *Proceedings of the IEEE International Conference on Systems, Man and Cybernetics*, vol. 2 (2003), pp. 1648–1653
18. C. Castellini, P. Van Der Smagt, Surface EMG in advanced hand prosthetics. Biol. Cybern. **100**(1), 35–47 (2009)
19. S.K. Das, Adaptive physical human-robot interaction (PHRI) with a robotic nursing assistant. Ph.D. Thesis, University of Louisville (2019)
20. P.D. Labrecque, C. Gosselin, Variable admittance for pHRI: from intuitive unilateral interaction to optimal bilateral force amplification. Robot. Comput.-Integr. Manuf. **52**, 1–8 (2018)
21. A. Bicchi, M.A. Peshkin, J.E. Colgate, Safety for physical human–robot interaction. *Springer Handbook of Robotics* (Springer, Berlin, 2008) pp. 1335–1348
22. A.Q. Keemink, H. van der Kooij, A.H. Stienen, Admittance control for physical human–robot interaction. Int. J. Robot. Res. **37**(11), 1421–1444 (2018)
23. S. Haddadin, E. Croft, Physical human-robot interaction, in *Springer Handbook of Robotics* (Springer, Cham, 2016), pp. 1835–1874
24. I. Díaz, J.J. Gil, E. Sánchez, Lower-limb robotic rehabilitation: literature review and challenges. J. Robot. **2011**(i), 1–11 (2011)
25. A. Martínez, B.E. Lawson, M. Goldfarb, Preliminary assessment of a lower-limb exoskeleton controller for stroke rehabilitation in overground walking, in *IEEE International Conference on Rehabilitation Robotics* (2017)
26. A.C. Villa-Parra, D. Delisle-Rodriguez, J.S. Lima, A. Frizera-Neto, T. Bastos, Knee impedance modulation to control an active orthosis using insole sensors. Sensors **17**(12), 2751 (2017)
27. M. Sanchez-Manchola, D. Gomez-Vargas, D. Casas-Bocanegra, M. Munera, C.A. Cifuentes, Development of a Robotic lower-limb exoskeleton for gait rehabilitation: AGoRA exoskeleton, in *2018 IEEE ANDESCON Conference Proceedings* (2018), pp. 1–6
28. M.D. Sánchez Manchola, L.J. Arciniegas Mayag, M. Múnera, C.A. Garcia, Impedance-based backdrivability recovery of a lower-limb exoskeleton for knee rehabilitation, in *4th IEEE Colombian Conference on Automatic Control: Automatic Control as Key Support of Industrial Productivity, CCAC 2019 - Proceedings, (Medellin, Colombia)* (2019), pp. 1–6

29. T. Poliero, C. Di Natali, M. Sposito, J. Ortiz, E. Graf, C. Pauli, E. Bottenberg, A. De Eyto, D.G. Caldwell, Soft wearable device for lower limb assistance: assessment of an optimized energy efficient actuation prototype, in *2018 IEEE International Conference on Soft Robotics, RoboSoft 2018* (2018), pp. 559–564
30. J. Schuy, A. Burkl, P. Beckerle, S. Rinderknecht, A new device to measure load and motion in lower limb prosthesis - tested on different prosthetic feet, in *2014 IEEE International Conference on Robotics and Biomimetics, IEEE ROBIO 2014* (2014), pp. 187–192
31. M. Molinari, M. Masciullo, F. Tamburella, N.L. Tagliamonte, I. Pisotta, J.L. Pons, Exoskeletons for over-ground gait training in spinal cord injury. Biosyst. Biorobot. **19**, 253–265 (2018)
32. T. Bacek, M. Moltedo, K. Langlois, G.A. Prieto, M.C. Sanchez-Villamañan, J. Gonzalez-Vargas, B. Vanderborght, D. Lefeber, J.C. Moreno, BioMot exoskeleton - towards a smart wearable robot for symbiotic human-robot interaction, in *IEEE International Conference on Rehabilitation Robotics* (2017), pp. 1666–1671
33. A.J. Young, D.P. Ferris, State of the art and future directions for lower limb robotic exoskeletons. IEEE Trans. Neural Syst. Rehabil. Eng. **25**(2), 171–182 (2017)
34. F. El Zahraa Wehbi, W. Huo, Y. Amirat, M.E. Rafei, M. Khalil, S. Mohammed, Active impedance control of a knee-joint orthosis during swing phase, in *IEEE International Conference on Rehabilitation Robotics* (2017), pp. 435–440
35. W.M. Dos Santos, A.A.G. Siqueira, Impedance control of a rotary series elastic actuator for knee rehabilitation, in *The International Federation of Automatic Control (IFAC)*, vol. 19 (2014), pp. 4801–4806
36. A. Taherifar, G. Vossoughi, A.S. Ghafari, Variable admittance control of the exoskeleton for gait rehabilitation based on a novel strength metric. Robotica **36**, 427–447 (2018)
37. V. Grosu, C. Rodriguez-Guerrero, S. Grosu, B. Vanderborght, D. Lefeber, Design of smart modular variable stiffness actuators for robotic-assistive devices. IEEE/ASME Trans. Mechatron. **22**(4), 1777–1785 (2017)
38. L.M. Mooney, H.M. Herr, Biomechanical walking mechanisms underlying the metabolic reduction caused by an autonomous exoskeleton. J. NeuroEng. Rehabil. **13**(1), 1–12 (2016)
39. N. Abhayasinghe, I. Murray, Human activity recognition using thigh angle derived from single thigh mounted IMU data, in *2014 International Conference on Indoor Positioning and Indoor Navigation (IPIN 2014)* (2014), pp. 111–115
40. D. Micucci, M. Mobilio, P. Napoletano, UniMiB SHAR: a dataset for human activity recognition using acceleration data from smartphones. Appl. Sci. **7**(10), 1101 (2017)
41. B. Barshan, M.C. Yüksek, Recognizing daily and sports activities in two open source machine learning environments using body-worn sensor units. Comput. J. **57**(11), 1649–1667 (2013)
42. M.N. Victorino, X. Jiang, C. Menon, *Wearable Technologies and Force Myography for Healthcare* (Elsevier, Amsterdam, 2018)
43. M.D. Manchola, M.J. Bernal, M. Munera, C.A. Cifuentes, Gait phase detection for lower-limb exoskeletons using foot motion data from a single inertial measurement unit in hemiparetic individuals. Sensors **19**(13), 2988 (2019)
44. M. Motor, Maxon flat EC motor. Maxon (2017). https://www.maxongroup.com/medias/sys_master/root/8831018893342/2018EN-270.pdf
45. J.F. Veneman, R. Kruidhof, E.E.G. Hekman, R. Ekkelenkamp, E.H.F.V. Asseldonk, H. van der Kooij, Design and evaluation of the LOPES exoskeleton robot for interactive gait rehabilitation. IEEE Trans. Neural Syst. Rehabil. Eng. **15**(3), 379–386 (2007)
46. A. Ortlieb, M. Bouri, R. Baud, H. Bleuler, An assistive lower limb exoskeleton for people with neurological gait disorders, in *2017 International Conference on Rehabilitation Robotics (ICORR)* (2017), pp. 441–446
47. S.O. Schrade, K. Dätwyler, M. Stücheli, K. Studer, D.A. Türk, M. Meboldt, R. Gassert, O. Lambercy, Development of VariLeg, an exoskeleton with variable stiffness actuation: first results and user evaluation from the CYBATHLON 2016 Olivier Lambercy; Roger Gassert. J. NeuroEng. Rehabil. **15**(1), 1–18 (2018)
48. N. Hogan, Impedance control: an approach to manipulation, in *1984 American Control Conference* (1984), pp. 304–313

49. E.S. Barjuei, S. Toxiri, G.A. Medrano-Cerda, D.G. Caldwell, J. Ortiz, Bond graph modeling of an exoskeleton actuator, in *2018 10th Computer Science and Electronic Engineering Conference, CEEC 2018 - Proceedings* (2019), pp. 101–106
50. F.R.O. Andres, A. Lopez-Delis, A.F. da Rocha, *Upper and Lower Extremity Exoskeletons* (Elsevier, Amsterdam, 2018)
51. Z. Li, Z. Yin, Position tracking control of mass spring damper system with time-varying coefficients, in *Proceedings of the 29th Chinese Control and Decision Conference, CCDC 2017* (2017), pp. 4994–4998
52. A.A. Nikooyan, A.A. Zadpoor, Mass-spring-damper modelling of the human body to study running and hopping-an overview. Proc. Inst. Mech. Eng H J. Eng. Med. **225**(12), 1121–1135 (2011)
53. C. Bayón, O. Ramírez, F. Mollà, J. Serrano, M. Del Castillo, J. Belda-Lois, R. Poveda, R. Raya, T. Martín Lorenzo, I. Martínez Caballero, S. Lerma Lara, C. Cifuentes, A. Frizera, E. Rocon, CPWalker, robotic platform for gait rehabilitation and training in patients with cerebral palsy, in *2016 IEEE International Conference on Robotics and Automation (ICRA)* (2015), pp. 3736–3741
54. A.C. Villa-Parra, D. Delisle-Rodriguez, T. Botelho, J.J.V. Mayor, A.L. Delis, R. Carelli, A. Frizera Neto, T.F. Bastos, Control of a robotic knee exoskeleton for assistance and rehabilitation based on motion intention from sEMG. Res. Biomed. Eng. **34**(3), 198–210 (2018)
55. M.D. Sánchez Manchola, L.J. Arciniegas Mayag, M. Munera, C.A. Garcia, Impedance-based backdrivability recovery of a lower-limb exoskeleton for knee rehabilitation, in *4th IEEE Colombian Conference on Automatic Control: Automatic Control as Key Support of Industrial Productivity, CCAC 2019 - Proceedings* (2019), pp. 1–6
56. J.B. Webster, B.J. Darter, Principles of normal and pathologic gait, in *Atlas of Orthoses and Assistive Devices*, chap. 4, 5th edn. (Elsevier, Amsterdam, 2019), pp. 49–62.e1
57. T. Roskilly, R. Mikalsen, Closed-loop stability, in *Marine Systems Identification, Modeling and Control* (Elsevier, Amsterdam, 2015), pp. 97–122
58. K. Wen, D. Necsulescu, J. Sasiadek, Haptic force control based on impedance/admittance control. IFAC Proc. Vol. **38**(1), 427–432 (2005)
59. A. Fortin-Coté, P. Cardou, C. Gosselin, An admittance control scheme for haptic interfaces based on cable-driven parallel mechanisms, in *Proceedings - IEEE International Conference on Robotics and Automation* (2014), pp. 819–825
60. M. Bortole, A. Del Ama, E. Rocon, J.C. Moreno, F. Brunetti, J.L. Pons, A robotic exoskeleton for overground gait rehabilitation, in *Proceedings - IEEE International Conference on Robotics and Automation* (2013), pp. 3356–3361
61. S.K. Banala, S.H. Kim, S.K. Agrawal, J.P. Scholz, Robot assisted gait training with active leg exoskeleton (ALEX), in *Proceedings of the 2nd Biennial IEEE/RAS-EMBS International Conference on Biomedical Robotics and Biomechatronics, BioRob 2008*, vol. 17 (2008), pp. 653–658

Brain–Computer Interface for Controlling Lower-Limb Exoskeletons

9

Angie Pino, Nicolás Tovar, Patricio Barria, Karim Baleta, Marcela Múnera, and Carlos A. Cifuentes

9.1 Introduction

About 15 million people around the world suffer a stroke each year [1]. After a stroke episode, one or more effects may be triggered, such as muscle weakness, hemiparesis, hemiplegia, fatigue, and spasticity. Those affectations are related in turn to limitations in the execution of different activities of daily living, restriction in participation, and a high degree of dependency on third parties [2]. Therefore, stroke is one of the leading causes of physical disability directly affecting the quality of life [1].

Post-stroke rehabilitation is a patient-centered process to maximize patients' functional independence who have suffered a series of disabilities associated with the episode [3]. Assistive technologies for motor rehabilitation include exoskeletons and robotic orthoses, which may provide high motor intensity, repeatability, and

A. Pino · N. Tovar · M. Múnera · C. A. Cifuentes (✉)
Biomedical Engineering, Department of the Colombian School of Engineering Julio Garavito, Bogotá D.C., Colombia
e-mail: angie.pino-l@mail.escuelaing.edu.co; bryan.tovar@mail.escuelaing.edu.co; marcela.munera@escuelaing.edu.co; carlos.cifuentes@escuelaing.edu.co

P. Barria
Department of Electrical Engineering, University of Magallanes, Punta Arenas, Chile

Club de Leones Cruz del Sur Rehabilitation Center, Punta Arenas, Chile

Brain-Machine Interface Systems Lab, Systems Engineering and Automation Department, Miguel Hernández University of Elche, UMH, Elche, Spain
e-mail: pbarria@rehabilitamos.org

K. Baleta
Club de Leones Cruz del Sur Rehabilitation Center, Punta Arenas, Chile
e-mail: kbaleta@rehabilitamos.org

C. A. Cifuentes, M. Múnera, *Interfacing Humans and Robots for Gait Assistance and Rehabilitation*, https://doi.org/10.1007/978-3-030-79630-3_9

precision [4]. However, one of the most critical problems that must be solved for the clinical implementation of these developments is their control systems.

Conventional control includes tools as inertial sensors, direct contact operation, and external transducers. However, despite the effectiveness of traditional control systems, some authors insist that these methods ignore the patient's involvement with the system regarding neurofeedback progression [5]. Thus, robotics-based rehabilitation becomes a process that does not fully exploit the patient's ability to generate neuroplasticity progressively, since neurological intend is not directly implicated [5].

In this way, given the rise of the Brain–Computer Interfaces (BCI) paradigm, many developments have focused their applications on motor rehabilitation and language assistance [6]. Control of exoskeletons and orthoses BCI-based has been extensively studied. Some research affirms that physical therapies involving BCI in patients with neuromotor conditions may improve their neuroplasticity more effectively [7]. There are many paradigms and modalities of BCI used in research; one of the most approached is Motion Imagery (MI) analysis, which is based on the electrical activity of the motor cortex that occurs when there is a movement intention of the subject [7]. This strategy seeks to improve the patient's interaction with the therapeutic mechanisms that pursuit an evolution of the neuroplasticity, adequately including the use of the neuromotor abilities through the BCI system in the rehabilitation process [8].

Following the above, this chapter discusses the main concepts of designing a BCI system in the robot-assisted rehabilitation field. To do so, the chapter is organized into seven sections. Section 9.2 conceptualizes the BCI term and electroencephalography (EEG) signals. Section 9.3 aims to present the stages in the universal design of a BCI system. Section 9.4 focuses on a literary review about BCI systems in lower-limb rehabilitation. Section 9.5 introduces the integrating control system for an ankle exoskeleton, based on the analysis of EEG signals and their involvement in the locomotor system. Section 9.6 addresses a case study with a post-stroke patient to evaluate the operation of a BCI control system for the exoskeleton control. Finally, the last section of the chapter presents conclusions and future works.

9.2 Brain–Computer Interface and Electroencephalographic Signals

Brain–computer interface (BCI) is considered a relatively novel communication method between a user and a machine. This communication may work as a control system in which human mind thoughts are translated into real-world interactions. Some recent studies have shown a significant role in future technologies for assisting people with disabilities [9, 10]. In rehabilitation and assistance technologies, the ideal BCI system is when a device may be controlled as naturally as using a human body limb [11]. To do this, BCI relies on EMG signals denoting the sum of the neurons' action potentials throughout receiving and processing sensory inputs from other neurons or external stimuli [12]. That means the EEG technology may accurately measure brainwave activity [13]. The way to access this physiological data is through sensitive electrodes attached to the scalp. The most common recording technique is the application of 21 electrodes and an equal number of channels. Other techniques include 256 electrodes and a number up to 64 channels [12, 13].

Even when there are other methods for extracting the brain activity, for instance, electrocorticograms (ECoGs) [14], magnetoencephalograms (MEGs) [15], functional magnetic resonance imaging (fMRI) [16], and near-infrared spectroscopy (fNIRS) [17], the popularity of EEG makes it widely used due to its non-invasive action, compatibility, portability, and its high temporal resolution in comparison with the mentioned methods above. Nevertheless, the EEG has a weak signal and is prone to several artifacts and relatively low spatial resolution [12]. This type of signal is generally in the order of microvolts (μV) range. Moreover, many investigations have categorized the EEG signals in the frequency domain, and until now, these ranges are divided into five main categories, which consists of delta (δ) (0.5–4 Hz), theta (θ) (4–8 Hz), alpha (α) (8–13 Hz), beta (β) (13–30 Hz), and gamma (γ)(>30 Hz).

9.3 BCI Control System Design

Every BCI system has a basic structure. According to He et al. [5], there are four primary stages to construct a universal BCI system. The first one is the signal acquisition from the brain. The second one is the pre-processing stage of the signal mentioned above. The third stage refers to the processing, which includes feature extraction and decoding or translation. Finally, the last stage is the execution stage that puts the device into operation according to the human brain's intent (see Fig. 9.1).

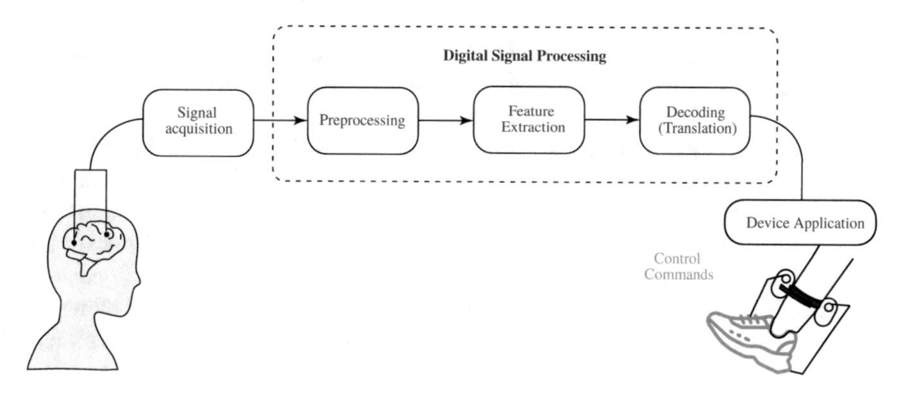

Fig. 9.1 BCI basic system diagram that includes signals acquisition and processing

It is essential to add that the basic diagram could change in some aspects, for instance, depending on the system design. The following subsections will describe the general objective of each stage.

9.3.1 Signal Acquisition

As was mentioned above, EEG is the preferred tool to extract brain activity from the user. However, the signal acquisition process may be executed in numerous ways depending on the suitable BCI modality. These modalities could be classified into two categories: exogenous and endogenous [18].

- **Endogenous Modalities:** In this case, EEG acquisition is produced independently from external stimulation. Namely it may entirely be managed voluntarily by the user. This modality is mainly applied to subjects who have neurological issues [18]. In this manner, the BCI could offer a more natural and spontaneous way of interaction. Neuroplasticity is a fundamental feature that may be improved with this modality [19].

  For instance, Event-Related Desynchronization/Synchronization (ERD/ERS) works based on the behavior of brain signals and motor intent. Frequency bands may show a power increasing or decreasing when a subject imagines or executes a lower movement [7]. Evidence has been shown that the methods to detect lower-limb motor imagery with ERD/ERS usually focus on the potency of the beta rebound band of the EEG cortical due to an abrupt increase in the signal power just when the movement of the lower-limb ends [20] (see Fig. 9.2). This behavior occurs similarly at the end of the imagination of a movement. Other strategies as the Movement-Related Cortical Potentials (MRCPs) are based on a set of power variations in the cortical activity before and after the movement execution [21].
- **Exogenous Modalities:** An exogenous BCI refers to the generation of external stimuli to add more effectiveness. There are many types of stimuli where the most common are auditory and visual [18]. This modality implies simple training strategies compared with the endogenous modalities, necessary for the subject to drive suitably the BCI system.

  External instruments as the Steady-State Visual Evoked Potential (SSVEP) are used to excite brain signals based on a set of multiple visual stimuli, such as LEDs or figures on a computer screen [22]. Likewise, the Event-Related Potential: P300 in BCI applications forces the subject to focus on the selected item on the screen and ignore the rest [23]. In this case, the positive deflection appears approximately 300 ms after presenting an attended stimulus. Nevertheless, one issue found in all these exogenous modalities is that the subject cannot manage the entire device independently and is dependent on external conditions.

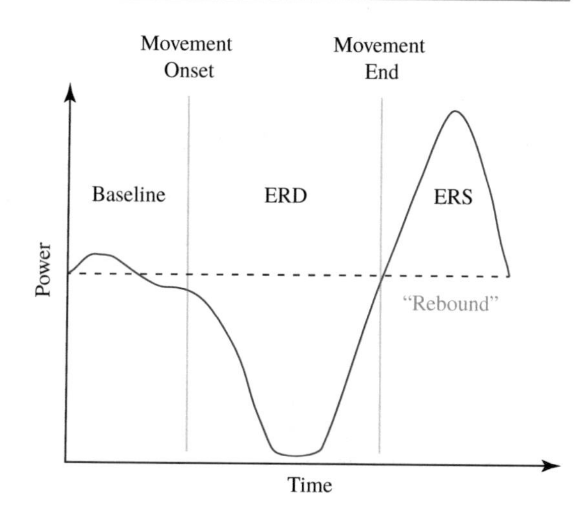

Fig. 9.2 Event-related desynchronization/synchronization (ERD/ERS) power behavior

In addition to the endogenous and exogenous modalities, the hybrid BCI modality (h-BCI) combines the unique advantages of two different systems or signals to make the BCI control more effective, accessible, and optimal [24]. Some signals could be added as hybrid BCI (h-BCI) systems or feedback sources to improve rehabilitation or assistance aspects [25, 26]. According to Hong et al. [27], there are three objectives for implementing an h-BCI. The first one is to enhance classification accuracy. The second is to increase the number of brain commands for control application. Finally, the third objective is to achieve a shortened brain-command detection time.

An example of a signal hybrid modality is the union of electromyographic (EMG) signals with EEG, a promising alternative for rehabilitation therapies [28]. EMG signals indicate muscles' electrical activity, which changes when a voluntary or not voluntary contraction appears. Therefore, the EMG signal confirms a detection system for muscular movement [27]. That said, the incorporation of these signals depends on the task the subject performs. However, in any case, EMG control is used as an additional control system in biomechanical action. For instance, the laterality detection [27] and the biomechanical freedom degrees detection [29] could be pretty complicated with only an EEG-based system.

Another h-BCI modality could combine both ERD/ERS and SSVEP systems. Bunner et al. [30] have achieved a high accuracy system carrying this proposal out. Other authors have done experiments with this implementation, concluding that this modality does not need an exhaustive training process. Moreover, this modality could reduce the non-legible population by 20% [31]. The above shows an improvement in the most significant disadvantage of the ERD/ERS individual modality, where generally one part of the population is not eligible due to the intrinsic users' characteristics against distractor factors on BCI performance [32].

9.3.2 Pre-processing

Once a set of signals are obtained, it is necessary to consider that this set is generally entirely raw and full of artifacts depending on the technology used for this objective, the environment, and the user's physical conditions (e.g., noise related to the hardware, electrode wear, interference and skin impedance fluctuations) [33]. Therefore, denoising and cleaning the data is a widely studied process that already has numerous advances. For instance, the Filter Bank Common Spatial Patterns (FBCSP) is the most used method in BCI systems due to its efficacy in pre-processing signals and further stages [33, 34]. On the other side, according to Tariq et al. [35], interference issues could also be managed through digital filters as a notch filter. Likewise, other methods such as independent component analysis (ICA), principal component analysis (PCA), non-linear adaptive filtering, and dipole analysis have been tested.

9.3.3 Feature Extraction

After the signal pre-processing, this stage oversees classifying as many features as the BCI system requires. Some BCIs are based on Motor Imagery (MI) [7]; this modality is related to specific frequency bands. Therefore, its use depends on the suitable method to characterize these specific ranges for future decoding to build a set of commands necessary to control the target device [36].

Numerous feature extraction methods have been studied in BCI systems. However, these strategies depend on the modality structure. According to Lotte et al. [37], the extraction methods could be divided into three categories: (1) Time-domain analysis, (2) Frequency-domain analysis, (3) Time–frequency-domain analysis, and (4) Spatial complimentary analysis. One of the most used methods is instantaneous statistics and autoregressive methods (AA) in a time-domain analysis. Likewise, frequency-domain methods include Fourier Fast Transform analysis, Short Time Fourier Transform (STFT), and Power Spectral analysis. Frequency-time techniques, for their part, may span Wavelet Transform and Hilbert–Huang Transform (HHT). The fourth classification is the Common Spatial Patterns and has been widely used [37].

9.3.4 Decoding

The feature extraction layer forms a set of classification that the decoding stage uses to identify the intent brain signals, namely, to manipulate the robotic device via machine-understandable commands for interfacing [18]. The system generally works by making a weighted class estimate, presented by a feature vector for mapping the desired driving application command. Some of these strategies are Linear Discriminant Analysis (LDA), Support Vector Machine (SVM), Gaussian Mixture Model (GMM), and Artificial Neural Network (ANN) [35].

9.4 Lower-Limb Exoskeletons with BCI Systems Review

This section introduces recent works on developing BCI Systems focused on the lower-limb robotic devices, the modality for signal acquisition, pre-processing strategies, and main results. Table 9.1 summarizes the principal data found in the research for each exoskeleton with a BCI integrated system.

9.4.1 Lokomat

Lokomat (Hocoma, Switzerland) is a robotic treadmill exoskeleton to automate locomotion training for spinal cord injured and stroke patients [38]. BCI system in *Lokomat* device was researched looking for the subject participation improvement. Initially, this orthosis worked in a training mode where the device influenced the subject's motion with a fixed gait pattern [38,39]. Even when some reports conclude this causes greater coordination of the muscles and the neuromotor system, BCI became an alternative to improving the device.

Donatti et al. [40] show some clinical assessments. Eight (8) chronic spinal cord injury (SCI) paraplegics were subjected to long-term training with a multi-stage BCI-based gait neurorehabilitation paradigm aimed at restoring locomotion. The BCI system modality was the MRCP. The Common Average Patterns method was used for pre-processing stage, including conventional digital filters for denoising. Moreover, the decoding process was based on Linear Discriminant Analysis (LDA) methods [40].

The methodology of these authors was composed of six parts, where the patient was addressed to familiarize the BCI with a tactile feedback system. Then this same familiarization continued in an orthostatic position supported by a stand-in table. Without using a BCI, they began conventional training with the *Lokomat* device, including body weight support (BWS) on the treadmill. Then a BWS training was continued without *Lokomat's* joint support. Finally, in sections the BCI was integrated into the gait training system supported by the tactile feedback system (on treadmill and overground, respectively).

After 1 year of training in the *Lokomat* BCI system, all eight subjects improved neurological motion and somatic sensation as pain and proprioceptive sensing. In terms of neuroplasticity, the research showed no significant differences between a desynchronization and synchronization of the beta wave from an event-related potential analysis at the onset of the training therapy period. However, after 10 months of therapy, these synchronization differences were observed in all patients. In terms of anatomic improvements, all patients exhibited a complete ROM of the joints and a maximal grade of lower-limb spasticity of 2 on the Ashworth scale. Furthermore, a test provided by *Lokomat* developers known as L-stiff was used. This test is in charge of quantifying the spasticity of hip and knee muscles for flexors and extensors. Thus, on average, all patients exhibited a reduced spasticity level by the end of 12 months.

Table 9.1 Comparison of the paradigms, characteristics, and results of the BCI system applied to lower-limb exoskeletons

Robotic device	Joints	Modality/paradigm	Pre-processing methods	Feature extraction methods	Decoding methods	Participants	Main results
Lokomat [38–40]	– Hip – Knee – Ankle	MRCP	– Common average patterns – Temporal filters	N/I*	LDA	8 SCI patients	– Neuroplasticity generation observed from significant beta wave de/synchronization – The patients exhibit complete ROM joint after therapy – Spasticity reduced
RoGO [41–43]	– Hip – Knee – Ankle	ERD/ERS – Serious games feedback	Own prediction channels method	– Frequency analysis – PSD – PDA	AIDA	1 Healthy subject 1 SCI patient	– 85% BCI commands accuracy – 0.812 correlation between natural movement and BCI MI detection – Average of 0.8 false alarms in BCI commands – No omissions in BCI commands
H2 [44, 45, 47]	– Hip – Knee – Ankle	h-BCI: – ERD/ERS – MRCP	– Z-scores method – Temporal filters	– Laplacian spatial filter – Common average patterns – PDA	SDA	3 healthy subjects 4 SCI patients	– 84% accuracy commands for a healthy subject – 77% accuracy for SCI patient – 55% false alarms commands or healthy subjects – 40% false alarms command for SCI patients

Rex [48–50]	– Hip – Knee – Ankle	MRCP	– Z-scores method – Temporal filters	Own method for delta wave isolation	MKL	1 healthy subject 1 SCI patient	– A rise of accuracy BCI commands from 60% to 90%
MAFO [51]	– Ankle	MRCP	– Spatial filters – Temporal filters	Local preserving projection	LDA	10 healthy subjects	– 73% accuracy reported – Feasible plasticity induction
H2 Foot-Ankle Orthosis [51]	– Ankle	– ERS/ERD – FES feedback	– Temporal filters	PSD	Own classi-fication method	5 healthy subjects	– 100% BCI-FES response (no omissions) –Only one subject had one false alarm

9.4.2 RoGo

RoGo (University of California, USA) is a robotic gait orthosis addressed to Spinal Cord Injury (SCI) patients [41]. This orthosis has been studied mainly with the BCI system control [41–43]. The BCI modality used in the investigation includes ERD/ERS induced by the kinesthetic motor imagination of the left hand, right hand, and feet. The pre-processing method is briefly described by Wang et al. [43] and includes an EEG prediction model that excludes those EEG channels with excessive artifacts. Two states were defined in the feature extraction method, Idling, and Walking states. Then this data was transformed in the domain frequency and their Power Spectral Densities. Moreover, a PCA algorithm was applied to reduce the data dimension. Finally, the researchers use the AIDA method to classify the commands to *RoGo*. Serious games have been implemented before a complete integration to the rehabilitation device. For instance, one experiment proposed to drive an avatar that expects to stop with a specific indication. The results gathered all the correct and wrong attempts and showed an 85% accuracy.

This system was assessed in a study by Do et al. [41] with a clinical assessment where patients with SCI impairments and one healthy subject were compared. The performance of this system was assessed by calculating the cross-correlation and latency between the computerized cues and BCI-RoGO response, and the omission and false alarm rates. The methodological protocol consisted of three divisions: (1) active walking (subject voluntarily walks while the *RoGO* servos are turned off), (2) cooperative walking (subject walks synergistically with the *RoGO*), and (3) passive walking (the subject is fully relaxed while the RoGO makes walking movements). Those different training stages were helpful in set baseline values for EMG and EEG. Finally, the accuracy of the EEG prediction model averaged 86.30% across both subjects. The cross-correlation between instructional cues and the BCI-RoGO walking epochs averaged across all subjects, and all sessions were 0.812. Also, there were, on average, 0.8 false alarms per session and no omissions.

9.4.3 H2 Exoskeleton

H2 (Technaid S.L., Spain) exoskeleton was developed in Spain and is addressed to stroke patients with gait impairments [44]. This device is aimed to assist and rehabilitate patients with suitable walking in a natural environment. According to the researchers, the exoskeleton has six joints, including the hip, knee, and ankle. Moreover, *H2* presents an open architecture that allows modifications in the control system [45, 46]. López-Laraz et al. [47] implemented a BCI control system with ERD/ERS-MRCP hybrid modality. Pre-processing methods are based on an automated procedure based on z-scores to eliminate the trials containing artifacts and conventional denoising filters. The ERD features were calculated after applying a small Laplacian filter to the frontocentral, central, and centroparietal EEG channels in terms of feature extraction. On the other hand, the Common

Average Patterns method was used for the MRCP modality. For the decoding process, a strategy named Sparse Discriminant Analysis (SDA) was used.

López-Laraz et al. [47] presented a clinical assessment where three (3) healthy subjects and four (4) SCI patients were tested. The basic system uses the BCI described above to trigger exoskeletons' assistive motion. Factors as fatigue and exertion level, usability, and user satisfaction were assessed. Results concluded for healthy subjects with approximately 84% of accuracy, and SCI subjects 77%. On average, 55 and 40% of the trials (for healthy subjects and patients, respectively) have suffered unexpected activations without the proposed control strategy.

9.4.4 Rex Exoskeleton

Rex(Rex Bionics Ltd., New Zealand) is an exoskeleton that aimed to assist rehabilitation and mobility for those with neurological and spinal injuries [48]. *Rex* has been developed for private users that can now perform tasks that are not possible when sitting in a wheelchair. Specifically, the exoskeleton aids the patient to improve git patterns and movement for standing and sitting [49]. A joystick system initially drove *Rex*, but some BCI systems were designed to include this device in the rehabilitation field [50] .

Zhang et al. [50] made an investigation with MRCP-based BCI implemented on the *Rex* exoskeleton. The authors included a filter in the 0.1–2 Hz range in terms of the pre-processing stage using a second-order Butterworth filter and standardized z-score method. In the feature extraction stage, isolation of the delta band was carried out. For the classification stage, a Multiple Kernel Learning (MKL) was used and compared with the SMV algorithm, where they conclude MKL was more suitable for the system. A clinical assessment was performed with two (2) subjects: one healthy subject and one with SCI impairment. Results conclude that the frontal/frontocentral regions were the most critical regions for classifying gait states of the tested subjects, consistent with the brain regions hypothesized to control lower-limb movements. Moreover, the classification accuracy increased, and the findings suggest cortical plasticity triggered by the BCI use.

9.4.5 Motorized Ankle–Foot Orthosis: MAFO

This study was carried out by Xu et al. [51], where a BCI system was applied to the *Motorized Ankle–Foot Orthosis (MAFO)*. The mentioned orthosis allows the assistance of the ankle dorsiflexion movement. The objective of this research was focused on the evaluation of the functionality of the BCI system commands and the verification of an increase in neuroplasticity in the subject. The paradigm chosen by the researchers was MRCP. Spatial filters and temporal filters were used as pre-processing. The processing means were defined by a Locality Preserving Projection (LPP) method in conjunction with the Linear Discriminant Analysis (LDA) decoding method.

The BCI system was evaluated through the manifestation of the subject on possible false commands or omissions. The rate of accurate detections was measured along with the rate of false commands per minute. In addition, subject monitoring was evaluated to verify her motor activity concerning MI. The results yielded 73% accuracy in the general system, weighting the values described. In terms of the induced plasticity, it was determined that there were significant differences before and after the tests that demonstrated induction of neuroplasticity in the subjects' cortical zone.

9.4.6 H2 Foot-Ankle Orthosis

This research led by Do et al. [52] does not include a robotic orthosis directly, but it is part of the research project that wants to improve the *H2* orthosis mentioned above. However, according to the research carried out, reports of this integration have not yet been carried out. Thus, a BCI system was integrated into a Functional Electrical Stimulation (FES) system, which potentially allows a robotic orthosis to be controlled in its dorsiflexion movement and, therefore, act like one. The digital processing of the signals was not described in detail, so they were limited to showing the acquisition process and the tests with the subject.

Five healthy subjects were evaluated, executing ten repetitions interspersed between dorsiflexion and relaxation. The BCI commands were intended to trigger the assistance caused by the FES system. The subject received signals to perform MI or remain at rest, with which the results of the system's functionality were observed. The results showed a correlation between the commands and the signals given to the subject of 0.77. Latencies were measured between the ranges of 1.4–3.1 s. Furthermore, no omissions were evidenced and only one subject had one false alarm.

9.5 BCI System Integration with T-FLEX

BCI Integration with *T-FLEX* (Colombian School of Engineering, Colombia) [53] emerges as a proposal that considers the patient's involvement with the system control through imaginary dorsiflexion movements. Thus, when the BCI system detects the activation, the user receives active movement through the robotic orthosis. In general terms, the integration system consists of the EEG signal acquisition system and the *T-FLEX* ankle exoskeleton (see Fig. 9.3). However, the process to command the device through EEG signals requires specific steps based on the theoretical concepts presented in the previous sections. Moreover, additional strategies are necessary for communication between systems.

Fig. 9.3 Setup for BCI system integration with *T-FLEX*

9.5.1 Signal Acquisition

Electroencephalography (EEG)-based endogenous BCI is selected with an ERD/ERS modality. The objective is to extract characteristics located in the activity of the beta wave rebound power, whose frequency range is from 16 to 24 Hz. The acquisition system is achieved through *Enobio 20 Hardware* (Neroelectrics, Spain). This hardware is linked through the *NIC 2.0* (Neroelectrics, Spain) to allow the motor cortex recording with a Laplacian montage. This type of setup uses multiple electrodes at once as a reference. In this way, a single output channel is related to the neighbor electrode average of a specific electrode. In this case, the acquisition protocol takes as reference the Cz electrode of the international system 10–20. Thus, the output is the average of the acquisition channels C1, C2, FCz, and CPz.

Once the raw signal is acquired through the software, it is connected to a local server capable of transferring this data to the pre-processing, feature extraction, and decoding system. OpenVibe (Inria Rennes, France) is an open-use program that allows the implementation of different BCI modalities. According to the creators, the interfaces in OpenVibe reach a speed of up to 1 selection per 5 s with a selection accuracy of up to 70% in motor imagery [54].

9.5.2 Pre-processing

The pre-processing method was divided into two stages. The first one is applying a Laplacian filter, and the second one is based on a temporal filter. These implementations are described in more detail below.

- **Laplacian Spatial Filter:** This spatial filter calculates the second derivative of the instantaneous spatial voltage distribution for each electrode, and therefore focuses the activity originating from radial sources immediately below the electrode [55]. This tool highlights localized activity and reduces poorly defined activity. Moreover, this filter can create the best possible linear combination of the electrodes used to obtain a signal with less noise and maximized utility in the data [56].
- **Temporal Filter:** According to Clerc et al. [20], this filter is applied conventionally as a Butterworth-type band-pass filter, order 100 and with a 0.5 dB band ripple. Consequently, with the frequency range of the beta wave, whose behavior is essential for applying the ERS/ERD paradigm, the lower and upper cutoff frequencies were 16 and 24 Hz, respectively.

9.5.3 Feature Extraction

The feature extraction is based on the methodology proposed by Clerc et al. [20] to obtain a signal as straightforward as possible and represent the beta wave behavior in its motor synchronization and desynchronization periods. The process is performed in four steps:

1. Filtered signal decomposition into 1 s long epochs with an overlap of 100 ms between two consecutive epochs.
2. Signal square operation.
3. Signal's average calculation over the input epoch. The average of the signal is calculated for each interval of 1 s received from the previous step.
4. Signal crop to a minimum value. The minimum value was obtained by averaging and adding 3 times the standard deviation of the signal acquired during the 5-min calibration period.

9.5.4 Decoding

As mentioned above, the last step of the feature extraction consists of a calibration process that defines a potential threshold of the extracted beta rebound power. Thus, the potentials detected below the threshold value are taken as zero, while those that exceed it would be considered potentials of motion intend. Consequently, there is a proportionality between the intensity of movement and the potential magnitude generated by the beta rebound.

9.5.5 Communication Between Systems: BCI-T-FLEX

T-FLEX is a wearable ankle exoskeleton whose main objective is to assist patients with impairment in the foot-ankle complex. This ankle exoskeleton comprises an

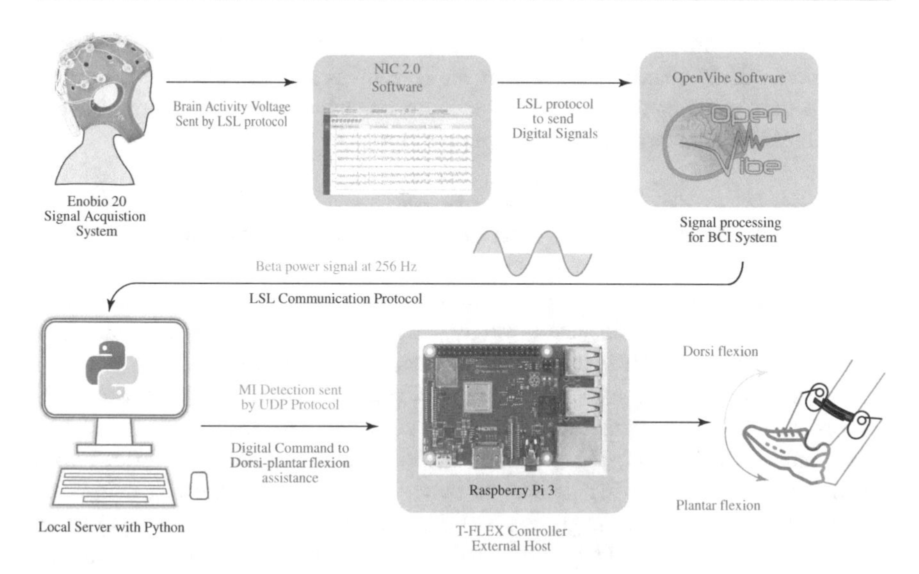

Fig. 9.4 Communication protocol diagram involved in the BCI-T-FLEX integration system

actuator system with bioinspired tendons commanded by a Raspberry Pi 3 to assist gait or perform dorsi-plantarflexion repetitions in stationary therapy (see Chap. 6). Considering the above, communication between BCI System and *T-FLEX* is carried out directly from OpenVibe sending data continuously to the Raspberry Pi 3. However, this communication requires two sections: (1) output of the OpenVibe software to a local server and (2) a delivery of data from the local server to the external server of the robot controller (see Fig. 9.4).

1. **OpenVibe to Local Server Connection:** For this data extraction, the Lab Streaming Layer (LSL) protocol is used. OpenVibe uses a native LSL system, in which it is necessary to specify a name of a transmission channel and the type of signal to be sent. Once this channel was configured in the OpenVibe box system, a local server was created in Python, whose objective is to receive the transmission channel (i.e., an array of variables for each sample of the EEG signal, which includes sample number, time in seconds, channel, encoding type, and magnitude) (see Fig. 9.4).
2. **Local Server to *T-FLEX* Controller:** Once the data arrives continuously through the LSL channel which has a frequency of 256 Hz, this data will be processed to detect the exceeding of the previously defined threshold. This implies that the calibration process must be appropriately associated with the local server created in Python. In this way, every time a threshold is exceeded, the data will be sent as a logical “1” to the Raspberry Pi 3. This will cause an action equivalent to the dorsiflexion assisted by the robot. However, once a drop below the beta rebound signal threshold is detected, the local server will send a logical “0” to the controller, and it will remain in plantarflexion.

The communication protocol used to send this data was the User Datagram Protocol (UDP) connection that uses the IP address data of both parties to carry out a data exchange. In this case, it is an open-loop system that only sends unidirectional data to the *T-FLEX* (see Fig. 9.4).

9.6 Case Study: BCI System Control Assessment with T-FLEX

This case study presents the results of the BCI operation carried out with a post-stroke patient (age: 55 years, weight: 84 Kg, and height: 173 cm) with right hemiparesis laterality. This study seeks to evaluate the operation of a BCI control system for the *T-FLEX* exoskeleton and its preliminary effect on neurological activity. In addition to considering the EEG signal acquisition system and the *T-FLEX* ankle exoskeleton, the proposed system includes a visual interface with a full-screen that guides the actions to be carried out during the test using text instructions. The system integration test for the development of this case study was carried out in Club de Leones Cruz del Sur Rehabilitation Center with its corresponding Ethics Committee approval.

9.6.1 Experimental Procedure

The following procedure is based on experimental designs found in the literature [57]. The experiment is developed under three stages: During the first stage, 5-min calibration is performed while the participant remains statically in a chair with a 90° knee flexion. The second stage corresponds to a stationary therapy (ST) [58] where the EEG signal is recorded while the patient receives alternating dorsi-plantar flexion motion for 10 s using the *T-FLEX* robotic orthosis. Afterward, the last stage considers motor imagination with visual stimulation (MIV) to trigger the *T-FLEX* robotic device. The above implies EEG signal records while the patient imagines alternating dorsi-plantar flexion movement for 10 s while observing an image showing the desired command. The second and third stages alternate with 10 s-periods of rest until reaching a 5-min test.

Both experimental conditions (ST and MIV) involve the use of the *T-FLEX* exoskeleton. Therefore, capturing EEG records is essential to present a posterior comparative analysis in the brain activation frequency band (8–32 Hz). In this way, the quantitative characterization of the BCI system is performed employing the following variables at the end of the test:

- Accuracy rate: The data associated with the motor imagery attempts correctly detected by the BCI will be collected in the 10-s periods in which the patient is asked to imagine movement.
- Acquisition of EEG signals from a cortical zone: Continuous signals will be acquired at each test interval from the cortical zone, using channels Fcz, C1, Cz, C2, and Cpz International System. In this way, signal processing will be

carried out to compare and conclude if there are significant differences in the brain-motor activity of the patient when using the *T-FLEX* device integrated into the BCI system and in its absence. To conclude in this regard, the Event-Related Potential (ERP) methodology will be used.

9.6.2 Results of the Study

The result in terms of the accuracy level of motor imagery detection made by the patient was 53.33%. This result is a consequence of the difficulty of some people to perform motor imagination without prior training. As previously mentioned, this is one of the disadvantages regarding the ERD/ERS modality [32]. In this way, future studies should implement long training to guarantee better performance and control of the system. Figure 9.5 shows the temporal response for the isolate frequency band during one of the 10-s periods when the user was required to perform motor imagination.

Meanwhile, as can be seen in Table 9.0, the MIV test has a higher associated Power Spectral Density related with the Event-Related Potentials (ERP) in the Cz, C2, and Cpz channels vs. the therapy mode of the *T-FLEX* device, in which the patient was not required to generate motor imagery (ST).

The above is corroborated in the EEG topographies that tend to be oriented to increased brain activity on visual stimulation tests (see Fig. 9.6). These significant differences in brain activity between these tests may indicate the significant difference in the conventional therapy mode and the use of integrated BCI. Therefore, these results are helpful since they show a preliminary added utility in the proposed integration concerning the conventional use of *T-FLEX*. However, to generate a significant difference in motor imagery-related brain activity, the most viable paradigm for future research must include another kind of stimulation besides the visual.

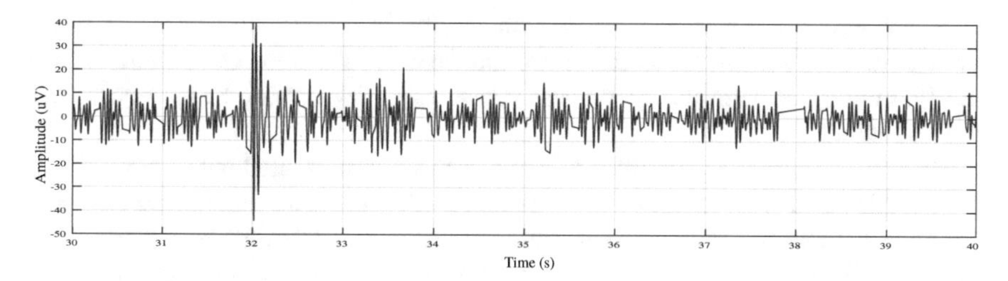

Fig. 9.5 Cz channel filtered signal on the band 8–32 Hz in motion imagination detection state with visual stimulation to command the ankle exoskeleton

Table 9.0 Power spectral density (PSD) associated with each channel in the ST and MIV test

Test	Fcz	C1	Cz	C2	Cpz
ST PSD (dB/Hz)	6.81	6.46	6.84	7.60	6.30
MIV PSD (dB/Hz)	7.75	7.66	7.61	8.83	7.19

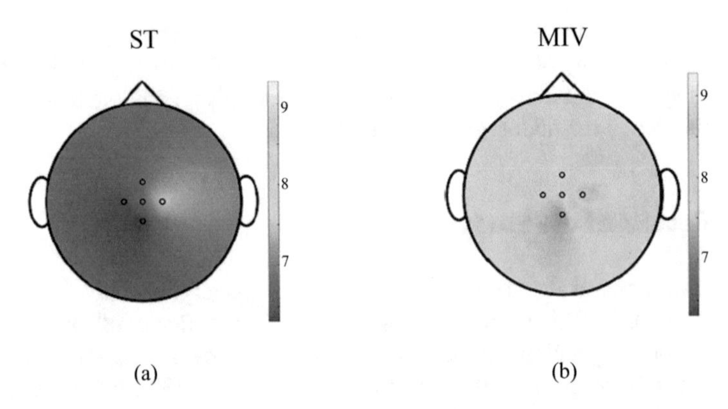

Fig. 9.6 EEG topographies of power spectral density associated with the event-related potentials for tests with and without motor imagination. (**a**) Stationary therapy. (**b**) Motor imagination with visual stimulation

It is essential to mention that these results are partially beneficial for generating neuroplasticity in post-stroke patients [59,60]. However, this single-session test does not prove that neuroplasticity was generated or induced in the patient using this interface. According to long-term research, this can be demonstrated in therapies with BCI systems and exoskeletons lasting between 10 to 12 months with a weekly intensity session [40]. Therefore, this case study is limited to achieving immediate results related to partially beneficial brain activity.

9.7 Chapter Conclusions

State of the art and conceptualization carried out in this chapter made it possible to compile the basic concepts of BCI systems, modalities, and EEG signal analysis to detect motor imagination. Moreover, the review applied to exoskeletons aimed at lower-limb rehabilitation with BCI shows the long-term advantages of using these systems in rehabilitation therapy to induce neuroplasticity. On the other hand, the BCI control system design process considers multiple features to acquire and process signals.

Finally, the experimental procedure and case study presented with a post-stroke patient allows us to conclude that stimulation methods or long training are essential to induce patients to generate movement imagination in BCI systems. Nevertheless, the BCI-TFLEX system provides a better neuronal response than conventional therapy performed with *T-FLEX* independently.

References

1. K.K. Ang, C. Guan, Brain-computer interface in stroke rehabilitation. Comput. Sci. Eng. **7**, 139–146 (2013)
2. B.A.P.B. Carvalho-Pinto, C.D.C.M. Faria, Health, function and disability in stroke patients in the community. Brazil. J. Phys. Therapy **20**, 355–366 (2016)
3. S. Whitehead, E. Baalbergen, Post-stroke rehabilitation. S. Afr. Med. J. **109**(2), 81–83 (2019)
4. A. Zeiaee, R. Soltani-Zarrin, R. Langari, R. Tafreshi, Design and kinematic analysis of a novel upper limb exoskeleton for rehabilitation of stroke patients, in *2017 International Conference on Rehabilitation Robotics (ICORR)* (IEEE, Piscataway, 2017), pp. 759–764
5. Y. He, D. Eguren, J.M. Azorín, R.G. Grossman, T.P. Luu, J.L. Contreras-Vidal, Brain–machine interfaces for controlling lower-limb powered robotic systems. J. Neural Eng. **15**(2), 021004 (2018)
6. C. Wang, K.S. Phua, K.K. Ang, C. Guan, H. Zhang, R. Lin, K.S.G. Chua, B.T. Ang, C.W.K. Kuah, A feasibility study of non-invasive motor-imagery BCI-based robotic rehabilitation for stroke patients, in *2009 4th International IEEE/EMBS Conference on Neural Engineering* (IEEE, Piscataway, 2009), pp. 271–274
7. M. Ortiz, L. Ferrero, E. Iáñez, J. M. Azorín, J.L. Contreras-Vidal, Sensory integration in human movement: a new brain-machine interface based on gamma band and attention level for controlling a lower-limb exoskeleton. Front. Bioeng. Biotechnol. **8**, 735 (2020)
8. W. Wang, J.L. Collinger, M.A. Perez, E.C. Tyler-Kabara, L.G. Cohen, N. Birbaumer, S.W. Brose, A.B. Schwartz, M.L. Boninger, D.J. Weber, Neural interface technology for rehabilitation: exploiting and promoting neuroplasticity. Phys. Med. Rehabil. Clin. **21**(1), 157–178 (2010)
9. S. Xie, W. Meng, *Biomechatronics in Medical Rehabilitation* (Springer, Berlin, 2017)
10. A.J. McDaid, S. Xing, S.Q. Xie, Brain controlled robotic exoskeleton for neurorehabilitation, in *2013 IEEE/ASME International Conference on Advanced Intelligent Mechatronics* (IEEE, Piscataway, 2013), pp. 1039–1044
11. M. Zhuang, Q. Wu, F. Wan, Y. Hu, State-of-the-art non-invasive brain–computer interface for neural rehabilitation: a review. J. Neurorestoratol. **8**(1), 12–25 (2020)
12. K. Najarian, R. Splinter, *Biomedical Signal and Image Processing* (Taylor & Francis, Oxfordshire, 2012)
13. R.M. Rangayyan, *Biomedical Signal Analysis* (Wiley, Hoboken, 2015)
14. T. Yanagisawa, M. Hirata, Y. Saitoh, A. Kato, D. Shibuya, Y. Kamitani, T. Yoshimine, Neural decoding using gyral and intrasulcal electrocorticograms. Neuroimage **45**(4), 1099–1106 (2009)
15. J.R. Wolpaw, D.J. McFarland, T.M. Vaughan, Brain-computer interface research at the Wadsworth Center. IEEE Trans. Rehabil. Eng. **8**(2), 222–226 (2000)
16. N. Weiskopf, K. Mathiak, S.W. Bock, F. Scharnowski, R. Veit, W. Grodd, R. Goebel, N. Birbaumer, Principles of a brain-computer interface (BCI) based on real-time functional magnetic resonance imaging (FMRI). IEEE Trans. Biomed. Eng. **51**(6), 966–970 (2004)
17. S. Nayak, R.K. Das, Application of artificial intelligence (AI) in prosthetic and orthotic rehabilitation, in *Service Robotics* (IntechOpen, London, 2020),
18. M.S. Al-Quraishi, I. Elamvazuthi, S.A. Daud, S. Parasuraman, A. Borboni, EEG-based control for upper and lower limb exoskeletons and prostheses: a systematic review. Sensors **18**(10), 3342 (2018)
19. L. Alonso-Valerdi, M. Arreola-Villarruel, J. Argüello-García, Interfaces cerebro-computadora: conceptualización, retos de rediseño e impacto social. Revista mexicana de ingeniería biomédica **40**(3), 1–18 (2019)

20. A. Andreev, A. Barachant, F. Lotte, M. Congedo, *Brain-Computer Interfaces 2: Technology and Applications* (Wiley, Hoboken, 2016)
21. A. Moran, M. Campbell, P. Holmes, T. MacIntyre, Mental imagery, action observation and skill learning, in *Skill Acquisition in Sport: Research, Theory and Practice*, vol. 94 (2012)
22. N. Ahmad, R.A.R. Ghazilla, M.Z.H.M. Azizi, Steady state visual evoked potential based BCI as control method for exoskeleton: a review. Malaysian J. Public Health Med. **16**(suppl. 1), 86–94 (2016)
23. M. Arvaneh, I.H. Robertson, T.E. Ward, A p300-based brain-computer interface for improving attention. Front. Hum. Neurosci. **12**, 524 (2019)
24. J. Choi, K.T. Kim, J.H. Jeong, L. Kim, S.J. Lee, H. Kim, Developing a motor imagery-based real-time asynchronous hybrid BCI controller for a lower-limb exoskeleton. Sensors **20**(24), 7309 (2020)
25. G. Schalk, D.J. McFarland, T. Hinterberger, N. Birbaumer, J.R. Wolpaw, Bci2000: a general-purpose brain-computer interface (BCI) system. IEEE Trans. Biomed. Eng. **51**(6), 1034–1043 (2004)
26. F. Duan, D. Lin, W. Li, Z. Zhang, Design of a multimodal EEG-based hybrid BCI system with visual servo module. IEEE Trans. Auton. Mental Develop. **7**(4), 332–341 (2015)
27. K.-S. Hong, M.J. Khan, Hybrid brain–computer interface techniques for improved classification accuracy and increased number of commands: a review. Front. Neurorobot. **11**, 35 (2017)
28. J.S. Brumberg, A. Nieto-Castanon, P.R. Kennedy, F.H. Guenther, Brain–computer interfaces for speech communication. Speech Commun. **52**(4), 367–379 (2010)
29. S. Balasubramanian, E. Garcia-Cossio, N. Birbaumer, E. Burdet, A. Ramos-Murguialday, Is EMG a viable alternative to BCI for detecting movement intention in severe stroke? IEEE Trans. Biomed. Eng. **65**(12), 2790–2797 (2018)
30. B.-J. Choi, S.-H. Jo, Hybrid SSVEP/ERD BCI for humanoid navigation, in *2013 13th International Conference on Control, Automation and Systems (ICCAS 2013)* (IEEE, Piscataway, 2013), pp. 1641–1645
31. B.Z. Allison, C. Brunner, C. Altstätter, I.C. Wagner, S. Grissmann, C. Neuper, A hybrid ERD/SSVEP BCI for continuous simultaneous two dimensional cursor control. J. Neurosci. Methods **209**(2), 299–307 (2012)
32. L.-W. Ko, S. Ranga, O. Komarov, C.-C. Chen, Development of single-channel hybrid BCI system using motor imagery and SSVEP. J. Healthcare Eng. **2017**, 3789386 (2017)
33. N. Elsayed, Z.S. Zaghloul, M. Bayoumi, Brain computer interface: EEG signal preprocessing issues and solutions. Int. J. Comput. Appl. **169**(3), 975–8887 (2017)
34. D. Delisle-Rodriguez, V. Cardoso, D. Gurve, F. Loterio, M.A. Romero-Laiseca, S. Krishnan, T. Bastos-Filho, System based on subject-specific bands to recognize pedaling motor imagery: towards a BCI for lower-limb rehabilitation. J. Neural Eng. **16**(5), 056005 (2019)
35. M. Tariq, P.M. Trivailo, M. Simic, EEG-based BCI control schemes for lower-limb assistive-robots. Front. Hum. Neurosci. **12**, 312 (2018)
36. J. Gomez-Pilar, R. Corralejo, L.F. Nicolas-Alonso, D. Álvarez, R. Hornero, Neurofeedback training with a motor imagery-based BCI: neurocognitive improvements and EEG changes in the elderly. Med. Biol. Eng. Comput. **54**(11), 1655–1666 (2016)
37. F. Lotte, L. Bougrain, A. Cichocki, M. Clerc, M. Congedo, A. Rakotomamonjy, F. Yger, A review of classification algorithms for EEG-based brain–computer interfaces: a 10 year update. J. Neural Eng. **15**(3), 031005 (2018)
38. S. Jezernik, G. Colombo, T. Keller, H. Frueh, M. Morari, Robotic orthosis Lokomat: a rehabilitation and research tool. **Neuromodulation: Technology at the neural interface **6**(2), 108–115 (2003)
39. R. Riener, Technology of the robotic gait orthosis Lokomat, in *Neurorehabilitation Technology* (Springer, Berlin, 2016), pp. 395–407
40. A.R. Donati, S. Shokur, E. Morya, D.S. Campos, R.C. Moioli, C.M. Gitti, P.B. Augusto, S. Tripodi, C.G. Pires, G.A. Pereira, et al., Long-term training with a brain-machine interface-based gait protocol induces partial neurological recovery in paraplegic patients. Sci. Rep. **6**(1), 1–16 (2016)

41. A.H. Do, P.T. Wang, C.E. King, S.N. Chun, Z. Nenadic, Brain-computer interface controlled robotic gait orthosis. J. Neuroeng. Rehabil. **10**(1), 1–9 (2013)
42. P.T. Wang, C. King, L.A. Chui, Z. Nenadic, A. Do, BCI controlled walking simulator for a BCI driven FES device, in *Proceedings of RESNA Annual Conference* (RESNA, Arlington, 2010)
43. P.T. Wang, C.E. King, L.A. Chui, A.H. Do, Z. Nenadic, Self-paced brain–computer interface control of ambulation in a virtual reality environment. J. Neural Eng. **9**(5), 056016 (2012)
44. J.L. Contreras-Vidal, M. Bortole, F. Zhu, K. Nathan, A. Venkatakrishnan, G.E. Francisco, R. Soto, J.L. Pons, Neural decoding of robot-assisted gait during rehabilitation after stroke. Am. J. Phys. Med. Rehabil. **97**(8), 541–550 (2018)
45. J.A. Gaxiola-Tirado, E. Iáñez, M. Ortíz, D. Gutiérrez, J.M. Azorín, Effects of an exoskeleton-assisted gait motor imagery training in functional brain connectivity, in *2019 41st Annual International Conference of the IEEE Engineering in Medicine and Biology Society (EMBC)* (IEEE, Piscataway, 2019), pp. 429–432
46. M. Bortole, A. Venkatakrishnan, F. Zhu, J.C. Moreno, G.E. Francisco, J.L. Pons, J.L. Contreras-Vidal, The h2 robotic exoskeleton for gait rehabilitation after stroke: early findings from a clinical study. J. Neuroeng. Rehabil. **12**(1), 1–14 (2015)
47. E. López-Larraz, F. Trincado-Alonso, V. Rajasekaran, S. Pérez-Nombela, A.J. Del-Ama, J. Aranda, J. Minguez, A. Gil-Agudo, L. Montesano, Control of an ambulatory exoskeleton with a brain–machine interface for spinal cord injury gait rehabilitation. Front. Neurosci. **10**, 359 (2016)
48. A.D. Gardner, J. Potgieter, F.K. Noble, A review of commercially available exoskeletons' capabilities, in *2017 24th International Conference on Mechatronics and Machine Vision in Practice (M2VIP)* (IEEE, Piscataway, 2017), pp. 1–5
49. A. Schütz, Robotic exoskeleton: for a better quality of life. Maxon motor (2012). https://www.maxongroup.com/maxon/view/application/Robotic-exoskeleton-For-a-better-quality-of-life
50. Y. Zhang, S. Prasad, A. Kilicarslan, J.L. Contreras-Vidal, Multiple kernel based region importance learning for neural classification of gait states from EEG signals. Front. Neurosci. **11**, 170 (2017)
51. R. Xu, N. Jiang, N. Mrachacz-Kersting, C. Lin, G.A. Prieto, J.C. Moreno, J.L. Pons, K. Dremstrup, D. Farina, A closed-loop brain–computer interface triggering an active ankle–foot orthosis for inducing cortical neural plasticity. IEEE Trans. Biomed. Eng. **61**(7), 2092–2101 (2014)
52. A.H. Do, P.T. Wang, C.E. King, A. Schombs, S.C. Cramer, Z. Nenadic, Brain-computer interface controlled functional electrical stimulation device for foot drop due to stroke, in *2012 Annual International Conference of the IEEE Engineering in Medicine and Biology Society* (IEEE, Piscataway, 2012), pp. 6414–6417
53. D. Gomez-Vargas, F. Ballen-Moreno, P. Barria, R. Aguilar, J.M. Azorín, M. Munera, C.A. Cifuentes, The actuation system of the ankle exoskeleton t-flex: first use experimental validation in people with stroke. Brain Sci. **11**(4), 1–17, Article 412 (2021)
54. Y. Renard, F. Lotte, G. Gibert, M. Congedo, E. Maby, V. Delannoy, O. Bertrand, A. Lécuyer, OpenViBE: an open-source software platform to design, test, and use brain–computer interfaces in real and virtual environments. Presence Teleop. Virt. Environ. **19**(1), 35–53 (2010)
55. D.J. McFarland, L.M. McCane, S.V. David, J.R. Wolpaw, Spatial filter selection for EEG-based communication. Electroencephalogr. Clin. Neurophysiol. **103**(3), 386–394 (1997)
56. L. Bradshaw, J. Wikswo, Spatial filter approach for evaluation of the surface Laplacian of the electroencephalogram and magnetoencephalogram. Ann. Biomed. Eng. **29**(3), 202–213 (2001)
57. A. Vourvopoulos, O.M. Pardo, S. Lefebvre, M. Neureither, D. Saldana, E. Jahng, S.-L. Liew, Effects of a brain-computer interface with virtual reality (VR) neurofeedback: a pilot study in chronic stroke patients. Front. Hum. Neurosci. **13**, 210 (2019)
58. D. Gomez-Vargas, M.J. Pinto-Betnal, F. Ballén-Moreno, M. Múnera, C.A. Cifuentes, Therapy with t-flex ankle-exoskeleton for motor recovery: a case study with a stroke survivor, in *2020 8th IEEE RAS/EMBS International Conference for Biomedical Robotics and Biomechatronics (BioRob)* (2020), pp. 491–496

59. F. Cincotti, F. Pichiorri, P. Aricò, F. Aloise, F. Leotta, F. de Vico Fallani, J.D.R. Millán, M. Molinari, D. Mattia, EEG-based brain-computer interface to support post-stroke motor rehabilitation of the upper limb, in *2012 Annual International Conference of the IEEE Engineering in Medicine and Biology Society* (IEEE, Piscataway, 2012), pp. 4112–4115
60. P. Langhorne, F. Coupar, A. Pollock, Motor recovery after stroke: a systematic review. Lancet Neurol. **8**(8), 741–754 (2009)

Control Strategies for Human–Robot–Environment Interaction in Assisted Gait with Smart Walkers

10

Sergio D. Sierra M., Mario F. Jiménez, Anselmo Frizera-Neto, Marcela Múnera, and Carlos A. Cifuentes

10.1 Introduction

Recent advances and developments in rehabilitation engineering have been focused on the design and implementation of control strategies that allow natural, safe, and compliant interaction between users, the smart walker, the environment, and the healthcare professionals [1, 2]. In particular, multiple research projects have been oriented to develop innovative tools to assist older people, neurological patients, and people with cognitive impairments. Among these, the *AGoRA Smart Walker* [3], the *UFES Smart Walker* [4], the *GUIDO Smart Walker* [5], and the *MOBOT Platform* [6] are found.

These strategies have gained considerable popularity in rehabilitation and everyday scenarios, owing to their positive usability outcomes, proper users and medical acceptance, and the increasing cooperation between engineers, medical staff, and patients [7–9]. Specifically, a key issue in such interdisciplinary collaborations is related to the fact that they focus on generating solutions with a particular focus on the user, that is to say, strategies and prototypes centered on the users' requirements [10].

S. D. Sierra M. · M. Múnera · C. A. Cifuentes (✉)
Biomedical Engineering, Department of the Colombian School of Engineering Julio Garavito, Bogotá D.C., Colombia
e-mail: sergio.sierra@escuelaing.edu.co; marcela.munera@escuelaing.edu.co; carlos.cifuentes@escuelaing.edu.co

M. F. Jiménez
School of Engineering, Science and Technology, Universidad del Rosario, Bogotá D.C., Colombia
e-mail: mariof.jimenez@urosario.edu.co

A. Frizera-Neto
Graduate Program in Electrical Engineering, Federal University of Espírito Santo, Vitória, Brazil
e-mail: frizera@ieee.org

C. A. Cifuentes, M. Múnera, *Interfacing Humans and Robots for Gait Assistance and Rehabilitation*, https://doi.org/10.1007/978-3-030-79630-3_10

Considering the particular case of smart walkers, in the previous chapters, it has been stated that these interaction strategies require the rehabilitation device to count with appropriate sensory architectures, precise actuation interfaces, and sufficient communication interfaces with the user. In general, the selection of such components is aimed at providing three types of interaction in smart walkers: (1) Human–Robot Interaction (HRI), (2) Robot–Environment Interaction (REI), and (3) Human–Robot–Environment Interaction (HREI). In this sense, this chapter seeks to describe the most common architectures that can be used to provide these types of interaction during walker-assisted gait and demonstrate some case studies and provide insights into their implementation in real devices.

10.2 Design Considerations for Control Strategies

During the design process of an interaction strategy for walker-assisted gait, several milestones should be attained by researchers, healthcare professionals, and stakeholders. For instance, this process involves (1) identification of patients' requirements, (2) co-design of robotic solutions (involving engineers, clinicians, patients, and relatives), (3) implementation in healthy patients, (4) validation in clinical scenarios, and (5) analysis of effects [11, 12]. This is not a straightforward process but a continuous loop of development, integration, and testing. Moreover, depending on the type of interaction that is sought to be developed, there are several baseline design criteria, such as safety, compliance, and comfort, among others. Table 10.1 describes these underlying concepts for HRI, REI, and HREI.

As it can be inferred from Table 10.1, it is a common denominator in the design criteria that the smart walker behavior is safe, intuitive, compliant, and appropriate for the users' specific requirements [1]. Some of these criteria have been widely reported in literature reviews focused on smart walkers [2, 13, 14]. Additional constraints might also include (1) the smart walker motion should be smooth and only triggered by the users' intentions, (2) the control strategies should not induce hazardous situations neither for the users nor for the environment, (3) the healthcare professional should always be involved in the interaction loop, either for monitoring or for active participation, and (4) the smart walker should track and store the users' progress and session's performance [1].

Finally, the control architectures for walker-assisted gait should include several minimal modules to interact with users. Particularly, Fig. 10.1 illustrates a standard control diagram in a walker-assisted gait application, where the involvement of the user, the smart walker, and the clinician is showcased. Moreover, this architecture aims to implement the design criteria defined previously (see Table 10.1). Several smart walkers reported in the literature have already proposed control strategies that can be framed within this control diagram [3–6, 15–17]. This diagram offers a generalization of the control strategies, and thus some smart walkers might not include all of the proposed modules.

Multiple control strategies will be described in the following sections, categorizing them as strategies for HRI, REI, or HREI. Moreover, some of such strategies

Table 10.1 Description of general design criteria for the development of control strategies in walker-assisted gait

Interaction type	Design criteria	Example
Human–Robot Interaction (HRI)	Recognition of physical interaction with the user	Sensing of forces, torque, and pressure exerted by the user
	Recognition of cognitive interaction with the user	Voice processing modules. Gestures recognition
	Estimation of user's navigation commands	Rule-based algorithms. Admittance controllers
	Estimation of user's gait parameters	Sensing gait with IMUs, pressure insoles, ranging devices, etc.
	User monitoring	Hearth rate estimation
	Safety management	Detection of proper user's support and posture. Emergency braking
	Implementation of compliant control strategies	Natural and intuitive interaction. Personalized behaviors
Robot–Environment Interaction (REI)	Smart walker motion control	Low-level controllers to generate desired velocity
	Implementation of autonomous navigation	Localization and mapping. Path planning. Obstacle detection. Guidance
	Social interaction ability	Detection of surrounding people. Motion adaption to avoid intruding into personal spaces
	Safety management	Redundant systems. Remote control
Human–Robot–Environment Interaction (HREI)	Implementation of shared control strategies	Adaptation of control authority. Modulation of user's participation. The user triggers motion
	Clinician participation	Close accompanying teleoperation and remote monitoring
	Feedback of environment information to the user	Haptic, auditory, and visual feedback

have already been validated with users, and thus, the following section presents two particular smart walkers used for these purposes: the *AGoRA Smart Walker* and the *UFES Smart Walker*.

10.3 Robotic Platforms

The *AGoRA Smart Walker* and the *UFES Smart Walker*, illustrated in Fig. 10.2, are two active robotic walkers that have been used to implement control strategies for HRI, REI, and HREI [3, 4]. In this way, their internal components and main characteristics are described as follows:

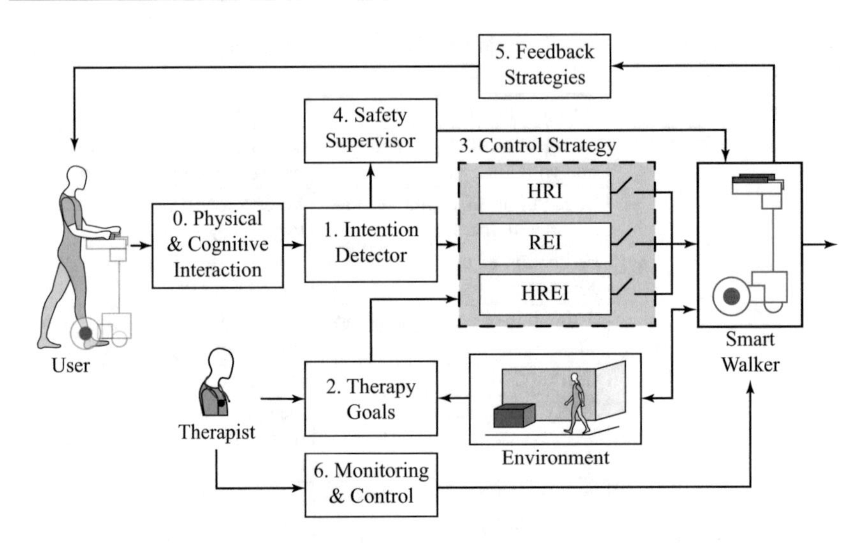

Fig. 10.1 Standard control architecture for a walker-assisted gait application

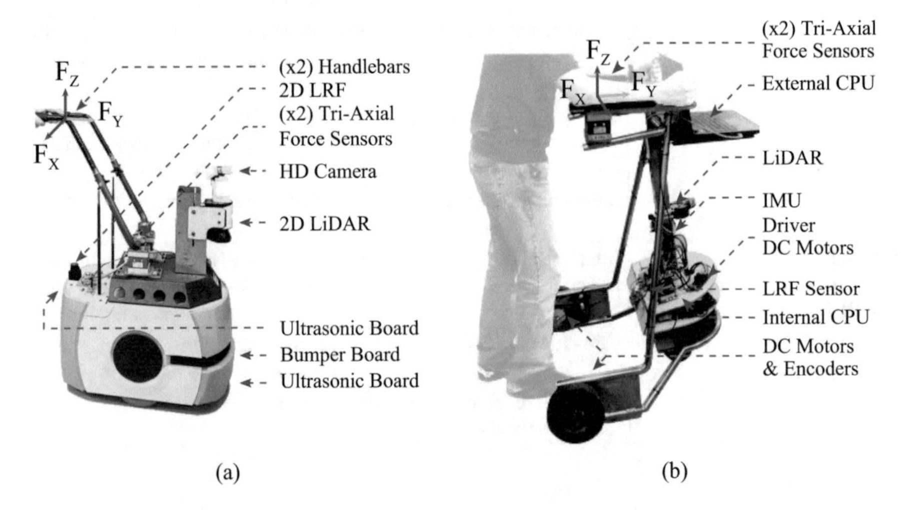

Fig. 10.2 Illustration of two standard robotic walkers for Human–Robot–Environment Interaction. (**a**) AGoRA Smart Walker. (**b**) UFES Smart Walker

10.3.1 AGoRA Smart Walker

The *AGoRA Smart Walker* is a robotic walker mounted on a commercial robot (Pioneer LX, Omron Adept, USA), emulating an assistive smart walker's structural frame and functionality (see Fig. 10.2a).

The platform is equipped with (1) two motorized wheels and four caster wheels, (2) two encoders, one Inertial Measurement Unit (IMU), and two hall sensors to measure walker's overall position and speed, (3) a 2D Light Detection and Ranging (LiDAR) sensor (S300 Expert, SICK, Germany) for environment sensing, (4) two ultrasonic boards for detection of users and low-rise obstacles, (5) a bumper panel to stop the platform under collisions, (6) two tri-axial load cells (MTA400, FUTEK, USA) to estimate the user's navigation commands, (7) a camera (LifeCam Studio, Microsoft, USA) to sense people in the environment, and (8) a 2D laser range finder (LRF) (URG-04LX, Hokuyo, Japan) for user's gait parameters estimation [3].

The device's onboard CPU runs a Linux distribution to support the Robotic Operating System (ROS) framework and the software requirements [3]. Moreover, to ensure efficient processing resources, an external computer is used to off-load non-critical modules. The platform's Ethernet and WiFi modules allow communication with the external CPU [3].

10.3.2 UFES Smart Walker

The *UFES Smart Walker*, developed at the Federal University of Espírito Santo, Brazil, is an active three-wheeled walker that provides gait rehabilitation and assistance. The platform is depicted in Fig. 10.2b, as well as its sensory and actuation interfaces.

The smart walker is based on a differential drive configuration with one front caster wheel and two rear motorized wheels. The device is equipped with (1) an encoder at each motorized wheel (H1, US Digital, USA) to estimate the wheel's position and movement, (2) an Inertial Measurement Unit (BNO055, Adafruit, USA) to estimate the platform's orientation, (3) two tri-axial force sensors at each forearm support handlebar (MTA400, Futek, USA) to estimate physical interactions between the user and the platform, (4) a laser rangefinder (LRF) sensor (URG-04LX, Hokuyo, Japan) located in front of the user's legs to obtain user's gait spatiotemporal parameters and distance to the platform, and (5) a 2D Light Detection and Ranging (LiDAR) sensor (RPLIDAR A1, SLAMTEC, CHN) pointing towards the front for environment sensing.

Additionally, the platform is equipped with an onboard computer (PC/104-Plus Standard, 1.67 GHz Atom N450, 2GB RAM). This computer is configured to run a real-time architecture based on the *Matlab Simulink Real-Time xPC Target Toolbox*. An external computer is used for programming purposes of the onboard computer and for experimental data storage.

10.4 Control Strategies for HRI

One of the significant improvements that have brought the emergence of smart walkers is their ability to acquire and process physical and cognitive interaction with users. Considering that each user may have different health conditions, the smart walkers are often equipped with a wide range of sensors and actuators to meet the particular assistance requirements of the users (see Chap. 2). This section describes several interaction strategies that have been proposed in the literature to provide natural and compliant HRI.

To follow the design criteria outlined in Table 10.1 and the control architecture illustrated in Fig. 10.1, the first module in an HRI strategy is related to estimating the user's intentions. This is a crucial issue, considering that the outputs of this module are in charge of triggering the smart walker motion, so that it is compliant with the user's motivational demands. A common source of this information is the physical interaction between the user and the smart walker. This interaction is often quantified employing force and pressure sensors on the device's forearm supports and handlebars (see Fig. 10.2). These sensors output force and torque signals that can be used to estimate the user's intentions.

10.4.1 Estimation of Physical Interaction

As shown in the *AGoRA Smart Walker* and the *UFES Smart Walker* sensory interfaces, the forces are acquired from the sensors placed on the left and right forearm supports. In particular, tri-axial force sensors can obtain magnitudes along the x, y, and z axes. In this way, to compute the final exerted force and torque by the user, the following equations are used:

$$\begin{aligned} \mathbf{F}_Y &= (\mathbf{F}_{LY} + \mathbf{F}_{RY}) * \frac{1}{2}, \\ \mathbf{F}_Z &= (\mathbf{F}_{LZ} + \mathbf{F}_{RZ}) * \frac{1}{2}, \end{aligned} \tag{10.1}$$

$$\tau = (\mathbf{F}_{LY} - \mathbf{F}_{RY}) * \frac{d}{2}. \tag{10.2}$$

In particular, Eq. 10.1 shows that the resulting impulse force $\mathbf{F_Y}$ can be estimated by averaging the forces along the y-axis on both sensors, i.e., $\mathbf{F_{LY}}$ and $\mathbf{F_{RY}}$, and it provides information about the users' intention to start walking. Similarly, the support force $\mathbf{F_Z}$ can be estimated using the forces along the z-axis on both sensors, i.e., $\mathbf{F_{LZ}}$ and $\mathbf{F_{RZ}}$. This support force is useful to detect the oscillatory components of gait and the posture of the user. Note that the force component along the x-axis is discarded, as it does not provide any additional or relevant information about the user's intentions.

Regarding the torque τ, it provides information about the turning intentions of the users. Equation 10.2 shows that it can be estimated using the difference between the forces along the y-axis, i.e., $\mathbf{F_{LY}}$ and $\mathbf{F_{RY}}$, and the sensors' separation distance d. In this case, the vertical forces are not used to calculate another torque signal, indicating the user's intention to roll the device about the x-axis.

10.4.2 Signals Processing

Current commercial force and torque sensors can extract clear signals, containing meaningful information about the user's support, propulsion, and turning intentions. However these signals also contain information about the oscillatory patterns of the users' gait and high-frequency noise related to vibrations produced by the floor [3, 18]. Therefore, before implementing a control strategy based on such interaction forces, a filtering and gait parameters extraction process is required. Consequently, the estimation of the user's intentions of movement and the user's navigation commands could be achieved with ease and fewer probabilities to misinterpretations [3].

According to the above, there are several alternatives to achieve such a filtering process. Two of them are briefly introduced as follows:

1. **Low-Pass Filters:** These types of filters have been proposed to remove high-frequency noise components. Among these, Gaussian, Blackman, Moving Average, and multiple-pass filters are commonly found. The Moving Average filter is one of the most common techniques in digital signal processing and one of the simplest in terms of implementation and formulation. This type of filter acts as a low-pass filter, although it has poor ability to handle frequency-domain responses. The operation of this filter takes M input points, estimates the average of those points, and finally produces a single output point. In a force signal processing application, Eq. 10.3 describes the formulation of the Moving Average filter:

$$\mathbf{F}'_{\mathbf{Y}}[i] = \frac{1}{M} \sum_{k=0}^{M-1} \mathbf{F}_{\mathbf{Y}}[i+k]. \tag{10.3}$$

   $\mathbf{F}'_{\mathbf{Y}}[i]$ is the filtered force signal, and M indicates the number of points to average [19]. Considering that this filter is intended to remove random noise, it might not remove oscillatory patterns related to gait. Moreover, given that the estimation of each filtered point requires M input samples, this filter induces an amount of delay that increases with the M value. Regarding Gaussian and Blackman variations of this type of filter, they have better stopband attenuation than the Moving Average filter itself. The Gaussian filter, for instance, sets smaller amplitudes near the ends of the averaging window, thus producing smoother results [19].
2. **Adaptive Filters Based on Gait Parameters**: It is well known that the gait pattern exhibits a natural oscillatory behavior, which is commonly related to movements of the human trunk and center of mass in the sagittal plane [3]. In walker-assisted gait applications, the force sensors also capture such movements of the users' upper body [3]. Thus, the frequency of the gait components that contaminate the force signals is often related to the gait cadence [20].

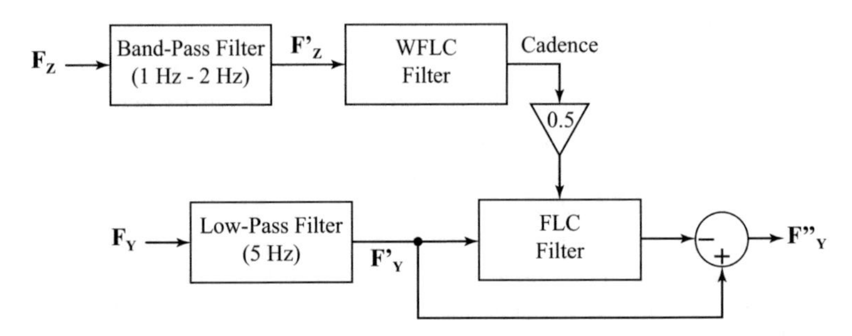

Fig. 10.3 Illustration of the adaptive filtering process using Weighted Fourier Linear Combiner (WFLC) and Fourier Linear Combiner (FLC)

In this way, as proposed in [20], an appropriate filtering process of the force signals requires estimating the gait cadence. This process relies on the implementation of adaptive filters, such as (i) the Weighted-Fourier Linear Combiner (WFLC) and (ii) the Fourier Linear Combiner (FLC), which allow the online tracking of quasi-periodic signals [20]. The mathematical formulation of these filters has been previously described in Chap. 5, and thus, their implementation for force signal filtering is outlined here.

As illustrated in Fig. 10.3, the filtering process consists of several steps. On the one hand, the resulting support force $\mathbf{F_Z}$ (see Eq. 10.1) is passed through a band-pass filter with cutoff frequencies of 1 Hz and 2 Hz to remove the signal's offset and high-frequency noise [3]. Several studies have validated these frequencies in walker-assisted gait applications [3, 17, 20]. Afterward, the first harmonic of the filtered force signal $\mathbf{F'_Z}$, i.e., the gait cadence, is estimated by the WFLC.

On the other hand, the resulting impulse force $\mathbf{F_Y}$ (see Eq. 10.1) is filtered by a fourth-order *Butterworth* low-pass filter with a cutoff frequency of 5 Hz. In parallel, the FLC is fed with the cadence from the WFLC and the filtered signal $\mathbf{F'_Y}$. Finally, the FLC outputs the cadence signal $\mathbf{F'_{Y\_CAD}}$, which is subtracted from the $\mathbf{F'_Y}$ to get the final $\mathbf{F''_Y}$ filtered signal.

The above-mentioned filtering processes are helpful to process and remove undesired components from the force signals acquired from the sensory interfaces of the smart walkers. Note that one can obtain a filtered torque signal if these processes are not carried out with the resulting force signals $\mathbf{F_Y}$ and $\mathbf{F_Z}$, but with the independent signals of each sensor, $\mathbf{F_{RY}}$, $\mathbf{F_{LY}}$, $\mathbf{F_{RZ}}$, and $\mathbf{F_{LZ}}$.

At this point, it is still necessary to obtain the users' intentions of movement to set appropriate behaviors (i.e., velocities) on the smart walker. To this end, the following section describes one of the most common methods employed to extract velocities from force and torque signals.

10.4.3 Motion Intention Detector

As stated in the design criteria, the smart walkers should compliantly respond to users' motivational demands, to guarantee safety and acceptance [3,4]. In this sense, admittance controllers have been widely implemented in smart walkers, as they allow users to control the device by exerting forces and torques on the handlebars or supporting themselves on the devices' forearms [1, 3, 4]. The main idea with these controllers is that the users require less effort to handle the smart walker than to control the device in a passive configuration [1].

In general, the admittance controllers are dynamic models that generate linear and angular velocities from users' intentions [3, 4]. These controllers model the smart walker as two first-order *mass-damper* systems, whose inputs are the resulting impulse force $\mathbf{F_Y}$[1] and the resulting torque $\boldsymbol{\tau}$. The outputs of these controllers are the linear (v) and angular (ω) velocities, meaning the user's navigation commands.

To estimate the linear velocity $\mu(t)$ from the exerted force $\mathbf{F_Y}(t)$, the first-order system shown in Eq. 10.4 is used:

$$\mu(t) = \frac{\mathbf{F_Y}(t) - m\dot{\mu}(t)}{b_\mu}, \tag{10.4}$$

where m is a virtual mass and b_μ is the damping constant. Similarly, this system can also be represented in terms of the following transfer function:

$$L(s) = \frac{\mu(s)}{\mathbf{F_Y}(s)} = \frac{\frac{1}{m}}{s + \frac{b_\mu}{m}}. \tag{10.5}$$

The torque $\boldsymbol{\tau}$ is used to obtain the angular velocity for the smart walker. Using the first-order *mass-damper* system, the angular velocity $\omega(t)$ can be calculated as shown in the equation below:

$$\omega(t) = \frac{\boldsymbol{\tau}(t) - J\dot{\omega}(t)}{b_\omega}, \tag{10.6}$$

where J represents virtual inertia and b_ω is the damping constant. Similarly, this system can be represented in the frequency domain, as follows:

$$A(s) = \frac{\omega(s)}{\boldsymbol{\tau}(s)} = \frac{\frac{1}{J}}{s + \frac{b_\omega}{J}}. \tag{10.7}$$

In addition to the above, the quality and type of interaction are strongly related to the values of the controllers' constants. In particular, during the selection of

[1] For simplicity, the filtered force is referred to as $\mathbf{F_Y}$.

these parameters, it is possible to provide different assistance levels by changing the general virtual stiffness of the platform [3,4]. On the one hand, experimental studies with the *AGoRA Smart Walker* reported that using $m = 0.5$ kg, $b_\mu = 4$ N.s/m, $J = 2.1$ kg.m^2/rad, and $b_\omega = 2$ N.m.s/rad, the controllers provided the most effortless and lightest interaction.

On the other hand, experimental studies have also reported that it might be helpful to make the smart walker oppose the users' intentions, i.e., for muscular and gait training purposes. In patients in later stages of rehabilitation, it could be helpful to set the controllers' parameters to render a heavier and more challenging maneuvering experience [21].

To accomplish this, it is assumed that people with higher Body Mass Index (BMI) values can exert higher force and torque values on the device. Therefore, a unique set of parameters is not suitable. In this sense, to provide a resistive mode, the virtual mass should be at least ten times greater than the virtual mass of the previous configuration. The value of the virtual inertia remains unchanged to avoid increasing the risk for falls. An experimental study with the *AGoRA Smart Walker* reported the following values: $m = 10$ kg, $b_\mu = \beta$ N.s/m, $J = 2.1$ kg.m^2/rad, and $b_\omega = 7$ N.m.s/rad. The calculation of the damping constant (β) employs the user's weight, as follows:

$$\beta = 0.375 * weight - 12.5, \tag{10.8}$$

The values of the model presented in Eq. 10.8 were estimated empirically, in such a way that a subject with a maximum weight of 120 kg or a minimum weight of 55 kg could move the device with moderate resistance [21].

10.4.4 HRI Strategy: Case of Study

As an illustration of the previously explained modules, a simple task is proposed. A user is asked to follow an L-shaped path, while the admittance controllers are in charge of generating linear and angular velocities from the force and torque signals (see Fig. 10.4). In this case, the user was asked to walk at the preferred speed, and the first set of constants was used (i.e., $m = 0.5$ kg, $b_\mu = 4$ N.s/m, $J = 2.1$ kg.m^2/rad, and $b_\omega = 2$ N.m.s/rad).

Moreover, for safety purposes, the motion of the smart walker is only allowed if the user is appropriately supporting on the device's forearm supports. This can be achieved by employing the information obtained from the support force $\mathbf{F_Z}$ and setting a simple threshold. Similarly, another implementation of this safety constraint can be made by using ranging sensors pointing towards the user's legs and setting a distance threshold. If the distance threshold is exceeded, the smart walker stops.

Figure 10.5 illustrates the outcomes of the HRI case study. In particular, Fig. 10.5a shows the raw and filtered signals of the resulting impulse force and the resulting torque. In this case, the filtered force and torque signals in dark red

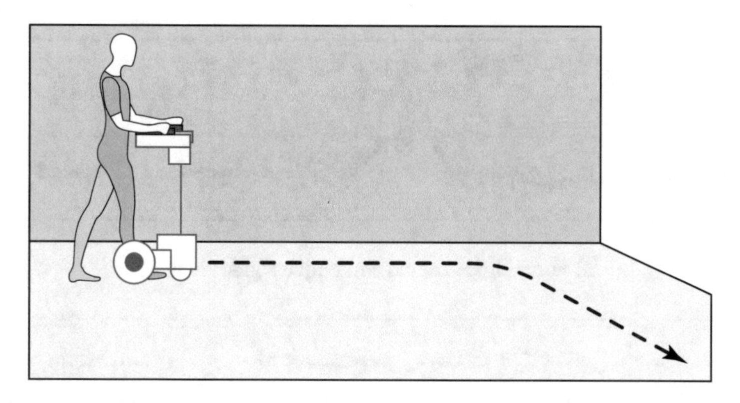

Fig. 10.4 Description of a simple case of study with the control strategy for HRI

and dark blue, respectively, were obtained by filtering the independent force signals of each sensor. As it can be noted, the filtering process removes both the high-frequency noise and the cadence components. Regarding the generation of linear and angular velocities, Fig. 10.5b shows the obtained velocities from the filtered force and torque signals. These velocities were generated using the admittance controllers described in the previous section.

In general, these outcomes highlight how the admittance controllers can extract the users' intentions to move. Particularly, increases in the impulse force are commonly associated with increases in the linear velocity. Analogously, increases or decreases in the torque exerted by the users are associated with turning intentions. Moreover, the turning intentions are also accompanied by a slight decrease in the impulse force. This behavior is explained by the fact that users prefer to perform soft turns, rather than turns around their axis (i.e., 90 degree turns) [3, 4].

10.5 Control Strategies for REI

Smart walkers are usually deployed in complex and dynamic environments, such as homes, hospitals, and rehabilitation centers. Likewise, smart walkers are often required to provide cognitive support to the user by assisting them in moving tasks. In this sense, the control strategies for REI are designed to provide guidance and path following capabilities. To this end, the smart walkers navigate autonomously and effectively while avoiding static and dynamic obstacles in the environment. According to the above, this section presents the following components for REI: (1) position control, (2) path following control, (3) autonomous navigation, and (4) low-level safety constraints.

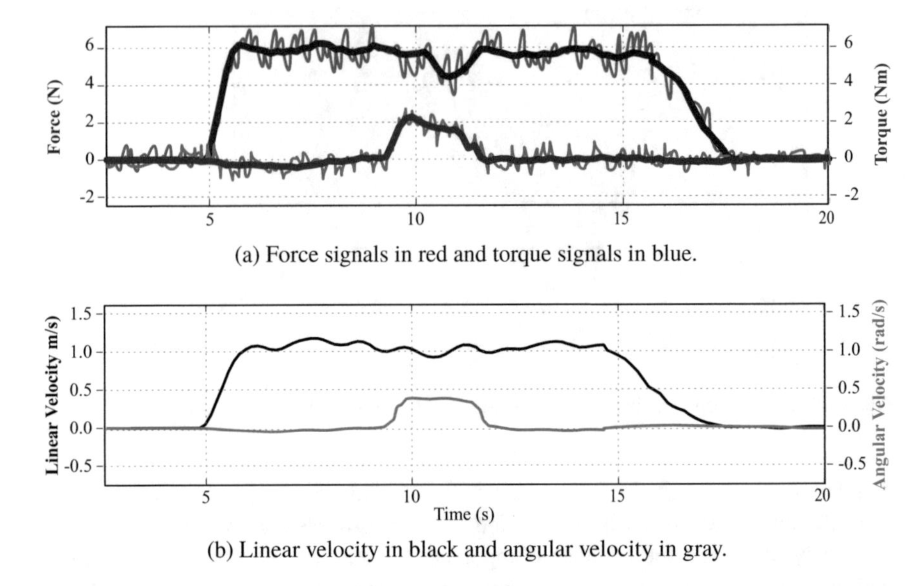

Fig. 10.5 Outcomes of the HRI case study. (**a**) Illustration of raw (light colors) and filtered (dark colors) force and torque signals. (**b**) Illustration of the linear and angular velocities generated by the admittance controllers

10.5.1 Positioning Control

One of the simplest ways to interact with the environment is to use a positioning strategy. This type of strategy allows a robotic walker to be taken from one point to another without defining a particular trajectory. In this scenario, two strategies are outlined below.

10.5.1.1 Non-linear Position Controller

In a positioning strategy, the controller should safely and naturally take the smart walker to a desired point. Let us describe this problem as shown in Fig. 10.6a, where the kinematic unicycle model is presented in polar coordinates $(e,\ \theta)$. In this scenario, the smart walker is located at the position defined by $(x_R,\ y_R)$, and the desired point is the origin of the inertial reference frame $(X_I,\ Y_I)$.

As explained in Chap. 2, the unicycle kinematic model is described by Eq. 10.9, and the control variables of the robot are μ and ω:

$$\begin{bmatrix} \dot{x_R} \\ \dot{y_R} \end{bmatrix} = \begin{bmatrix} \mu\ \cos(\varphi) \\ \mu\ \sin(\varphi) \end{bmatrix}. \tag{10.9}$$

The conversion to polar coordinates is described by Eq. 10.10, where e is the error distance between the smart walker and the goal, and θ is the orientation with respect to the global reference frame:

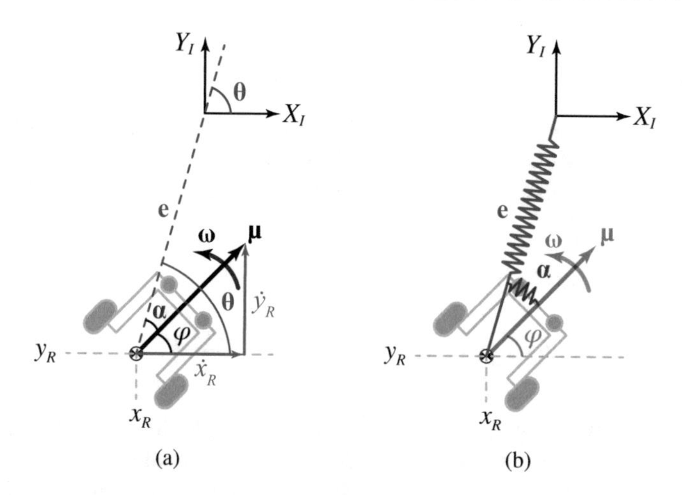

Fig. 10.6 (**a**) Polar coordinates for the unicycle model in positioning problem. The goal is located at the origin of the global reference frame. (**b**) Formulation of the positioning problem as a mechanical system

$$\begin{bmatrix} X_I \\ Y_I \end{bmatrix} = \begin{bmatrix} e\ \cos(\theta) \\ e\ \sin(\theta) \end{bmatrix}. \tag{10.10}$$

Considering that $e^2 = X_I^2 + Y_I^2$ and that the orientation error is defined by $\alpha = \theta - \varphi$, the kinematic model is replaced by

$$\begin{bmatrix} \dot{e} \\ \dot{\alpha} \\ \dot{\theta} \end{bmatrix} = \begin{bmatrix} -\mu\ \cos(\alpha) \\ -\omega + \mu\frac{\sin(\alpha)}{e} \\ \mu\frac{\sin(\alpha)}{e} \end{bmatrix} \tag{10.11}$$

At this point, the problem is focused on finding a control law that guarantees that $e \to 0$, $\alpha \to 0$, and $\theta \to 0$, asymptotically. To this end, as described in [22], the Lyapunov-based control method can be applied. In particular, the basic idea of Lyapunov's direct method is to find the mathematical extension of a physical observation for a given system [23]. In general, this method states that if a mechanical or electrical system's energy is continuously dissipated, it must eventually stabilize to an equilibrium point [23].

In this way, consider the alternative formulation of the positioning problem shown in Fig. 10.6b. Specifically, the unicycle model is described as a mechanical system that comprises two springs, which are in charge of taking the system to the desired goal. Thus, the candidate Lyapunov function for this system can be formulated by examining the total system energy, i.e., the sum of the energy of the springs, as shown in the equation below:

$$V(e,\alpha) = V_1 + V_2 = \frac{1}{2}e^2 + \frac{1}{2}\alpha^2. \tag{10.12}$$

Moreover, the rate of energy variation of the system is obtained by taking the time derivative of $V(e,\alpha)$, as expressed in Eq. 10.13. Physically, this implies that the system will stabilize at the natural length of the springs, i.e., at the desired goal [23].

$$\begin{aligned} \dot{V} &= \dot{V}_1 + \dot{V}_2 = e\dot{e} + \alpha\dot{\alpha} \\ \dot{V} &= e(-\mu\ \cos(\alpha)) + \alpha\left(-\omega + \mu\frac{\sin(\alpha)}{e}\right). \end{aligned} \tag{10.13}$$

The Lyapunov energy-like function $V(e,\alpha)$ should be positive definite and should have a continuous first partial derivative to guarantee the stability of the system's equilibrium point stability. Similarly, $\dot{V}$ should be negative semi-definite. Furthermore, if $\dot{V}$ is locally negative definite, the stability is asymptotic. The rigorous formulation of this theorem can be found in [23].

According to this, $\dot{V}_1$ can be made non-positive by choosing:

$$\mu = \lambda e \cos(\alpha),\ \lambda > 0 \tag{10.14}$$

which yields that

$$\dot{V}_2 = \alpha(-\omega + \lambda \sin(\alpha)\cos(\alpha)), \tag{10.15}$$

and consequently, $\dot{V}_2$ can also be made non-positive by choosing

$$\omega = k\alpha + \lambda \sin(\alpha)\cos(\alpha),\ k > 0. \tag{10.16}$$

Then, replacing the equations from e and α, it gives

$$\dot{V} = -\lambda e^2 \cos^2(\alpha) - k\alpha^2, \tag{10.17}$$

which finally implies that $e(t), \alpha(t) \to 0$ with $t \to \infty$.

At this point, the equations of the position controller have been defined by Eqs. 10.14 and 10.16. However, to obtain a safe behavior for μ and ω, a saturation strategy can be added. Particularly, to avoid motor saturation, the linear and angular velocities can be truncated, by saturating the error e with the hyperbolic tangent. Thus, the controller is now defined by

$$\begin{bmatrix} \mu \\ \omega \end{bmatrix} = \begin{bmatrix} \lambda\ \tanh(e)\cos(\alpha) \\ k\alpha + \lambda\frac{\tanh(e)}{e}\sin(\alpha)\cos(\alpha) \end{bmatrix}. \tag{10.18}$$

As an illustration of the behavior of this control strategy, Fig. 10.7 shows the outcomes of several positioning tasks with a healthy user. Remarkably, three positioning tasks from the same initial pose were executed (see Fig. 10.7a). Moreover, to demonstrate the asymptotic behavior of the distance error, Fig. 10.7b shows the distance error e for each goal. Finally, Fig. 10.7c presents the behavior of the steering error α for each goal. In this case, α did not exhibit a strict asymptotic behavior. However, it stabilized around an equilibrium point near 0.

It should be noted that this controller does not consider the users' intentions to make the smart walker move. Thus, to guarantee users' safety, this controller can be coupled with a motion triggering system, so that the smart walker only moves if the user is exerting a minimal impulse force.

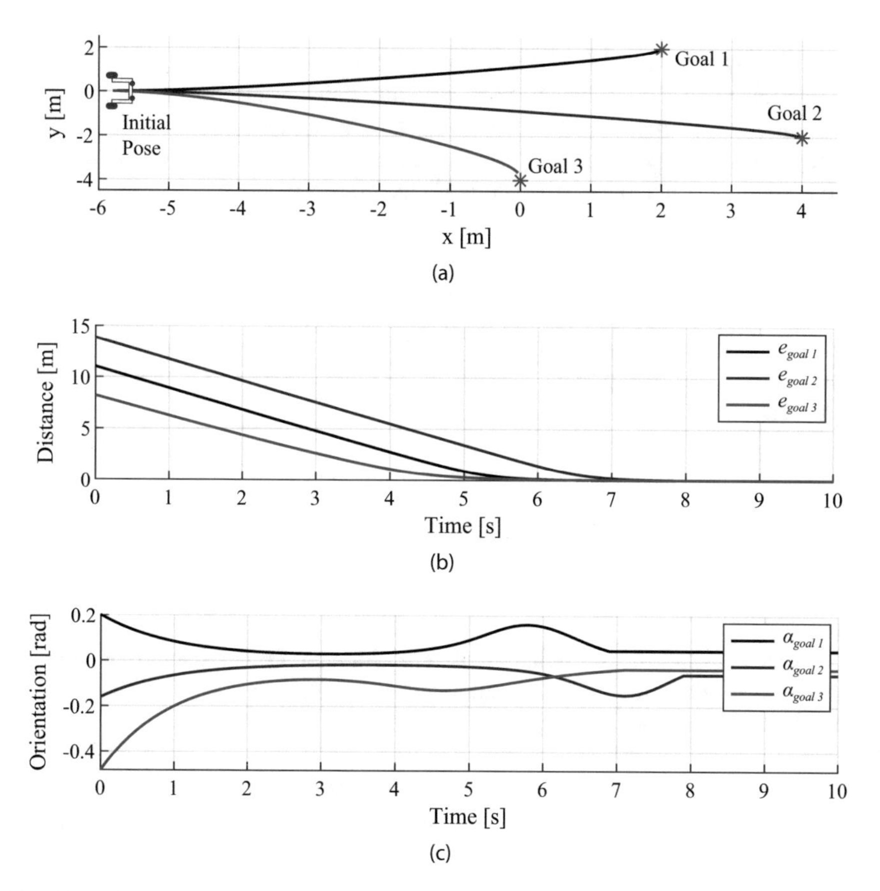

Fig. 10.7 Example of three positioning tasks. (**a**) describes the initial position of the smart walker, the performed trajectory, and the final position. (**b**) describes the behavior of the distance error e between the smart walker and each goal. (**c**) describes the behavior of the orientation error α

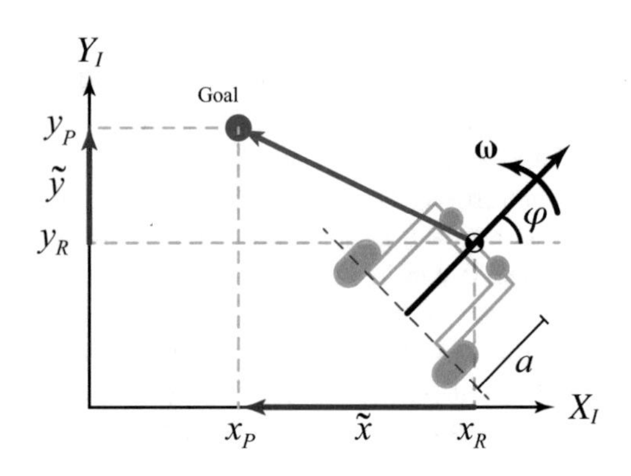

Fig. 10.8 Simple formulation of the positioning problem for the displaced kinematic model

10.5.1.2 Proportional Position Controller

An alternative to the previously described non-linear controller is related to the formulation of a simple proportional controller. In this case, consider the position problem described in Fig. 10.8, where the smart walker is modeled with the displaced kinematic model presented in Chap. 2 (see Eq. 10.19).

$$\begin{bmatrix} \dot{x} \\ \dot{y} \end{bmatrix} = \begin{bmatrix} \mu \cos(\varphi) - a\,\omega \sin(\varphi) \\ \mu \sin(\varphi) + a\,\omega \cos(\varphi) \end{bmatrix}. \tag{10.19}$$

This model can be expressed in terms of the kinematic matrix C (also referred to as Jacobian matrix $\mathbf{J}$) as

$$\begin{aligned} \begin{bmatrix} \dot{x} \\ \dot{y} \end{bmatrix} &= \begin{bmatrix} \cos(\varphi) & -a \sin(\varphi) \\ \sin(\varphi) & a \cos(\varphi) \end{bmatrix} \begin{bmatrix} \mu \\ \omega \end{bmatrix} \\ \begin{bmatrix} \dot{x} \\ \dot{y} \end{bmatrix} &= C \begin{bmatrix} \mu \\ \omega \end{bmatrix}. \end{aligned} \tag{10.20}$$

In this context, the formulation of a position controller should provide an expression for the required linear and angular velocities to reach the desired goal. Thus, it can be obtained from Eq. 10.20 that

$$\begin{aligned} \begin{bmatrix} \mu \\ \omega \end{bmatrix} &= \begin{bmatrix} \cos(\varphi) & \sin(\varphi) \\ -\frac{1}{a} \sin(\varphi) & \frac{1}{a} \cos(\varphi) \end{bmatrix} \begin{bmatrix} \dot{x} \\ \dot{y} \end{bmatrix} \\ \begin{bmatrix} \mu \\ \omega \end{bmatrix} &= C^{-1} \begin{bmatrix} \dot{x} \\ \dot{y} \end{bmatrix}, \end{aligned} \tag{10.21}$$

where C^{-1} is the inverse kinematic (or Jacobian) matrix. At this point, the equations of a proportional controller can be used to define $[\dot{x}, \dot{y}]^T$, as follows:

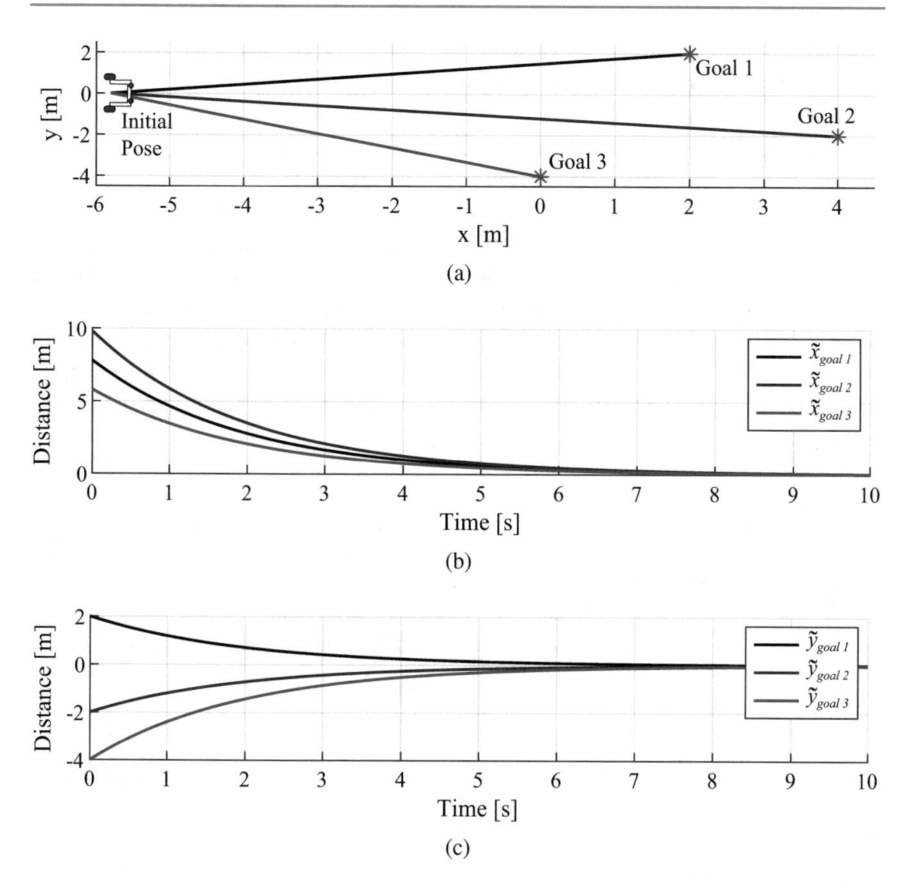

Fig. 10.9 Example of three positioning tasks with the proportional controller. (**a**) describes the initial position of the smart walker, the performed trajectory, and the final position. (**b**) describes the behavior of the error between the x coordinates of the smart walker and the desired goal. (**c**) describes the behavior of the error between the y coordinates of the smart walker and the desired goal

$$\begin{bmatrix} \dot{x} \\ \dot{y} \end{bmatrix} = \begin{bmatrix} K\tilde{x} \\ K\tilde{y} \end{bmatrix}. \tag{10.22}$$

K is a proportional gain, while $\tilde{x}$ and $\tilde{y}$ are the distance errors between the x and y coordinates of the smart walker and the desired goal, respectively. An illustration of the behavior of this controller with a healthy user is shown in Fig. 10.9. The same three goals were used to compare this controller with the previous non-linear strategy (see Fig. 10.9a). Moreover, Fig. 10.9b and c shows the behavior of $\tilde{x}$ and $\tilde{y}$, respectively.

For this controller, it should also be noted that the users' intentions are not taken into account; thus, it is recommended to integrate a motion triggering system. Additionally, the results presented in Fig. 10.9 might suggest that the proportional controller exhibits a faster and more asymptotic response than the outcomes presented in Fig. 10.7 for the non-linear controller.

However, let us analyze the behavior of the linear and angular velocities generated by these two controllers, for one of the proposed goals (see Fig. 10.10). Regarding the linear velocities illustrated in Fig. 10.10a, the proportional controller generates larger velocity magnitudes, as it does not integrate any saturation strategy. This behavior might lead to sudden unsafe movements of the smart walker in an actual application. The angular velocity generated by the non-linear controller (see Fig. 10.10b) exhibits a soft behavior that does not impose considerable effort on the robot's actuators. Nevertheless, the angular velocity generated by the proportional controller exhibits more aggressive behavior that could induce unsafe movements in the smart walker (see Fig. 10.10c). Furthermore, this controller saturates the robot's actuators.

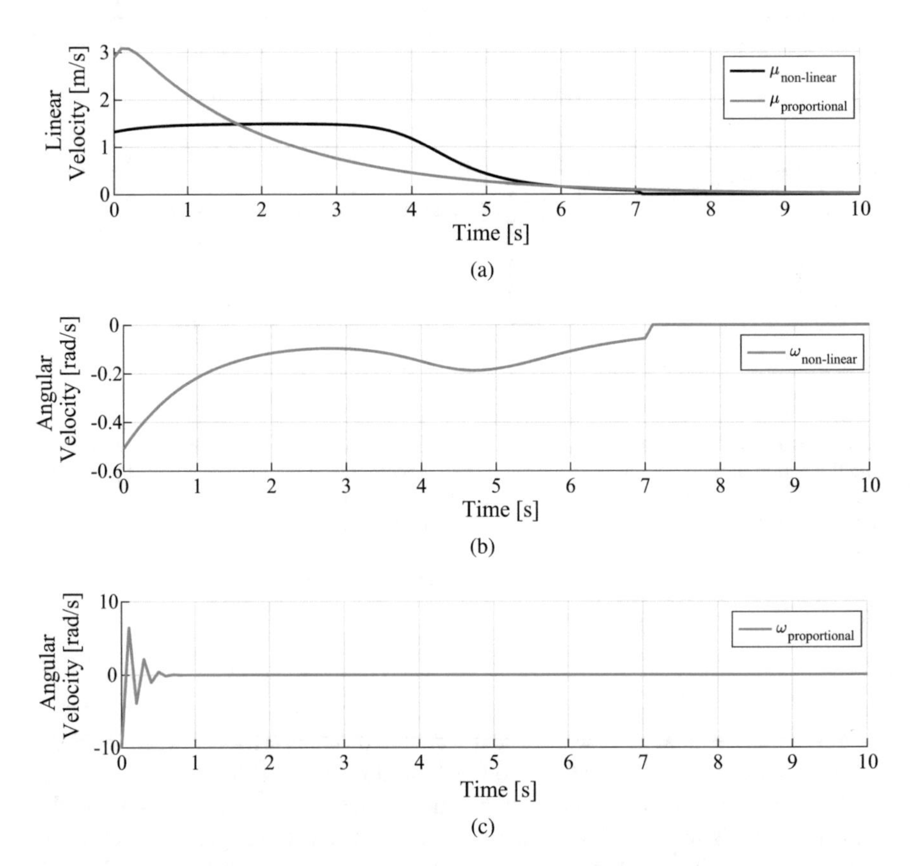

Fig. 10.10 Comparison of linear and angular velocities generated by the two positioning strategies. (**a**) Linear velocities generated by the two controllers. (**b**) Angular velocity generated by the non-linear controller. (**c**) Angular velocity generated by the proportional controller

10.5.2 Path Following Control

As stated in the previous chapters, a valuable functionality of smart walkers is related to guiding users with cognitive and physical impairments. In general, a guiding task aims at taking the user through the desired route, consisting of several predetermined poses or goals. A standard solution for path following in *wheeled mobile robots* consists of proposing a controller that generates the required linear and angular velocities to achieve the desired route.

10.5.2.1 Non-linear Path Following Controller

Let us consider the path following problem described in Fig. 10.11, where the smart walker is also modeled with the displaced kinematic model, previously described in Eq. 10.19.

This model can also be expressed in terms of the kinematic matrix C, as shown in Eq. 10.20. In this context, the inverse kinematic matrix C^{-1} is again used to obtain expressions for the required linear and angular velocities to follow a particular path (see Eq. 10.21).

Thus, finding the equations of the path following controller relies on defining proper expressions for $[\dot{x},\ \dot{y}]^T$. In particular, Andaluz *et al.* proposed a set of equations for this formulation, which have been widely used in mobile robotics applications [24]. For every path pose, the closed-loop equation of the controller proposed in [24] is represented as follows:

$$\begin{bmatrix} \dot{x} \\ \dot{y} \end{bmatrix} = \nu_p + \nu_a, \qquad (10.23)$$

where ν_p is the desired velocity vector on the path and ν_a is an attraction vector to the path. In this sense, Eq. 10.23 can be further expressed as

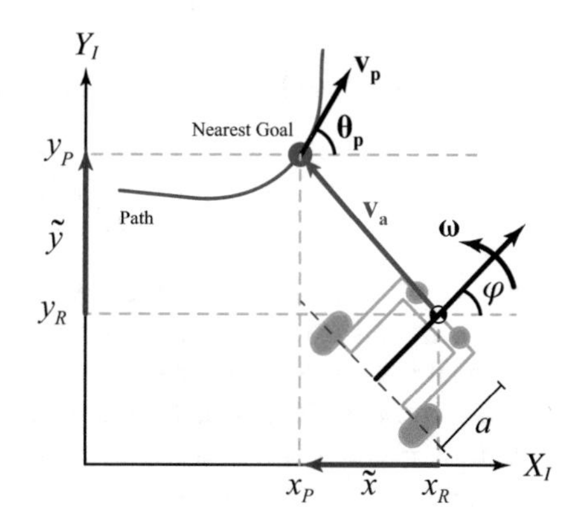

Fig. 10.11 Illustration of the path following problem, where the smart walker should reach a desired point on the route

$$\begin{bmatrix} \dot{x} \\ \dot{y} \end{bmatrix} = \begin{bmatrix} v_r \cos(\theta_p) + l_x \tanh\left(\frac{k_x}{l_x}\tilde{x}\right) \\ v_r \sin(\theta_p) + l_y \tanh\left(\frac{k_y}{l_y}\tilde{y}\right) \end{bmatrix}, \tag{10.24}$$

where v_r is the magnitude of the desired velocity on the path; θ_p is the reference orientation of the path, defined by the tangent of the nearest point to the path; l_x and l_y determine the saturation limits of the position error; k_x and k_y are constant gains that establish the linear zone of the position error; and $\tilde{x}$ and $\tilde{y}$ are the position errors of the smart walker with respect to the path [4, 24].

With these equations, it is possible to implement a path following strategy in any wheeled mobile robot that can be modeled as shown in Eq. 10.19. Note that in the case of walker-assisted gait applications, this solution does not consider the users' intentions, and thus it assumes that the user will follow the smart walker's motion. However, to avoid unsafe situations, this strategy can be coupled with a simple motion triggering system, so that the motion of the smart walker is only enabled if the user exerts a minimum impulse. Such impulse force will indicate that the user is ready to start walking with the device.

As an illustration of this control strategy, Fig. 10.12 describes the outcomes of a path following task with a healthy user, where the desired route was configured as a lemniscate curve. Figure 10.12a also shows the initial pose of the smart walker and the pose after 5 s (t_5). The blue asterisk indicates the desired goal on the route at t_5. Moreover, Fig. 10.12b shows the distance errors $\tilde{x}$ and $\tilde{y}$ during the execution of the task.

10.5.2.2 Proportional Path Following Controller

Another simple yet valuable solution to following a desired route is based on the formulation of a proportional controller. Let us assume the same formulation shown in Fig. 10.11, modeling the smart walker with the displaced kinematic model described by Eq. 10.20. Once again, the equations for the linear and angular velocities can be obtained from Eq. 10.21, and it is required to formulate an expression for $[\dot{x},\ \dot{y}]^T$.

In this case, the closed-loop equation of the controller can be defined as

$$\begin{bmatrix} \dot{x} \\ \dot{y} \end{bmatrix} = \begin{bmatrix} \Delta x + K\tilde{x} \\ \Delta y + K\tilde{y}' \end{bmatrix} \tag{10.25}$$

where Δx and Δy correspond to the difference of the x and y coordinates between the current and the next desired point on the route, respectively. Likewise, $\tilde{x}$ and $\tilde{y}$ are the position errors between the smart walker and the desired goal on the path, and K is a proportional constant. Similar to the previous controller, this formulation does not consider the users' intention to move. Therefore, a motion triggering system is required only to make the smart walker move, when the user exerts a minimum impulse force. An illustration of this control strategy is shown in Fig. 10.13.

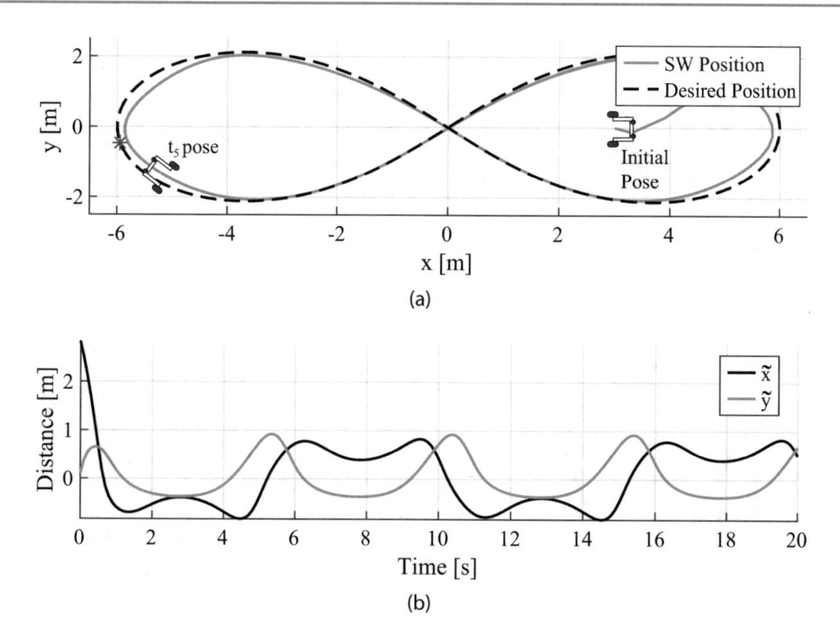

Fig. 10.12 Example of a path following task using the non-linear path controller proposed by Andaluz et al. [24] (**a**) describes the position of the smart walker during the task. (**b**) shows the distance errors during the task

Figure 10.13a compares the desired route and the position of the smart walker during the task. It also shows the initial pose of the smart walker and the pose after 5 s (t_5). The blue asterisk indicates the desired goal on the route at t_5. Moreover, Fig. 10.13b shows the distance errors $\tilde{x}$ and $\tilde{y}$ during the execution of the task.

In this case, the errors between the smart walker and the route are larger, and thus the smart walker does not follow the exact desired route. Moreover, considering that the controller is based on a proportional equation, it might generate large velocities when the error is not zero. This can cause saturation on the smart walker's motor and abrupt movement of the device. In this way, this controller could be helpful to provide path following; however, it should be softened with saturation prevention strategies. Likewise, the formulation of this controller can be extended to versions including integral and derivative gains.

10.5.3 Autonomous Navigation

Navigation during walker-assisted gait is mainly focused on safety provision while guiding the users through different environments. Such guiding might respond to the users' intentions, e.g., "*I want to go to my room*," or be used in rehabilitation scenarios to perform circuit-based tasks. The concept of autonomous navigation often differs from the previous position and path following controllers, as these controllers

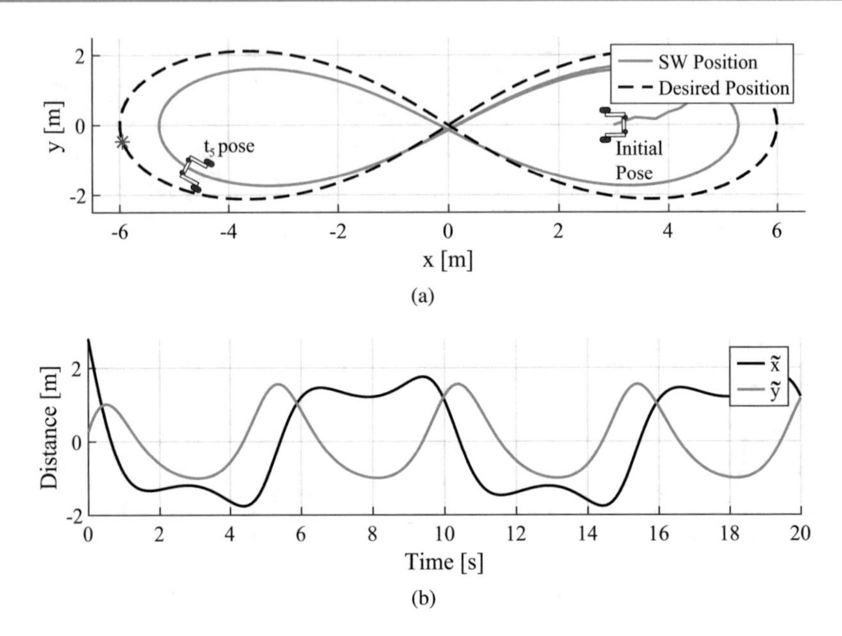

Fig. 10.13 Example of a path following task using a simple proportional controller a. The desired route and the position of the smart walker during the task b. The distance errors during the execution of the task

often require a static environment with controlled conditions. Moreover, one of the significant benefits of autonomous navigation is related to the online estimation and adaption of routes based on the environment's dynamic characteristics. Overall, the fundamentals of autonomous navigation are based on four concepts [25].

1. **Perception**: This refers to the ability of the robotic device to acquire information from the environment through sensory interfaces and extract meaningful information. This is often accomplished by ranging sensors (e.g., ultrasonic sensors and laser rangefinders), cameras, and depth cameras, among others.
2. **Localization**: This concept states that the robot must determine and track its position (i.e., odometry) in known and unknown environments. Commonly, this concept is also related to the robot's ability to create and update maps of its environment.
3. **Decision-Making**: Given a set of mission commands or goals, the navigation system must decide how to achieve such objectives. Similarly, this concept also states that the robot should determine if a particular goal is feasible or not and to determine when to perform recovery strategies, e.g., the localization is distorted. Moreover, this also refers to making plans and updating them if the environment conditions change.

4. **Motion Control**: Finally, with a particular plan or path to follow, the robot must determine the appropriate motor commands to achieve such a plan. This concept is very similar to the path following and position controllers described previously.

In this context, the navigation systems in smart walkers employ the same navigation's concepts for wheeled mobile robots. It comprises map building, autonomous localization, obstacle avoidance, and path following strategies [3]. Moreover, for simplicity, these concepts will be introduced in a high-level context, and their implementation will be based on the functionalities of the Robot Operating System (ROS) navigation stack.

10.5.3.1 Requirements for Navigation

Several hardware and software considerations should be met so that a wheeled mobile robot (or a smart walker) can integrate a navigation system, such as the one provided by ROS.

First, the robot should have a **transform configuration** system that provides information about the relationships between the coordinate frames of the robot. Typically, every sensor, the robot's decks and the robot's base have a coordinate frame. Thus, the transform configuration defines offsets in translation and rotation between different coordinate frames [26]. Second, the device must implement a reliable **odometry system** to estimate its position, orientation, and velocity [27]. Third, the robot must equip and properly acquire information from **range sensors**. Preferably, laser rangefinders or LiDARs are recommended as they provide sufficient information about the surrounding obstacles [28]. Finally, the robot must integrate a low-level **base controller** in charge of converting the linear and angular velocities generated by the navigation system to independent motors' commands [29].

The following sections describe the main functional components required for an autonomous navigation system, focused on the ROS navigation stack.

10.5.3.2 Localization and Map Building

The problem of robot's localization has been an active research area in the last decades, since several methodologies for both indoors and outdoors applications have been proposed [25]. In general, every localization strategy requires an environment map that could be made by hand or automatically with ranging and odometry sensors.

For mapping, the ROS navigation stack offers an implementation of a 2D map building algorithm based on the Simultaneous Localization and Mapping (SLAM) technique [3]. This strategy has emerged considering that robots often equip sensors with limited ranges. Therefore the robots are obligated to create maps while exploring and localizing themselves in an unknown environment [25]. Specifically, the *GMapping* ROS package aims at creating a static map of the complete interaction environment [30]. The static map is made offline and focuses on defining the main constraints and characteristics of the environment [3]. For instance, in a walker-assisted gait application in clinical environments, the map of the environment should be built before any interactions with users or patients.

Once the desired map is obtained, it is also used for the robot's online localization. To this end, the navigation stack implements the Adaptive Monte Carlo Localization Approach (AMCL) [31]. This is a probabilistic strategy that uses a particle filter to estimate all the possible poses of the robot within the map. With the robot's motion, the filter starts to converge to a single localization [32].

A common problem related to these strategies is often associated with zones such as stairs, elevator entrances, corridor railings, and glass walls. These zones are defined as non-interaction unsafe zones, and the ranging sensors such as LiDARs cannot completely sense their physical surface. Moreover, these zones are also restricted to the robot, mainly due to the risk of collisions and falls [1]. In this sense, and considering that a gray-scale image often represents the environment's map, these restrictions are achieved by editing the resulting static map [3].

It is worth mentioning that robots can also perform autonomous navigation without having a map of the environment. In this case, the robot can only plan and execute motion tasks that are limited to the field of view of its ranging sensors.

10.5.3.3 Path Planners and Cost Maps

One of the major functionalities of a navigation system is to receive a mission command or a desired goal and plan an optimal route to attain such an objective. In this sense, to perform path planning, the ROS navigation system implements three main concepts: (1) cost maps, (2) a global planner, and (3) a local planner.

A cost map consists of an occupancy grid, where every detected obstacle is represented as a cost, and it is elaborated from the previous edited map. To define the numerical value of an obstacle's cost, several aspects are considered. For instance, the distance that the robot is allowed to approach the obstacles, and this process is called obstacles inflation [3]. There is a global cost map, as well as a local cost map. The global cost map is generated by inflating the obstacles on the edited map. The local cost map is generated online by inflating obstacles detected by the field of view of the robot's sensors. These cost maps are also capable of semantically separating the obstacles in several layers [33]. Frequently, a navigation system employs a *static map layer*, an *obstacle layer*, a *sonar layer*, and an *inflation layer* [33]. Moreover, if the robot can detect people in the environment, a *social layer* can also be used to integrate proxemics and social zones.

Regarding the global planner, it executes two tasks. First, the global planner checks if the desired goal is feasible; if not, a near feasible goal is automatically proposed. Second, it seeks to find a global route with a minimum cost between the start point (i.e., current robot's position) and the endpoint (i.e., desired goal). To this end, ROS provides implementations of path planning algorithms such as Dijkstra's and A* algorithms [32].

Regarding the local planner, it provides a controller that drives the robot in the environment. Using the local cost map, the local planner creates a kinematic trajectory for the robot to get from a start point to a goal location [3]. Specifically, the Trajectory Rollout and the Dynamic Window Approach (DWA) plan local paths based on environmental data and sensory readings [32, 34]. To determine the required linear and angular velocities to execute the local plan, the DWA performs forward simulations from the robot's current state to predict possible velocities and trajectories [32]. Each simulation is scored by metrics, such as proximity to obstacles, proximity to the goal, proximity to the global path, and speed [32]. The trajectories that collide with obstacles are considered illegal and thus discarded. Finally, the highest-scoring trajectory is chosen and the associated velocity is sent to the robot's motion controller [32].

10.5.3.4 Considerations for Smart Walkers

Accordingly, a navigation system is of great relevance in walker-assisted gait applications for guiding and path following tasks [1,3]. However, as in the previous controllers, this system does not consider the user's movement intentions. In this sense, for the navigation system to be safe and intuitive, the walker must move only when the user wants it.

There are several alternatives to solve this. One option is to leave control of the smart walker's linear speed to the user through an admittance controller, while the navigation system controls the angular speed. This allows the users to follow the planned route at their speed. Another option is to implement a triggering system, as mentioned in the previous controllers. In this case, the movement of the smart walker is only allowed if the user is exerting a minimum impulse force. Also, for this, a maximum navigation speed must be configured to be comfortable for the user.

In addition to the above, it is essential to configure the navigation system so that the robotic walker does not make turns on its axis. In other words, a minimum turning radius must be established to prevent users from stumbling and thus reduce the risk of falls.

10.5.4 Low-Level Safety Supervisor

This module is in charge of guaranteeing users' safety in case of malfunctioning of the above-described control strategies. In particular, the *AGoRA Smart Walker* and the *UFES Smart Walker* have reported several safety rules that constraint the walker's movement when hazardous situations are detected [3,4]. As proposed in Fig. 10.1, the safety supervisor should be implemented to override the control strategies, if required. In general, the supervisor monitors two main safety conditions.

1. **User condition:** In this case, the device movement is only allowed if the user supports himself/herself on the walker handlebars and properly stands behind it. This information can be obtained from the force or pressure sensors on the forearm supports and from ranging sensors pointing towards the user.
2. **Warning zone condition:** Using the information gathered from the ranging sensors mounted on the device, the walker's speed can be constrained when surrounding obstacles are detected. In this sense, an area of interest must be defined around the robotic walker, for which obstacles will be taken into account. This is known as a warning zone. There is no definitive warning zone since, depending on the application, the context, or the user's requirements, one zone or

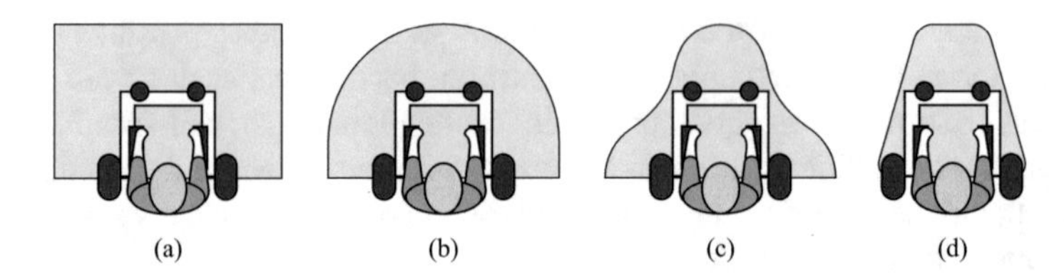

Fig. 10.14 Illustration of possible warning zones to detect surrounding obstacles and constraint the smart walker's motion. (**a**) Square zone. (**b**) Semi-circular zone. (**c**) Gaussian zone. (**d**) Conic zone

another may be chosen. In particular, Fig. 10.14 shows some applicable warning zones. Regardless of the shape of the waning zone, the speed limitation goes as follows [3]:

a. The readings from ranging sensors are processed to detect surrounding obstacles. Clustering algorithms can be helpful for such processing.
b. Only the obstacles in the warning zone are taken into account.
c. The smart walker's velocity is constrained proportionally to the distance between the smart walker and the obstacle.
d. A stopping distance is defined, so that if the obstacle is at this distance or closer, the robot comes to a complete stop. In this case, only angular velocities are allowed, to let the user avoid the obstacle.

On the other hand, smart walkers are constantly monitored by healthcare professionals or researchers. In this sense, if a device malfunctioning occurs, they can remotely disable, fix, or stop the device. The supervisor's safety restrictions should always be redundant. That is to say, they are executed from the onboard computer, as well as from external computers. In case of communication loss with the external computer, the device can continue running the safety supervisor autonomously.

10.6 Conclusions

The ability of robotic walkers to interact with the user or the environment is due to their actuators, sensory interface, and the implementation of control strategies. These strategies allow obtaining specific behaviors such as responding to the user's movement intentions, guiding a user between two points or along a trajectory.

In this sense, this chapter presented some of the control strategies that are most commonly used in robotic walkers to ensure Human–Robot interaction (HRI), Robot–Environment interaction (REI), and Human–Robot–Environment Interaction. It is essential to clarify that some of these strategies have already been implemented in wheeled mobile robots, which are not necessarily related to rehabilitation. Thus, in these sections, concepts of applying these strategies in walker-assisted gait applications were given.

References

1. S. Sierra, L. Arciniegas, F. Ballen-Moreno, D. Gomez-Vargas, M. Munera, C.A. Cifuentes, Adaptable robotic platform for gait rehabilitation and assistance: design concepts and applications, in *Exoskeleton Robots for Rehabilitation and Healthcare Devices* (Springer, 2020), pp. 67–93
2. T. Mikolajczyk, I. Ciobanu, D.I. Badea, A. Iliescu, S. Pizzamiglio, T. Schauer, T. Seel, P.L. Seiciu, D.L. Turner, M. Berteanu, Advanced technology for gait rehabilitation: An overview. Adv. Mech. Eng. **10**, 1–19 (2018)
3. S.D. Sierra M., M. Garzón, M. Múnera, C.A. Cifuentes, Human–Robot–environment interaction interface for smart walker assisted gait: AGoRA walker. Sensors **19**, 2897 (2019)
4. M.F. Jiménez, M. Monllor, A. Frizera, T. Bastos, F. Roberti, R. Carelli, Admittance controller with spatial modulation for assisted locomotion using a smart walker. J. Intell. Robot. Syst., 1 (2018)
5. G.J. Lacey, D. Rodriguez-Losada, The evolution of guido. IEEE Robot. Autom. Mag. **15**(4), 75–83 (2008)
6. E. Efthimiou, S.-E. Fotinea, T. Goulas, A.-L. Dimou, M. Koutsombogera, V. Pitsikalis, P. Maragos, C. Tzafestas, The MOBOT platform – Showcasing multimodality in human-assistive robot interaction, in *Universal Access in Human-Computer Interaction. Interaction Techniques and Environments* (Springer, 2016), pp. 382–391
7. S. Sierra, M. Jimenez, M. Munera, T. Bastos, A. Frizera-Neto, C. Cifuentes, A therapist helping hand for walker-assisted gait rehabilitation: A pre-clinical assessment, in *4th IEEE Colombian Conference on Automatic Control: Automatic Control as Key Support of Industrial Productivity, CCAC 2019 - Proceedings* (2019)
8. P. Boissy, S. Briere, H. Corriveau, A. Grant, M. Lauria, F. Michaud, Usability testing of a mobile robotic system for in-home telerehabilitation, in *2011 Annual International Conference of the IEEE Engineering in Medicine and Biology Society* (IEEE, 2011), pp. 1839–1842
9. Y. Koumpouros, A systematic review on existing measures for the subjective assessment of rehabilitation and assistive robot devices. J. Healthcare Eng. **2016**, 1–10 (2016)
10. J. Varela, R.J. Saltaren, L.J. Puglisi, J. López, M. Alvarez, J.C. Rodríguez, User centred design of rehabilitation robots, in *Advances in Automation and Robotics Research in Latin America. Lecture Notes in Networks and Systems* (Springer, Cham, 2017), pp. 97–109
11. A.A. Ramírez-Duque, L.F. Aycardi, A. Villa, M. Munera, T. Bastos, T. Belpaeme, A. Frizera-Neto, C.A. Cifuentes, Collaborative and inclusive process with the autism community: A case study in Colombia about social robot design. Int. J. Soc. Robot. **13**, 153–167 (2021)
12. D. Casas-Bocanegra, D. Gomez-Vargas, M.J. Pinto-Bernal, J. Maldonado, M. Munera, A. Villa-Moreno, M.F. Stoelen, T. Belpaeme, C.A. Cifuentes, An open-source social robot based on compliant soft robotics for therapy with children with ASD. Actuators **9**, 91 (2020)
13. M.M. Martins, C.P. Santos, A. Frizera-Neto, R. Ceres, Assistive mobility devices focusing on Smart Walkers: Classification and review. Robot. Auton. Syst. **60**(4), 548–562 (2012)
14. M. Martins, C. Santos, A. Frizera, R. Ceres, A review of the functionalities of smart walkers. Med. Eng. Phys. **37**, 917–928 (2015)
15. M.F. Jiménez, R.C. Mello, T. Bastos, A. Frizera, Assistive locomotion device with haptic feedback for guiding visually impaired people. Med. Eng. Phys. (2020)
16. A. Wachaja, P. Agarwal, M. Zink, M.R. Adame, K. Möller, W. Burgard, Navigating blind people with walking impairments using a smart walker. Autonomous Robots **41**, 555–573 (2017)
17. C.A. Cifuentes, A. Frizera, *Human-Robot Interaction Strategies for Walker-Assisted Locomotion*, vol. 115 of *Springer Tracts in Advanced Robotics* (Springer International Publishing, Cham, 2016)
18. M.A.D. Brodie, T.R. Beijer, C.G. Canning, S.R. Lord, Head and pelvis stride-to-stride oscillations in gait: validation and interpretation of measurements from wearable accelerometers. Physiological Measurement **36**, 857–872 (2015)

19. S.W. Smith, Moving average filters, in *The Scientist and Engineer's Guide to Digital Signal Processing*, ch. 15, 2nd edn. (California Technical Publishing, San Diego, 1999), pp. 277–284
20. A. Frizera, J. Gallego, E. Rocon de Lima, A. Abellanas, J. Pons, R. Ceres, Online cadence estimation through force interaction in walker assisted gait, in *ISSNIP Biosignals and Biorobotics Conference 2010* (Vitória, 2010), pp. 1–5
21. S.D. Sierra M., M. Munera, T. Provot, M. Bourgain, C.A. Cifuentes, Evaluation of physical interaction during walker-assisted gait with the AGoRA walker: Strategies based on virtual mechanical stiffness. Sensors (In revision) (2021)
22. S.G. Tzafestas, Mobile robot control I, in *Introduction to Mobile Robot Control* (Elsevier, 2014), pp. 137–183
23. J. Slotine, J. Slotine, W. Li, *Applied Nonlinear Control* (Prentice Hall, 1991)
24. V.H. Andaluz, F. Roberti, J.M. Toibero, R. Carelli, B. Wagner, Adaptive dynamic path following control of an unicycle like mobile robot, in *Intelligent Robotics and Applications*, ch. 56 (Springer Berlin Heidelberg, 2011), pp. 563–574
25. R. Siegwart, I.R. Nourbakhsh, D. Scaramuzza, *Introduction to Autonomous Mobile Robots*, 2nd edn. (MIT Press, 2011)
26. ROS, Setting up your robot using tf (2021)
27. ROS, Publishing odometry information over ROS (2019)
28. ROS, Publishing sensor streams over ROS (2012)
29. ROS, Setup and configuration of the navigation stack on a robot (2018)
30. G. Grisetti, C. Stachniss, W. Burgard, Improved techniques for grid mapping With Rao-Blackwellized particle filters. IEEE Trans. Robot. **23**, 34–46 (2007)
31. D. Fox, W. Burgard, F. Dellaert, S. Thrun, Monte Carlo localization: efficient position estimation for mobile robots, in *AAAI-99*, no. Handschin, vol. 1970, pp. 343–349 (1999)
32. K. Zheng, ROS navigation tuning guide (2016)
33. D.V. Lu, D. Hershberger, W.D. Smart, Layered costmaps for context-sensitive navigation, in *2014 IEEE/RSJ International Conference on Intelligent Robots and Systems* (IEEE, 2014), pp. 709–715
34. D. Fox, W. Burgard, S. Thrun, The dynamic window approach to collision avoidance. Robot. Autom. Mag., 1–23 (1997)

Socially Assistive Robotics for Gait Rehabilitation

11

Marcela Múnera, Luis F. Aycardi, Nathalia Cespedes, Jonathan Casas, and Carlos A. Cifuentes

11.1 Introduction

Gait is a rehabilitation process that involves physical and cognitive parameters [1]. Rehabilitation may need to be done in a cognitively stimulating context to maximize its impact on neuroplasticity and cognition [2]. Engagement, motivation, and adherence during the process have shown a high impact on the patient's performance. Social Robots have been used to assist the patient physically and cognitively [3] through factors like robot embodiment, social, emotional intelligence, and socio-cognitive skills [4]. Socially Assistive Robotics (SAR) focuses on achieving specific goals in rehabilitation, training, or education [5].

In the first section of this chapter, the basic concepts of social robotics and the importance of the cognitive process are presented. In the second section, the parameters considered for developing patient–robot interfaces based on SAR for gait rehabilitation are described. The application of these concepts is presented through an example in neurological rehabilitation.

M. Múnera · L. F. Aycardi · C. A. Cifuentes (✉)
Biomedical Engineering, Department of the Colombian School of Engineering Julio Garavito, Bogotá D.C., Colombia
e-mail: luis.aycardi-c@mail.escuelaing.edu.co; carlos.cifuentes@escuelaing.edu.co

N. Cespedes
Centre for Advanced Robotics at Queen Mary University of London, London, England
e-mail: n.cespedesgomez@qmul.ac.uk

J. Casas
Department of Mechanical and Aerospace Engineering, Syracuse University, Syracuse, NY, USA
e-mail: jacasasb@syr.edu

C. A. Cifuentes, M. Múnera, *Interfacing Humans and Robots for Gait Assistance and Rehabilitation*, https://doi.org/10.1007/978-3-030-79630-3_11

11.2 Social Interaction

To understand social robotics is essential to have a clear meaning of social interaction. The main objective of a social robot is to assist the patient not only physically but also cognitively. Over time, social interaction has been studied, and it has been represented through a variety of theories. However, a general definition of social interaction from a sociology approach is as follows: "social interaction is a dynamic, changing sequence of social actions between individuals or groups" [6]. As a product of social interaction, the partners can modify their actions and reactions.

In this context, social robots have several ways to change and share actions. The channels commonly used for social robots are verbal and nonverbal communication [7]. Verbal communication is considered the exchange of symbols that can be spoken or written [8], and nonverbal communication can be produced through gestures and gaze [6]. For long-term periods, this interaction is expected to be more robust and very similar to the human–human social interaction. Currently, SAR applications for long-time experiences still represent a challenge. Factors such as robot embodiment, social, emotional intelligence, and socio-cognitive skills [4] must be considered during the design of social robots and their applications.

11.2.1 Relevant SAR Characteristics During Social Interaction

Some characteristics differentiate SAR from other forms of social interaction like virtual agents, affecting the relationship with humans in different scenarios [9]. The parameters described here will be social robots' embodiment, social-emotional intelligence, and socio-cognitive skills (Fig. 11.1).

11.2.1.1 Social Robots' Embodiment

Social robots are developed to interact with users in a human-centric way. The robot embodiment is not always the same; robots can have various external appearances (e.g., human-like [10], animal-like [11], or abstract designs [12], see Fig. 11.2), but they share the aim of engaging users in an interpersonal manner [4]. Despite the several social robots, people tend to have a greater acceptance of anthropomorphic robots [13]. This preference occurs as humans attribute their mental stages (e.g., thoughts, emotions, and desires) to this kind of robot [14]. The design of the robot depends on its final application. It is crucial to include whole-body motion proxemics, facial expression, linguistic vocalization, and touch-based communications in some areas. To achieve the correct embodiment features is vital to use methodologies as an inclusive-participatory design [15], where the participants contribute to the decision-making process to increase the acceptance and effectiveness of the impact caused by the robot.

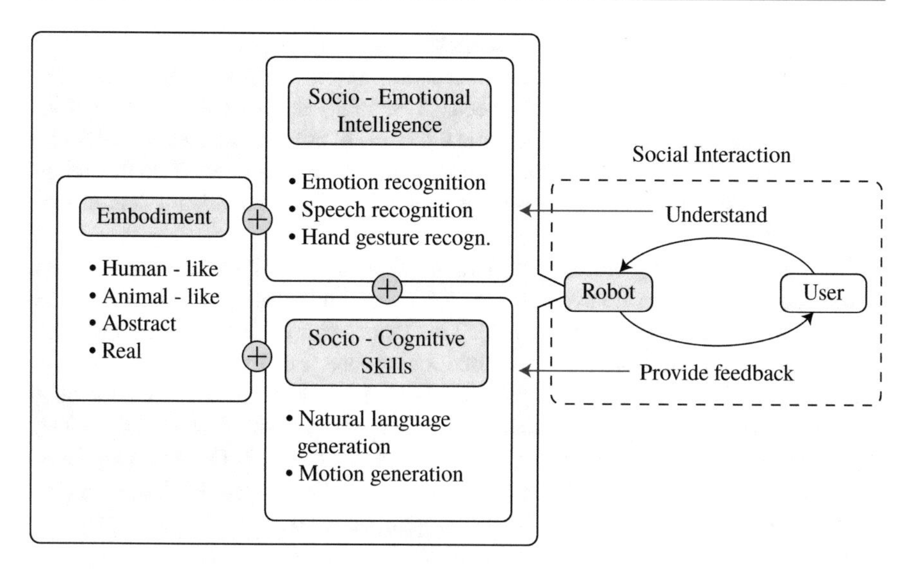

Fig. 11.1 Parameters that differentiate SAR from other forms of social interaction

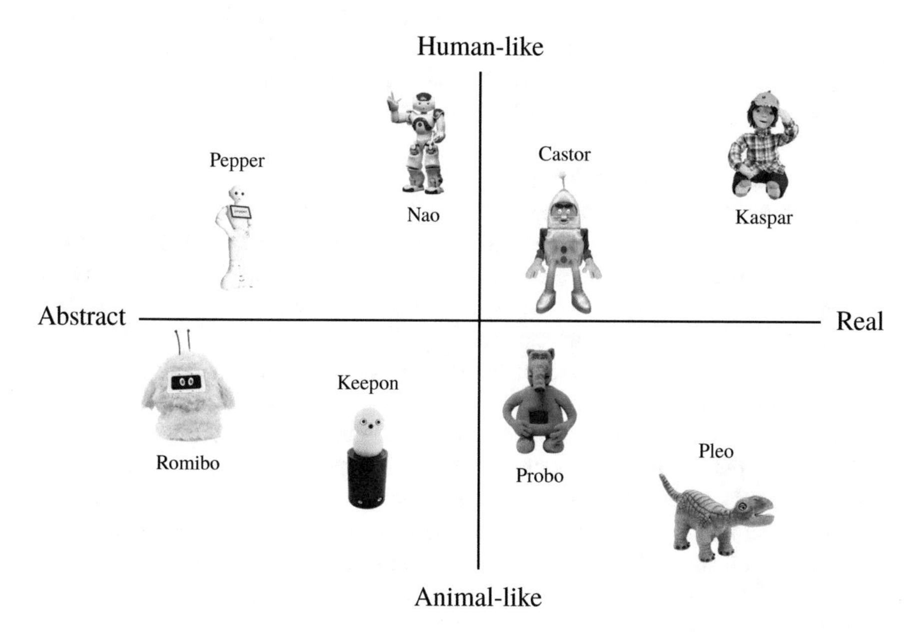

Fig. 11.2 Socially Assistive Robots classification. In this chapter, we consider two main categories: real/abstract referring to their similarity to living beings and human/animal referring to their similarity with humans or in contrast their similarity with animals

11.2.1.2 Socio-emotional Intelligence

Human communication and social interaction often integrate compelling and emotive cues. Thus, social robots need to be able to recognize and interpret affective signals from the users. Theoretical models of emotions for social robots are currently being developed to derive coherent computational models. Two theoretical models are mainly used in social robotics: *appraisal theory model* and *dimensional theory model*.

The *appraisal theory* emphasizes a connection between cognition and emotion. In this model, emotions are evoked from personal significance events (e.g., individual beliefs, desire, and intentions) [16]. This theory can be described as a discrete model, where an emotional event causes a response. For example, the Artificial Intelligence (AI) with if-then rule codes is based on this kind of model. On the other hand, the *dimensional theory* is based on continuous dimensional space [17], where the user's emotional state can be represented in a 3D space. PAD models are based on this theory [18]. PAD models are represented by P (i.e., pleasure/valence), A (i.e., arousal/intensity), and D (i.e., dominance/coping potential).

Emotional Empathy is another factor relevant in order to achieve long-term interactions between robots and humans. Empathy can be broadly defined as an "affective response more appropriate to someone else's situation than the one's own" [19]. Several works are currently focused on empathy approaches to enhance the social robots' capabilities [20]. Most of these studies use mimicking user's affective states to endorse the effects of social robotics [21].

11.2.1.3 Socio-cognitive Skills

Social robots must understand and predict human behaviors. Therefore, robots have to be aware of people's goals and intentions, so the robot's behaviors can be adjusted to help the users in terms of their goals and needs [22]. In this way, several strategies are used. The most common features used in robots are memory (i.e., face recognition) [23, 24] and communicative skills (i.e., speech recognition) [25].

A key challenge in this kind of interaction recalls critical past events during conversations and activities [26]. *Episodic memory* is a core concept to define this challenge. The *episodic memory* stores the data related to past events and adds perspective to the robot to choose actual and emotional events and preserve temporal labels to use them in future referencing. Several applications consider the use of automatic speech recognition (ASR) to produce casual communication and social exchanges [27]. However, this remains a challenge. Limitations on the environmental characteristics and the voice properties are highlighted in various research studies [28, 29].

11.2.2 Importance of the Cognitive Approach in Rehabilitation

Gait rehabilitation is a process that involves a multidisciplinary approach. Several medical specialties are included (i.e., physiatry, internal medicine, and orthopedics), physical therapy, occupational therapy, speech pathology, social work, clinical psychology, neuropsychology, orthotics/prosthetics, nutrition, and recreational therapy [1]. A basic premise of rehabilitation medicine is that optimal patient recovery requires the concerted efforts of some combination of each of these treatment disciplines [1]. It has been proposed that rehabilitation may need to be done in a cognitively stimulating context to maximize its impact on neuroplasticity and cognition [2]. Physical and cognitive training on their own are helpful to some extent for improving cognition, but there may be added benefits to combining the two into a single activity [30]. Social cognitive and system formulations can help revise how we attempt to deliver comprehensive rehabilitation efforts [1]. In this context, Bandura's social cognitive theory of human behavior and cognition suggests that environmental factors, internal factors, and behavioral outcomes combine to shape and direct human learning, cognition, and behavior [31].

The integration of a cognitive approach in physical rehabilitation has been done through different studies. The study by Dhami et al. in 2015 proposed dancing as an alternative to physical therapies as used in neurorehabilitation. This produced a positive impact on physical functioning and cognitive perception, due to the combined, or multimodal framework in therapies, which incorporate simultaneous physical and cognitive activities in a stimulating environment [30]. This can also be achieved through SAR. SAR shares with assistive robotics to assist human users, but SAR constraints that assistance through non-physical social interaction. SAR focuses on achieving specific convalescence, rehabilitation, training, or education goals [5]. Integrating a socially assistive robot can help provide one-on-one support to the patient [3]. It can facilitate the healthcare staff to focus on patients' individual needs, immediately detect any complications during the session [32], analyze the patient's progress within the program in more detail [33], and provide a more tailored plan [34]. Unlike virtual agents [35], socially assistive robots present a physical embodiment, which improves likeability [36, 37], user engagement and motivation [38], adherence [39, 40], and task performance [38], which are essential in long-term healthcare programs. This subsection will further discuss the importance of motivation and adherence and gait rehabilitation and how it can be improved using a cognitive approach.

11.2.2.1 Intrinsic Motivation

Motivation is the most challenging part of the work of the therapeutic profession. Motivated patients are believed to perform better in rehabilitation activities and make more gains than those described as less enthusiastic for treatment [41]. Yet, motivation is recognized as the most significant challenge in physical rehabilitation and training [42]. The more an individual is motivated and engaged in the learning activity, the better the learning outcome [41].

SAR technology can provide novel means for monitoring, motivating, and coaching [42]. Socially assistive robots have been shown to improve user motivation and engagement in several studies in rehabilitation [5, 43–47]. A complementary application where robots are used to motivate and increase the adherence in long-term therapies and medical self-care is diabetes mellitus treatments, where robots play personal assistants in the adult [48] and children [49] population, showing potential results within motivational aspects and treatment engagement.

11.2.2.2 Adherence

Improving adherence to therapy is a critical component of advancing outcomes and reducing rehabilitation costs [50]. Rehabilitative robotics has the potential to enhance adherence to rehabilitation recommendations, which is known to be difficult for those with chronic health conditions [51]. Research suggests that poor adherence compromises health outcomes [52], while high levels of therapy practice optimize motor recovery [53], underscoring the importance of strategies and technologies that bring rehabilitation support into patients' homes. Different studies have shown positive results of social robotics regarding this factor. *Gadde et al.* evaluated an interactive personal robot trainer in the early stages to monitor and increase exercise adherence in older adults [54]. The system was tested with 10 participants, initially showing positive response and a favorable interaction. In another study by White et al., using focus groups, all participants favorably endorsed the potential utility of a socially assistive robot that functioned as a personal coach. They identified three areas in which such a system would be helpful for (1) adherence to therapy recommendations, (2) organizing and remembering things to do, and (3) locating and supporting participation in social and recreational activities [50].

11.3 Patient–Robot Interfaces Based on SAR

Natural human-to-human interaction is performed using senses (e.g., vision, touch, taste, smell, and touch) that facilitate perception of the environment and the ability to communicate employing diverse information channels [55]. This information serves as the input of cognitive processes that are conformed by a sequence of tasks, including reasoning, planning, and execution of a given situation [56, 57]. Unlike human beings, who use their senses to perceive the world, computers and robotic systems implement interfaces composed of a set of sensors that provide the required data to perceive the environment, process the information to define a plan, and perform a determined behavior according to the context [58]. Hence, aiming to generate an effective interaction between the user and the robot, it is of relevance to provide multiple communication channels from different sources. In other words, these interfaces should be multimodal to allow interaction as naturally as possible [56]. For this reason, in most of the Human–Robot Interaction (HRI) systems, there are considered not only humans and robots but also multimodal interfaces that work as an intermediary between both agents [58]. Classic Human–Computer interfaces commonly conform to such interfaces (HCis), such as graphical computer interfaces in conjunction with visual interfaces (e.g., camera-based vision and recognition interfaces) and sensors. Among the most used sensors are Inertial Measurement Units (IMUs), laser rangefinders (LRFs), or wearable devices associated with different communication modalities that are integrated within the HCi [56].

The way an HRI is developed is critical to achieving a natural interaction that can potentiate the intervention with SAR, and over the years, researchers have used different methods to plan and produce these interfaces. A method that has shown promising results is the participatory design. The design of a Patient–Robot interface based on SAR following this methodology is presented in this section. The process is done in a generic rehabilitation program where there is a component of gait rehabilitation. The core activities for which an HCi can be developed are the ones carried out on a treadmill.

11.3.1 Participatory Design

Participatory design (PD) is a well-known method to develop products and services for a target population. The process intends to empower the people involved in a specific activity or situation (users, designers, and stakeholders) by providing them space and a voice to contribute to the decision-making [59]. This way, the real needs, desires, and expectations of the population are met in the final products or services.

PD was initially used in areas like industrial design, but given the effectiveness of designs based on participatory practices, researchers' attention has gained attention in different fields. PD techniques are up-and-coming when transferring knowledge and developments from research to the real world, especially when interacting with humans is vital in the final product or service. Additionally, implementing PD constitutes an opportunity to understand and gain knowledge about the target community's context. An occasion to show the benefits of technological tools and build a relationship based on trust and confidence between researchers and the community.

In health care, PD has been used in the design of social robots for (i) Autism Spectrum Disorder (ASD) [15, 60] and (ii) a children's hospital [61]. In all these contributions, PD methods have been implemented to recognize the target populations and their environment (families, society, groups of allies, and friends) as partners with experience that can be a part of the solution. They are no longer only a source to obtain information and requirements to produce results [62]. All the actors in the project are acknowledged as valuable contributors, which plays a crucial role in ethical, political, and social considerations of the development. The philosophy behind PD is not to provide a step-by-step list of the activities or phases to develop the final product or service, as there are multiple possible ways to implement it. It is up to the researchers to plan and design an appropriate methodology based on the population, context, variables, and objective. A general diagram of the main phases to consider during a PD is presented in Fig. 11.3.

When the participatory process is correctly implemented, it comprises different stages that could lead researchers to:

- obtain contextual information that successfully establishes the needs, interests, preferences, fears, desires, and priorities related to the product or service's functionality,

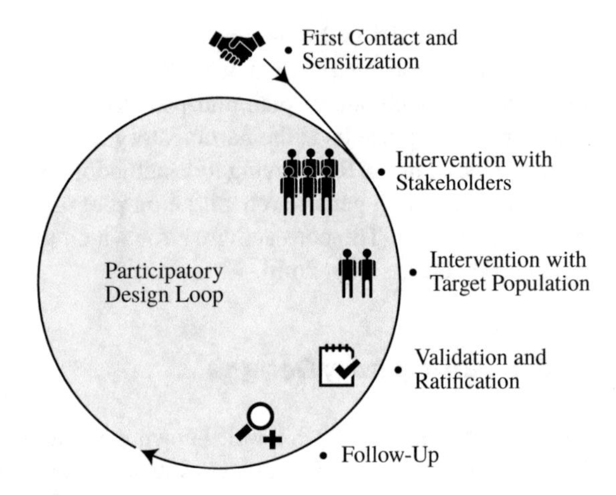

Fig. 11.3 General diagram of participatory design phases to design product and services

- validate or refute the insights gained in previous studies or developments to design the different products or services, and
- generate ideas and creative solutions through reflecting on experiences from the various participants.

11.3.2 Design Criteria

The contextual information found when applying PD leads to the establishment of design criteria or requirements. A natural way to understand and address them in an ordered fashion can be by grouping them according to their characteristics.

Observations from the process that follow the PD steps in Fig. 11.3 set the requirements that the HRI must accomplish. These requirements can be broadly classified into three main groups:

- **Variables:**
  In most rehabilitation scenarios, three types of variables are required to be measured by the system: (i) spatiotemporal, (ii) physiological, and (iii) exercise intensity variables. Spatiotemporal variables include the measurement of items as speed (mph) of the band, cadence (Hz), which is the step frequency of the patient (amount of steps per second), and step length (m), which refers to the distance between legs on each step during exercise. The clinicians typically request the measurement of these variables to monitor the patient's movement. Additionally, the cadence and the step length measurement are used to determine the patients' walking speed. Physiological variables control the patients' physical condition, usually employing the heart rate, blood pressure, and posture while walking. Finally, exercise intensity can be monitored employing the Borg scale and configured with the treadmill inclination.

- **Interactivity:**
  This requirement is provided through a Graphical User Interface (GUI) that allows visual interaction and provides corrections to avoid risk during the session. Similarly, a social robot must be integrated to provide a more natural and social interaction with the system and monitor and motivate patients during exercise.

- **Follow-up:**
  The third requirement is associated with the data management and follow-up of the program. Hence, a database must be included to provide a record of the events generated on each session and record each parameter of the sessions to allow the clinical staff to perform analysis on the patient's evolution.

  The requirements are summarized in Table 11.1.

11.3.3 Patient–Robot Interface Structure

After recognizing the need for the different modules presented in Table 11.1, the structure of the HRI is evident. The system must accomplish a continuous measuring and recording of variables while providing visual interactivity (employing a GUI). These functionalities that comprise the variables, follow-up, and the GUI regarding the interactivity requirement are considered the HCi. Similarly, the robotic social agent can address the interactivity requirements associated with social interaction,

Table 11.1 Requirements for the design of an HRI based on SAR

	Module	Feature
Variables	Sensor interface	*Spatiotemporal*
		Speed (mph)
		Cadence (Hz)
		Step length (m)
		Physiological
		Heart rate (bpm)
		Blood pressure (mmHg)
		Posture correction
		Exercise Intensity
		Treadmill inclination (°)
		Borg scale
Interactivity	Graphical User Interface	Visual interaction
	Social Robotic Agent	Social interaction
		Monitoring
		Motivation
Follow-up	Database	Events recording
		Parameters recording

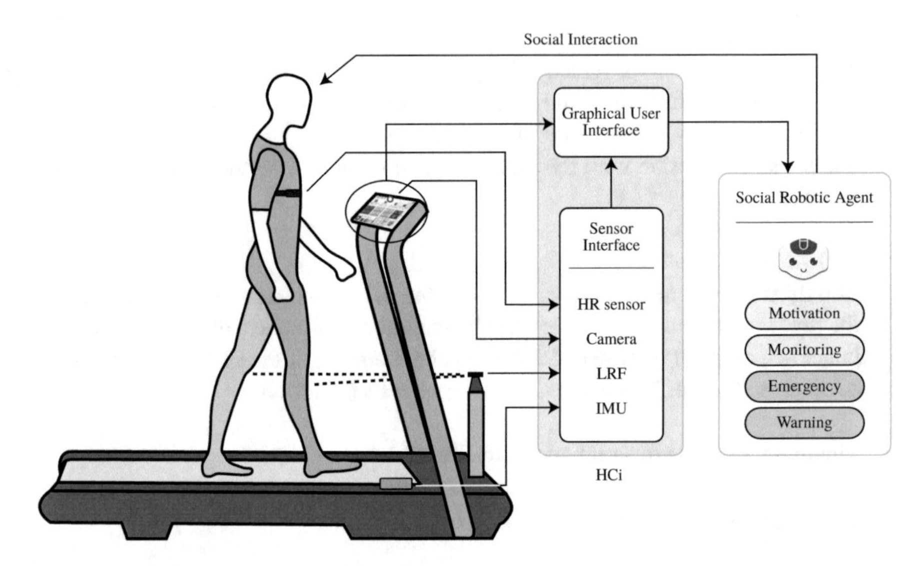

Fig. 11.4 Modules to consider in the design of a Patient–Robot Interface

monitoring, and motivation. Both systems, in conjunction, conform to the Patient–Robot Interface illustrated in Fig. 11.4.

A Patient–Robot Interface focuses on three main properties: acquisition of sensory data, computer interaction, and social interaction between the patient and the system. As depicted in Fig. 11.4, the HCi handles variables described in Table 11.1 utilizing a *sensor interface* and the user requests through the *GUI*. The therapy info is processed in the HCi and sent to the *social robotic agent*. The robot analyzes this information, and based on the result, the state of the therapy and the behavior that must be adopted are determined (i.e., motivation, monitoring, emergency, and warning). These behaviors are established according to the risks associated with the therapy. Hence, with this control loop, the patient's health condition is monitored and controlled, reducing the probability of risk occurrences. While at the same time, the robot can provide feedback and motivation through social interaction.

11.3.3.1 Sensor Interface

The sensor interface measures three types of variables usually selected by the medical staff to monitor the patient's status during the therapy, presented in Sect. 11.3.2. This interface integrates the measurement from a heart rate (HR) monitor, an IMU (reporting the treadmill inclination), an LRF (to estimate gait parameters), a camera (recording the patient's posture), and periodic results from the Borg scale. The system must be designed to present the three primary metrics and examples of the technology that can be involved are as follows:

Gait Spatiotemporal Parameters

As these parameters require tracking the displacement of the patient's legs during exercise, the selected sensor must locate the patient in the band and measure the leg difference distance (LDD). Additionally, the number of steps per second, namely the cadence, must be achieved by the exact measurement. Moreover, the sensor must accomplish the measurement at a frequency higher than the gait frequency. However, gait frequency is low compared to electronic devices. Hence, one sensor that meets all previous requirements could be the Hokuyo-URG 04LX-UG01 (Hokuyo, USA) [63]. This is a laser rangefinder (LRF) used to measure areas using an infrared electromagnetic wave (a wavelength of 785nm), and the distance measurement principle is based on the light phase difference. Similarly, this sensor allows measuring in a range of 240 degrees with a maximum distance of 4m. However, for this application, the measurement range must be limited to 60 degrees to limit the measurement of the treadmill band area. The sensor can perform a scan composed of 683 measurements in 0.1 s, which indicates a sampling frequency of 10 Hz, which is suitable for measuring gait spatiotemporal parameters. As shown in Fig. 11.4, an LRF sensor reports measurements used to estimate the patient's cadence, step length, and speed. The estimation of these parameters was proposed and validated in previous work [56, 64].

Physiological Parameters

The appropriate sensor to measure the HR must meet three main requirements: (1) it must allow physical activity while performing the measurement; in other words, the sensor must resist movement perturbations. (2) This sensor must allow online data transmission since the heart rate must be monitored in real-time during therapy. Finally, (3) the sensor must provide the processed data; namely, the sensor has to measure the signal and provide the heart rate value without requiring any additional processing. Hence, a suitable sensor for this application could be the heart rate monitor Zehpyr HxM BT (Zehpyr, USA) [65]. This sensor is located on the user's chest and reports a wireless and continuous measurement of the heart rate using Bluetooth communication [66].

Additionally, cervical posture corrections (the flexion of lower cervical vertebrae and its inclinations [67]) can be measured with the front camera of the tablet (Microsoft, USA) placed on the treadmill screen. A gaze estimator algorithm can be used. During the exercise, a proper cervical posture is set when the patient looks straight. In most therapies performed on a treadmill, the proper posture is essential to avoid dizziness, falls, and nausea. This measure represents the counting of a binary ("look-straight, not look-straight") value extracted from a gaze vector.

Exercise Intensity Parameters

Two different metrics are used to measure the physical activity difficulty: the inclination of the treadmill and the reported difficulty of the exercise. As the inclination most often cannot be accessed directly from the treadmill, an additional sensor must be installed. This sensor must be capable of measuring inclination angles in a range of 0 to 5 degrees (slope available on the treadmill), and as with the other sensors, it must allow online data transferring. Hence a sensor that meets these requirements is an IMU that will be placed on the treadmill so that one of its rotation angles corresponds to the central rotation axis of the treadmill. This way, changes in the measured IMU angle are equal to changes in the treadmill slope. For example, the MPU9150 IMU (Invensense, USA) [68] is an embedded system that combines a 3-axis gyroscope, one 3-axis accelerometer, a 3-axis magnetometer, and a digital motion processor.

11.3.3.2 Graphical User Interface

The GUI can run in a tactile computer monitor (i.e., Surface Pro-Microsoft, USA). This interface must present basic information and control panels regarding the status of the therapy (e.g., current user, session time, start/stop panel, emergency status, and biofeedback display) (see Fig. 11.5). As was presented in Fig. 11.4, the system receives the sensory data to be processed, stored, and displayed on the screen. With this information, the patient has access to visual feedback provided by the HCi. Hence, the graphical interface should report the synchronized and processed data from the sensors and allow the user to interact and respond to the requests generated by the system or the robot. Additionally, the interface must estimate the patient's fatigue level or related values, which can be captured employing the Borg scale (a qualitative measurement that estimates the perceived exertion of the patient, 6 for low intensity and 20 for very high intensity [69]). This value has to be periodically requested by the system or the robot.

Figure 11.5 presents an example of the main window (i.e., *MainTherapyWindow*) displayed during the therapy time. However, the system can contain additional functionalities and forms that allow the medical staff to register users, log in to the therapy session, and set therapy configuration parameters. Additionally, the system may allow the user to select different modalities of the therapy. In the

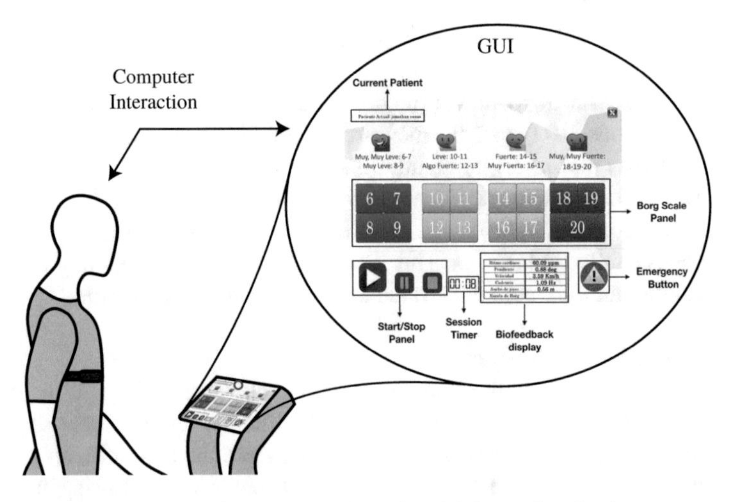

Fig. 11.5 Graphical User Interface to assess the patients' fatigue, view the therapy parameters as a form of feedback, configure the robot, monitor sensors, and control the therapy performance

first modality, the system could only work with the HCi; namely, the system only measures performance through the sensor interface and stores it in the database. Additionally, the GUI could request the Borg scale, even if feedback is not displayed on the screen. This modality is meant to measure a patient's performance without providing any feedback or social interaction and can be used for validation purposes with a control group defined in the baseline. The second modality could incorporate the social robot to provide social interaction, motivation, and monitoring. Similarly, the GUI should provide feedback regarding the state of the measured parameters (biofeedback display, see Fig. 11.5).

11.3.3.3 Social Robotic Agent

The robot module is focused on the interaction between the user and the robot. This interaction is provided through three robot roles: (i) motivational support, (ii) performance monitoring, and (iii) online feedback (emergency and warning). A typical therapy with the robot starts with an initial greeting, where the robot made an introduction of its functionalities during the rehabilitation program. Then, when the patient starts the exercise on the treadmill, the performance monitoring state is activated. During this state, sensory information is analyzed. Depending on the values given by each sensor, the current state can turn to the online feedback state or remain in the same state. If the online feedback is activated, two robot behaviors can be triggered (emergency or warning) when the system detects an increment in the physiological parameters such as training heart rate, Borg scale, and cervical posture.

11.4 Case Study: A Social Robot for Gait Rehabilitation with Lokomat

Patients who suffer from neurological disorders as spinal cord injuries, dementia, and cerebrovascular diseases like stroke are usually recommended to enter a physical rehabilitation (PR) program. PR is an active process to achieve a full recovery or an optimal physical, mental, and social potential to integrate the person appropriately back into society [70]. This is done through a combination of a (1) physiological treatment (e.g., cardiovascular, aerobic, and muscle control training) and (2) cognitive rehabilitation (e.g., language, perception, motivation, attention, and memory training) according to the patient's condition [70, 71]. There are different methods to perform PR. The conventional method is based on the guidance and manual assistance of the therapist in repetitive exercises that are used to improve the patient's performance [72]. In this method, the results depend merely on the expertise of the physiotherapist and the intensity of the exercises [73]. Alternatively, robot-assisted PR therapies combine a body weight support (BWS) system with a lower-limb exoskeleton to train physiological gait patterns on treadmills. Such is the case of PR with Lokomat, a robotic orthotic device that includes a BWS system to retrain gait [74].

Several studies have demonstrated the benefits of robot-assisted PR with Lokomat over the conventional therapies, including improvements in cardiovascular parameters [75], motor control [76], and balance [77], among others. However, even with robotic aids, PR is complex, and its second component, i.e., cognitive rehabilitation, has not been fully integrated. In healthcare, including social robots to rehabilitation procedures has shown progress regarding adherence to the treatments, assistance, and perception [5, 78]. This section presents the design and implementation of a patient–robot interface using SAR during Lokomat therapies.

11.4.1 Patient–Robot Interface

Following the approach presented in Sect. 11.3.3, Cespedes et al. [79, 80] developed and tested the interface. The system was composed of three main modules: (i) the sensory module, which allowed the acquisition and processing of sensory data, (ii) the Graphical User Interface, which was used for the computer interaction and monitoring, and (iii) the social robot module, in charge of the social interaction and the assistance of the patients. Figure 11.6 shows the patient–robot interface proposed for the Lokomat gait rehabilitation therapy using SAR.

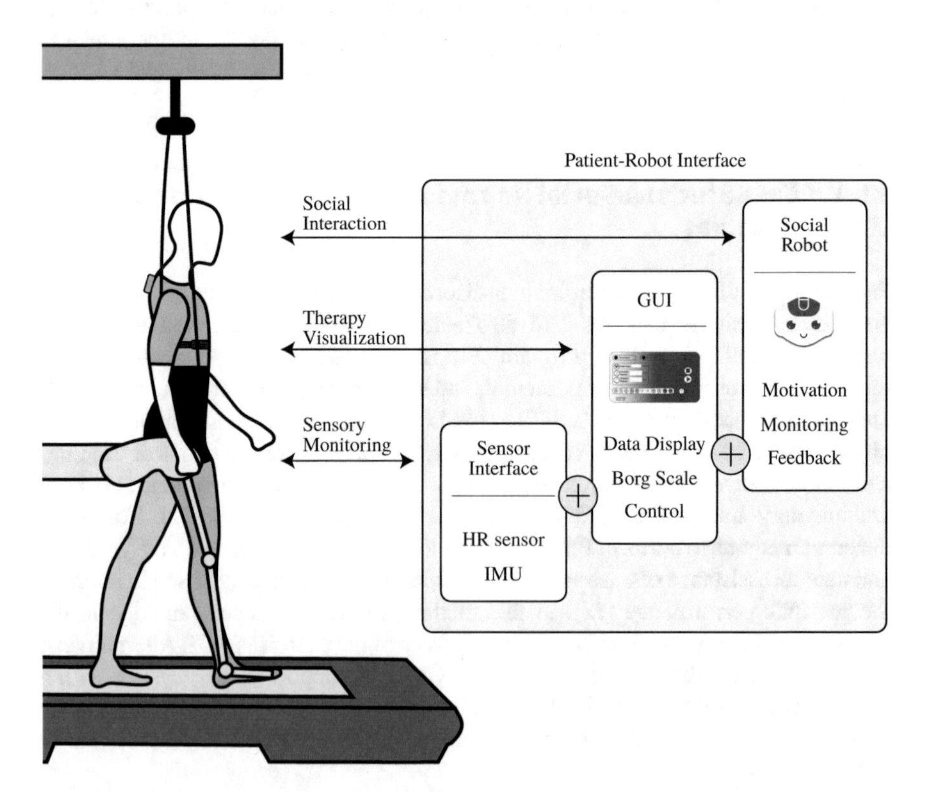

Fig. 11.6 Patient–robot interface for the Lokomat gait rehabilitation therapy using SAR

Sensory Module
The system acquired and processed the following physiological variables:

- The spinal posture (cervical and thoracic postures), measured with an IMU BNO055 (Adafruit, USA).
- The heart rate, measured with a Zephyr HxM sensor (Medtronic, New Zealand).
- The patients' perceived exertion during the exercise, measured with the Borg scale.

Graphical User Interface
The GUI was in charge of visualizing the therapy's data and controlling the session flow. Additionally, it allowed the therapists to interact with the patient and manage the session. A tablet Surface Pro (Windows, USA) was used to display the interface.

Social Robot Module
The robot's role was to provide feedback to the patients regarding their physiological parameters and motivate them during the therapy development. It supported the therapists while performing other tasks during the session, enabling physical distancing between the clinicians and the patient. The feedback given by the robot included nonverbal (imitation of healthy postures) and verbal gestures. The nonverbal gestures and the conversation scheme designed for the robot were developed with a rule-based algorithm. It depended on the events that took place during the sessions and the types of feedback presented. An NAO V6 robot (Softbank Robotics, France) was used to achieve the interaction.

11.4.2 Setup and Results

The study took place at the Mobility Group Rehabilitation Center located in Bogota, Colombia, where patients received Neurological Rehabilitation with Lokomat. A total of 10 patients were recruited to perform the rehabilitation assisted by the robot during 15 sessions. A session was conducted per week and lasted around 50 min. In the end, only 60% of the patients finished rehabilitation with Lokomat. Two conditions were established: (i) a *Control condition* and (b) a *Robot condition*. In the *Control condition*, the participants performed a conventional session of neurological rehabilitation with Lokomat. During the *Robot condition*, the participants performed the sessions assisted by the social robot. Patients were monitored in both conditions through the sensory module and the GUI and received support and additional feedback from the healthcare staff. Test sessions, where physiological parameters were measured, were performed at the beginning, in the middle, and at the end of the study. Afterward, the patients were assigned randomly to start with one of the conditions during six sessions. Finally, considering the start condition, the patients changed the scenario during another six sessions.

Two types of variables were analyzed to evaluate the robot assistance. The first one was quantitative variables, including the unhealthy posture time, the Borg scale, and the heart rate at training. The second group was qualitative variables from a questionnaire to observe the patient's perceptions of the robot's role. A Wilcoxon Signed-Rank test was applied to compare the patient's progress in both conditions. This is a non-parametric test used to compare two related samples and assess their difference [81]. A descriptive analysis was performed for the closed questions in the qualitative parameters, and a textual data analysis test was performed for the open items.

The study provided promising results regarding the inclusion of SAR in long-term PR and expanded the boundaries of robotic-assisted PR:

(i) In the case of patients who started the study with the *Control condition*, the percentage of unhealthy posture time regarding both the cervical and the thoracic postures (for both planes of motion) decreased when performing the session with the robot. The heart rate was maintained in a healthy range considering the exercise performed during the session, and the Borg scale was perceived at a low level.

(ii) Similarly, in the case of patients who started the*Robot condition* study, the percentage of unhealthy posture time regarding both postures was lower with the robot than without it. However, the patients seemed to maintain the posture after the robot intervention (the unhealthy posture time was lower compared to the previous group of patients), indicating that the patient learns how to control the cervical posture on the sagittal plane. Both the heart rate and the Borg scale were performed in healthy ranges.

(iii) Statistical differences in the different measurements between the *Robot* and the *Control condition* were found. For example, the percentages of unhealthy posture time were lower in the *Robot condition* than in the *Control condition*. Contrarily, the heart rate and the Borg scale parameters did not show differences between conditions.

(iv) During the robot condition, many benefits were evidenced. First, the feedback given by the robot allowed the patients to maintain a healthy posture and promote full gait rehabilitation. Patients considered that the system was safe and secure as they were continuously monitored. At the same time, this monitoring gave the medical team the possibility of performing other tasks during the session, which enriched the therapy sessions.

(v) An essential contribution of this study is how a patient–robot interface can enhance the methods in therapy by integrating different sources of information. For instance, the heart rate is not measured in conventional therapies. With the system and the robot's interaction, the clinicians could be warned by the robot and take action during the therapy if the patients had a high heart rate. Additionally, the inclusion of the Borg scale provided the clinicians with precise information regarding the performance of the patients. Altogether, clinicians could evaluate the patient's progress, not only in the gait behavior but also in their cardiovascular functioning and the exertion perceived during each session.

(vi) Although, in general, the robot's sociability was perceived as low by the users, they highlighted the platform's potential in PR with Lokomat. Fluid speech and conversation with the robot is the next step towards better social interaction. At the end of the sessions, most of the patients suggest using the robot with other patients. Clinicians' overall perception was also positive and in accordance with recent findings that evidence the need for social and cognitive support during PR [82]. These results showed the potential of SAR in gait rehabilitation as a tool to enhance the conventional sessions.

11.5 Conclusions

Gait rehabilitation is a primary component of physical rehabilitation processes for many patients with neurological disorders. Robot-assisted methods to perform this rehabilitation therapy have shown many physical benefits for patients. Nevertheless, these methods can still improve substantially as their cognitive component is explored. Social Assistive Robotics have been used in the last years in therapy to include cognitive aspects such as patient motivation and engagement. Starting from the basic concepts of social interaction, relevant characteristics of social robotics and their importance in rehabilitation processes were presented. After that, the methodology to design a Patient–Robot interface based on social robots was guided through a real-world rehabilitation scenario on a treadmill. The impact and promising future of including SAR in physical rehabilitation were at last shown in a case study of long-term gait rehabilitation with Lokomat.

References

1. L.L. Mullins, J.R. Keller, J.M. Chaney, A systems and social cognitive approach to team functioning in physical rehabilitation settings. Rehabilitation Psychology **39**(3), 161–178 (1994)
2. K. Fabel, G. Kempermann, Physical activity and the regulation of neurogenesis in the adult and aging brain. NeuroMolecular Medicine **10**, 59–66 (2008)
3. D. Feil-Seifer, M. Mataric, in *9th International Conference on Rehabilitation Robotics, 2005. ICORR 2005* (IEEE, Chicago, IL, USA, 2005), pp. 465–468
4. C. Breazeal, K. Dautenhahn, T. Kanda, Social robotics, in *Springer Handbook of Robotics* (Springer International Publishing, 2016), pp. 1935–1971
5. M.J. Matarić, J. Eriksson, D.J. Feil-Seifer, C.J. Winstein, Socially assistive robotics for post-stroke rehabilitation. J. NeuroEng. Rehab. **4**, 5 (2007)
6. Understanding Social Interaction|Boundless Sociology, https://courses.lumenlearning.com/boundless-sociology/chapter/understanding-social-interaction/. [Online; accessed 23-July-2019]
7. D. Feil-Seifer, M.J. Matarić, Toward Socially Assistive Robotics For Augmenting Interventions For Children With Autism Spectrum Disorders, Tech. rep., 2008
8. R. Krauss, Psychology of verbal communication. Int. Encyclopedia Soc. Behav. Sci., 16161–16165 (2004)
9. J. Casas, N. Cespedes, M. Múnera, C.A. Cifuentes, *Chapter One - Human-Robot Interaction for Rehabilitation Scenarios* (Academic Press, 2020)

10. D. Casas-Bocanegra, D. Gomez-Vargas, M.J. Pinto-Bernal, J. Maldonado, M. Munera, A. Villa-Moreno, M.F. Stoelen, T. Belpaeme, C.A. Cifuentes, An open-source social robot based on compliant soft robotics for therapy with children with ASD. Actuators **9**(3), 91 (2020)
11. K. Goris, J. Saldien, I. Vanderniepen, D. Lefeber, *The Huggable Robot Probo, a Multi-disciplinary Research Platform*, vol. 33 (IEEE, 1970)
12. A. Shick, *Romibo Robot Project: An Open-Source Effort to Develop a Low-Cost Sensory Adaptable Robot for Special Needs Therapy and Education* (IEEE, 2013)
13. Y.H. Wu, C. Fassert, A.S. Rigaud, Designing robots for the elderly: Appearance issue and beyond. Archiv. Gerontol. Geriat. **54**, 121–126 (2012)
14. J. Fink, Anthropomorphism and human likeness in the design of robots and human-robot interaction, in *Lecture Notes in Computer Science (Including Subseries Lecture Notes in Artificial Intelligence and Lecture Notes in Bioinformatics)*, vol. 7621 LNAI (Springer, Berlin, Heidelberg, 2012), pp. 199–208
15. A.A. Ramírez-Duque, L.F. Aycardi, A. Villa, M. Munera, T. Bastos, T. Belpaeme, A. Frizera-Neto, C.A. Cifuentes, Collaborative and inclusive process with the autism community: A case study in Colombia about social robot design. Int. J. Soc. Robot. **13**, 153–167 (2021)
16. R.S. Lazarus, *Emotion and Adaptation* (Oxford University Press, 1991)
17. J.A. Russell, Core affect and the psychological construction of emotion. Physchol Rev. **110**(1), 145–172 (2003)
18. A. Mehrabian, J.A. Russell, *An Approach to Environmental Psychology* (MIT Press, 1974)
19. M.L. Hoffman, Toward a comprehensive empathy-based theory of prosocial moral development., in *Constructive & Destructive Behavior: Implications for Family, School, & Society* (American Psychological Association, 2004), pp. 61–86
20. I. Leite, S. Mascarenhas, C. Martinho, R. Prada, A. Paiva, The influence of empathy in human-robot relations. Int. J. Human Comput. Stud. **71**(3), 250–260 (2012)
21. B. Gonsior, S. Sosnowski, C. Mayer, J. Blume, B. Radig, D. Wollherr, K. Kuhnlenz, Improving aspects of empathy and subjective performance for HRI through mirroring facial expressions, in *Proceedings - IEEE International Workshop on Robot and Human Interactive Communication* pp. 350–356 (2011)
22. C. Breazeal, Cynthia, Social robots: From research to commercialization, in *Proceedings of the 2017 ACM/IEEE International Conference on Human-Robot Interaction - HRI '17*, pp. 1–1 (2017)
23. Z. Kasap, N. Magnenat-Thalmann, Towards episodic memory-based long-term affective interaction with a human-like robot, in *Proceedings - IEEE International Workshop on Robot and Human Interactive Communication*, pp. 452–457 (2010)
24. T. Belpaeme, P.E. Baxter, R. Read, R. Wood, H. Cuayáhuitl, B. Kiefer, S. Racioppa, I. Kruijff-Korbayová, G. Athanasopoulos, V. Enescu, R. Looije, M. Neerincx, Y. Demiris, R. Ros-Espinoza, A. Beck, L. Cañamero, A. Hiolle, M. Lewis, I. Baroni, M. Nalin, P. Cosi, G. Paci, F. Tesser, G. Sommavilla, R. Humbert, Multimodal child-robot interaction: Building social bonds. J. Human Robot Interact. **1**(2), 33–53 (2013)
25. E. Tsardoulias, A.L. Symeonidis, P.A. Mitkas, An automatic speech detection architecture for social robot oral interaction, in *AM '15* (2015)
26. Z. Kasap, N. Magnenat-Thalmann, Building long-term relationships with virtual and robotic characters: The role of remembering. Visual Computer **28**, 87–97 (2012)
27. K. Jokinen, G. Wilcock, Multimodal open-domain conversations with robotic platforms, in *Multimodal Behavior Analysis in the Wild: Advances and Challenges* (Elsevier, 2018), pp. 9–26
28. L. Yang, H. Cheng, J. Hao, Y. Ji, Y. Kuang, A survey on media interaction in social robotics, in *Lecture Notes in Computer Science (including subseries Lecture Notes in Artificial Intelligence and Lecture Notes in Bioinformatics)*, vol. 9315 (Springer, 2015), pp. 181–190
29. T. Salter, I. Werry, F. Michaud, Going into the wild in child-robot interaction studies: Issues in social robotic development. Intell. Serv. Robot. **1**(2), 93–108 (2008)
30. P. Dhami, S. Moreno, J.F.X. DeSouza, New framework for rehabilitation - fusion of cognitive and physical rehabilitation: the hope for dancing. Front. Psychol. **5**, 1478 (2015)

31. A. Bandura, *Social Foundations of Thought and Action: A Social Cognitive Theory*. Prentice-Hall Series in Social Learning Theory (Prentice-Hall, Englewood Cliffs, NJ, 1986)
32. B. Irfan, N.C. Gomez, J. Casas, E. Senft, L.F. Gutierrez, M. Rincon-Roncancio, M. Munera, T. Belpaeme, C.A. Cifuentes, *Using a Personalised Socially Assistive Robot for Cardiac Rehabilitation: A Long-Term Case Study* (IEEE, 2020)
33. J. Casas, E. Senft, L.F. Gutierrez, M. Rincon-Rocancio, M. Munera, T. Belpaeme, C.A. Cifuentes, Social assistive robots: assessing the impact of a training assistant robot in cardiac rehabilitation. Int. J. Soc. Robot., 1–15 (2020)
34. N. Céspedes, B. Irfan, E. Senft, C.A. Cifuentes, L.F. Gutierrez, M. Rincon-Roncancio, T. Belpaeme, M. Múnera, A socially assistive robot for long-term cardiac rehabilitation in the real world. Front. Neurorobot. **15**, 633248 (2021)
35. M. Bautista, C.A. Cifuentes, M. Munera, Conversational agents for healthcare delivery: Potential solutions to the challenges of the pandemic, in *Internet of Medical Things*, 1st edn. (CRC Press, 2021), p. 26
36. J. Fasola, M. Mataric, A socially assistive robot exercise coach for the elderly. J. Human Robot Interact. **2**, 3–32 (2013)
37. J. Li, The benefit of being physically present: A survey of experimental works comparing copresent robots, telepresent robots and virtual agents. Int. J. Human Comput. Stud. **77**, 23–37 (2015)
38. V. Vasco, C. Willemse, P. Chevalier, D. De Tommaso, V. Gower, F. Gramatica, V. Tikhanoff, U. Pattacini, G. Metta, A. Wykowska, Train with me: A study comparing a socially assistive robot and a virtual agent for a rehabilitation task, in *Social Robotics*, vol. 11876, ed. by M.A. Salichs, S.S. Ge, E.I. Barakova, J.-J. Cabibihan, A.R. Wagner, A. Castro Gonzalez, H. He (Springer International Publishing, Cham, 2019), pp. 453–463
39. T.W. Bickmore, R.W. Picard, Establishing and maintaining long-term human-computer relationships. ACM Trans. Comput. Human Interact. **12**, 293–327 (2005)
40. C.D. Kidd, C. Breazeal, A robotic weight loss coach, in *Proceedings of the 22nd National Conference on Artificial Intelligence - Volume 2*, AAAI'07 (AAAI Press, 2007), p. 1985–1986
41. N. Maclean, P. Pound, A critical review of the concept of patient motivation in the literature on physical rehabilitation. Soc. Sci. Med. **50**, 495–506 (2000)
42. A. Tapus, M. Mataric, B. Scassellati, Socially assistive robotics [grand challenges of robotics]. IEEE Robot. Autom. Mag. **14**, 35–42 (2007)
43. K.I. Kang, S. Freedman, M.J. Matarić, M.J. Cunningham, B. Lopez, A hands-off physical therapy assistance robot for cardiac patients, in *9th International Conference on Rehabilitation Robotics, 2005. ICORR 2005*, pp. 337–340 (2005)
44. R. Gockley, A. Bruce, J. Forlizzi, M. Michalowski, A. Mundell, S. Rosenthal, B. Sellner, R. Simmons, K. Snipes, A. Schultz, Jue Wang, Designing robots for long-term social interaction, in *2005 IEEE/RSJ International Conference on Intelligent Robots and Systems* (IEEE, 2005), pp. 1338–1343
45. J. Fasola, M.J. Matarić, Using socially assistive human-robot interaction to motivate physical exercise for older adults. Proc. IEEE **100**(8), 2512–2526 (2012)
46. S. Sabanovic, C.C. Bennett, Wan-Ling Chang, L. Huber, PARO robot affects diverse interaction modalities in group sensory therapy for older adults with dementia, in *2013 IEEE 13th International Conference on Rehabilitation Robotics (ICORR)* (IEEE, Seattle, WA, 2013), pp. 1–6
47. K. Swift-Spong, E. Short, E. Wade, M.J. Mataric, Effects of comparative feedback from a socially assistive robot on self-efficacy in post-stroke rehabilitation, in *2015 IEEE International Conference on Rehabilitation Robotics (ICORR)* (IEEE, Singapore, Singapore, 2015), pp. 764–769
48. R. Looije, F. Cnossen, M. Neerincx, Incorporating Guidelines for Health Assistance into a Socially Intelligent Robot, Tech. rep., 2006
49. I. Baroni, M. Nalin, P. Baxter, C. Pozzi, E. Oleari, A. Sanna, T. Belpaeme, What a robotic companion could do for a diabetic child, in *The 23rd IEEE International Symposium on Robot and Human Interactive Communication* (IEEE, 2014), pp. 936–941

50. M. White, M.V. Radomski, M. Finkelstein, D.A.S. Nilsson, L.I.E. Oddsson, Assistive/socially assistive robotic platform for therapy and recovery: Patient perspectives. Int. J. Telemed. Appl. **2013**, 1–6 (2013)
51. P.L. Wilbourne, E.R. Levensky, Enhancing client motivation to change, in *Clinical Strategies for Becoming a Master Psychotherapist* (Elsevier, 2006), pp. 11–36
52. W.H. Organization, Adherence to long term therapies: Evidence for action. *WHO* (2013)
53. D.X. Cifu, J.S. Kreutzer, S.A. Kolakowsky-Hayner, J.H. Marwitz, J. Englander, The relationship between therapy intensity and rehabilitative outcomes after traumatic brain injury: a multicenter analysis11no commercial party having a direct financial interest in the results of the research supporting this article has or will confer a benefit upon the author(s) or upon any organization with which the author(s) is/are associated. Archiv. Phys. Med. Rehab. **84**, 1441–1448 (2003)
54. P. Gadde, H. Kharrazi, H. Patel, K.F. MacDorman, Toward monitoring and increasing exercise adherence in older adults by robotic intervention: A proof of concept study. J. Robot. **2011**, 1–11 (2011)
55. R. Sharma, V.I. Pavlovic, T.S. Huang, Toward multimodal human-computer interface. Proc. IEEE **86**(5), 853–869 (1998)
56. C.A. Cifuentes, A. Frizera, Human-robot interaction strategies for walker-assisted locomotion. Springer Tracts Adv. Robot. **115**(September), 105 (2016)
57. E. Bruno, S. Oussama, K. Frans, E.B. Siciliano, O. Khatib, F. Groen, *Springer Tracts in Advanced Robotics*, vol. 26 (Springer, 2003)
58. M.A. Goodrich, A.C. Schultz, Human-robot interaction: A survey. Found. Trends® Human Comput. Interact. **1**(3), 203–275 (2008)
59. M.L. Guha, A. Druin, J.A. Fails, Cooperative inquiry revisited: Reflections of the past and guidelines for the future of intergenerational co-design. Int. J. Child Comput. Interact. **1**(1), 14–23 (2013)
60. A.A. Ramírez-Duque, T. Bastos, M. Munera, C.A. Cifuentes, A. Frizera-Neto, Robot-assisted intervention for children with special needs: A comparative assessment for autism screening. Robot. Autonom. Syst. **127**, 103484 (2020)
61. N. Vallès-Peris, C. Angulo, M. Domènech, Children's imaginaries of human-robot interaction in healthcare. Int. J. Environ. Res. Public Health **15**(5), 970 (2018)
62. S. Merter, D. Hasırcı, A participatory product design process with children with autism spectrum disorder. Int. J. CoCreat. Des. Arts **14**(3), 170–187 (2018)
63. Y. Maeda Kamitani, Scanning Laser Range Finder Corrector Amended Reason, pp. 1–4 (2009)
64. A. Aguirre, S.D. Sierra M., M. Múnera, C.A. Cifuentes, Online system for gait parameters estimation using a LRF sensor for assistive devices. IEEE Sensors J., 1 (2020)
65. Zephyr Technology, HXM Bluetooth API Guide, pp. 1–21 (2011)
66. J.S. Lara, J. Casas, A. Aguirre, M. Munera, M. Rincon-Roncancio, B. Irfan, E. Senft, T. Belpaeme, C.A. Cifuentes, *Human-Robot Sensor Interface for Cardiac Rehabilitation* (IEEE, 2017)
67. R. Shafer, *Chapter 7: The Cervical Spine*, 2nd edn. (Wiliams & Wilkins, 1987)
68. Invensense, MPU-9150 Datasheet, vol. 1(408), pp. 1–50 (2013)
69. J. Scherr, B. Wolfarth, J.W. Christle, A. Pressler, S. Wagenpfeil, M. Halle, Associations between Borg's rating of perceived exertion and physiological measures of exercise intensity. Eur. J. Appl. Physiol. **113**, 147–155 (2013)
70. WHO, T. Dua, A. Janca, A. Muscetta, Public health principles and neurological disorders. Neurol. Disord. Public Health Challen. **2**, 7–25 (2006)
71. S.B. O'Sullivan, T.J. Schmitz, G.D. Fulk, *Physical Rehabilitation* (F.A. Davis, 2013)
72. L.L. Stanley Fisher, A. Trasher, Robot-Assisted Gait Training for Patients with Hemiparesis Due to stroke (2011)
73. G.L. Shahid Hussain, S. QuanXie, Robot assisted treadmill training: Mechanisms and training strategies. Med. Eng. Phys. **33**, 527–533 (2011)
74. M. Munera, A. Marroquin, L. Jimenez, J.S. Lara, C. Gomez, S. Rodriguez, L.E. Rodriguez, C.A. Cifuentes, *Lokomat Therapy in Colombia: Current State and Cognitive Aspects* (IEEE, 2017)

75. A. Mayr, M. Kofler, E. Quirbach, H. Matzak, K. Fröhlich, L. Saltuari, Prospective, blinded, randomized crossover study of gait rehabilitation in stroke patients using the Lokomat gait orthosis. Neurorehab. Neural Repair **21**(4), 307–314 (2007)
76. R. Banz, M. Bolliger, G. Colombo, V. Dietz, L. Lunenburger, Computerized visual feedback: An adjunct to robotic-assisted gait training. Physical Therapy **88**, 1135–1145 (2008)
77. D.-H. Bang, W.-S. Shin, Effects of robot-assisted gait training on spatiotemporal gait parameters and balance in patients with chronic stroke: A randomized controlled pilot trial. NeuroRehabilitation **38**(4), 343–349 (2016)
78. J.A. Casas, N. Céspedes, C.A. Cifuentes, L.F. Gutierrez, M. Rincón-Roncancio, M. Múnera, Expectation vs. reality: Attitudes towards a socially assistive robot in cardiac rehabilitation. Appl. Sci. (Switzerland) **9**(21), 4651 (2019)
79. N. Céspedes, D. Raigoso, M. Múnera, C.A. Cifuentes, Long-term social human-robot interaction for neurorehabilitation: Robots as a tool to support gait therapy in the pandemic. Front. Neurorob. **15**, 10 (2021)
80. N. Céspedes, M. Múnera, C. Gómez, C.A. Cifuentes, Social human-robot interaction for gait rehabilitation. IEEE Trans. Neural Syst. Rehab. Eng. **28**(6), 1299–1307 (2020)
81. F. Wilcoxon, Individual comparisons by ranking methods. Biometrics Bulletin **1**, 80 (1945)
82. D. Raigoso, N. Céspedes, C. Cifuentes, A.J. del Ama, M. Munera, A survey on socially assistive robotics: Clinicians' and patients' perception of a social robot within gait rehabilitation therapies. Brain Sciences **11**(6), 738 (2021)

Serious Games in Robot-Assisted Rehabilitation Therapy for Neurological Patients

12

Angie Pino, Marcela Múnera, and Carlos A. Cifuentes

12.1 Introduction

Robotic rehabilitation therapy has evolved as an innovative solution to overcome motor impairments caused by neurologic injuries such as stroke or spinal cord disease. Robotic systems and exoskeletons in rehabilitation favor acquiring lost motor skills, improving both repetition and intensity of training [1]. These tools assist the patient in reaching his functionality level and supporting the therapist to increase the time for an effective rehabilitation process [2].

The technological strategies involve sensors, actuators, and control systems that give information to the robot to learn and optimize the therapy. However, the interaction environment is essential in the human–robot interaction in a two-way exchange of information. In this way, the user transmits information to the robotic device and receives information from the system once the action has been executed [3].

Although motor assistance robots, like exoskeletons, are not of a social or purely interactive nature, it has been shown the importance of implementing additional tools that, together with it, provide information about the user's performance during the therapy process. Thus, after the reasoning, planning, and execution of an action, the participant retains an informational component that allows the user to correct or improve the performance [4]. Notwithstanding the above, rather than the informational component, patient compliance, adherence, and motivation are some of the most determinative factors that remain challenges in rehabilitation therapies with robotic devices [2]. During physical therapy, there is evidence that the lack of

A. Pino · M. Múnera · C. A. Cifuentes (✉)
Biomedical Engineering, Department of the Colombian School of Engineering Julio Garavito, Bogotá D.C., Colombia
e-mail: angie.pino-l@mail.escuelaing.edu.co; marcela.munera@escuelaing.edu.co; carlos.cifuentes@escuelaing.edu.co

C. A. Cifuentes, M. Múnera, *Interfacing Humans and Robots for Gait Assistance and Rehabilitation*, https://doi.org/10.1007/978-3-030-79630-3_12

positive feedback is one of the main factors related to non-compliance. In this way, poor user engagement affects the success of the motor restoration process [5,6]. The biomedical robot's design features can significantly influence the motivation and adherence of patients to robot-aided treatments in the ease of adaptation to training. Motor task's difficulty level, modes of interaction, and the quantity and quality of feedback presented to the patient show how robotic technology is directly involved in maintaining patients' interest high during motor rehabilitation [7]. In this context, serious games are video games based on visual feedback strategies designed to carry out a learning process from an entertainment environment. These strategies prove to be solutions to long and repetitive conventional therapies [8–10].

This chapter aims to present strategies based on serious games to encourage user's motivation and commitment to therapeutic rehabilitation. To do so, it is necessary to understand the principles of serious game design and its relationship with motor learning and neurorehabilitation. Likewise, the design of a serious game for lower-limb therapeutic purposes using an exoskeleton of variable stiffness is presented.

This chapter is organized into six sections that include relevant concepts about adherence, feedback, and gaming experiences in rehabilitation environments. Section 12.2 presents the importance of motivation and adherence in rehabilitation therapies focusing on literary reports demonstrating its success in motor therapy. Section 12.3 defines a serious game based on the game and the therapy component that characterize it. Section 12.4 addresses the main components to consider when designing a serious game for therapeutic purposes. Section 12.5 focuses on the latest developments in serious games for lower-limb rehabilitation. Some of the most used strategies and the common therapeutic objective of all the games are presented. Section 12.6 presents the first version of a serious game designed for motor rehabilitation using an ankle exoskeleton. Its functionality, configuration, and form of access are presented. Finally, Sect. 12.7 presents the conclusions and recommendations for future works in this field and the challenges of serious games in the rehabilitation context.

12.2 Motivation and Adherence to Rehabilitation Therapies

Recent studies traditionally show that long-term rehabilitation therapies produce, in a habitual way, demotivation and even gradual desertion of the physical therapy program by part of the user due to a tedious process. Low motivation is a prominent concern that is significantly affecting therapy success [11]. A motivational character is associated with a persistent behavioral tendency to work independently to achieve goals [12]. In a rehabilitation program, this is achieved through active engagement toward the treatment, leading to greater satisfaction and better therapy outcomes [7].

That said, it is generally thought that motivation is related to the person's personality traits. However, it is a factor that depends for the most part on the interaction between the individual and the environment. Biomedical robot's design features can significantly influence the motivation and adherence in robot-aided treatments. These aspects are focused on the way the rehabilitation practice is presented to the patient. That is, the difficulty of the motor task, the ease of adaptation to training, and the quantity and quality of feedback presented that go beyond the simple execution of a repetitive task. On the other hand, there is evidence that a patient's perception of therapy can influence motivation and improve exercise adherence [7]. Although most early robotic devices lacked engaging interfaces, nowadays, the interactivity and virtual reality elements are being developed to challenge patients and provide additional motivation [2].

12.3 Serious Games for Rehabilitation Therapy

Serious games are video games designed to achieve a particular therapeutic purpose through participation in an interactive experience [8]. This type of strategy and the ones that include virtual reality systems are promising developments to avoid discouragement problems and limit lack of enthusiasm in long-term therapy. Several researchers have proved that its application in lower-limb physical therapy is safe and effective in improving gait rehabilitation and motor function [13, 14].

This type of augmented feedback operates as an external source that enhances the interaction between the patient and the therapy. Through intrinsic motivation, the user believes in his abilities and challenges himself to improve his performance and rehabilitation outcomes [15]. In general, video games can incorporate different motivation levels essential to the task and elements that satisfy plasticity and motor learning principles. Following the above, serious games contain both learning and entertainment components to engage user attention through a cognitive and motor process [9].

12.3.1 Game Component

A serious game has its nature as a component of entertainment and fun, which is considered essential to motivate the patient. The game's playability is described as the degree to which the game is experienced as enjoyable, where emotions are regarded as fundamental during the game experience. Although the aspiration to achieve the goals can motivate the player to continue playing, the experienced emotions determine whether the player finds the experience pleasant and continues to play. Curiosity, virtuosity, and sociality are some of the emotions that might carry out a pleasurable experience. However, considering that our attention span is limited, interactive tools must be a mechanism that eludes this limitation and maintains the user's attention [16].

The automation of movements and striking graphics has proven to be attractive to keep the player's attention. Notably, after an automatization task is fully achieved, the performance may increase efficiency and speed, allowing for a more controlled task [16]. These computational environments enable motor learning thanks to the ease with which users can adapt to the extrinsic visual, auditory, or even haptic feedback [15].

12.3.2 Therapy Component

Neurological patients have suffered from an injury to the central nervous system like stroke or spinal cord disease. The therapy treatment usually considers a multidisciplinary rehabilitation program specializing in neurorehabilitation and motor learning [17]. In the same way, serious games consider a therapeutic objective to recover lost motor skills at a neurological and motor level.

12.3.2.1 Neurological Rehabilitation

According to the World Health Organization, neurorehabilitation is considered an active process where individuals with an injury or illness can achieve an integral recovery, mainly at a neurological level. The facilitation of adaptive learning based on experience and learning processes is the main focus of this type of treatment [18]. How the brain reorganizes its structure, functions, and connections to respond to learning processes or gradual recovery from brain damage is known as neuroplasticity and is the principal objective of any neurorehabilitation system related to the brain's ability to change in response [1, 18, 19].

In neurological rehabilitation, motivation, attention, and skill acquisition promote optimal learning and are also vital factors in the success of therapies to induce neuroplasticity [20, 21]. It has been shown that enhancing neuroplasticity during poststroke rehabilitation might help patients overcome their motor impairments [17, 22].

12.3.2.2 Motor Rehabilitation

Regaining the skills practiced depends mainly on processes that involve cognitive and motor methods related to practice or experience [15]. Motor learning implies the acquisition of skilled movements through practice [23]. It leads to changes at neurological or performance levels, supported in part by the implicit memory system or by the effect of explicit information in the form of feedback. The type of practice and its intensity is related to exercise-dependent neuroplasticity, task-specific practice, and motivation to optimize motor learning and recovery. An improvement in motor learning has been found due to the perception of self-control during training employing physiotherapy practice, feedback, and physical assistive devices [15].

12.4 Serious Games Design

Serious game design involves a broad range of requirements to include entertainment and learning components on the same platform. As seen in Fig. 12.1, the game's operation with robotic devices meets a constant loop. The process begins

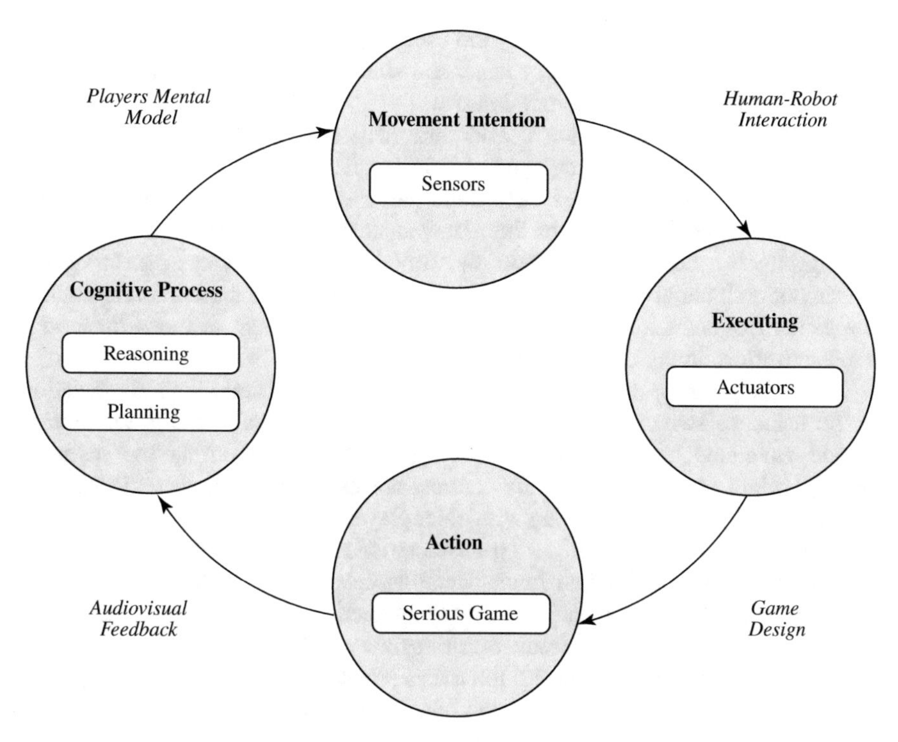

Fig. 12.1 Learning loop in serious games design through assistive robotic technology for rehabilitation

with establishing a mental model in which the player reasons, plans, and makes decisions. Then, considering the robotic assistance technology systems, the player's intention to move is detected, and the movement is executed through the actuators. The action carried out is interpreted under the rules and the design of the game, which informs the consequence of the executed action. In this way, the player receives automatic feedback to modify his cognitive process to advance the game and start the loop again. This is when the entertainment level must be high enough to allow people to meet both therapeutic and game goals. Remarkably, these games are directly related to visual feedback, where the subject performs an action and sees the result of it immediately on the screen. However, different strategies could be used to feedback the user, including haptic systems, virtual reality, and even brain–computer interface.

Within the game strategies, the use of characters to command the game objectives is one of the most useful in this field. Neurological patients need to know their position in the game, and the usage of cartoon persons generates the feeling of “property” and improves interaction with visual strategy [24]. Besides, the psychological color of game characters is even more critical. Blue and green tones convey well-being, and red colors capture the players' attention and convey danger [25]. Gonçalves et al. [24], include, within the visual strategies, the integration of sidebars to relate the effort made by the user with the effort necessary to advance in the game. This strategy is very beneficial since it informs the game's needs to advance and inspires the person to improve.

The therapeutic intervention's tasks and goals must be individualized to the learner's motor abilities and enjoyment [15]. In this way, the serious game design must include a calibration task before the game according to the control performance. This seems to be one of the most valuable requirements in the design of videogames for rehabilitation. Short and straightforward activities measuring the maximum abilities of each user are essential to evaluate the calibration to control the game. Besides establishing the difficulty for each patient, it looks for a non-excessive effort during the gameplaying.

In general, during the game, the program must have a real-time computation performance to switch according to the player's input. Additionally, the system should guarantee good screen resolution and refresh rate during the sessions. Finally, bright colors that generate contrast are essential to capture the player's attention even more when dealing with older players.

In addition to the general design characteristics mentioned above, the design of serious games includes four fundamental aspects for the development of entertainment technologies in rehabilitation: (1) interaction technologies, for the active participation of the user; (2) feedback strategies, which can fulfill the function of informing, motivating, or focusing the user's attention; (3) incentive strategies, to keep the player committed to therapy; and (4) the user performance evaluation, valuable metrics to inform both the player and the therapist about the results of the gaming experience.

12.4.1 Interaction Technologies

Generally, serious games in rehabilitation therapy are associated with using a robotic device in stationary therapy and different control interfaces for motion intent detection. These strategies favor the interaction between the user and the therapy to actively operate movement-assistive devices [26]. Therefore, the player interaction with the games is essential to associate the execution made by the user with the feedback on the screen. To do that, the use of sensor interfaces is indispensable to carry out movement intention detection based on neural signals from the central nervous system, muscle contraction forces, or relative joint rotations and motion of body segments. Several sensors are involved in this field: electroencephalography sensors, electromyography sensors, force sensors, inertial sensors, soft wearable motion sensors, strain sensors, among others [26]. However, the study carried out by Farjadian et al. [27] showed that the control over position seems to cause less fatigue and intensity level than the one over force. In this way, movement detection from the force exerted by the user is not commensurate. Other strategies that involve changes in the position or joint flexion angle (i.e., inertial sensors, soft wearable motion sensors) seem to be better options to interact with the system. That said, despite motion intention can also be derived from explicit commands of the user (e.g., by pressing command switches or handling joysticks), these are not relevant in the motor rehabilitation field [26].

12.4.2 Feedback Strategies

In human performance systems, the term feedback refers to the information about the movement outcomes or movement performance. Mainly, serious games are extrinsic audiovisual feedback, also known as augmented feedback, due to the information that comes from an external source. In general, this information could be focused to (1) produce a motivation to increase effort, (2) provide information about errors as a basis for corrections, or (3) direct the learner's attention toward a movement goal [21]:

- **Motivational Properties:** An essential function of this type of feedback is to act as a "stimulant" to improve the learners' effort to take on tasks. Rewards, challenges, and even casual commentaries using keywords to highlight correct performance or indicate progress motivate to keep going a more extensive practice [21].
- **Informational Properties:** An essential part of motor learning feedback is its information about the action pattern. This feedback is about error guides for improving future performance [21]. Although this information can come directly from the therapist, audiovisual resources reporting on performance immediately or at the end of the session (i.e., metric evaluation results) represent an additional tool to support the informational process of the interactive experience [13].
- **Attentional-Focusing Properties:** The main objective of this feedback is to improve performance and learning. This is accomplished by directing the subject's attention to achieving the movement's goal [21]. As mentioned above, game characters and brightly colored didactic interfaces are examples of the strategies that make it possible to capture the participant's attention.

Any of the strategies mentioned above can be used during a serious game and supported by feedback from other sensory channels. However, it is crucial to consider the frequency in which this is supplied since it is sought not to overload the player with information. In this way, considering neural patients, feedback content should be adjusted for the user's learning stage in faded feedback frequency. The early stages of learning must be more frequent, but the feedback should be reduced gradually to accomplish permanent skill learning as the player gains experience. Besides, it must consider when the practitioner internalizes his proprioceptive feelings of the performed skill [1, 28, 29].

Feedback can be provided positively or negatively. Positive feedback enhances or amplifies the person's performance, while negative feedback corrects and regulates its execution. Therefore, to deal with negative feedback carefully, it has been shown that it is crucial to handle failure positively in rehabilitation. Positive failure is how the player fails, but his attitude continues being influenced by the enjoyment of the challenge more than nominal success [13]. Therefore, patients would not feel failure from their physical limitations, and they would be more likely to remain engaged [30].

12.4.3 Incentive Strategies

Besides the feedback, there is evidence suggesting that clear goals, rewards, and an optimal challenge are indispensable in game design to increase engagement and adherence. These strategies are linked with successful rehabilitation where the player connects himself with the game to self-motivate during the experience [13, 15, 31].

12.4.3.1 Clear Objectives

Developing clear objectives and instructions has been shown to increase user motivation and avoid frustration or confusion when interacting with the platform [13]. The therapeutic intervention's tasks and goals must be individualized to the learner's motor abilities and enjoyment [15]. In this way, self-improvement is associated with goals or instructions to complete a specific assignment specified in a task-by-task tutorial. Patients who are unclear about the therapeutic goals during the game may develop an opposite effect to what is being sought with the interactive tool. Situations like these and even those in which the tasks or game objectives are complex for the patient achieve a lead in a decisive way to a low motivation [13].

12.4.3.2 Reward

Gameplaying is motivated by rewarding and challenging experiences where dopamine is associated with feelings of enjoyment, learning, and motivation [13]. The reward is often given after correcting or correctly completing an assignment. Several pieces of research included the reward in terms of the score or according to the amount of the collected items, seeking to encourage the patient to achieve his highest [24]. Scores are incredibly beneficial in maintaining the patient's motivation high throughout the session, improving the video-game experience where a higher score indicates better performance.

12.4.3.3 Challenge

It is necessary to avoid boredom or frustration by always giving a scaffold practice from easy to difficult. Multilevel serious games offer the best solution to keeping the proper level of challenge throughout the interactive experience. This concept of challenge and difficulty should be administered according to the therapy dosage to lead to a more significant skill transfer [13]. Patients must be able to fulfill the serious game's objectives to avoid frustration, but at the same time, the game must have a component of difficulty to maintain adherence to the therapeutic process and to demand the player indirectly to improve his performance.

12.4.4 User Performance Evaluation

In rehabilitation, evaluating the individual player execution and experience in serious games is crucial during player–game interaction. This process depends on the kind of serious game design. However, there are different measurements to analyze user efficiency during the interactive experience. One of the most effective is based on the individual player skill, where the progress in achieving goals or evolution over time is measured [32]. This quantification of the performance/skill level can be achieved through performance metrics like completion time, the number of hits, trajectories features, etc. [27]. Parameters as adaptability and progress in time enable one to know the learner's progress and actions within the game. Besides, strategies like learning, gaming experience, and usability are some other subjective attributes that should be assessed through surveys, scenario analysis, psychometrics, video observation, or interviews to inform how the practice impacts the participant [33, 34].

In the analysis of the user's performance, many studies evaluated the execution based on the data thrown by the robotic device or according to the game results. Data from the device mainly included a range of motion and muscle strength. Data from the game was related to the score and failure (e.g., number of collected items, number of collisions) [24, 27].

12.5 Lower-Limb Rehabilitation with Serious Games

The application of serious games in rehabilitation has been mainly influenced by the motor recovery of the upper limb. However, gradually audiovisual strategies have been developed to support lower-limb therapy, along with lower-limb exoskeletons or robotic platforms. In the field of stationary rehabilitation, several and different game environments have been developed. Some of them were designed in collecting items [24], jumping over obstacles [25], tracking trajectories [35], and reaching targets [27, 36]. However, what all these developments have in common is the therapy component that seeks to regain lost mobility of the ankle after a neural injury.

After a stroke, the loss or reduction of descending input to the spinal centers is the primary deficit that results in the incapability to activate muscles and reduce muscle force voluntarily [37]. The ankle joint usually shows an increase in stiffness and insufficient dorsiflexion movement due to muscle fibers' changes in the gastrocnemius or the tibialis anterior. In this case, the spastic ankle tends to show over 50% more resistance than the average ankle in the displacement [38]. Therefore, the therapeutic objective for neural patients aims at the ankle complex's critical movements (i.e., plantar flexion, dorsiflexion, ab-/adduction, and inversion–eversion) to improve the range of motion of the joint.

Within the serious game experience, different strategies have been found that favor the rehabilitation process. Farjadian et al. [27] included auditory feedback in which collecting the goal was rewarded with a particular sound while colliding with walls was penalized with a different audio signal. Asín-Prieto et al. [35] besides the visual feedback of the game, bring in perturbations via a haptic adaptive feedback approach based on the user performance.

On the other hand, Ren et al. [39] were focused on the therapist's active participation through the game settings, including the challenge during the therapy. Other studies set the game's difficulty according to facial expression recognition [25], or through an automatic system that analyses the performance in real-time to increase challenge [35].

12.6 Jumping Guy: Ankle Rehabilitation Therapy with T-FLEX

To improve the experience during ankle rehabilitation after stroke with a robotic orthosis, an interface was proposed to follow and challenge the user's motor rehabilitation process. In this way, this strategy was thought to be a rehabilitation supplement that, together with interdisciplinary work, contributes to better therapeutic results. *Jumping Guy: Ankle Rehabilitation with T-FLEX* is a serious game designed to improve the results already obtained in the rehabilitation process with the variable stiffness device *T-FLEX* (Colombian School of Engineering, Colombia) (see Chap. 7) contribute to the user's commitment to it. This game's virtual environment and functionalities were performed with Unity software in version 2.3.1 in C# language and developed to be executed in a Windows 10 operating system. The sprites, sounds, and graphic resources were taken from a free retro-type game called "Jumping Guy" [40]. From this existing version of the game, the alternate version is provided with therapeutic objectives. Most of the original game features were modified to provide the patient with an entertaining experience during motor recovery.

This serious game tries to vary the therapy intensity throughout audiovisual feedback. The game involves two main characters: the avatar and the enemies (see Fig. 12.2). The avatar is the character over which the user has control to trigger a jump. This action is carried out through combined subject movements of plantar and dorsal flexion. The enemies, on their behalf, are characters that the user faces, and in front of them, the player must make the jump to avoid hitting them. The more enemies are avoided, the higher the player's score will be.

The challenge is to achieve the highest number of points in 20 min sessions for each level, considering conventional therapy. In total, there are three levels, differentiated not only by the game environment but also by the speed at which the enemies come out, which increases as the levels advance and leads the player to perform more dorsi-plantar flexion movements. In this way, while the person seeks to achieve the most significant number of points and overcome oneself, the serious game seeks to induce neuroplasticity through repetitions. Thus, the brain promotes the neural systems reorganization to recover the lost motor function.

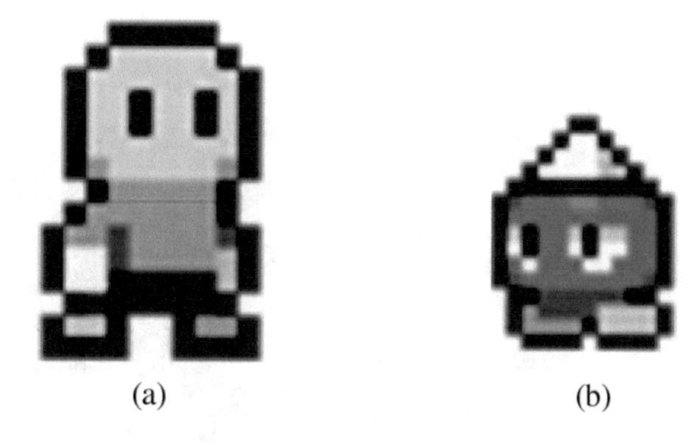

Fig. 12.2 Jumping guy characters. (**a**) Avatar. (**b**) Enemy

The system configuration, serious game design strategies (i.e., for motion intention detection, providing feedback, evaluating performance), and the final interactive experience are presented below.

12.6.1 System Setup

The proposed system for serious game ankle rehabilitation includes the following three critical components (see Fig. 12.3):

1. ***T-FLEX* Exoskeleton:** Its principal function is to assist the dorsiflexion and plantarflexion movements to control the game's avatar.
2. **Graphic Interface:** It presented a serious game to engage the user in the therapy process.
3. **IMU:** Inertial sensor (100 Hz, BNO055, Bosch, Germany), located at the tip of the paretic foot. It allowed estimating the user movement intention.

As seen in Fig. 12.3, the participant must be seated in a chair with 90° knee flexion. The lower member where the device is located must be raised, avoiding contact with the ground. Moreover, the orthosis must be used in its therapy mode [6]. The *T-FLEX* electronic system in this operational mode integrates a Raspberry Pi 3 Board running in Debian operating system where the sensor acquisitions, algorithms, and controllers are deployed to command the actuators. In this way, the device executes repetitions and assists the user's dorsi-plantarflexion movements (see Chap. 7).

The connection and feedback of the system are carried out using the model in Fig. 12.4. The system consists of an input signal corresponding to the user movement assisted with the robotic device. Specifically, these plantar and dorsal flexion movements are identified with the inertial sensor data (i.e., the angular velocity along the sagittal plane). The data are taken from the sensor pass through a pre-processing to perform the detection of movement intent. Only a "jump" command is transmitted through a TCP/IP model to communicate with the game.

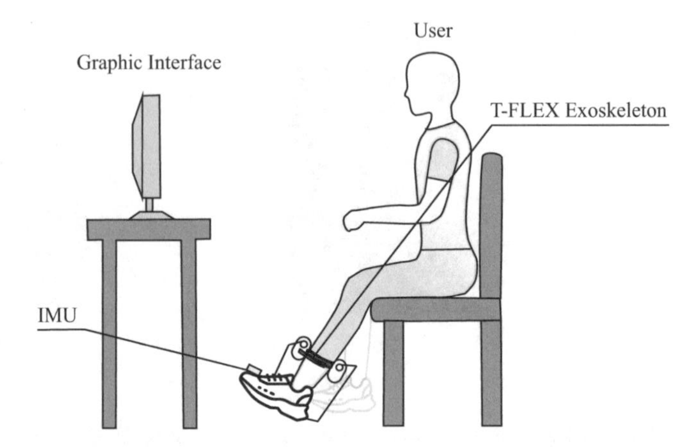

Fig. 12.3 Setup for the interaction between the serious game and the variable stiffness orthosis

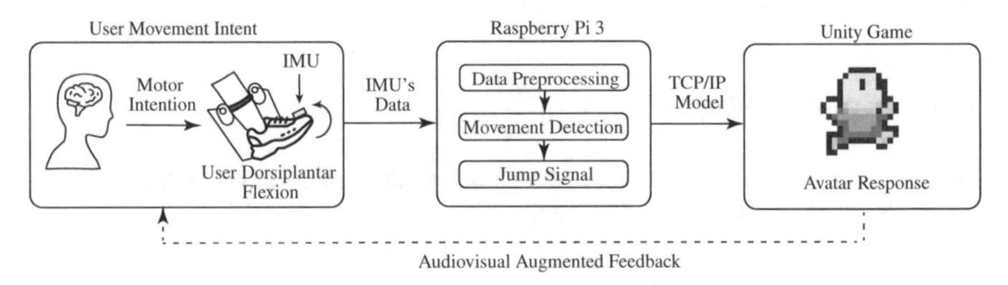

Fig. 12.4 General motion control model during the game considering the data acquisition and processing on the Raspberry Pi 3, and the graphical game response in unity

The TCP communication is configurated through a pair of sockets for each connector. One works as a server listening to the incoming messages, and the other one as a client connecting to the applications. In this sense, the data exchange considers the server in the Unity interface and the *T-FLEX*'s Raspberry Pi as the client. Thus, the sockets are opened in both cases with the IP direction and the port, and only when the game receives the string "jump" the avatar jump will be executed. In this way, the feedback, according to the movement performed, allows the person to correct their movement strategy subsequently.

12.6.2 Movement Intention Detection Strategy

The module runs in the Robotic Operating System (ROS) on a Raspberry Pi 3 since the controllers for the orthosis actuators operate in this framework. As mentioned previously, all the IMU's data enter the pre-processing stage, where a 4th-order Butterworth low-pass filter (cutoff of 6 Hz) removes the electromagnetic noise. After filtering, the data is analyzed in code to detect in real-time if there is a movement intention. To do this, data higher than one rad/s (radian per second) are analyzed to determine action (see Fig. 12.5). This motion intention detection process is constantly performed in three different stages of the game: calibration stage, tutorial, and game stage.

In the calibration stage, a threshold value is set with the angular velocity average of five movements' intentions performed by the user and detected by the system. This stage only ends when the number of movements requested is achieved. Once the threshold has been established, the system proceeds to either the game stage or the tutorial stage since both works under the same threshold terms. The tutorial only differs in considering additional time to teach the desired performance. However, in both, the filtered data are compared in real-time with the threshold value calculated before, and the participant is asked to perform his best dorsiflexion movement. In this way, the avatar jumps, and the exoskeleton executes a repetition only when the angular velocity is higher than the threshold (see Fig. 12.5). Therefore, the result observed on the screen is the one that provides feedback on whether or not the execution made by the participant was sufficient to move the avatar.

12.6.3 Feedback Strategies

The increase in the user's commitment to rehabilitation and their learning and participation was fostered through augmented and positive feedback. In general, the term "losing" was not implemented in the game's design because the main idea was to allow the user to reach the maximum score during the established time. In this way, the reward of the game was always optimistic and looking to avoid frustration.

In addition to the above, the different feedback strategies mentioned in Sect. 12.4.2 are implemented throughout the audiovisual experience of the serious game. Mainly, the attentional-focusing feedback is constantly presented through the graphics, the avatar, and the gameplay:

- **Before the Game (Informational Feedback):** Before starting the game, the instructions and the interactive interface's user position are presented using a tutorial. In this way, through a short practice similar to the in-game interface, the tutorial stage shows step by step the directed movements to reach the game's primary objective.
- **During the Game (Informational and Motivational Feedback):** Throughout the interactive experience, the user is subjected to informative and motivational feedback, seeking to improve commitment to therapy.
  - Informational Feedback: The display modality, besides being visual, is also auditory. In this sense, the user knows the immediate result of his action because both sensory channels are stimulated. The reward in terms of the score is continuously displayed on the screen since the user achieves the game's objectives.
  - Motivational Feedback: During the game, performance-dependent comments are also visually implemented, considering direct and straightforward feed-

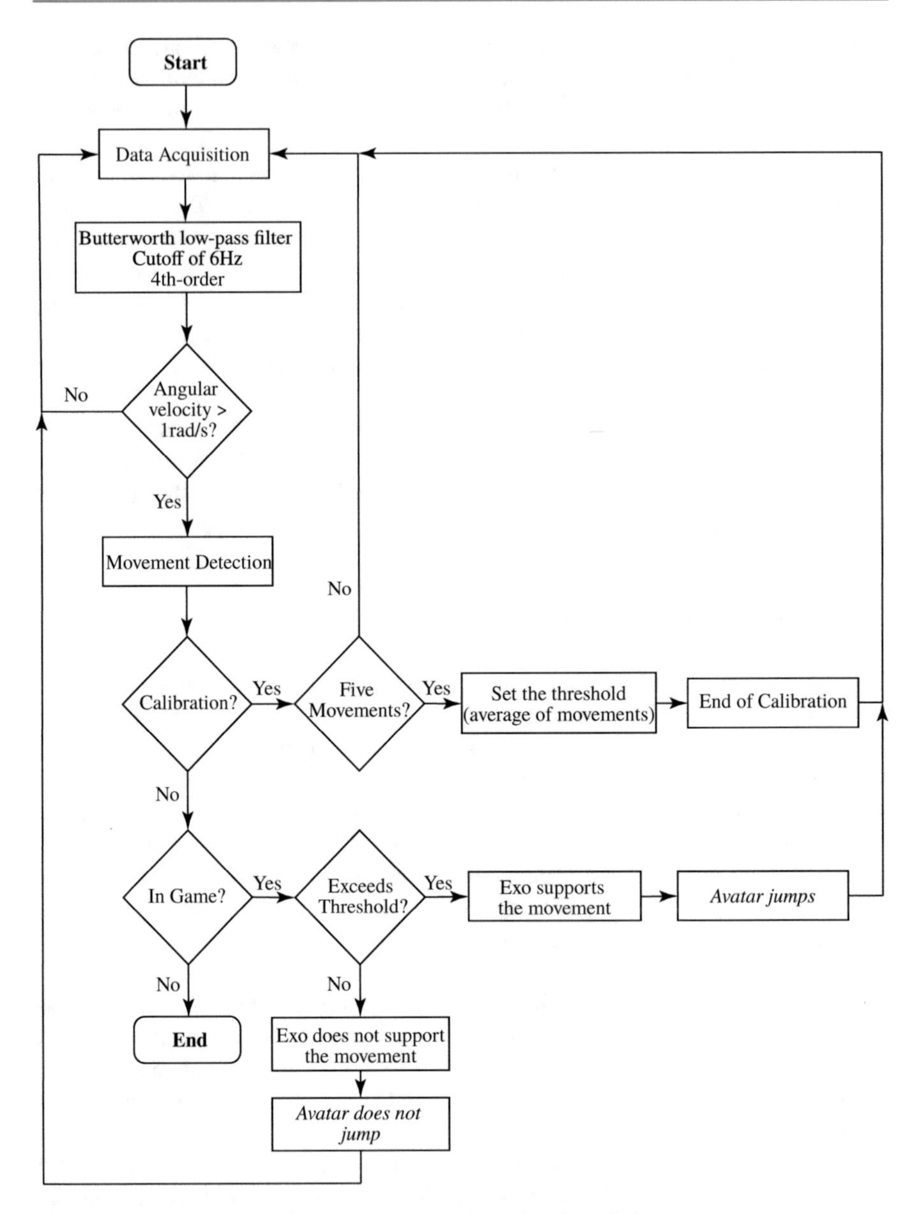

Fig. 12.5 Blocks diagram of the user movement intention through the game

back. There are employed keywords to highlight correct performance or indicate progress (e.g., Good Job! You beat the record!) according to the score reached for time intervals during the game. This motivation uses fade feedback, which is more frequent during the initial learning stage (i.e., during the first levels).

- **After the Game (Informational Feedback):** Finally, with a database, additional augmented feedback is included with the Knowledge of Results (KR), where the user knows and compares his results into the broader context of the game [21]. The above was thought to fulfill the informative function to improve motor learning for both therapist and patient.

12.6.4 Performance Evaluation

The evaluation of the user's progress during the game was based on the number of jumps and missed avoiding the enemies, the percentage of precision during the entire session, and the type of response in front of each enemy. In this last parameter, it was empirically evaluated when the action was effected to achieve the point. An ideal skip was counted as one in which the avatar passes without approaching the enemy. From this, the anticipated or a delayed time response were those in which the enemy was gently closer in his back or front, respectively (see Fig. 12.6). The system detects the moment for which the jump is made and classifies, through trajectory colliders, the type of response of the player when jumping on the enemy (i.e., from the jump start zone to the end zone). In this way, "Early" was classified as a jump 0.15 s before the ideal jump, and "Late" was a jump 0.34 s after the ideal. On the other side, the system also evaluated the user's precision recorded over time (i.e., user accuracy every 20 s) to evaluate the user's adaptability during the session. However, as the speed increases throughout the levels, all the evaluation metrics results are maintained for each game stage.

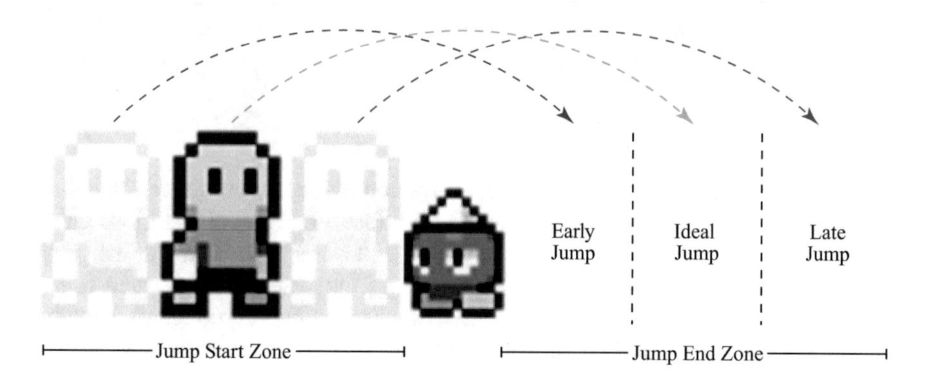

Fig. 12.6 Possible player responses to the enemy: ideal, early, or late

12.6.5 Serious Game Experience

This first version of *Jumping Guy: Ankle Rehabilitation Therapy with T-FLEX* consists of a 2D adventure game with a single player. The serious game is designed to be compiled on desktop devices with a Windows platform and with a minimum architecture of 32 bits. By default, the game is full screen, and its resolution is 640x360. However, the system allows to select the resolution of the game with a display resolution dialog and the option of a resizable window. This option was provided and configured in Unity to display initially, right after running the game. One of the advantages of this mechanism is that it facilitates interaction with any Windows device, where the user can select the graphic quality of the game and its screen layout according to his comfort.

Once the game is executed after the display resolution dialog, a first welcome interface appears, giving details about the connection with the game. As mentioned in the past, the game connects to the robotic device through a TCP/IP model, demanding an IP address and a port. In this specific case, the game is designed to automatically acquire the computer IP address on which it is running with a 3014 connection port number. On the other hand, this interface also informs that the game results will be gathered in a text file inside the folder where the executable is located.

Important It is critical to ensure that all robotic, sensing, and assistive devices are connected under the same *T-FLEX* Network, the game is running on. Moreover, the Raspberry Pi must function as a client that commands game communication with the same IP address and port. The client performs precisely by sending through sockets the encoded "jump" message every time the lower-limb movement intention is detected during the calibration, the tutorial, and the in-game stage.

A login interface is directly presented by continuing in the game, including the game's name and the user-password system (see Fig. 12.7). The serious game works with a local MySQL database storing the player's basic data and his game results throughout all sessions. However, in the demo version this system is not enabled; therefore, to continue, it is not necessary to fill out everything or access the registration option. It is only requested to enter the player's name in the "Username" space and continue pressing the "Login" button. The name placed here will be the one that will differentiate the results text file.

12.6.5.1 Before the Game

Before starting the game, the user is faced with a calibration stage that is necessary to establish the threshold of surpassing each player. The user is asked to perform five ankle flexion movements (i.e., dorsiflexion and plantarflexion movements). Visually, the interface shows both the avatar jump when detecting an intention of movement

Fig. 12.7 Login interface. The indicated section corresponds to the only one the player must fill out to enter the game

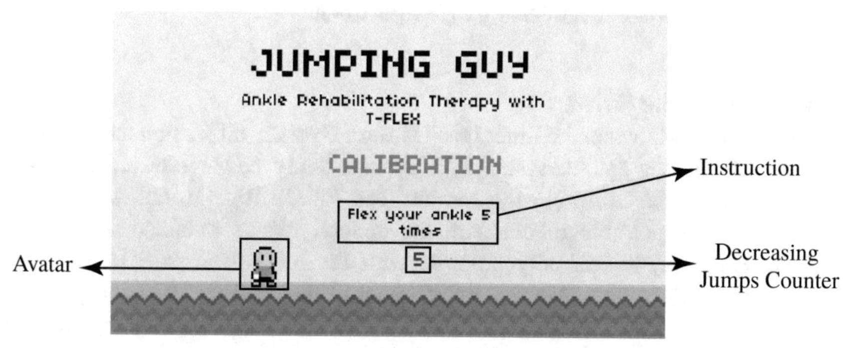

Fig. 12.8 Calibration interface. Request the execution of 5 movements of plantar and dorsal flexion. When the system detects a movement, the avatar jumps, and the counter decreases

and the countdown of the movements made (see Fig. 12.8). Thus, when the user completes them all, he can already start playing or taking a short tutorial session. The tutorial is based on a similar experience to the game. However, it works at a much slower speed, with instructions, and in it, the game does not progress until the player executes the action requested.

12.6.5.2 During the Game

Once the game is entered, the player finds the avatar located in the lower-left part of the screen, and a series of enemies generated one by one from the lower right part of the screen and that advance toward the avatar. These are the enemies, which, as mentioned previously, must be avoided using the avatar jump. As shown in Fig. 12.9 the interface includes the game points (given for each jump that avoids an enemy), the time of the remaining session, and a series of visual and motivational messages, which occasionally appear with the player's evolution.

The user can pause the game by pressing the space bar on the computer keyboard during the game. There it can be chosen whether to continue the game or to quit. If the continue option is selected, the panel disappears, and the game continues its course. However, if the exit option is selected, the game saves the process carried out so far in the text file, and the game closes automatically.

Fig. 12.9 Visual interface design during the game (first level)

12.6.5.3 After the Game

The game ends only when the timer reaches zero. By then, all session data is stored in the text document, and the system is directed to show the game results regarding the user's performance during the session (see Fig. 12.10). Overall, through the drop-down, it is possible to compare the general results obtained in previous sessions. However, this is only possible with the local database. On the other hand, by clicking on the "More Details" button, the interface graphically shows adaptability over time. This graph is related to the user's accuracy record evaluated every 20 s during the whole game session.

Jumping Guy: Ankle Rehabilitation Therapy with T-FLEX is a serious game whose demo version is available from the *T-FLEX* public repository at https://github.com/GummiExo/t_flex. This repository, in addition to containing the details for the configuration, installation, and connection of the device, includes a set of files-folders with an executable to start the game. To download it, you must access the folder titled T-FLEX GAME and run the application titled "T_FLEX_Game," which has the logo of the game.

The demo version, under the *T-FLEX* Network, allows access to all the game levels for 5 min each. This version does not make available to access the user-password system, nor to the in-game results of the previous sessions. Nevertheless, as mentioned above, enter and store the data is possible without any hassle.

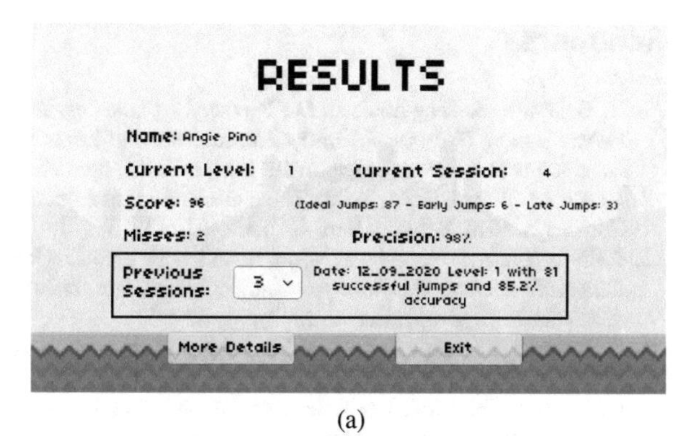

(a)

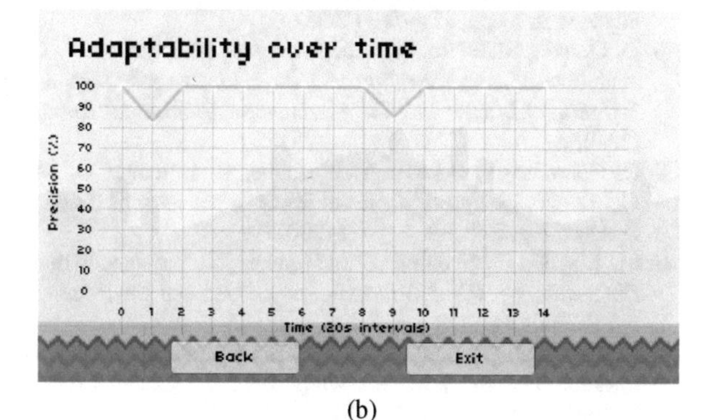

(b)

Fig. 12.10 Interface results model. (**a**) Interface with the general results of the session. The designated session does not show results from past sessions. (**b**) Interface with adaptability results over time. Precision graph at 20 s intervals

12.7 Chapter Conclusions

This chapter has presented serious games as a promising tool in solving patients' low motivation and commitment to therapy. It has been shown that its application in the rehabilitation field can considerably improve user participation toward the treatment. Besides, the importance of both therapy and game components has been presented to design a learning and entertainment experience.

This chapter also showed through *Jumping Guy: Ankle Rehabilitation Therapy with T-FLEX* that strategies such as clear objectives, rewards, challenges, feedback strategies, and performance evaluation are essential for designing interactive games in motor rehabilitation. Likewise, the game functionality with robotic devices requires sensing methods for movement detection and a necessary calibration stage to individualize the learner's motor abilities.

References

1. N. Bejarano, S. Maggioni, L. De Rijcke, C. Cifuentes, D. Reinkensmeyer, *Robot-Assisted Rehabilitation Therapy: Recovery Mechanisms and Their Implications for Machine Design*. Biosystems & Biorobotics (Springer, Berlin, 2016), pp. 197–223
2. J. Laut, M. Porfiri, P. Raghavan, The present and future of robotic technology in rehabilitation. Curr. Phys. Med. Rehabil. Rep. **4**(4), 312–319 (2016)
3. S. Sierra Marín, L. Arciniegas Maya, F. Ballen Moreno, D. Gomez, M. Munera, C.A. Cifuentes, *Adaptable Robotic Platform for Gait Rehabilitation and Assistance: Design Concepts and Applications* (Springer, Berlin, 2020), pp. 67–93
4. L. Bueno, F. Brunetti, A. Frizera, J.L. Pons, J.C. Moreno, E. Rocon, J.M. Carmena, E. Farella, L. Benini, *Human–Robot Cognitive Interaction*, chap. 4 (Wiley, Hoboken, 2008), pp. 87–125
5. E.M. Sluijs, G.J. Kok, J. van der Zee, Patient compliance is of considerable cause treatment effects partly depend importance in physical therapy be- on it. The efficacy of therapeutic exer. Phys. Ther. **73**(1), 771–786 (1993)
6. D. Gomez, M. Pinto, F. Ballen Moreno, M. Munera, C. Cifuentes G., Therapy with t-flex ankle-exoskeleton for motor recovery: a case study with a stroke survivor, in *The 8th IEEE RAS/EMBS International Conference on Biomedical Robotics & Biomechatronics BIOROB* (2020)
7. R. Colombo, F. Pisano, A. Mazzone, C. Delconte, S. Micera, M.C. Carrozza, P. Dario, G. Minuco, Design strategies to improve patient motivation during robot-aided rehabilitation. J. NeuroEng. Rehabil. **4**, 1–12 (2007)
8. M. Ma, K. Bechkoum, Serious games for movement therapy after stroke, in *Conference Proceedings - IEEE International Conference on Systems, Man and Cybernetics* (2008), pp. 1872–1877
9. P. Rego, P.M. Moreira, L.P. Reis, Serious games for rehabilitation: a survey and a classification towards a taxonomy, in *Proceedings of the 5th Iberian Conference on Information Systems and Technologies, CISTI 2010* (2010)
10. A. Pino, D. Gomez, M. Munera, C.A. Cifuentes, Visual feedback strategy based on serious games for therapy with t-flex ankle exoskeleton, in *The International Symposium on Wearable Robotics (WeRob2020) and WearRAcon Europe* (Springer, Berlin, 2020)
11. N. Barrett, I. Swain, C. Gatzidis, C. Mecheraoui, The use and effect of video game design theory in the creation of game-based systems for upper limb stroke rehabilitation. J. Rehabil. Assist. Technol. Eng. **3**, 205566831664364 (2016)
12. R. Dishman, W. Ickes, Self-motivation and adherence to therapeutic exercise. J. Behav. Med. **4**, 421–38 (1982)
13. K. Lohse, N. Shirzad, A. Verster, N. Hodges, H.F. Van Der Loos, Video games and rehabilitation: using design principles to enhance engagement in physical therapy. J. Neurol. Phys. Ther. **37**(4), 166–175 (2013)
14. S.C. Howes, D.K. Charles, J. Marley, K. Pedlow, S.M. McDonough, Gaming for health: systematic review and meta-analysis of the physical and cognitive effects of active computer gaming in older adults. Phys. Ther. **97**(12), 1122–1137 (2017)
15. M.F. Levin, H. Sveistrup, S.K. Subramanian, Feedback and virtual environments for motor learning and rehabilitation. Schedae **1**, 19–36 (2010)
16. N.C. Nilsson, S. Serafin, R. Nordahl, Gameplay as a source of intrinsic motivation for individuals in need of ankle training or rehabilitation. Presence Teleop. Virt. Environ. **21**(1), 69–84 (2012)
17. M.A. Dimyan, L.G. Cohen, Neuroplasticity in the context of motor rehabilitation after stroke. Nat. Rev. Neurol. **7**(2), 76–85 (2011)
18. L. Carey, A. Walsh, A. Adikari, P. Goodin, D. Alahakoon, D. De Silva, K.L. Ong, M. Nilsson, L. Boyd, Finding the intersection of neuroplasticity, stroke recovery, and learning: scope and contributions to stroke rehabilitation. Neural Plast. **2019**, 1–15 (2019)

19. L.M. Muratori, E.M. Lamberg, L. Quinn, S.V. Duff, Applying principles of motor learning and control to upper extremity rehabilitation. J. Hand Ther. **26**(2), 94–103 (2013)
20. A. Kliem, A. Wiemeyer, Comparison of a traditional and a video game based balance training program. Int. J. Comput. Sci. Sport **9**(2010), 80–92 (2010)
21. R.A. Schmidt, T.D. Lee, *Motor Learning and Performance*. Human Kinetics, 5th edn (2014)
22. S.C. Cramer, J.D. Riley, Neuroplasticity and brain repair after stroke. Curr. Opin. Neurol. **21**(1), 76–82 (2008)
23. H. Masaki, W. Sommer, Cognitive neuroscience of motor learning and motor control. J. Phys. Fitness Sports Med. **1**(3), 369–380 (2012)
24. A.C.B. Gonçalves, W.M. Dos Santos, L.J. Consoni, A.A. Siqueira, Serious games for assessment and rehabilitation of ankle movements, in *SeGAH 2014 - IEEE 3rd International Conference on Serious Games and Applications for Health, Books of Proceedings* (2014)
25. A.M. Salazar, A.B. Ortega, K.G. Velasco, A.A. Pliego, Mechatronic integral ankle rehabilitation system: ankle rehabilitation robot, serious game, and facial expression recognition system, in *Advanced Topics on Computer Vision, Control and Robotics in Mechatronics* (Springer, Berlin, 2018), pp. 291–320
26. J. Lobo-Prat, P.N. Kooren, A.H. Stienen, J.L. Herder, B.F. Koopman, P.H. Veltink, Non-invasive control interfaces for intention detection in active movement-assistive devices. J. NeuroEng. Rehabil. **11**(168), 1–22 (2014)
27. A.B. Farjadian, M. Nabian, A. Hartman, S.C. Yen, B. Nasseroleslami, Visuomotor control of ankle joint using position vs. force. Eur. J. Neurosci. **50**(8), 3235–3250 (2019)
28. K. Carr, N. Zachariah, P. Weir, N. McNevin, An examination of feedback use in rehabilitation settings. Crit. Rev. Phys. Rehabil. Med. **23**(1–4), 147–160 (2011)
29. P. van Vliet, G. Wulf, Extrinsic feedback for motor learning after stroke: what is the evidence?. Disabil. Rehabil. **28**(13–14), 831–840 (2006)
30. J.W. Burke, M.D. McNeill, D.K. Charles, P.J. Morrow, J.H. Crosbie, S.M. McDonough, Optimising engagement for stroke rehabilitation using serious games. Vis. Comput. **25**(12), 1085–1099 (2009)
31. S.T. Smith, D. Schoene, The use of exercise-based videogames for training and rehabilitation of physical function in older adults: current practice and guidelines for future research. Aging Health **8**(3), 243–252 (2012)
32. L. Nacke, A. Drachen, S. Gobel, Methods for evaluating gameplay experience in a serious gaming context. Electron. J. e-Learn. **10**(2), 172–184 (2012)
33. J. Moizer, J. Lean, E. Dell'Aquila, P. Walsh, A.A. Keary, D. O'Byrne, A. Di Ferdinando, O. Miglino, R. Friedrich, R. Asperges, L.S. Sica, An approach to evaluating the user experience of serious games. Comput. Edu. **136**, 141–151 (2019)
34. I. Mayer, G. Bekebrede, C. Harteveld, H. Warmelink, Q. Zhou, T. Van Ruijven, J. Lo, R. Kortmann, I. Wenzler, The research and evaluation of serious games: toward a comprehensive methodology. British J. Edu. Technol. **45**(3), 502–527 (2014)
35. G. Asín-Prieto, A. Martínez-Expósito, F.O. Barroso, E.J. Urendes, J. Gonzalez-Vargas, F.S. Alnajjar, C. González-Alted, S. Shimoda, J.L. Pons, J.C. Moreno, Haptic adaptive feedback to promote motor learning with a robotic ankle exoskeleton integrated with a video game. Front. Bioeng. Biotechnol. **8**, 1–15 (2020)
36. S.N. Jeon, J. H. Choi, The effects of ankle joint strategy exercises with and without visual feedback on the dynamic balance of stroke patients. J. Phys. Ther. Sci. **27**(8), 2515–2518 (2015)
37. N. Arene, J. Hidler, Understanding motor impairment in the paretic lower limb after a stroke: a review of the literature. Top. Stroke Rehabil. **16**(5), 346–356 (2009)
38. A.F. Thilmann, S.J. Fellows, H.F. Ross, Biomechanical changes at the ankle joint after stroke. J. Neurol. Neurosurg. Psychiatry **54**(2), 134–139 (1991)
39. Y. Ren, Y.N. Wu, C.Y. Yang, T. Xu, R.L. Harvey, L.Q. Zhang, Developing a wearable ankle rehabilitation robotic device for in-bed acute stroke rehabilitation. IEEE Trans. Neural Syst. Rehabil. Eng. **25**(6), 589–596 (2017)
40. Academia de Unity | Escuela de videojuegos | Hektor Profe. https://www.hektorprofe.net/

13 Assessment of Robotic Devices for Gait Assistance and Rehabilitation

Luis F. Aycardi, Felipe Ballen-Moreno, David Pinto-Fernández, Diego Torricelli, Carlos A. Cifuentes, and Marcela Múnera

13.1 Introduction

In the last decades, the development of robotic devices for gait assistance and rehabilitation has shown ongoing growth [1, 2]. As these technologies have expanded and matured, the need for accurate assessment and understanding of how users perform with the robotic devices has become evident and has been a convergence point for multiple technology designers. Even if robotic technology's potential was and is indisputable, demonstrating its value on a quantitative basis has been challenging. Trying to address this general concern, many research studies have started to evaluate robotic devices' performance, resulting in an abundant and highly diverse compilation of methods, variables, and protocols. The enormous amount of information led the robotics community to increase interest in benchmarking to scientifically assess and compare robotic devices' performance for gait assistance and rehabilitation. Even though benchmarks have been long used to verify and compare the readiness level of different technologies in many domains, not long ago, the primary approach to compare devices like exoskeletons was only through competitions, such as Cybathlon [3]. The big challenge of unifying a benchmark is even more difficult for the specific case of assistive and rehabilitation devices. The intrinsic interaction of these devices with the subjects complicates finding

L. F. Aycardi · F Ballen-Moreno · C. A. Cifuentes (✉) · M. Múnera
Biomedical Engineering, Department of the Colombian School of Engineering Julio Garavito, Bogotá D.C., Colombia
e-mail: luis.aycardi-c@mail.escuelaing.edu.co; felipe.ballen@mail.escuelaing.edu.co; carlos.cifuentes@escuelaing.edu.co; marcela.munera@escuelaing.edu.co

D. Pinto-Fernández · D. Torricelli
Neural Rehabilitation Group, Cajal Institute, Spanish National Research Council (CSIC), Madrid, Spain
e-mail: david.pinto@cajal.csic.es; diego.torricelli@csic.es

C. A. Cifuentes, M. Múnera, *Interfacing Humans and Robots for Gait Assistance and Rehabilitation*, https://doi.org/10.1007/978-3-030-79630-3_13

appropriate metrics to measure their performance. Hence, studies in this area generally have to be accompanied by performance studies of the subject and not just the robots.

The foundations to build such standards have been laid by recent efforts in the field of benchmarking bipedal locomotion to consolidate a unified scheme for humanoids, wearable robots, and humans [4]. Subsequently, work has been done attempting to organize the available assessment information and identify performance indicators that could be converted into practical benchmarks [5, 6].

This chapter presents an overview of the most promising and used measures, experimental procedures, equipment, sensors, and tools so far identified in the literature to assess gait robotic assistive and rehabilitation devices. The chapter starts with the introduction to the basic concepts to understand the implications and ways to assess the performance of an activity. Thereafter, the different modules towards a correct assessment are explained.

13.2 Motor Skills, Abilities, and Performance

The assessment of robotic devices for gait assistance and rehabilitation is a multidisciplinary area. Engineers and clinicians of different backgrounds have to agree on common nomenclature and classification systems to conceive standards in the assessment process. Inspired by the approach by Magill [7], further organized and discussed by Torricelli et al. [4], three basic concepts are often used to understand the area: motor skill, motor ability, and motor performance (see Fig. 13.1).

A motor skill, also called action in the motor learning and control research literature, refers to an activity or task that has a specific goal to achieve. However, not all activities with a goal are considered motor skills. To be studied as one, it needs to have other characteristics as: (i) be performed voluntarily, (ii) require the movement of joints and body segments, and (iii) be learned or relearned (as it usually happens in the field of rehabilitation) [7]. The most basic motor skill in this book is walking, but several others will be contemplated in the following sections.

Highly related to the concept of skill is the one of abilities. Motor abilities can be referred to as the general traits of an individual that are a determinant of his achievement potential for the performance of specific skills [7]. Let the skill be walking. The abilities may refer to stability, coordination, compliance, and any other characteristic needed to walk.

The last concept is motor performance, defined as the level of achievement of the goal, i.e., how well the goal established in the skill is achieved. The performance of any motor skill is influenced by (a) characteristics of the skill itself, (b) the environment in which the skill is performed, and (c) the person performing the skill [7], as presented in Fig. 13.1. The person is the agent in charge of learning and adapting the skill through the observation and perception of the performance.

Measuring the level of achievement of a skill is not a straightforward process. Many ways to assess motor performance have been defined over time. These different measuring methods are called the performance indicators (PI) [4] and can

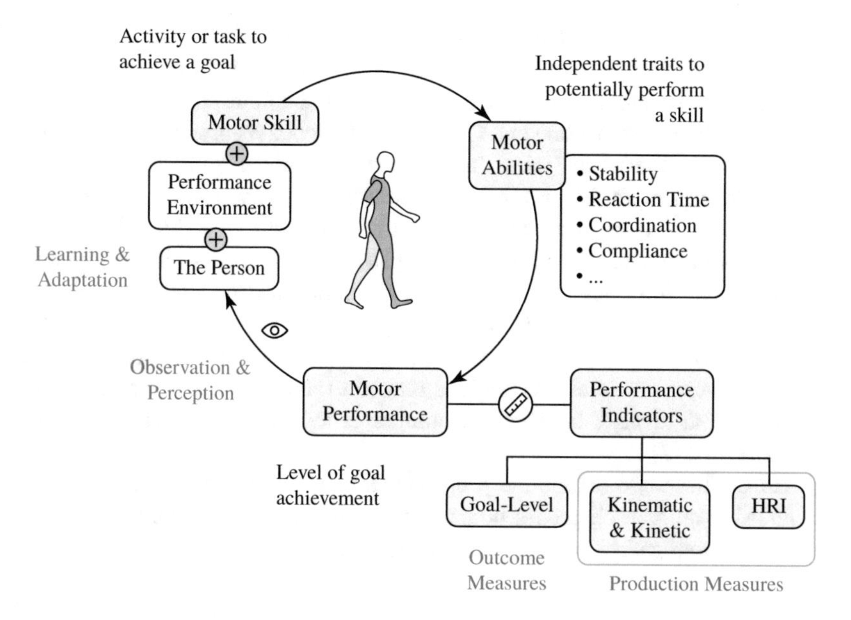

Fig. 13.1 Basic concepts in the assessment of robotic devices that interact with humans

be grouped into two main categories according to Magill. In the first one are the performance outcome indicators, which indicate the result of performing a motor skill (e.g., how far or fast a person walked). They provide information where the primary concern is whether or not the goal of the skill was accomplished. In the second category are the performance production indicators, which indicate how the different human systems (e.g., the nervous system, the muscular system, and the movement of the limbs or joints) function during the performance of a motor skill [7]. This category includes both kinematic/kinetic measures and the ones defined as Human–Robot Interaction (HRI) measures [4], which will be addressed in further sections.

13.3 Classifying Motor Skills

A complete understanding and characterization of related motor skills is crucial to correctly assess the performance of robotic devices used in gait assistance and rehabilitation. Classifying the motor skills for which these devices are developed and the possible variations and conditionals involved is the first step in the assessment process. Several proposals like the ones by Gentile [8] and Fleishman et al. [9] successfully classify motor skills and motor abilities and are commonly used in physical therapy and psychology.

Similar to what was established to influence a motor performance in Fig. 13.1, Gentile classified motor skills according to two general items. The first one is the environment, which he divided into: (i) unaltered motion and (ii) with the presence of inter-trial variability or unexpected disturbances. The second item corresponds to the function of the motor skill, which is classified according to: (i) the motion of the body (posture or transport) and (ii) the simultaneous manipulation of an object during the execution of the task. Furthermore, Fleishman proposed a list of the "fewest independent ability categories which might be most useful and meaningful in describing performance in the widest variety of tasks" [9]. In addition to the abilities previously established as an example for the skill of walking, significant motor abilities from that list are inter-limb coordination, static and dynamic strength, limb flexibility, gross body equilibrium, reaction time, speed of limbs, and control precision [4]. However, Fleishman's lists should not be considered exhaustive inventories of all the abilities related to motor skill performance, as the objective was to identify the smallest number of abilities that would describe the tasks performed [7].

Based on those two taxonomic proposals, a benchmark for bipedal locomotion was created to unify a scheme for humanoids, wearable robots, and humans [4]. The motor skill classification presented here maintains the conventions defined there.

13.3.1 Walking

Walking is undoubtedly the core motor skill to be assessed and the main focus of the robotic devices described in this book. However, since motor skills can be further classified according to: (i) environment variability and (ii) the presence of external disturbances, the relevant motor skills for robotic devices are:

- Walking in a **static** environment with a **constant or absent** disturbance:
  This includes walking on flat ground, constant slopes, ascending or descending stairs, and backward walking.
- Walking in a **static** environment with a **variable** disturbance:
  This includes walking on variable slopes, irregular terrains, and slippery surfaces.
- Walking in a **moving** environment with a **constant or absent** disturbance:
  This includes walking on a constant treadmill, a constant soft ground, and walking while bearing additional weight.
- Walking in a **moving** environment with a **variable** disturbance:
  This includes walking on a variable treadmill, a variable soft ground, when pushed, overcoming obstacles and slalom or turning.

13.3.2 Standing

Even if most of the efforts when designing a robotic device are devoted to walking, standing (maintaining an upright posture) is critical motor skill to assess. Standing is evaluated employing the same two previous variables:

- Standing in a **static** environment with a **constant or absent** disturbance:
  This includes standing on a horizontal surface and an inclined surface.
- Standing in a **static** environment with a **variable** disturbance:
  This includes standing on uneven terrains and during manipulation.
- Standing in a **moving** environment with a **constant or absent** disturbance:
  This includes standing while bearing additional weight and while periodic tilts or moving ground.
- Standing in a **moving** environment with a **variable** disturbance:
  This includes standing in the presence of pushes and while irregular tilts or rough translations.

13.3.3 Others

Finally, other skills related to the assessment of robotic devices and not included in either of the aforementioned categories can be of value and are covered. This includes activities, where the environment is static and there are no or constant disturbances, such as: lateral stepping, crouching or kneeling, changing from sitting-to-standing or from standing-to-sitting, and running.

A complete illustrated scheme presented in an interactive application, with the first step being the selection of a motor skill from the previously listed skills, is available in the official *Benchmarking Locomotion Website* [10].

Once the skill is fully determined and characterized, the following action towards the assessment corresponds to selecting the desired measures to be taken when performing it.

13.4 Performance Indicators

As mentioned before, there are multiple ways to measure motor skill performance. A first and useful way to organize them is by grouping them into two categories defined in Fig. 13.1 that relate to the different levels of performance observations, as suggested by Magill [7]. The first type, the performance outcome indicators, received another name in Pinto-Fernandez et al. [5] and will be the one adopted in this chapter. They label them as Goal-Level variables or measurements. In the second category, the same authors identified two different subgroups that will also be used further on. On one side are the kinematic and kinetic indicators focusing on the limbs, head, or body movements that lead to the observed outcomes. On the other side are the HRI measurements that relate more to the variables that might influence the intrinsic interaction between the user and the robotic device. The PIs that correspond to each of these categories will now be addressed.

13.4.1 Goal-Level Performance Indicators

To indicate the results of performing a motor skill, different variables can be considered. The following are the most commonly used Goal-Level PI in the field of the assessment of robotic devices for gait assistance and rehabilitation:

- **Time indicators**
  This category includes various time-related measurements. One of the preferred metrics for performance evaluation is the minimum time or the maximum speed achieved to correctly complete a task. However, another important indicator under this category is the reaction time (RT), which indicates how long it takes for a person to prepare and initiate a movement. Time indicators are mostly calculated during clinical tests, such as the 10 Meter Walking Test (10MWT), the 6 Minute Walking Test (6MWT), and the Timed Up and Go (TUG) test [5], and are measured in time units (e.g., sec, min).
- **Error indicators**
  Metrics related to errors have a prominent place in human performance research and in everyday living activities (assistance and rehabilitation). Multiple ways of reporting errors are accepted and it is up to the researchers to decide if they correspond to a study of accuracy either spatial, temporal, or both. Error indicators can be in the form of: (i) the amount of error in performing criterion movement, e.g., absolute error (AE), constant error (CE), or variable error (VE), or (ii) the number or percentage of errors [7].
- **Distance**
  The distance covered when performing a motor skill with a device is frequently used as PI. In exoskeletons, the 6MWT is found to be the preferred PI in this category [5].
- **Stability (to external disturbances)**
  Stability can be understood as the ability to maintain equilibrium over the support base during the motor skill execution [4]. The PIs in this category include: maintaining the center of mass (CoM) above the polygon of support (what Fleishman on his list referred to as gross body equilibrium), forefoot and rearfoot loading, length of the motion path, or confidence ellipse area [5].
- **Endurance**
  This PI generally refers to the ability to perform long periods of functioning or multiple cycles of work to test the robot's skills (also in the benchmark proposal). Nevertheless, it can also apply to other robotic devices as it is usually measured by the power development per joint, joint stiffness, and battery usage [5].
- **Repetitions**
  PIs under this category are measured with integer numbers and one of the simplest to recognize. Good examples are the number of successful attempts and the number of trials or repetitions to complete the task.
- **Versatility**
  Versatility is here understood as the ability of the robotic devices to cope with

different motor skills in the same run [5]. It is mainly used in cases where an exoskeleton takes part. This PI can be implemented together with the last category by measuring the number of successful transitions between tasks, or independently, with step width adaptability criteria.

Goal-Level PIs, especially time, error, and distance indicators, are very popular and globally accepted indicators. They are relatively simple and practical to use, making them particularly useful during competitions in the area (Cybathlon [3], for example). However, they can be rather insufficient to validate or quantify the robotic systems' performance [5], as robotics in rehabilitation and assistance are highly conditioned to the subject's performance. Given that these PIs are not very reproducible, other types of PIs are usually needed.

13.4.2 Kinematic and Kinetic Performance Indicators

Addressing the robotic device's performance during the motor skill by measuring the production indicators includes many more parameters to consider than the outcome indicators. To capture the complexity of the action to be performed more closely, they require specific instruments and equipment, as presented in the following sections.

Kinematic and kinetic PIs include many of the most common indicators used to assess robotic devices [5]. They are traditionally associated with biomechanics and refer to descriptors of motion without concern for its cause and force as a cause of motion, respectively [7]. Under this category are the following PIs for the assessment of robotic devices in the field:

- **Spatiotemporal Parameters**
  They correspond to parameters of distance (spatial) and time (temporal) during gait. They are considered standard metrics that can grasp the kinematic performance's main features in basic locomotion tasks [5]. The spatial parameters are related to the step and stride length but can include others like the number of steps. On the other hand, temporal parameters comprise the cadence, walking speed and the complete cycle, and individual phase time.
- **Kinematic indicators**
  As previously stated, kinematic indicators are a description of motion without regard to force or mass. As PIs, they portray the displacement, velocity, and acceleration of the human and robotic joints. This includes: joint trajectories, range of motions (ROMs), speed, and CoM position along the three principal planes of motions (sagittal, frontal, and transverse).
- **Kinetic indicators**
  In kinetic indicators, force is the main parameter to consider in the analysis of joints. Therefore, these PIs are of joints torques, force, power, and work, global forces, and power and ground reaction forces (GRF).

- **Symmetry**
  The symmetry indexes are the percentage of symmetry between the right and left gait cycle regarding their curve of acceleration or pelvic angles. Pelvic angles are the tilt, obliquity, and rotation, according to the plane of motion. As the indexes approach 100, the more symmetry there is along the trial [11].
- **Coordination**
  Coordination PIs come after the previously explained spatiotemporal parameters. For a cyclic movement, like gait, an indicator of coordination between two limb segments is the relative phase. This index calculates the phase angles for each limb segment or limb at a specific point in time and then subtracts one phase angle from the other [7].

The first three PIs presented in this section are very popular in assessing exoskeletons as they can grasp the entire complexity of limb dynamics. However, the kinematic and kinetic indicators are often difficult to compare and replicate as there are no typical standard setups, data labeling, or experimental protocols. Symmetry and coordination, on the other hand, are still poorly used in the evaluation of exoskeletons' performance [5].

13.4.3 Human–Robot Interaction Performance Indicators

The second type of performance production indicators comprises all the measurements that characterize the synergy between the user and the robotic device. Given the nature of this group of indicators, HRI PIs include both quantitative and qualitative variables. The first ones evaluate the user's physical parameters, while the others reveal subjective levels of acceptance of the technology by the user during the interaction. The main PIs in this category are:

- **Metabolic cost**
  Metabolic cost is a way to describe the intensity of an activity or motor skill. Many indicators can be used to that end. The most frequent PIs are: heart rate, blood lactate concentration, oxygen consumption, carbon dioxide production, metabolic power, biological power, work, and calorimetry [5, 12].
- **Muscle activity**
  This type of indicator is the most commonly employed variable for the assessment of HRI. It is generally measured by electromyography (EMG), in which the intention of movement is captured through muscles' electrical activity. EMG recordings are relevant to motor learning and control issues as they can indicate when a muscle begins and ends activation [7] and can be used to quantify the effects of a robot on muscle fatigue.
- **Brain activity**
  Research on the relationship between brain activity and performance has led to rapid brain assessment technology implementation on motor rehabilitation.

Similar to the previous indicators, brain activity is usually measured by electroencephalography (EEG) recordings.

- **Interaction forces**
  This category does not need any extra information other than the fact that it is measured through three PIs: the power delivered to the robot, the interface transmitted forces, and the interaction forces themselves.
- **Comfort**
  Comfort is defined in this document as the user's perception of the HRI. This one corresponds to the qualitative variables previously mentioned and has many ways of being measured. Among the most relevant indicators are pain scales, skin irritation, sore spots, spasticity, clinical questionnaires, and user sense of comfort [5].
- **Ergonomics**
  Ergonomics refers to the design and arrangement of things people use to make the interaction the most efficient and safe possible. The main PIs used in this category are HR relative position, interface displacements, anthropometric database percentiles, and adaptability to different height ranges [5].
- **Safety**
  This indicator assesses the condition of being protected from harm or other non-desirable outcomes. Safety PIs are a mix of both quantitative and qualitative indicators. Quantitative PIs are the number of falls, blood pressure, and heart rate. Qualitative PIs include the skin, spine, and joint status after using the robot, and clinical questionnaires similar to those implemented for comfort.

Some of the most expected performance outcomes and production measurements here are included in the official *Benchmarking Locomotion Website* [10]. They correspond to the second step of selecting the organization of the currently available metrics and protocols to assess bipedal function into a meaningful taxonomy.

Keeping in mind the provided overview of the motor skills and PI, the only unexplored and missing area to fully understand how to assess robotic devices in gait assistance and rehabilitation is the section of the required equipment and sensors.

13.5 Equipment and Sensors

By equipment and sensors, one should understand in this chapter all the set of tools, devices, and kits, assembled to measure and capture the different PIs for the chosen motor skills. Regarding their location, the equipment and sensors can: (i) be mounted or fixed in the testing environment and record from strategic points of the activity or the specific events, or (ii) be wearable, which means that the user wears them during the performance of the motor skill. The first type is considered the gold standard in accuracy for walking kinematics [13], but their main disadvantages are the price and their limitation to indoor use with a very controlled environment [14]. On the contrary, wearable sensors have become popular due to their affordability and flexibility of use, together with shorter donning/doffing times [15].

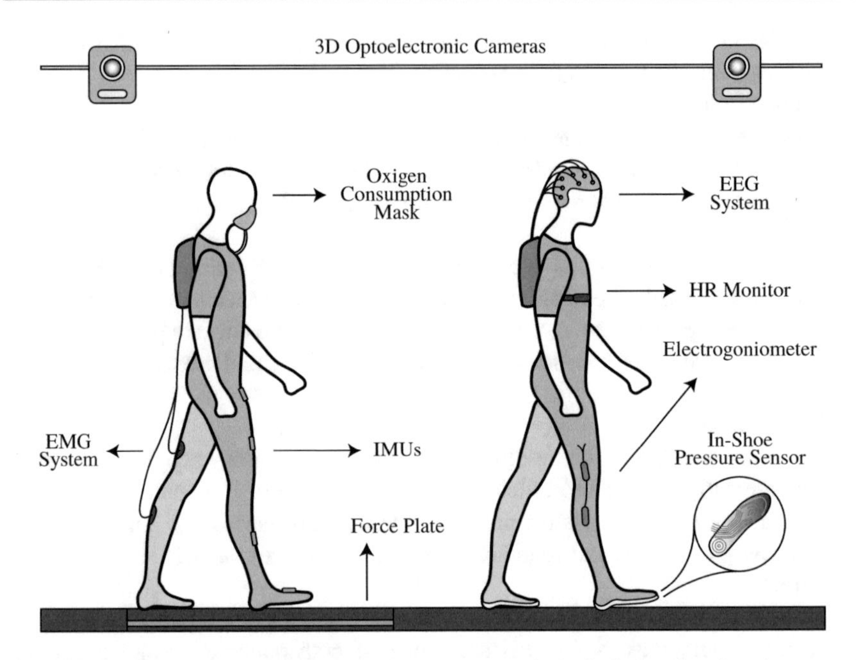

Fig. 13.2 Equipment and sensors used for the assessment of robotic devices in gait assistance and rehabilitation

This section presents a non-exhaustive catalog of the leading equipment and sensors used to assess the motor skills' performance, as mentioned earlier, employing the desired PI. Most of them are depicted in Fig. 13.2. They are grouped in the same categories used to classify the PIs. Given the purpose of the Goal-Level PIs and their intention to measure outcomes, most of the metrics are not complex and with simple equipment like timers, counters, and rulers can be calculated. Therefore, no further details are presented regarding this kind of PI, except possibly for stability, which can be addressed with the equipment of other kinematic and kinetic PIs.

13.5.1 Equipment and Sensors for Kinematic and Kinetic Performance Indicators

The extraction of most kinematic PIs (including spatiotemporal, symmetry, and coordination) was first done with portable sensors called electrogoniometers. Afterward, the measurements evolved to 2D and 3D video systems, which need to be placed in the performance environment, depend on specific laboratory conditions and imply complex protocols and high economic costs. Nowadays, the extraction is moving back to relay on wearable sensors like inertial motion units (IMUs).

13.5.1.1 Electrogoniometers

Electrogoniometers are electromechanical devices that span the joint to be measured by attaching to the proximal and distal limb segments. They measure the joint's angular change by providing an output voltage proportional to the change and assuming that the attachment segments move with the limb segment's midline [16]. The two significant advantages of these devices are ease of use and low cost. However, a significant limitation in using them is that the angles are only acquired in a single motion plane [17].

13.5.1.2 Video Systems

Video systems are based on a computer vision approach, in which the main goal is to extract gait patterns from sequential images [18]. There are both 2D and 3D configurations and it depends on the complexity of the motor skill and the chosen PI, which of them to implement. 2D Systems, as electrogoniometers, can record joint angles in only one plane of motion. 3D Systems, through the inclusion of depth, can extract joint angles in all three planes. The use of active (LED markers that are pulsed sequentially) or passive (lightweight reflective markers) markers is widespread when implementing this kind of system, even though some have worked their way out of the markers. The leading video systems used in the field are:

- **2D Systems**
  The Kinect is the most used exemplar of this technology. First developed by Microsoft, in 2010, collects information from RGB cameras, infrared projectors, and detectors that mapped depth to perform real-time gesture recognition and skeletal body detection, among others. In this sense, a biomechanical model based on rigid segments can be implemented to acquire human motion data [19]. As the human body is modeled, joint angles are acquired while performing a motor skill. As previously mentioned, it can only record angles in one plane of motion, and, in this case, users are not required to wear any markers. Additionally, this equipment is portable (easy to relocate) and low cost. The main drawback is that it is no longer produced as by 2018, Microsoft discontinued all Kinect hardware for video games. Moreover, for those who still can get their hand on them, specific lighting and space conditions (controlled or laboratory conditions) are required.

  Other alternatives to this system include motion tracking software, based on recordings by a 2D camera (and possible reflective markers) to calculate almost all kinematic parameters. *MaxTRAQ 2D* (Innovision Systems, USA) [20] includes tools and analysis of angles, distances, the center of mass, and more.
- **3D Systems**
  3D optoelectronic camera systems for motion capture are often regarded as the gold standard in acquiring biomechanical parameters, given their robustness [21]. They detect light and use it to estimate the 3D position of reflective markers via time-of-flight triangulations. To correctly place markers on the user and allow an optimal estimation many protocols have been developed. The accuracy of these

systems is dependent on the different details of the experimental setup: (a) the location of each of the cameras relative to the others, (b) the distance between the cameras and the markers, (c) the position, number, and type of the markers implemented, and (d) the motion of the markers within the capture volume [22].

Systems of this type are based on fixed cameras, which means they can only acquire data in a restricted area [23]. The number of cameras, their field of view, and the space between them condition the total volume in which the skill can be performed and captured. The most extensive measured range reported, to the authors' knowledge, is 824 m^2, obtained with a Vicon MX13 (UK) measurement system [6]. To capture this range, a total of 24 cameras were required.

Among the major drawbacks of these systems are high costs, lack of portability, constant need for calibration and synchronization, high labor in the organization and processing of trials, and high sensitivity to alterations in setup. By increasing the number of cameras increases the level of all of these items. Further limitations of the system are the necessity of line-of-sight, which means that the data output will be interrupted when the cameras lose sight of the markers [6, 24], and the need for dark areas (indoors), as bright sunlight interferes with the measurements [6].

Important and widely used manufacturers of this technology include: Vicon (UK) [25–27], Motion Analysis (USA) [17, 28, 29], Qualisys (Sweden) [20], and BTS Bioengineering (Italy) [30].

An extensive review of vision-based systems that have been proposed for tracking human motion in the past years can be found in Moeslund et al. [31].

13.5.1.3 Inertial Unit System (IMU)

An inertial measurement unit (IMU) is a sensor composed of the fusion of three other sensors: gyroscopes, accelerometers, and magnetometers. Through this combination of components, the unit can acquire gravitational acceleration and rotational velocity, to estimate the velocity, acceleration, and orientation of the element they are attached to. In a person's lower limbs, they are usually positioned on the waist, thigh, shank, and foot instep [32–34]. To estimate complex PIs, multi-sensor arrangements are widely used to assess a specific task. Several studies used a multi-sensor to estimate and compare the efficacy and precision, analyzing signal patterns of body segments in different locations [35–37]. They are of relatively low cost and provide an alternative to 3D systems as they do not require specific light and space conditions to function properly. Nevertheless, signal processing can be challenging as it involves the fusion of three sensors and the presence of cumulative drift error and the growth of quadratic or cubic error [38], which can distort the measured parameters. There are many commercially available IMUs on the market. From sophisticated modules like Xsens (Netherlands) to simple units from manufacturers as Bosch (Germany) [39].

A complete analysis of the accuracy of the three previously presented systems for the capture and assessment of human motion (aimed but not strictly to sports applications) can be found in the work by van der Kruk and Reijne [40].

13.5.1.4 Ground Reaction Force (GRF) Sensors

To calculate kinetics indicators for each of the joints involved in the motor skill, dedicated software based on inverse kinematics analysis has been developed. The most prominent exponent is *C-motion* (Visual3D, USA) [20, 41, 42]. The basic inputs for this software are: (i) kinematic PI, obtained by any of the motion capture systems shown before, (ii) ground reaction forces (GRF), and (iii) segmental mass distribution models. GRF can be measured with two main types of sensors:

- **Force Platforms or Plates**
  A force platform can be understood as a pair of plates, one over another with force transducers between them at the corners [43]. There are several types of force plates on the market and they are classified either by how many pedestals (single-pedestal or multi-pedestal) or by the type of transducer they employ. The types of transducers commonly found in force platforms are: strain gauge, piezoelectric sensor, capacitance gauge, Hall effect, and piezoresistive sensor, each with the advantages and drawbacks inherent in their nature. For gait analysis, force platforms with three or four pedestals are used to permit forces that migrate across the plate [44]. They are usually synchronized with 3D optoelectronic camera systems to provide a simultaneous analysis of the different PIs [25, 28, 45].
- **Pressure Mapping Systems**
  Pressure mapping systems quantify the interface pressure between two contacting surfaces. They can come in different forms, from walking mats or strip of carpet-like sensors, to a completely wireless thin insole (in-shoe technology). These systems use a larger number of sensors (typically in the hundreds, depending on the size) to capture the pressure distribution and profiles in the foot, and the position and trajectories of the center of pressure (CoP) during stance phases of gait. Nonetheless, they have also been used to measure force profiles during many activities. For example, the F-Scan (TekScan, USA) in-shoe pressure mapping system has been effectively used to measure GRF during able-bodied walking [46, 47], and the Pedar-X mobile (Novel Gmbh, Germany) in-shoe system was used for collecting GRF using a lower limbs exoskeleton [48].

13.5.2 Equipment, Sensors, and Tools for Human–Robot Interaction Performance Indicators

As mentioned in the HRI PI characterization, this type of measurement includes quantitative and qualitative variables. Two big groups of equipment and sensors, which refer to the user's physical parameters, describe the majority of the quantitative HRI indicators. The qualitative PIs are clustered in one independent group in this chapter.

13.5.2.1 Metabolic Cost Systems

Metabolic cost encloses a variety of PIs, as it was presented in Sect. 13.4.3. Authors have measured it in numerous ways and with different types of equipment. Some of the sensors and calculations that best exemplify this are:

(i) Malcom et al. measured the metabolic cost of subjects walking with an exoskeleton through respiratory gas analysis. They analyzed respiratory gasses with a computerized O_2–CO_2 analyzer flow meter (Oxycon Pro, Germany) and estimated metabolic cost with the formula from Brockway [20, 49]. **(ii)** Lee et al. equipped elder exoskeleton users with a facemask connected to a computerized portable cardiopulmonary metabolic system (Cosmed K4B2, Italy), to measure breath-by-breath metabolic costs. They also measured the heart rate via a wireless chest-strap heart rate monitor [29]. **(iii)** Award et al. measured the energy cost of walking in individuals in the chronic phase of stroke recovery using an exosuit. They defined it as mass normalized oxygen consumption per meter ambulated (mlO_2/kg/m) measured with indirect calorimetry (Cosmed K4B2, Italy) and normalized by body weight (kg) and walking speed (m/min) [42]. Finally, **(iv)** Arazpour et al. evaluated the physiological cost index (PCI) of walking (a proxy measure of energy consumption) in a group of subjects with poliomyelitis. They used a Polar Heart Rate monitor (Polar, USA) to evaluate the PCI through a calculation including heart rate at steady-state walking (HRss) and heart rate at rest (HRar) [50].

13.5.2.2 EMG and EEG Systems

Muscle and brain activity and their corresponding subindicators are measured using the electrical signals associated with each human system as mentioned in Sect. 13.4.3. For researchers to achieve non-invasive and painless EMG and EEG recordings, surface electrodes are attached to the skin over muscles (known as surface EMG or sEMG) or a person's scalp. Typically, electrodes are placed on standard locations on the muscles and scalp to measure the voltage fluctuations. In the EEG case, the electrodes are usually contained in an elastic cap in their appropriate locations on the scalp to measure the activity of thousands or millions of neurons immediately beneath them [7]. sEMG systems are widely used to assess muscle activity PI during gait [28, 29, 51].

13.5.2.3 Clinical Scales and Evaluations

This last group comprises all measurements that cannot be captured or characterized with sensors or equipment as the ones explained before. This section intends not to list all the existing tools to assess qualitative PIs, as it would be extensive, but rather to give examples that have been used in the literature.

A detailed description and compilation of more than 500 measures of clinical protocols, scales, indexes, and questionnaires are found in the *Rehabilitation Measures Database Website* of the Shirley Ryan AbilityLab [52]. Additionally, in Chapter 14: *Experiences of Clinicians Using Rehabilitation Robotics*, some of the most used standardized questionnaires to evaluate user's ergonomics, comfort, and safety are presented.

Regarding practical examples of the clinical scales used to assess HRI PI in the field, the following are some of the reported studies:

(i) Visual analog scales (VAS) are used to assess features like user fatigue, pain, and comfort [53]. del-Ama et al. implemented a VAS consisting of a 10 centimeter rectangle. With that scale, the user was asked to rate the pain perception by placing a mark inside the rectangle, rating from no pain at the left edge of the rectangle, to intolerable pain at the right edge of the rectangle [54]. **(ii)** The Ashworth scale (AS) and the Modified Ashworth scale (MAS) are utilized to evaluate spasticity [54, 55], and spasm frequency and severity are quantified using the Penn Spasm Frequency Scale (PSFS) [55]. **(iii)** To evaluate all aspects of patients' health and assess if there has been an improvement or decline in clinical status, the patient's global impression of change (PGIC) is used [55]. Finally, **(iv)** to assess the static balance and fall risk the Berg Balance Scale (BBS) is usually implemented [56].

13.6 Conclusions

The assessment of robotic devices' performance for gait assistance and rehabilitation is a multidisciplinary area that involves the mastering of many different concepts. Recent efforts to benchmark bipedal locomotion have settled the basis to understand the various considerations when classifying a motor skill and measuring its performance. The overview presented hopes to have organized and explained the key components one needs to consider when assessing gait robotic assistive and rehabilitation devices. According to the focus given to the performance, a reasonably detailed description of the implemented measures was achieved through the characterization of the existing PI. Additionally, the inclusion of practical information of their use and application in research intends to favor future studies, where standardized nomenclature, parameters, and benchmarking, in general, are included. Finally, some of the most popular equipment, sensors, and tools used in the literature and commercially available to measure motor performance were described. The knowledge and understanding of all the components presented are fundamental in the process of accurately assessing technology towards better assistance and rehabilitation of patients.

References

1. N. Koceska, S. Koceski, Review: Robot devices for gait rehabilitation. Int. J. Comput. Appl. **62**(13), 1–8 (2013)
2. M.M. Martins, C.P. Santos, A. Frizera-Neto, R. Ceres, Assistive mobility devices focusing on Smart Walkers: Classification and review. Robot. Autonom. Syst. **60**, 548–562 (2012)
3. R. Riener, The Cybathlon promotes the development of assistive technology for people with physical disabilities. J. NeuroEng. Rehab. **13**, 49 (2016)
4. D. Torricelli, J. Gonzalez-Vargas, J.F. Veneman, K. Mombaur, N. Tsagarakis, A.J. Del-Ama, A. Gil-Agudo, J.C. Moreno, J.L. Pons, Benchmarking bipedal locomotion: A unified scheme for humanoids, wearable robots, and humans. IEEE Robot. Autom. Mag. **22**(3), 103–115 (2015)

5. D. Pinto-Fernandez, D. Torricelli, M.d.C. Sanchez-Villamanan, F. Aller, K. Mombaur, R. Conti, N. Vitiello, J.C. Moreno, J.L. Pons, Performance evaluation of lower limb exoskeletons: A systematic review. IEEE Trans. Neural Syst. Rehab. Eng. (2020)
6. J. Spörri, C. Schiefermüller, E. Müller, Collecting kinematic data on a ski track with optoelectronic stereophotogrammetry: A methodological study assessing the feasibility of bringing the biomechanics lab to the field. PLOS ONE **11**(8), 1–12 (2016)
7. R.A. Magill, D.I. Anderson, *Motor Learning and Control: Concepts and Applications* (McGraw-Hill, New York, 2007)
8. A. Gentile, Skill acquisition: Action, movement, and neuromotor processes, in *Movement Science: Foundations for Physical Therapy in Rehabilitation* (Aspen Publishers Inc., MD, 1987), pp. 93–154
9. M.K. Fleishman, E.A. Quaintance, *Taxonomies of Human Performance*, vol. 7 (Academic Press, Orlando, FL, 1984)
10. C. Neural Rehabilitation Group, Cajal Institute, KNOW the benchmarking scheme – Benchmarking locomotion (2021)
11. L.F. Aycardi, C.A. Cifuentes, M. Múnera, C. Bayón, O. Ramírez, S. Lerma, A. Frizera, E. Rocon, Evaluation of biomechanical gait parameters of patients with Cerebral Palsy at three different levels of gait assistance using the CPWalker. J. NeuroEng. Rehab. **16**(1), 1–9 (2019)
12. N. Postol, S. Lamond, M. Galloway, K. Palazzi, A. Bivard, N.J. Spratt, and J. Marquez, The metabolic cost of exercising with a robotic exoskeleton: A comparison of healthy and neurologically impaired people. IEEE Trans. Neural Syst. Rehab. Eng. **28**(12), 3031–3039 (2020)
13. A. Miller, Gait event detection using a multilayer neural network. Gait Posture **29**(4), 542–545 (2009)
14. F. Attal, Y. Amirat, A. Chibani, S. Mohammed, Automatic recognition of gait phases using a multiple-regression hidden Markov model. IEEE/ASME Trans. Mechatron. **23**(4), 1597–1607 (2018)
15. R. Caldas, M. Mundt, W. Potthast, F. Buarque de Lima Neto, B. Markert, A systematic review of gait analysis methods based on inertial sensors and adaptive algorithms. Gait Posture **57**(June), 204–210 (2017)
16. E.L. Bontrager, M. Bontrager, Instrumented gait analysis systems, in *Gait Analysis in the Science of Rehabilitation* (Diane Pub Co, 1998), p. 112
17. K.E. Gordon, G.S. Sawicki, D.P. Ferris, Mechanical performance of artificial pneumatic muscles to power an ankle-foot orthosis. J. Biomech. **39**(10), 1832–1841 (2006)
18. M. Nieto-Hidalgo, F.J. Ferrández-Pastor, R.J. Valdivieso-Sarabia, J. Mora-Pascual, J.M. García-Chamizo, A vision based proposal for classification of normal and abnormal gait using RGB camera. J. Biomed. Inf. **63**, 82–89 (2016)
19. J.P. Silva Cunha, A.P. Rocha, H.M. Pereira Choupina, J.M. Fernandes, M.J. Rosas, R. Vaz, F. Achilles, A.M. Loesch, C. Vollmar, E. Hartl, S. Noachtar, A novel portable, low-cost kinect-based system for motion analysis in neurological diseases, in *2016 38th Annual International Conference of the IEEE Engineering in Medicine and Biology Society (EMBC)*, pp. 2339–2342 (Aug 2016)
20. P. Malcolm, W. Derave, S. Galle, D. De Clercq, A simple exoskeleton that assists plantarflexion can reduce the metabolic cost of human walking. PLoS ONE **8**(2), 1–7 (2013)
21. S. Corazza, L. Mündermann, E. Gambaretto, G. Ferrigno, T. Andriacchi, Markerless motion capture through visual hull, articulated ICP and subject specific model generation. Int. J. Comput. Vis. **87**, 156–169 (2010)
22. L.P. Maletsky, J. Sun, N.A. Morton, Accuracy of an optical active-marker system to track the relative motion of rigid bodies. J. Biomecha. **40**(3), 682–685 (2007)
23. M. Begon, F. Colloud, V. Fohanno, P. Bahuaud, T. Monnet, Computation of the 3D kinematics in a global frame over a 40m-long pathway using a rolling motion analysis system. J. Biomech. **42**(16), 2649–2653 (2009)
24. A. Panjkota, I. Stancic, T. Šupuk, Outline of a qualitative analysis for the human motion in case of ergometer rowing (2009)

25. A.S.-L. Hung, H. Guo, W.-H. Liao, D.T.-P. Fong, K.-M. Chan, Shulphqwdo 6Wxglhv Rq. Lqhpdwlfv Dqg. Lqhwlfv Ri, vol. 2011(June), pp. 45–50, 2011
26. D. Gomez-Vargas, F. Ballen-Moreno, P. Barria, R. Aguilar, J.M. Azorín, M. Munera, C.A. Cifuentes, The actuation system of the ankle exoskeleton T-FLEX: First use experimental validation in people with stroke. Brain Sciences **11**(4), 412 (2021)
27. S. Sierra, M. Múnera, T. Provot, M. Bourgain, C. Cifuentes, Evaluation of physical interaction during walker-assisted gait with the AGoRA Walker: Strategies based on virtual mechanical stiffness. Sensors (Under Review) **21**(9), 3242 (2021)
28. E.C. Ranz, E. Russell Esposito, J.M. Wilken, R.R. Neptune, The influence of passive-dynamic ankle-foot orthosis bending axis location on gait performance in individuals with lower-limb impairments. Clinical Biomechanics **37**, 13–21 (2016)
29. J. Lee, K. Seo, B. Lim, J. Jang, K. Kim, H. Choi, Effects of assistance timing on metabolic cost, assistance power, and gait parameters for a hip-type exoskeleton, in *2017 International Conference on Rehabilitation Robotics (ICORR)* (IEEE, 2017), pp. 498–504
30. A.M. Alsubaie, E. Martinez-Valdes, A.M. De Nunzio, D. Falla, Trunk control during repetitive sagittal movements following a real-time tracking task in people with chronic low back pain. J. Electromyography Kinesiol. **57**, 102533 (2021)
31. T.B. Moeslund, A. Hilton, V. Krüger, A survey of advances in vision-based human motion capture and analysis. Comput. Vis. Image Understand. **104**(2), 90–126 (2006)
32. J. Rueterbories, E.G. Spaich, O.K. Andersen, Gait event detection for use in FES rehabilitation by radial and tangential foot accelerations. Med. Eng. Phys. **36**(4), 502–508 (2014)
33. M. Yuwono, S.W. Su, Y. Guo, B.D. Moulton, H.T. Nguyen, Unsupervised nonparametric method for gait analysis using a waist-worn inertial sensor. Appl. Soft Comput. J. **14**, 72–80 (2014)
34. P. Catalfamo, S. Ghoussayni, D. Ewins, Gait event detection on level ground and incline walking using a rate gyroscope. Sensors **10**(6), 5683–5702 (2010)
35. M. Rabuffetti, G.M. Scalera, M. Ferrarin, Effects of gait strategy and speed on regularity of locomotion assessed in healthy subjects using a multi-sensor method. Sensors (Switzerland) **19**(3), 513 (2019)
36. H.B. Menz, S.R. Lord, R.C. Fitzpatrick, Acceleration patterns of the head and pelvis when walking are .... Gait Posture **18**, 35–46 (2003)
37. S.M. Rispens, M. Pijnappels, K.S. van Schooten, P.J. Beek, A. Daffertshofer, and J.H. van Dieën, Consistency of gait characteristics as determined from acceleration data collected at different trunk locations. Gait Posture **40**(1), 187–192 (2014)
38. C. Cao, *Development of a low-cost wearable prevention system for MSDS using IMU systems and electrically conductive materials via additive manufacturing*. Ph.D. thesis, Universidad de Navarra, 2020
39. M.D. Sánchez Manchola, M. Pinto, M. Munera, C. Cifuentes, Gait phase detection for lower-limb exoskeletons using foot motion data from a single inertial measurement unit in hemiparetic individuals. Sensors **19**, 2988 (2019)
40. E. van der Kruk, nd M.M. Reijne, Accuracy of human motion capture systems for sport applications; state-of-the-art review. Eur. J. Sport Sci. **18**(6), 806–819 (2018)
41. Z.F. Lerner, D.L. Damiano, T.C. Bulea, A robotic exoskeleton to treat crouch gait from cerebral palsy: Initial kinematic and neuromuscular evaluation, in *Proceedings of the Annual International Conference of the IEEE Engineering in Medicine and Biology Society, EMBS*, vol. 2016-Octob, pp. 2214–2217, 2016
42. L.N. Awad, J. Bae, K. O'Donnell, S.M. De Rossi, K. Hendron, L.H. Sloot, P. Kudzia, S. Allen, K.G. Holt, T.D. Ellis, C.J. Walsh, A soft robotic exosuit improves walking in patients after stroke. Sci. Transl. Med. **9**(400), eaai9084 (2017)
43. F. Bonde-Petersen, A simple force platform. Eur. J. Appl. Physiol. Occupat. Physiol. **34**, 51–54 (1975)
44. D. Robertson, G. Caldwell, J. Hamill, G. Kamen, S. Whittlesey, *Research Methods in Biomechanics* (2004)

45. A.J. Ikeda, J.R. Fergason, J.M. Wilken, Effects of altering heel wedge properties on gait with the intrepid dynamic exoskeletal orthosis. Prosthetics Orthotics Int. **42**(3), 265–274 (2018)
46. A.L. Randolph, M. Nelson, S. Akkapeddi, A. Levin, R. Alexandrescu, Reliability of measurements of pressures applied on the foot during walking by a computerized insole sensor system. Archiv. Phys. Med. Rehab. **81**(5), 573–578 (2000)
47. D.B. Fineberg, P. Asselin, N.Y. Harel, I. Agranova-Breyter, S.D. Kornfeld, W.A. Bauman, A.M. Spungen, Vertical ground reaction force-based analysis of powered exoskeleton-assisted walking in persons with motor-complete paraplegia. J. Spinal Cord Med. **36**(4), 313–321 (2013)
48. H. gon Kim, J. won Lee, J. Jang, S. Park, C. Han, Design of an exoskeleton with minimized energy consumption based on using elastic and dissipative elements. Int. J. Control Autom. Syst. **13**(2), 463–474 (2015)
49. J. Brockway, Derivation of formulae used to calculate energy expenditure in man. Human Nutrition. Clin. Nutrition **41**(6), 463–471 (1987)
50. M. Arazpour, M.A. Bani, M. Samadian, M.E. Mousavi, S.W. Hutchins, M. Bahramizadeh, S. Curran, M.A. Mardani, The physiological cost index of walking with a powered knee-ankle-foot orthosis in subjects with poliomyelitis: A pilot study. Prosthetics Orthotics Int. **40**(4), 454–459 (2016)
51. J. Park, H. Park, J. Kim, Performance estimation of the lower limb exoskeleton for plantarflexion using surface electromyography (sEMG) signals. J. Biomech. Sci. Eng. **12**(2), (2017)
52. S.R. AbilityLab, Rehabilitation Measures Database (2021)
53. G. Zeilig, H. Weingarden, M. Zwecker, I. Dudkiewicz, A. Bloch, A. Esquenazi, Safety and tolerance of the ReWalk^{TM} exoskeleton suit for ambulation by people with complete spinal cord injury: A pilot study. J. Spinal Cord Med. **35**(2), 96–101 (2012)
54. A.J. Del-Ama, Á. Gil-Agudo, J.L. Pons, J.C. Moreno, Hybrid gait training with an overground robot for people with incomplete spinal cord injury: A pilot study. Front. Human Neurosci. **8**(MAY), 1–10 (2014)
55. G. Stampacchia, A. Rustici, S. Bigazzi, A. Gerini, T. Tombini, S. Mazzoleni, Walking with a powered robotic exoskeleton: Subjective experience, spasticity and pain in spinal cord injured persons. Neurorehabilitation **39**(2), 277–283 (2016)
56. T. Yoshimoto, I. Shimizu, Y. Hiroi, M. Kawaki, D. Sato, M. Nagasawa, Feasibility and efficacy of high-speed gait training with a voluntary driven exoskeleton robot for gait and balance dysfunction in patients with chronic stroke: Nonrandomized pilot study with concurrent control. Int. J. Rehab. Res. **38**(4), 338–343 (2015)

Experiences of Clinicians Using Rehabilitation Robotics

14

Marcela Múnera, Maria J. Pinto-Bernal, Nathalie Zwickl, Angel Gil-Agudo, Patricio Barria, and Carlos A. Cifuentes

14.1 Introduction

While there is a consensus that it is essential to involve users in developing rehabilitation technology, there are few examples of how to do this, and no studies of which techniques are most effective [1]. In recent years, many useful robotic devices have been used in daily therapeutic life. The experience shows that the devices could not always be used successfully. Some impracticability factors such as being time-consuming, complicated usage, and though wearing, were the reason for the device's failure. There is a growing recognition that if medical devices are of real value, their users' need and capabilities must be considered [2]. In the case of rehabilitation robotics, "User" covers both the patients treated with a device, and the

M. Múnera · M. J. Pinto-Bernal · C. A. Cifuentes (✉)
Escuela Colombiana de Ingeniería Julio Garavito, Bogotá, Colombia
e-mail: marcela.munera@escuelaing.edu.co; maria.pinto@mail.escuelaing.edu.co; carlos.cifuentes@escuelaing.edu.co

N. Zwickl
Zürcher Hochschule für Angewandte Wissenschaften, ZHAW, Institut für Physiotherapie, Winterthur, Switzerland
e-mail: nathaliezwickl@hotmail.com

A. Gil-Agudo
Hospital Nacional de Parapléjicos, Finca De, Carr. de la Peraleda, S/N, Toledo, Spain
e-mail: amgila@sescam.jccm.es

P. Barria
Department of Electrical Engineering, University of Magallanes, Punta Arenas, Chile

Club de Leones Cruz del Sur Rehabilitation Center, Punta Arenas, Chile

Brain-Machine Interface Systems Lab Systems Engineering and Automation Department, Miguel, Hernández University of Elche UMH, Elche, Spain
e-mail: pbarria@rehabilitamos.org

C. A. Cifuentes, M. Múnera, *Interfacing Humans and Robots for Gait Assistance and Rehabilitation*, https://doi.org/10.1007/978-3-030-79630-3_14

staff responsible for using the device to treat them [1]. This shows the importance of involving the clinicians and patients in the development process. In this chapter, the concept of the users' involvement in the development process is developed: the parameters assessed, the stages for this involvement, and the tools used. In the final sections, the application of these concepts is shown in three case studies.

14.2 Parameters Evaluated in the User's Input

In the development process [3, 4], the device's acceptance and practicability should be considered. Therefore, involving clinicians and patients who are the end- user group who will work with the device daily is essential. If the clinicians are not convinced about a device, its success could be significantly doubted [3, 5, 6]. Some reasons why the clinicians can reject the devices could be the difficulty and time of the donning/doffing of the devices, the complications in its handling, or if the device is triggering fear in the patient. Therefore, it is crucial to analyze social or physical robots' acceptability and practicability, more specifically [7]. In the following paragraphs, the key points of the clinicians' view for the application of robots in a daily therapeutic environment are presented.

14.2.1 Practicability

Practicability is a critical point for a clinician to use a physical device in their therapeutic sessions [7]. Practicability is the quality of being able to be done, or of being likely to be successful. Here, we describe this aspect through three important parameters found in the literature. First, we have to consider **specific target patient group considerations**. The clinician knows the target patient group which will use the device, and they know where specific problems may occur [3, 8]. The robotic device which helps physically is usually aimed at patients with neurological or muscular difficulties. Each target patient group might have specific needs and limitations [3, 4]. For example, a device targeted for patients who usually are in a wheelchair needs to be easily donned in a sitting position. Such limitations should be considered when designing the device, and the clinicians are a very reliable source for these design considerations [9]. For this parameter, in the case of social robotic in rehabilitation, the group should be analyzed with the clinicians' help. Different points should be clarified and considered, such as age (are the patients comfortable with using electronic devices?), weight (are the activities appropriate to the patients' weight group?), fatigue (are the patients motivated to have a certain level of activities?). Second, the devices should have **simple handling**. The clinicians are the end-users and should feel comfortable when working with the device. The usability of the device should be simple and should not have complicated and deep technical steps. Typically, the end-users are healthcare personnel without technical robotic knowledge [3, 8, 10]. Therefore, the possible technical adverse effect should be explained in a manual [9]. Third, the device should consider the **session process**

communication. The clinicians know how to explain the therapeutic session and the health-relevant problems in a patient-friendly language [3]. In each session, it is vital to create a safe and pleasant environment for the patient. Ideally, the patient should understand how the device works and what interactions with the device will happen. Having the process and interactions communicated to the patient helps the clinician achieve better compliance [6, 11]. In the use of social robots, patients have more direct contact with the robots. The clinicians' role is essential to communicate to the patients that they are not being treated solely by a robot but rather by a combination of human and machine. Having that communicated, the clinician is also influential in helping the patient to understand, cooperate with, interact with, and comply with the robot [3, 6, 11].

For physical robots, there is an additional parameter to take into account: **time**. The clinician knows the timeframe in which the device should be used and how long the donning/doffing should take to have a practical therapeutic session. They also know which steps are more time-consuming. Ideally, the device should be easy and not time-consuming to put on. The device donning should be straightforward with rather a small number of steps and the patient should not need to go through repetitive actions because this could cause the patient's loss of energy before the session starts [9].

14.2.2 Acceptability

Acceptance is defined as a phenomenon that reflects to what extent potential users are willing to use a specific system [12]. The difference between acceptance and acceptability is that acceptance is described as the respondent's attitudes, including their behavioral responses, after introducing a measure, and acceptability as the coming judgment before such future introduction [13]. Hence, acceptability is linked closely to usage, and acceptance will depend on how user needs are integrated into the system's development. Previous studies [14–17] showed that physiotherapists generally had a positive attitude to robotic devices' potential and a lack of knowledge about the systems currently being developed. Those studies indicated concerns about patient confidentiality and the cost and usability of robotic systems. For instance, a survey demonstrated which features of robotic devices physiotherapists considered to be desirable around the areas of safety, positioning, movement control, patient feedback, and display of and access to information [18]. As shown in the practicability, for the acceptability there are also different parameters for the physical robots and the social robots.

There are four critical parameters to define the acceptance of physical robots. First, the physical robot should cause **no harm**, which is a fundamental design principle. The clinicians have a good understanding of the human anatomy and the possible points of contact between the robotic device and the human body [3, 4]. Therefore, they can help understand specific implications of working with the device and potential harm points in the force transference. In this sense, an example of pressure points and the danger zones for skin integrity is shown in [9]. Second, the device should have a **Familiarization phase**. Clinicians know how long the patient needs to familiarize with the new device [8]. The goals of the use of the device as well as the limits of the device should also be clarified with the patients [3]. This would help to avoid future deception, insufficient compliance and disappointment. Third, the robotic device should **avoid fear**. Having a safe feeling along with a sense of comfort is desirable for the patients. With a proper knowledge of the device, clinicians should accurately use the device in their sessions [4, 11]. The role of the clinician is critical to provide the feeling of being safe and avoiding fear. Clinicians provide empathy due to the experience of handling patients with different diseases. Having a safe feeling during the therapy session is essential to achieve compliance. A patient who lacks information (e.g., how the assistive forces operate in a physical robot) starts to develop fear and will be rarely convinced to continue to work with the device [9]. The fourth parameter is **relevancy**. The clinician can decide if the device has relevance to the therapy. The device should help the clinician to achieve a better therapeutic result [3–5].

In the use of social robotics, it is crucial to clarify the **Role of the device** additionally. The clinician can discern the needs in therapy that can be helped by the device and how to implement it in the therapeutic environment. It should be clarified to the clinicians that the device is aimed to complement their role rather than replacing them [3]. For that, the clinician should understand how the social robot works, what its advantage is, and what its limits are [8, 10, 11]. In the category of social robots, the insecurity of the role of the clinicians is very high. Therefore, it is crucial to clarify the vital role of the clinicians next to the device. Generally, the robot can do a part of the clinician's work, which gives the clinician the liberty to perform other tasks [4, 11].

As the field of robotics develops, acceptance levels may rise, but physiotherapists, and rehabilitation in general, need to be in a position to make the best use of this by stepping out of established comfort zones. For this, they should recognize potential benefits to the patients and a broader assessment of cost/benefit that includes initial cost, storage, maintenance, training, and improved, efficient outcomes [19].

14.3 Stages for Clinical and Patient Input

Different tools have been used to collaborate between patients, the health staff, and the device's designers and developers. Those tools can help at different stages of the research in the use of new robotic devices. Figure 14.1 presents some of the inputs, those tools can offer in designing, implementing, and assessing a new robotic device.

14.3.1 Planning Stage

The study of the practicability of a device can be started in planning at the beginning of the project. From the beginning, a clinician can improve the device's

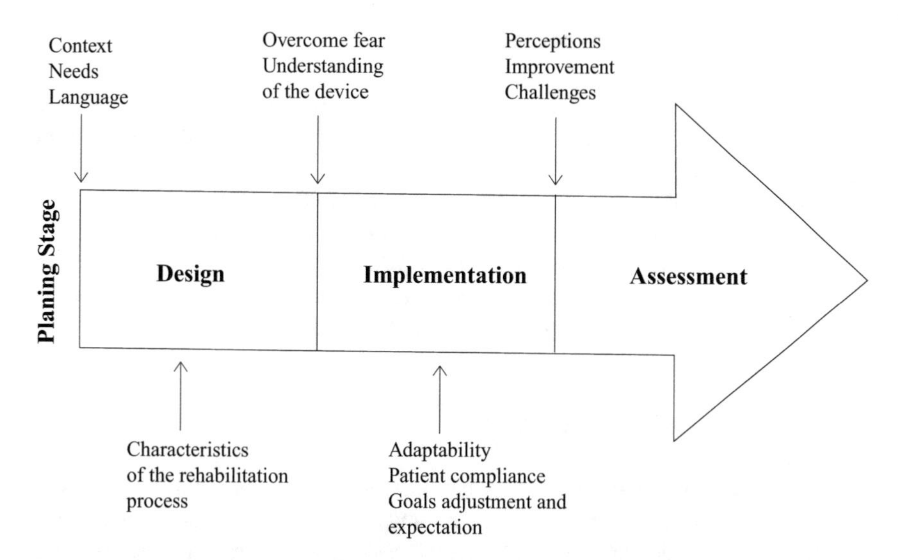

Fig. 14.1 Flow diagram of technology acceptance and perception: different aspects that can be obtained in the different stages of a study

practicability (e.g., make a wearable robot easier to donning and doffing the device). Working together with the engineers, they can save time through safe usefulness steps. The cooperation between clinicians and engineers can make the device more practical for clinicians and patients. Approaching the stakeholders during this stage can also give information about the context, specific needs, and specific language of the community.

14.3.2 During the Building Process

Involving clinicians in the planning and producing process support the building process and the stakeholder's integration. In this sense clinicians understand technical factors involved into the process, trade-offs, and limits. Engineers understand the clinical process and the challenges of handling patients. This approach also allows engineers to understand the clinical process and the challenges of handling patients. Moreover, a clinician involved in the building process can improve compliance with other clinicians. This factor will simplify and accelerate the implementation of the device in the therapy. Additionally, including clinicians in device design helps to overcome many clinicians' common fear, which is the job replacement. Concluding, clinicians will ideally tend to see their relationship with the new devices as cooperation rather than competition.

14.3.3 During the Implementation

The clinicians' role is crucial in providing a safe environment for patients when new devices are being implemented. Clinicians can better explain the device to the patient (e.g., how the forces work in a physical robot, and when wrong functioning is generated). The clinician can also define, together with the patient, reasonable goals and explain the devices. The role of the clinician is also vital in avoiding dangers. For example, in gait rehabilitation with robotic devices, the patient might get overly motivated, and therefore the risk of a fall might increase. Participants might lose their understanding of their physical limits during the interventions.

14.3.4 After the Implementation

After implementing a new device in a rehabilitation process, clinicians' and patients' opinions can be collected to view the community's perception of the device clearly. This can help to understand their idea of usefulness, difficulty, efficacy, and other parameters. Additionally, this can help the designers and engineers understand the challenges in using the device and the improvement possibilities. In the second section of this chapter, some tools are described. Those tools are divided into questionnaires, interviews, and focal groups. In the third section ,some specific examples in rehabilitation are presented. In those specific case studies, it is shown the use of those tools only at the beginning, only at the end, and at the beginning and the end of the process.

14.4 Perception Studies and Survey

Ergonomics and comfort are some of the most relevant aspects of user-machine interaction [20]. Those parameters are often measured using subjective scales. The idea of comfort for a robotic device in rehabilitation can be seen from different points of view. It can be related to the physical interface, its usefulness, and its safety among others. Additionally, the comfort can be related not only to the patient but also to the health staff that is part of the rehabilitation activity. Concerning the health staff, the comfort using the device can be related to how this device contributes to the therapy's development or can increase its difficulty. All those parameters can be grouped in the idea of acceptance and perception. This is usually obtained through qualitative data that is subjective, difficult to analyze consistently, and open to interpretation, but provides much richer information, explaining and giving context to the quantitative responses [1].

Some studies discuss the importance of balancing the device's functional requirements as defined by the potential users and the technical requirements from an engineering perspective. They also highlight the need for suitable strategies to gather information within a particular population to avoid problems associated with the trial of an unfamiliar device. They concluded that feedback from potential users is essential for device design [21].

In this section, a summary of different techniques to measure acceptance and perception will be developed. In this case, three methodologies will be explained (questionnaires, interviews, and focus groups). It seems probable that the most appropriate method and level of involvement to measure this perception will depend upon the nature of the device, stage of development, and the nature of the users involved [1].

14.4.1 Questionnaires

Standardized questionnaires have been proposed to provide a more reliable measure of people's perception [22]. These questionnaires have been standardized to have high reliability and validity measures, and they are compared based on their sensitivity degrees [23]. In this chapter, some questionnaires to use before the implementation of a robotic device (i.e., Knowledge, Attitude and Practice KAP questionnaire), during the intervention (i.e., Working Alliance Inventory WAI), and after the use of the robotic device (i.e., QUEST, UTAUT) are presented.

14.4.1.1 Knowledge, Attitude and Practice KAP

The Knowledge, Attitude, and Practice (KAP) questionnaire is a representative survey conducted on a particular population to identify the knowledge (K), attitudes (A), and practices (P) of a population on a specific topic [24]. In most KAP studies, data are gathered orally by an interviewer who uses a structured, predefined questions formatted in standardized questionnaires, making it a quantitative method that provides access to quantitative and qualitative information [24].

This questionnaire collects the data on the knowledge (i.e., what is known), attitudes (i.e., what is perceived), and practices (i.e., what is done) of a particular population [25]. In the first one, it is possible to measure the knowledge level regarding information acquired by a population and ensure that the tools used are appropriately suited to the people in question. In the attitudes, the gap between knowledge and practices is measured and shows the various restrictions people are bound. In the practices, the information about actual acts carried out by people in the situation, in their context, is measured [24].

KAP surveys are prevalent in health care because they provide helpful information and appear easy to design and execute [26]. KAP can be used before an intervention to reveal misconceptions or misunderstandings that may represent obstacles to the activities that would be implemented and potential barriers to behavior change [24]. It can also measure the impact of education programs [24] used at the beginning and the end, providing recommendations for implementing of future projects. KAP surveys burgeon when novel situations arise, such as the use of robotics in new clinical scenarios or during the COVID-19 pandemic, which has spawned several KAP studies in the population at large as well as in selected subpopulations, including health care workers [26].

Some reasons for conducting KAP Surveys are: (1) To identify the baseline knowledge, myths, misconceptions, attitudes, beliefs, and behaviors concerning a specific health-related topic, (2) To understand, analyze, and communicate about topics or situations of interest in the field, (3) To provide information on needs, issues, and barriers related to the development of effective, locally relevant public health interventions, and (4) To measure post-intervention changes, and thus, the effectiveness of intervention programs that were aimed at correcting and changing health-related knowledge, attitudes, behaviors, and practice.

Note that a KAP survey essentially records an "opinion" and is based on the "declarative" (i.e., statements). In other words, the KAP survey reveals what was said, but there may be considerable gaps between what is said and what is done [24]

14.4.1.2 Working Alliance Inventory (WAI)

The Working Alliance Inventory (WAI) is a method developed to evaluate some generic degree of success in counseling. This measurement by Hovarth et al. in 1989 is based on Bordin's pantheoretical tripartite conceptualization (i.e., bonds, tasks, and goals) [27]. In social robotics, it allows measuring the adaptation to the devices. These three subscales are assessed with a 36-item-self-report instrument. The *Bond* construct measures the degree to which the robot and the patient like and trust each other (e.g., "My relationship with the robot is important to me "); the *Task* construct evaluates the degree to which the robot and the patient agree on therapeutic tasks (e.g., "The things that the robot is asking to me do not make sense); and the *Goal* construct aimed to measure the degree to which the robot and the patient agree on the therapy goals (e.g., "The robot perceives accurately what my goals are").

This measurement has been analyzed and used in studies based on long-term interaction in social robotics, mainly based on the WAI proposed by Bickmore et al. in 2005 [28]. For example, to measure the robot interaction, the researchers use the WAI without the task construct in a study to measure the effects of anticipatory perceptual simulation on practiced human–robot tasks [29]. On the other hand, in Kidd et al. in 2008 [30], the interaction between the robot and the users in a long-term period scenario was measured. The researchers compare the WAI scores of a group who experienced the interaction with a relational robot with users who use a non-relational robot. The results show that the bond between the robot and the users was significantly better for the relational robot. Finally, Abdulrahman and Richards [31] modeled the therapeutic alliance using a user-aware embodied conversational agent that promotes treatment adherence. The researchers used the WAI to investigate the agent's influence on the adherence and therapeutic outcomes after 3 and 6 months of interaction.

14.4.1.3 Acceptance and Usability Assessment Based on UTAUT Test

Technology acceptance is commonly described as the favorable reception and ongoing use of newly introduced devices and systems [32]. Questionnaires used to assess this acceptance can be specific to an application or be universal. That means, they can be adapted to different forms of technology. For universal questionnaires, three criteria of ISO 9241-11 are the most taken into account: effectiveness, efficiency, and satisfaction [23].

The Unified Theory of Acceptance and Use of Technology (*UTAUT*) [33] was developed as an evolution to the Technology Acceptance Model (*TAM*) [34]. The TAM model is the basis for evaluating acceptance in different applications (e.g., e-commerce acceptance model (*EAM*), technology acceptance associated with mobile health devices [35]). However, the *TAM* has been criticized as it lacks precision and ignores influential factors such as the complexity of the technology, and user characteristics that are relevant on many applications [32]. The UTAUT model has been used in healthcare to evaluate of different devices and technologies, applications such as web-based devices and rehabilitation technologies [36]. This acceptance can be from the patients [37] or health care staff [38].

Based on the UTAUT and UTAUT2 models by [33, 39], and the questionnaire developed by Heerink et al. [40], an acceptance and usability questionnaire can be designed and adapted. Six categories are usually established in order to evaluate different perception constructs: Facilitating Conditions (*FC*), Performance and Attitude Expectancy (*PAE*), Effort expectancy and anxiety (*EEA*), Behavior Perception (*BP*), Trust (*TR*), and Attitude Towards Using Technology (*AT*). Each question is scored with a 5 points Likert scale (1 strongly disagree, 2 disagree, 3 neutral, 4 agree, and 5 strongly agree). The **Facilitating conditions** define if the user has the necessary knowledge to use the system or have previously used similar systems. The **Performance and Attitude Expectancy** asks if the user finds the device helpful, compelling or if it increases the task's performance. The **Effort expectancy and anxiety** ask about the fear, difficulties, or learning necessary to use the system. The **Behavior Perception** defined the perception of the user about the communication and understandability of the device. The **Trust** can use questions such as, "I would trust the system if it gave me advice" or "I would follow the advice that the system gives me." Finally, the **Attitude Towards Using Technology** asks about attitudes such as the fun or interest in using the system.

In social robots, they seek to interact as humans do [41] and this represents a difference with other technologies. Therefore, the perception models need some adaptations to meet the social robotics needs [42–45]. Heerink et al. found that the UTAUT model did not indicate that social abilities contribute accepting a social robot [40]. This work presents an adapted version of the UTAUT model incorporating social aspects relevant to assess social robotic agents [46]. They described user acceptance as "the demonstrable willingness within a user group to employ technology for the tasks it is designed to support." This model integrates several constructs that enable to know social factors influenced by a social robot (e.g., anxiety, attitude, facilitating conditions, social influence, intention to use, perceive adaptability, perceived enjoyment, perceived ease of use, perceived sociability, and perceived usefulness) [14, 47].

14.4.1.4 QUEST

The user perception can also be assessed after an interaction employing a Quebec User Evaluation of Satisfaction with Assistive Technology *QUEST* test [48]. The

QUEST was designed as an outcome measurement instrument to evaluate a person's satisfaction with a wide range of assistive technology (AT). The original QUEST survey comprises 27 questions related to participants' satisfaction concerning the robotic device [49]. The user is asked to indicate the degree of importance they attribute to each of the satisfaction variables and then to rate their degree of satisfaction ranging from 1 (not satisfactory at all) to 5 (very satisfactory) [50]. Satisfaction is defined as a person's critical evaluation of several aspects of a device [48]. This definition is based on the principle that each variable's relative importance needs to be determined by the consumer to interpret the satisfaction data [51]. It was intended as a clinical and research instrument. As a clinical tool, the rating scale provides practitioners with a means of collecting satisfaction data to document AT's real-life benefits and justify these devices' needs for these devices. QUEST test, as a research tool, can compare satisfaction data with other outcome measures such as clinical results, quality of life, functional status, cost factors, and comfort. It can also compare satisfaction results obtained with different user groups, settings, and countries [48].

In the second version of this tool (QUEST 2.0), the instrument is divided into two domains based on the results of factor analysis [48,52]. Some items concerning satisfaction are related to the assistive technology device ("Device" domain), while other items are related to the assistive technology services in which the assistive device is delivered ("Services" domain) [50]. In the work of Demers et al. in 2002 concerning the QUEST, test–retest reliability was high, with intra-class correlation coefficients (ICCs) of 0.82, 0.82, and 0.91 for the "Device" and "Services" domains and the total scores, respectively [48]. The questionnaire is designed for either self-administration or interview [50]

This tool has been used to assess different assistive technologies like wheelchairs, exoskeletons, orthesis, among others [50, 51, 53, 54]. In Wearable devices, the questions concerning characteristics about the device are: dimensions, weight, adjustments, safety, durability, simplicity of use, comfort, and effectiveness [54]. Questions concerning the service are: service delivery, repairs/servicing, professional service, and follow-up services [51].

14.4.2 Interviews

A key aspect to the planning of a research project in rehabilitation robotics is patients and their families and health staff likely to use the system within their routine practice [1]. A user interview is a User eXperience (UX) research method during which a researcher asks one user questions about a topic of interest (e.g., use of a system, behaviors, and habits) to learn about that topic. This technique can be a quick and easy way to collect user data. Interviews can give insights into what users think about a new robotic device in a rehabilitation process. They can point out what people feel is essential in the process and what ideas for improvement they may have.

Interviews can be used alone or combined with questionnaires and observations [1]. For the questionnaires, the interviews can also be done before the design process to know the context, inform the population; or at the end of a usability test to collect verbal responses related to observed behaviors. When done at the end of a usability test, it is best to defer the interview until after the behavioral observation segment of the usability study. If the perception questions are asked before the participant tries to perform tasks with the proposed design, it can bias the user to pay special attention to whatever features or issues were asked about.

Interviews have been used to obtain general impressions about the benefits and barriers of using robotic therapy devices for in-home rehabilitation [55], the involvement of health staff and members of the public in the design stages of an upper-limb robotic device [1, 19, 56], to investigate and prioritize the needs concerning the personal mobility domains and their attitudes towards assistive robots [57], to ask clinical therapists their perspectives on robotic stroke rehabilitation [58], among others.

Usually, the data collected at the interviews are analyzed to obtain the more frequent themes and concerns of the population about a specific goal [55] and thematic content analysis by underlying recurrent topics [57]. For this analysis is vital to set a goal for the interview and avoid leading, closed, or vague questions [55].

Some studies show through interviews the health staff opinion concerning ways to improve the handling of the robot, additional features that they would like to see, existing features that they considered unnecessary or undesirable, the type of patients they would use the system with, the benefits (if any) that they saw in using the robot; and the barriers (if any) that may limit the use of the robot [56]. In some cases, interviews alone are not sufficient to meet all the work/task analysis needs. It is vitally important to observe users doing work in their natural settings, and to gather and document examples of that work for designers to gain a thorough understanding of potential users' work (including its surrounding context) which an intended application [59].

14.4.3 Focus Groups

The focus groups are a video- or audio-taped small group discussion that explores topics selected by the researcher and is typically timed to last no more than 2 h [60]. Unlike user interviews, which are one-on-one sessions, focus groups involve 6 to 9 users [61]. As a qualitative method for gathering data, focus groups bring together several participants to discuss: (1) a topic of mutual interest to themselves and the researcher or (2) issues and concerns about the features of a user interface [60]. This enables the project team to take the user's perspective and argue from the user's point of view [62]. Moreover, it can help researchers to assess user needs and feelings both before interface design and long after implementation [61].

Focus group participants are usually led through the discussion by a moderator, often the researcher [60]. For participants, the focus group session should feel free-flowing and relatively unstructured, but in reality, the moderator must follow a pre-planned script of specific issues and set goals for the type of information to be gathered [61]. The data collected from focus group sessions are typically analyzed qualitatively [60]. In interactive systems development, the proper role of focus groups is not to assess interaction styles or design usability but to discover what users want from the system [61].

Focus groups not only give us access to certain kinds of qualitative phenomena that are poorly studied with other methods but also represent an essential tool for breaking down narrow methodological barriers [60]. Focus groups can serve a variety of purposes related to rehabilitation programs. Among these are to (a) obtain general background information about a program (b) generate program ideas that can be subsequently tested, (c) diagnose program problem areas (d) gather information about clients' impressions about a program, and (e) learn how clients talk about the program or topic of interest [63]. However, the information of the focus groups should be complemented with other techniques due to the inaccurate data that can be produced because users may think they want one thing when they need another [61]. Within the realm of qualitative methods, focus groups have much to offer as an adjunct to other qualitative techniques, such as informant interviewing and participant observation [60].

In rehabilitation, focus groups can be used to empower its conventional programs [64]. This technique has been used in e-Health for stroke rehabilitation [65], with potential users of exoskeletons for Spinal Cord Injury [66], virtual reality training systems [67], and home-based stroke rehabilitation [68]. In some cases, through Focal groups, it has been found that the system's requirements between patients/informal caregivers and health professionals differed on several aspects [65]. Therefore, involving the perspectives of all end-users in the design process of Rehabilitation programs are needed to achieve a user-centered design [65].

In the field of social robotics, focus groups have been used to introduce *SAR* and discuss their questions and concerns associated with the technology [69], and to create new application within the community [70]. Some changes in opinion and perception are found in the participants once the robotic application has been explained and they had the opportunity to witness in situ demonstrations [69, 71].

14.5 Clinician's Experiences and Perception of Robotics

According to the previous section's perception studies and surveys already mentioned, this section presents three studies previously performed. These studies show different measurements used to evaluate the patient's and clinician's perception and experience with the technology and their results. The first study contemplates the patient's and clinician's perception before using technology. The second study evaluated only the patient's and clinician's acceptability after used the technology. Finally, the third study evaluated the patient's and clinician's perception and expectations before using technology and their acceptability.

14.5.1 Expectations of Healthcare Professionals for Robots During COVID-19

The study of Sierra et al. [72] presents the design and implementation of a perception questionnaire to assess healthcare providers' level of acceptance and education towards robotic solutions for the COVID-19 pandemic. In this work remarkably, several questionnaires were proposed to evaluate the perception of medical robotics, as well as of three types of robotics platforms for COVID-19 mitigation and control: (DIS) Disinfection and cleaning robots; (ASL) Assistance, Service, and Logistic robots; and (TEL) Telemedicine and Telepresence robots.

The researchers designed a qualitative survey to assess health professionals' concepts, ideas, perceptions, and attitudes toward robotics in managing the COVID-19 pandemic through the KAP questionnaire. As illustrated in Sect. 14.4.1.1, this questionnaire collects the data on the knowledge (i.e., what is known), attitudes (i.e., what is perceived), and practices (i.e., what is done) of a particular population [25]. It is essential to highlight, as mentioned in Sect. 14.4.1.1, several KAP surveys on COVID-19 have been reported in the literature. However, they aimed to assess the overall perception of COVID-19 in patients and survivors, and not to evaluate robotics perception for COVID-19 outbreak management [73–75]. Therefore, the survey was designed taking into account three sections, as follows:

- **The first part** was designed using knowledge-oriented questions. These questions measure the level of awareness and understanding healthcare professionals have regarding robotic tools for DIS, ASL, and TEL.
- **The second part** was designed using attitude-oriented questions. These questions measure how healthcare professionals feel about robotic tools for DIS, ASL, and TEL, as well as any preconceived ideas or beliefs they may have about this topic.
- **The third part** was designed using practice-oriented questions. These questions provide insight into how healthcare professionals apply their knowledge and attitudes regarding robotic tools for DIS, ASL, and TEL through their everyday actions.

Overall, yes or no questions were rated using 1 and -1 scores, respectively; the questions asking to rate experience, knowledge about a topic, and questions formulated as statements were evaluated using a 5-point Likert scale, converted to a scale from -2 to 2 points.[1]

Summarizing, 41 (20 women and 21 men, $35.39\pm$ 8.48 y.o) healthcare professionals (e.g., nurses, doctors, biomedical engineers, among others) satisfactorily accomplished the surveys, assessing three categories: DIS, ASL, and TEL robots. Participants were asked to virtually fill out the perception questionnaires using the

[1]The results of these surveys are available in the following link https://doi.org/10.6084/m9.figshare.13373741 [72].

Google Forms online tool. At the beginning of the form, participants were presented with the informed consent, which they read carefully and accept before proceeding with the form. Afterward participants were asked for demographic information about their profession and their work environment. Preceding the questionnaires, a brief description of each type of robot was presented (i.e., DIS, ASL, and TEL) to homogenize the definition of such devices among the participants. Then, the questionnaires were applied [72].

KAP results (see Fig. 14.2) related to the three questionnaire constructs (i.e., knowledge, attitude, and practice) revealed:

i. There is a positive level of knowledge about medical robotics in general for the surveyed population.
ii. Concerning robots for disinfection (DIS), assistance (ASL), and telemedicine (TEL), participants indicated that they have a low level of knowledge and experience with these types of robots.
iii. 82.9% of participants reported a positive attitude towards robots' usefulness and benefits in managing and controlling the COVID-19 pandemic.
iv. 65.8% of clinicians recommend using ASL robots in the pandemic.
v. Approximately 60% of the participants assumed a neutral position when asked if they considered a replacement.

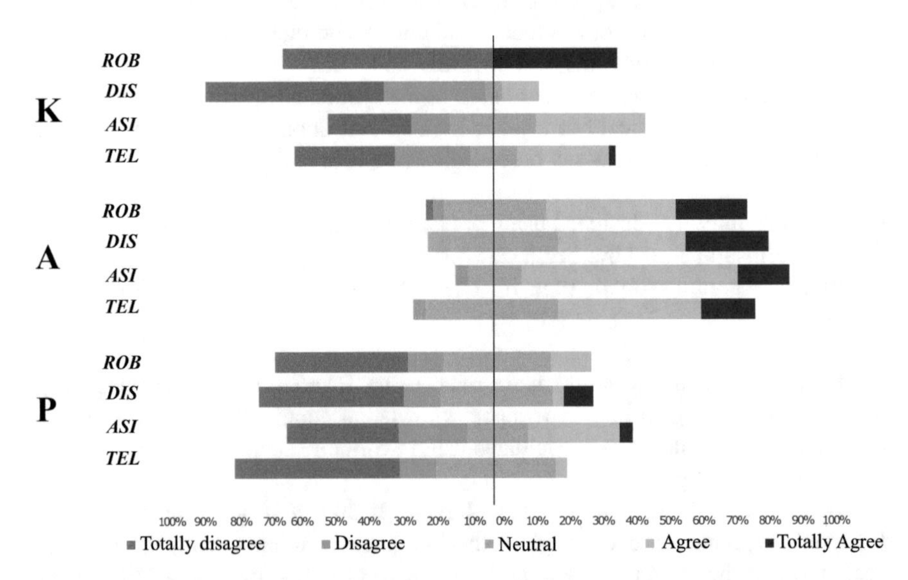

Fig. 14.2 Likert scale distribution for the KAP construct, K refers to knowledge, A to attitude and P to practice for the DIS, ASL, and TEL robots. At the same the results of general knowledge about robots, labeled ROBOT, are reported

The outcomes (i, ii, iii) showed that participants have a positive level of knowledge regarding medical robots in general. However, the clinicians' experience and knowledge regarding DIS, ASL, and TEL platforms are shallow. Consequently, the research suggested that the clinician's awareness and education have to be increased to understand these tools' opportunities, functions, and features [72]. Regarding the outcome (iv), i.e., the robot's role, clinicians prefer platforms capable of supporting logistic tasks, medication, and food delivery, and monitoring the environment. In the case of DIS and TEL platforms, a lower perception was presented. Hence, these technologies' efforts have to increase the clinicians' trust and develop comprehensive platforms capable of providing assistance and disinfection or teleoperation. Finally, concerning the result (v), this suggests that in the first instance, a familiarization stage is recommendable to increase healthcare personnel's trust and motivation as reported in the literature [76]. It is necessary to carry out education and awareness processes in the medical community [77], to strengthen the idea that robots can enhance and improve their work. However, they cannot replace healthcare professionals in fundamental activities. For instance, Coombs et al. [76] recommend performing a familiarization stage based on culture theory to understand individuals' social practices when interacting with the technology and their preferences within its usages. This culture theory will increase their motivation and trust towards technology, such as medical robotics.

14.5.2 Acceptance and Perception of Healthcare Staff in an Application of Social Robotics in Lokomat Therapy

In contrast to the study presented in the previous Sect. 14.5.1, this section presents the design and implementation of an acceptance questionnaire to assess patients and healthcare providers' level of acceptability after used a Social Assistive Robot (SAR) during Lokomat therapies. In this work by Raigoso et al. [78], before implemented the SAR during the therapies, a technology explanation was performed to inform the patients and clinicians about the possible robot's role during the rehabilitation procedure. Three robot assistance tasks were highlighted in this study: (1) clinicians support, e.g., the social robot give feedback to the patients about their cervical and thoracic posture; (2) patient's online monitoring; (3) corrections and motivations provided by the SAR. Overall, the robot was used to complement the therapist's tasks and motivate patients during therapy. It should be noted that, as mentioned in the previous section, several studies [79,80] recommend this first step (i.e., technology explanation) to understand better the technology dimensions (i.e., the robot's limitations and capabilities tasks). Afterward, the researches designed and implemented a questionnaire based on The Almere Model adapted from the Unified Theory of Acceptance and Use of Technology (UTAUT) questionnaire [81]. As illustrated in Sect. 14.4.1.3, this questionnaire assesses the perception of the participants through different constructs (e.g., *Psychological factor* (PF), *Social perception* (SP), *Entertainment Level* (EL), *Effort's Expectations* (EE), *Performance Expectations* (PE), and *Facility Conditions* (FC)).

Table 14.1 Acceptance questionnaire for lokomat therapy users. Adapted from [78]

Construct	No.	Questions
PF	1	I think that the robot will give me confidence.
	2	Using the robot will generate stress.
	3	I think that the robot express emotions during the sessions will be uncomfortable.
	4	I think that the robot will increase the concentration during the therapy.
SP	1	I think that using the robot in rehabilitation could be more enjoyable.
	2	I think that the interaction with the robot would be nice.
	3	Using the robot will give me satisfaction.
EL	1	I think that the therapy could turn boring with the use of the robot.
	2	I think that I will enjoy more the therapy with the robot.
	3	I think that the robot company will make the therapy more enjoyable.
EE	1	Following the robot's instructions would be difficult.
	2	I think that using the robot would improve Lokomat therapy.
	3	I think that use the robot will be easy.
PE	1	I think that the robot will be helpful during the rehabilitation process.
	2	I think that use the robot will make the therapies faster.
	3	I think that the presence of the robot will affect the engagement in the therapy.
	4	I think that the use of the robot will motivate the patients to perform better the rehabilitation.
FC	1	I consider that the robot can be challenging to control.
	2	I consider that the robot could be adapted to any scenario.
	3	I would like the robot to reduce the workload I have during the rehabilitation procedure.

The survey used in [78] consisted of 40 questions based on the constructs above. The questions are divided into 36 closed questions, 32 items are evaluated through a 5-point Likert scale (i.e., 1: strongly disagree to 5: strongly agree), four dichotomous type questions answered with three scores (i.e., Yes, No, Maybe); and four open questions. It is essential to highlight that to avoid the bias in their results; the researches implemented for the closed questions, positive (e.g., "The therapy is more enjoyable if a robot participates in it") and negative formulation (e.g., "The therapy can be boring using the robot"). An illustration of the implemented questionnaire is shown in Table 14.1.

A total of 88 healthcare professionals and patients involved in physical rehabilitation procedures based on Lokomat therapies in two different countries (Colombia and Spain) satisfactorily accomplished the surveys online using the Google Forms online tool. UTAUT results (see Fig. 14.3) related to the six questionnaire constructs proposed in this questionnaire revealed that the robot's perception is primarily positive (PF, 63.92%; SP, 82.5%; EL, 73.29%, and PE, 67.17%). However, a negative perception was found in the Effort's Expectation and Facility Conditions constructs (EE, 51.14%; FE, 43.63%) [78]. These results are interesting because most patients and clinicians think robot usage can be tricky (e.g., ease of use,

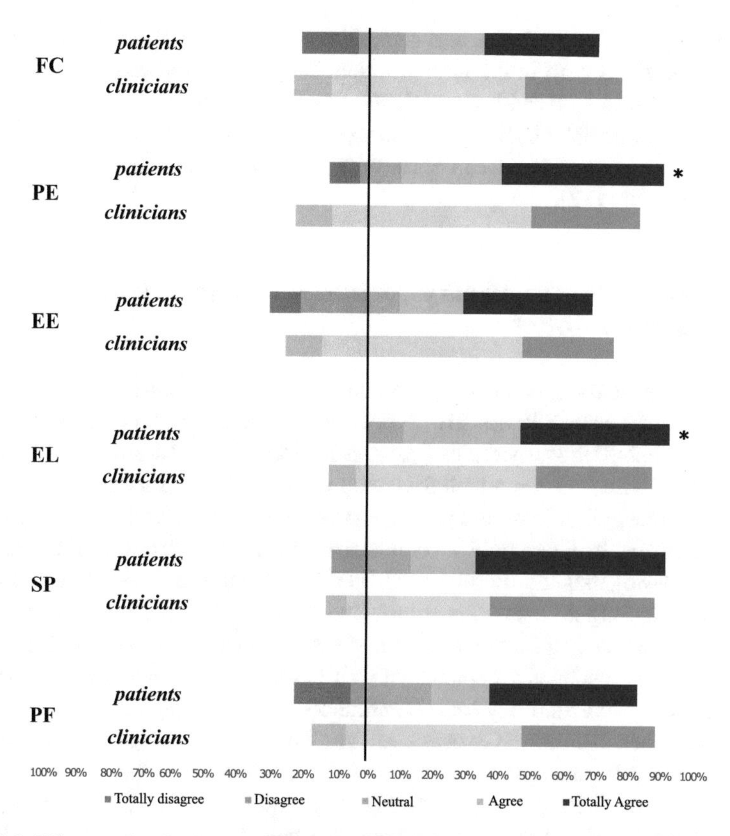

Fig. 14.3 Likert scale distribution for (FC), (PE), (EE), (EL), (SP), and (PF) construct of the acceptance and perception questionnaire applied to patients and clinicians. (*) Statistically significant differences between patients' and clinicians' groups

understand, and follow up the robot instructions, among others) considering the EE construct. In FC, the results show that the participants perceive that the robot role is exclusive for specific treatments and cannot be used in various tasks apart from the rehabilitation procedures [78]. This perception is expected as the robot's interaction is unknown for the users, suggesting that an introduction phase is needed to implement the robot in the future. In fact, in the literature, several studies [82, 83] recommend performing an initial stage where the participants could interact with the technology and understand it to increase the acceptance of the robot in the time.

Summarizing, the results are very encouraging as they highlight the positive perception of different kinds of participants (clinicians and patients) towards the robot in a physical rehabilitation scenario. More than 60% of the population evaluated accept the social robot in the PR with Lokomat. On the other hand, measuring the perception and acceptance in the first stage allows an initial perspective to the participants' needs and expectations. Moreover, the results also show that it is essential to perform a stage to present the robotic system's capabilities and introduce the technology (i.e., robustness and capabilities of the SAR system) to understand and integrate the system in the rehabilitation. The results allowed to build a social robotic interface to work with the patients. Results showed that the robot's support improves the patients' physiological progress by reducing their unhealthy spinal posture time, with positive acceptance. 65% of patients described the platform as helpful and secure [17].

14.5.3 Expectation vs. Reality: Attitudes Towards a Socially Assistive Robot

This section presents a user perception and acceptance questionnaire to assess the attitudes towards a socially assistive robot designed to support the outpatient phase of cardiac rehabilitation therapies.Casas et al. [47] designed and implemented a questionnaire based on the adapted version of the UTAUT model [33] for social robots to evaluate clinicians' and patients' perceptions before and after using a SAR during therapies. It is essential to emphasize that in the literature, there is some evidence [84–86] that the modified UTAUT is a reliable method to assess social acceptance, investigate users' reactions, and analyze the societal impact. In this context, this model has been successfully used in healthcare to evaluate various applications. For instance, acceptance of web-based aftercare devices [87], therapist acceptance of new technology for rehabilitation [5, 36], among others.

On the other hand, in Casas et al. similar to Raigoso et al presented in Sect. 14.5.2 participants had no previous experience with the robotic system. Hence, the researchers provided a technology explanation, i.e., participants were briefly contextualized about SAR systems, the benefits that they can provide, and the variables that are measured in this application, followed by the presentation of a video where the real cardiac scenario is displayed and the robot with its functionality can be appreciated [47]. The system used was comprised of a sensor interface, aiming to measure all relevant therapy variables such as cardiopulmonary parameters (e.g., heart rate and blood pressure), spatiotemporal parameters (e.g., speed, cadence, and step length), and exertion perception scale. Moreover, the robot behaviors were designed in three situations (motivation, warning, and emergency) to interact with the patient while monitoring its performance, and to communicate with the therapists if an event of emergency occurs during the therapy (e.g., heart rate over the maximum allowed level and dizziness) [47]. Afterward, to analyze patients' perception and attitudes towards incorporating this technology in clinical applications, such as cardiac rehabilitation, from both perspectives), two conditions were defined:

1. **Intervention condition:** Patients had a long-term interaction (more than 18 weeks) with the system and experienced the benefits and disadvantages.
2. **Control group**, where an interview was conducted for patients with no experience with the robot.

Besides, Casas et al. analyzed how clinicians are familiar with technology and the effects this might have on rehabilitation programs. Hence, a group of clinicians that work at the clinic in areas associated with cardiac rehabilitation were invited to participate in a focus group at the clinic. This focus group aimed to introduce SAR and discuss their questions and concerns about the technology [47].

Overall, the purpose of having three conditions in [47] was to contrast initial attitudes and expectations against a post-interaction period, to understand how the users can accept this technology more. Therefore, a total of participants performed the study, i.e., this questionnaire was administered to a group of 20 patients without experience with the robot (control group, male = 63.15%, female = 36.84%), eight patients (intervention group, male = 87.5%, female = 12.5%) who spent 18 weeks with the robot during therapy, and 15 clinicians (focus group, male = 6.66%, female = 93.33%, age 36.86 ± 8.78 years old, years of expertise years 11.13 ± 7.68) who work on the cardiac rehabilitation service.

Regarding the implemented UTAUT model in *Casas et al.* [47], it integrated several constructs (Usefulness (U), Utility and Advantages (U/A), Perceived Utility (PU), Safety (S), Perceived Trust (PT), Ease of Use (EU), Perceived Sociability (PS), and Social Presence (SP)), which provided insight into the social factors influenced by the SAR in cardiac rehabilitation scenarios. An example of the questionnaire used for the patients is illustrated in Table 14.2 and an example of the questionnaire implemented to the clinicians is in Table 14.3. The questions were based on a Likert scale. However, the questions were formulated only in a positive manner.

For the patient group (i.e., intervention and control condition), the UTAUT results (see Fig.14.4) allowed comparing the expectation and perception regarding a social robot's role in cardiac rehabilitation. Overall, the perception presented in both

Table 14.2 Perception questionnaire for patients. Adapted from [47]

Construct	No.	Questions
U	1	I consider that using robots it is a good tool to assist cardiac rehabilitation therapies.
	2	I consider that my interaction with the robot was comfortable.
	3	I am satisfied with the work that the robot did.
PU	1	I think that the use of the robot helps me to compromise to do a good job.
S	1	I consider it was easy to give information to the robot.
EU	1	I consider that the robot's instructions were clear.
PT	1	The robot made me confident.
	2	It gave me confidence that the robot guides my therapy.
PS	1	I find the robot pleasant to interact with.
	2	I think the robot is nice.
SP	1	When interacting with the robot I felt like I am talking to a real person.
	2	I can imagine the robot to be a living creature.
	3	I often think the robot is not a real person.

Table 14.3 Acceptance questionnaire for clinicians. Adapted from [47]

Construct	No.	Questions
U/A	1	I consider that using robots is a good tool to measure the HR and the BP during CR sessions.
	2	I consider that using robots it's a good tool to alert me if there is an abnormal heart rate.
	3	I consider that using robots can help me carry out my tasks faster.
	4	I consider that the verbal motivation given by the robot could help the patient to be more productive.
U	1	I might find the system easy to use.
	2	Learning to use the robot could be easy for me.
PU	1	I consider that using robots can bring benefits for the patients.
	2	I feel that the robot could replace me.
	3	I consider that using robots could aid me to evaluate the therapy better.
S	1	The robot would represent a risk to the patient's health.
PT	1	I would feel safe using the robot in the therapies.
	2	I would trust the robot to help me guide the therapy.
	3	I would be afraid to use a robot in therapy.
PS	1	I would like that the interaction between the patient and the robot can be pleasant.
	2	I would like the robot to act as a friendly companion.
	3	I would like to choose the program that the robot should perform during therapy.
SP	1	I consider that the interaction with the robot might feel like talking to a real person.
	2	I would consider good if the patient had the feeling that the robot will observe him in therapy.
	3	I consider patients would usually think that the robot is not a real person.

groups can be interpreted as positive; however, some categories showed differences between both conditions. For instance, the perceived trust (PT) is higher in the intervention group, than in the control group, which expressed low confidence in the robot. This is an expected reaction associated with the lack of experience and contact with the robot and suggested that the trust in the robot will influence the continuous use of the system in the future [88]. Moreover, the results showed for the utilitarian factor, which encloses ease of use (EU), perceived utility (PU), and usefulness (U), which are fundamental for the engagement in long-term relationships [88] differences between both conditions. Specifically, the following:

- **The (EU) construct** suggested that the intervention group perceives more ease of use than the control group; this is due to the time that these patients spent interacting with a robot where they had the opportunity to realize how complex the interaction with the robotic platform can be, in contrast to the patients in the control group who had limited knowledge of the system and its functionality, it is difficult to understand the complexity of the use of the platform.

- **The (PU) construct** showed a higher positive percentage in the intervention group than in the control group; although control patients perceive a high degree of utility, it can be evidenced that after the interaction, this expectation is overcome. This is because patients who had the opportunity to interact with the robot throughout the rehabilitation process expressed motivation and encouragement to perform better.
- **The (U) construct** was mainly focused on patients' perception of the system and its functionality (e.g., robot interventions, adaptability, manipulation, among others.). In this case, the same pattern as the previous categories was found. The intervention group attributes more usefulness to the system than the control group [47].

These results reflect the positive impact that the platform provided and the potential that it might have in future cardiac therapies. In general, the results of the utilitarian factor suggested the perception of the robot is better qualified for the group who interact with the robot more times as they are familiarized with it [14].

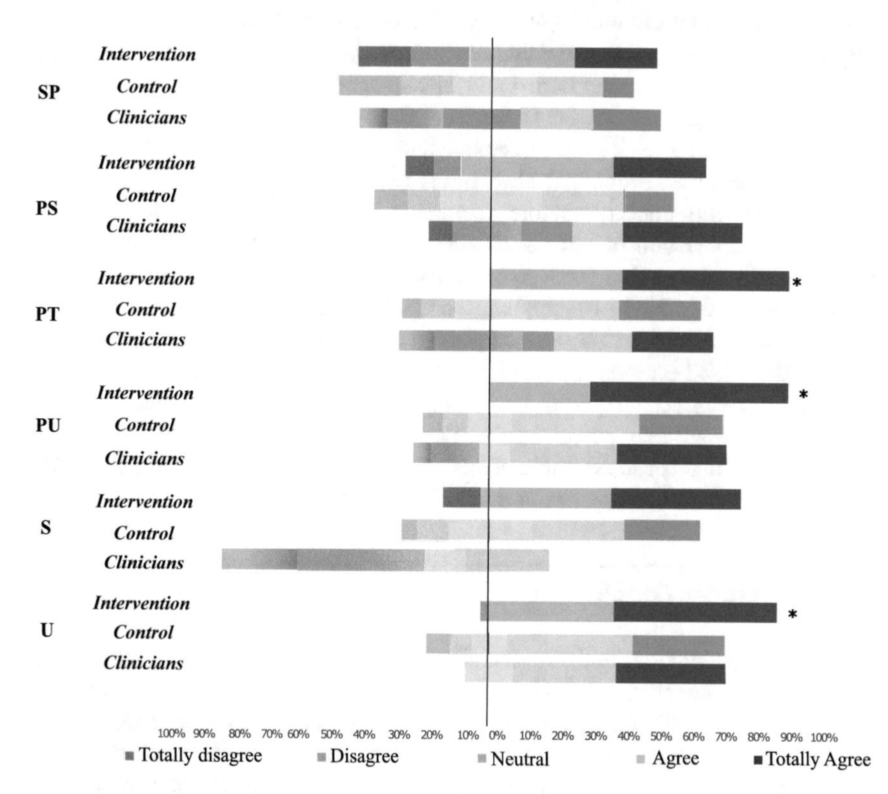

Fig. 14.4 Likert scale distribution for (SP), (PS), (PT), (PU), (S), and (U) construct of the acceptance and perception questionnaire applied to patients (i.e., control and intervention) and clinicians. (*) Statistically significant differences between patient groups

For the clinician group, the UTAUT results (see Fig.14.4) showed positive opinions regarding (U/A), (U), (PU), (PT), and (PS) categories, which means that clinicians think that the robot and the parameters measured are helpful and reliable in cardiac rehabilitation sessions [47]. The results reported:

- **The (S) construct** was scored negative; however, the research stated that it is due to the question formulation and that the results regarding this construct were positive as the clinicians did not consider the robot a risk for the patients.
- **The (SP) perception** showed a neutral response in general; this can be due to the robot's perception as a social agent before the focus group was performed. In this case, the responses related to this construct showed that clinicians think that the robot could not have social skills as the humans (e.g., emotion and living creature) due the robot is perceived like an object [47].

One of the essential aspects of these results was the clinician perception change, i.e., some clinicians perceived incorporating a social robot as a thread, as they regard the robot as a potential replacement. However, after the technology explication, the system's demonstrations and its objectives (e.g., the researchers emphasize that the robot must be considered as a tool that can improve its efficiency during therapy), the clinician's system perception turned into a positive one, where they showed interest and provided suggestions for the system improvements.

Summarizing, Casas et al. [47] demonstrated how the participants (patients and clinicians) present a lower expectation of the robot's usefulness, sociability, safety, and data reliability concerns before interacting with the SAR. However, after the technology explanation or after a considerable interaction time with the robotic platform, this expectation is overcome. Although there is a bias when people consider using this kind of technology, once they can become familiar with the social robot and interact for an adequate period, their attitudes and perception towards the SAR become more positive. The use of this interface has shown that patients felt more encouraged to perform physical activity and continue with the rehabilitation when they perceived that monitored and supervised by the system, demonstrating that it can be implemented as a reliable tool that would potentially leverage tasks carried out by health professionals [15].

14.6 Chapter Conclusions

Acceptance, perception, and the overall opinions of clinicians and patients can change how effectively a new device can be used in a clinical setting. Those concepts cover a series of different parameters for the community, like the ease of use, the time it adds to the therapy, and the physical and emotional comfort for the patient. Several techniques can be used to analyze these opinions and to quantify this qualitative information. Those techniques, like questionnaires, interviews, and focus groups, can be used before or during the design of the application, before or during the implementation of the device, or after for a post-treatment assessment.

The time of use of these techniques can give different information: the knowledge of the device before its application, the adaptability to its use, the challenges and improvement possibilities, among others. Finally, the case studies presented in this chapter show the overall positive perception of clinical staff and patients using robotics in the clinical process. It also showed some fears and challenges and how this can be overcome with the appropriate information about the application. This highlights the importance of the use of these techniques and the new opportunities for robotics in rehabilitation and clinical programs.

References

1. R. Holt, S. Makower, A. Jackson, P. Culmer, M. Levesley, R. Richardson, A. Cozens, M.M. Williams, B. Bhakta, User involvement in developing rehabilitation robotic devices: an essential requirement, in *2007 IEEE 10th International Conference on Rehabilitation Robotics* (2007), pp. 196–204
2. M.E. Wiklund, S.B. Wilcox, *Designing Usability into Medical Products* (CRC Press, Boca Raton, 2005)
3. C.C. Chen, R.K. Bode, Factors influencing therapists' decision-making in the acceptance of new technology devices in stroke rehabilitation. Am. J. Phys. Med. Rehabil. **90**(5), 415–425 (2011)
4. D. Conti, S. Di Nuovo, S. Buono, A. Di Nuovo, Robots in education and care of children with developmental disabilities: a study on acceptance by experienced and future professionals. Int. J. Soc. Robot. **9**(1), 51–62 (2017)
5. L. Liu, A.M. Cruz, A.R. Rincon, V. Buttar, Q. Ranson, D. Goertzen, What factors determine therapists' acceptance of new technologies for rehabilitation – a study using the unified theory of acceptance and use of technology (UTAUT). Disabil. Rehabil. **37**(5), 447–455 (2015). PMID: 24901351
6. S. Mazzoleni, G. Turchetti, I. Palla, F. Posteraro, P. Dario, Acceptability of robotic technology in neuro-rehabilitation: preliminary results on chronic stroke patients. Comput. Methods Programs Biomed. **116**(2), 116–122 (2014)
7. A. Lerdal, L.N. Bakken, S.E. Kouwenhoven, G. Pedersen, M. Kirkevold, A. Finset, H.S. Kim, Poststroke fatigue—a review. J. Pain Symptom Manage. **38**(6), 928–949 (2009)
8. E. Broadbent, R. Stafford, B. MacDonald, Acceptance of healthcare robots for the older population: review and future directions. Int. J. Soc. Robot. **1**(4), 319–330 (2009)
9. N. Zwickl, Evaluating feasibility of Myosuit in a physiotherapeutic rehabilitation environment, in *Masterarbeiten Master of Science in Physiotherapie (MScPT) Studiengang 2016*, ed. by Z.H. für Angewandte Wissenschaften, ch. 48 (Zurich University of Applied Sciences, School of Health Professions, Research and Development, Institute of Physiotherapy, Winterthur, 2019), p. 48
10. P. Lum, D. Reinkensmeyer, R. Mahoney, W.Z. Rymer, C. Burgar, Robotic devices for movement therapy after stroke: current status and challenges to clinical acceptance. Topics Stroke Rehabil. **8**(4), 40–53 (2002)
11. L. Liu, A. Miguel Cruz, A. Rios Rincon, V. Buttar, Q. Ranson, D. Goertzen, What factors determine therapists' acceptance of new technologies for rehabilitation–a study using the unified theory of acceptance and use of technology (UTAUT). Disabil. Rehabil. **37**(5), 447–455 (2015)
12. S. Vlassenroot, K. Brookhuis, V. Marchau, F. Witlox, Towards defining a unified concept for the acceptability of intelligent transport systems (ITS): a conceptual analysis based on the case of intelligent speed adaptation (ISA). Transp. Res. F **13**(3), 164–178 (2010)
13. J. Schade, B. Schlag, et al., *Acceptability of Urban Transport Pricing* (Valtion Taloudellinen Tutkimuskeskus, Helsinki 2000)

14. N. Céspedes, B. Irfan, E. Senft, C.A. Cifuentes, L.F. Gutierrez, M. Rincon-Roncancio, T. Belpaeme, M. Múnera, A socially assistive robot for long-term cardiac rehabilitation in the real world. Front. Neurorobot. **15**, 633248 (2021)
15. J. Casas, E. Senft, L.F. Gutiérrez, M. Rincón-Rocancio, M. Múnera, T. Belpaeme, C.A. Cifuentes, Social assistive robots: assessing the impact of a training assistant robot in cardiac rehabilitation. Int. J. Soc. Robot. (2020). https://doi.org/10.1007/s12369-020-00708-y
16. N. Céspedes, M. Múnera, C. Gómez, C.A. Cifuentes, Social human-robot interaction for gait rehabilitation. IEEE Trans. Neural Syst. Rehabil. Eng. **28**(6), 1299–1307 (2020)
17. N. Céspedes, D. Raigoso, M. Múnera, C.A. Cifuentes, Long-term social human-robot interaction for neurorehabilitation: robots as a tool to support gait therapy in the pandemic. Front. Neurorobot. **15**, 612034 (2021)
18. M. Lee, M. Rittenhouse, H.A. Abdullah, Design issues for therapeutic robot systems: results from a survey of physiotherapists. J. Intell. Robot. Syst. **42**, 239–252 (2005)
19. A. Stephenson, J. Stephens, An exploration of physiotherapists' experiences of robotic therapy in upper limb rehabilitation within a stroke rehabilitation centre. Disabil. Rehabil. Assist. Technol. **13**(3), 245–252 (2018). PMID: 28366037
20. J.L. Pons, *Wearable Robots: Biomechatronic Exoskeletons* (Wiley, Hoboken, 2008)
21. S.J. Mulholland, T.L. Packer, S.J. Laschinger, J.T. Lysack, U.P. Wyss, S. Balaram, Evaluating a new mobility device: feedback from women with disabilities in India. Disabi. Rehabil. **22**, 111–122 (2000)
22. K. Hornbaeck, Current practice in measuring usability: challenges to usability studies and research. Int. J. Hum. Comput. Stud. **65**(2), 79–102 (2006)
23. A. Assila, K. Marçal de Oliveira, H. Ezzedine, Standardized usability questionnaires: features and quality focus. Electron. J. Comput. Sci. Inf. Technol. **6**(1), (2016)
24. G. Fabienne, *Knowledge, Attitudes and Practices for Risk Education: How to Implement KAP Surveys* (Vassel graphique, Bron, 2009)
25. World Health Organization, Knowledge , attitudes , and practices (KAP) surveys during cholera vaccination campaigns: guidance for oral cholera vaccine stockpile campaigns "WORKING COPY ", Tech. Rep., World Health Organization (2014)
26. C. Andrade, V. Menon, S. Ameen, S.K. Praharaj, Designing and conducting knowledge, attitude, and practice surveys in psychiatry: practical guidance. Ind. J. Psychol. Med. **42**(5), 478–481 (2020). PMID: 33414597
27. A. Hovarth, L. Greenberg, Development and validation of the working alliance inventory. J. Counsel. Psychol. **36**(2), 223–233 (1989)
28. T.W. Bickmore, R.W. Picard, Establishing and maintaining long-term human-computer relationships. ACM Transa. Comput.-Hum. Interact. **12**, 293–327 (2005)
29. G. Hoffman, C. Breazeal, Effects of anticipatory perceptual simulation on practiced human-robot tasks. Auton. Robot. **28**(4), 403–423 (2010)
30. C. Kidd, C. Breazeal, Robots at home: understanding long-term human-robot interaction, in *2008 IEEE/RSJ International Conference on Intelligent Robots and Systems* (IEEE, Piscataway, 2008), pp. 3230–3235
31. J.M.K. Westlund, H.W. Park, R. Williams, C. Breazeal, Measuring young children ' s long-term relationships with social robots, in *IDC '18: Proceedings of the 17th ACM Conference on Interaction Design and Childre* (2018), pp. 207–218
32. K. Laver, S. George, J. Ratcliffe, M. Crotty, Measuring technology self efficacy: reliability and construct validity of a modified computer self efficacy scale in a clinical rehabilitation setting. Disabil. Rehabil. **34**, 220–227 (2012)
33. V. Venkatesh, M.G. Morris, G.B. Davis, F.D. Davis, User acceptance of information technology: toward a unified view. MIS Quart. **27**, 425–478 (2003)
34. F.D. Davis, R.P. Bagozzi, P.R. Warshaw, User acceptance of computer technology: a comparison of two theoretical models. Manage. Sci. **35**, 982–1003 (1989)
35. R. Schnall, T. Higgins, W. Brown, A. Carballo-Dieguez, S. Bakken, Trust, perceived risk, perceived ease of use and perceived usefulness as factors related to mhealth technology use.. Stud. Health Technol. Inf. **216**, 467–71 (2015)

36. M. Hatami Kaleshtari, I. Ciobanu, P. Lucian Seiciu, A. Georgiana Marin, M. Berteanu, Towards a model of rehabilitation technology acceptance and usability. Int. J. Soc. Sci. Hum. **6**, 612–616 (2016)
37. S. Hennemann, M.E. Beutel, R. Zwerenz, Drivers and barriers to acceptance of web-based aftercare of patients in inpatient routine care: a cross-sectional survey. J. Med. Internet Res. **18**(12), e337 (2016)
38. L. Liu, A. Miguel Cruz, A. Rios Rincon, V. Buttar, Q. Ranson, D. Goertzen, What factors determine therapists' acceptance of new technologies for rehabilitation – a study using the unified theory of acceptance and use of technology (UTAUT). Disabil. Rehabil. **37**, 447–455 (2015)
39. V. Venkatesh, J.Y.L. Thong, X. Xu, Consumer acceptance and use of information technology: extending the unified theory. MIS Quart. **36**(1), 157–178 (2012)
40. M. Heerink, B. Kröse, B. Wielinga, V. Evers, Measuring the influence of social abilities on acceptance of an interface robot and a screen agent by elderly users, in *Proceedings of the 23rd British HCI Group Annual Conference on People and Computers: Celebrating People and Technology, BCS-HCI'09, Swinton* (British Computer Society, London, 2009), pp. 430–439
41. A. Weiss, R. Bernhaupt,M. Lankes, M. Tscheligi, The USUS evaluation framework for human-robot interaction, in *AISB2009: Proceedings of the Symposium on New Frontiers in Human-Robot Interaction* (2009), pp. 158–165
42. D.-H. Shin, H. Choo, Modeling the acceptance of socially interactive robotics: social presence in human–robot interaction. Interact. Stud. **12**, 430–460 (2011)
43. M.M. de Graaf, S.B. Allouch, T. Klamer, Sharing a life with Harvey: exploring the acceptance of and relationship-building with a social robot. Comput. Hum. Behavior **43**, 1–14 (2015)
44. M. Heerink, B. Kröse, B. Wielinga, V. Evers, Measuring the influence of social abilities on acceptance of an interface robot and a screen agent by elderly users, in *Proceedings of the 23rd British HCI Group Annual Conference on People and Computers: Celebrating People and Technology*. 430–439 (2009)
45. M. Fridin, M. Belokopytov, Acceptance of socially assistive humanoid robot by preschool and elementary school teachers. Comput. Hum. Behavior **33**, 23–31 (2014)
46. M. Heerink, B. Kröse, V. Evers, B. Wielinga, Assessing acceptance of assistive social agent technology by older adults: the Almere model. Int. J. Soc. Robot. **2**, 361–375 (2010)
47. J.A. Casas, N. Céspedes, C.A. Cifuentes, L.F. Gutierrez, M. Rincón-Roncancio, M. Múnera, Expectation vs. reality: attitudes towards a socially assistive robot in cardiac rehabilitation. Appl. Sci. **9**(21), 4651 (2019)
48. L. Demers, R. Weiss-Lambrou, B. Ska, The Quebec user evaluation of satisfaction with assistive technology (QUEST 2.0): an overview and recent progress. Technol. Disabil. **14**(3), 101–105 (2002)
49. L. Demers, R. Weiss-Lambrou, B. Ska, Development of the Quebec user evaluation of satisfaction with assistive technology (QUEST). Assistive Technol. **8**, 3–13 (1996)
50. S.C. Chan, A.P. Chan, User satisfaction, community participation and quality of life among Chinese wheelchair users with spinal cord injury: a preliminary study. Occup. Ther. Int. **14**(3), 123–143 (2007)
51. A.L. Bergström, K. Samuelsson, Evaluation of manual wheelchairs by individuals with spinal cord injuries. Disabil. Rehabil. Assist. Technol. **1**(3), 175–182 (2006)
52. R.D. Wessels, L.P.D. Witte, Reliability and validity of the Dutch version of QUEST 2.0 with users of various types of assistive devices. Disabil. Rehabil. **25**, 267–272 (2003)
53. A.M. Karmarkar, D.M. Collins, A. Kelleher, R.A. Cooper, Satisfaction related to wheelchair use in older adults in both nursing homes and community dwelling. Disabil. Rehabil. Assist. Technol. **4**(5), 337–343 (2009)
54. D. Gomez-Vargas, F. Ballen-Moreno, P. Barria, R. Aguilar, J.M. Azorín, M. Munera, C.A. Cifuentes, The actuation system of the ankle exoskeleton t-FLEX: first use experimental validation in people with stroke. Brain Sci. **11**, 412 (2021)
55. C.O. Cherry, N.R. Chumbler, K. Richards, A. Huff, D. Wu, L. M. Tilghman, A. Butler, Expanding stroke telerehabilitation services to rural veterans: a qualitative study on patient

experiences using the robotic stroke therapy delivery and monitoring system program. Disabil. Rehabil. Assist. Technol. **12**(1), 21–27 (2017). PMID: 26135221

56. M.P. Dijkers, P.C. deBear, R.F. Erlandson, K. Kristy, D.M. Geer, A. Nichols, Patient and staff acceptance of robotic technology in occupational therapy: a pilot study. J. Rehabil. Res. Develop. **28**(2), 33–44 (1991)
57. L. Fiorini, M. De Mul, I. Fabbricotti, R. Limosani, A. Vitanza, G. D'Onofrio, M. Tsui, D. Sancarlo, F. Giuliani, A. Greco, et al., Assistive robots to improve the independent living of older persons: results from a needs study. Disabil. Rehabil. Assist. Technol. **16**(1), 92–102 (2021)
58. K. Lo, M. Stephenson, C. Lockwood, Adoption of robotic stroke rehabilitation into clinical settings: a qualitative descriptive analysis. JBI Evid Implement **18**(4), 36–390 (2020)
59. L.E. Wood, Semi-structured interviewing for user-centered design. Interactions **4**(2), 48–61 (1997)
60. D.L. Morgan, M.T. Spanish, Focus groups: a new tool for qualitative research. Qual. Soc. **7**, 253–270 (1984)
61. R. Krueger, M. Casey, *Focus Groups: A Practical Guide for Applied Research* (SAGE, Newcastle upon Tyne, 2014)
62. M. Richter, M. Flückiger, *User-Centred Engineering: Creating Products for Humans* (Springer, Berlin, 2014)
63. D.W. Stewart, P.N. Shamdasani, *Focus Groups: Theory and Practice*. Applied Social Research Methods Series, vol. 20 (Sage, Newbury Park, 1990)
64. K.E. Race, D.F. Hotch, T. Packer, Rehabilitation program evaluation. Eval. Rev. **18**, 730–740 (1994)
65. M. Wentink, L. van Bodegom-Vos, B. Brouns, H. Arwert, S. Houdijk, P. Kewalbansing, L. Boyce, T.V. Vlieland, A. de Kloet, J. Meesters, How to improve eRehabilitation programs in stroke care? A focus group study to identify requirements of end-users. BMC Med. Inf. Decis. Mak. **19**, 145 (2019)
66. A.W. Heinemann, D. Kinnett-Hopkins, C.K. Mummidisetty, R.A. Bond, L. Ehrlich-Jones, C. Furbish, E. Field-Fote, A. Jayaraman, Appraisals of robotic locomotor exoskeletons for gait: focus group insights from potential users with spinal cord injuries. Disabil. Rehabil. Assist. Technol. **15**, 762–772 (2020)
67. L. Schmid, A. Glässel, C. Schuster-Amft, Therapists' perspective on virtual reality training in patients after stroke: a qualitative study reporting focus group results from three hospitals. Stroke Res. Treat. **2016**, 1–12 (2016)
68. D.J. van der Veen, C.M.E. Döpp, P.C. Siemonsma, M.W.G.N. van der Sanden, B.J.M. de Swart, E.M. Steultjens, Factors influencing the implementation of home-based stroke rehabilitation: professionals' perspective. Plos One **14**, e0220226 (2019)
69. K. Winkle, P. Caleb-Solly, A. Turton, P. Bremner, Social robots for engagement in rehabilitative therapies: design implications from a study with therapists, in *Proceedings of the 2018 ACM/IEEE International Conference on Human-Robot Interaction* (2018), pp. 289–297
70. A.A. Ramírez-Duque, L.F. Aycardi, A. Villa, M. Munera, T. Bastos, T. Belpaeme, A. Frizera-Neto, C.A. Cifuentes, Collaborative and inclusive process with the autism community: a case study in Colombia about social robot design. Int. J. Soc. Robot. **13**, 153–167 (2021)
71. J. Casas, N. Cespedes, M. Múnera, C.A. Cifuentes, Human-robot interaction for rehabilitation scenarios, in *Control Systems Design of Bio-Robotics and Bio-Mechatronics with Advanced Applications* (Elsevier, Amsterdam, 2020), pp. 1–31
72. S.D. Sierra Marín, D. Gomez-Vargas, N. Céspedes, M. Múnera, F. Roberti, P. Barria, S. Ramamoorthy, M. Becker, R. Carelli, C.A. Cifuentes, Expectations and perceptions of healthcare professionals for robot deployment in hospital environments during the COVID-19 pandemic. Front. Robot. AI **8**, 102 (2021). https://doi.org/10.3389/frobt.2021.612746
73. M.Z. Ferdous, M.S. Islam, M.T. Sikder, A.S.M. Mosaddek, J.A. Zegarra-Valdivia, D. Gozal, Knowledge, attitude, and practice regarding COVID-19 outbreak in Bangladesh: an online-based cross-sectional study. Plos One **15**, e0239254 (2020)

74. IFRC Turkish Red Crescent, Knowledge, attitudes and practices (KAP) assessment on COVID-19 - community based migration programme, [EN/TR] - Turkey I ReliefWeb (2020)
75. REACH, COVID-19 knowledge, attitudes and practices (KAP) survey: Northwest Syria - August–September 2020 (Round 4) - Syrian Arab Republic I ReliefWeb (2020)
76. C. Coombs, Will COVID-19 be the tipping point for the intelligent automation of work? A review of the debate and implications for research. Int. J. Inf. Manage. **55**, 102182 (2020). https://doi.org/10.1016/j.ijinfomgt.2020.102182
77. P.-S. Goh, J. Sandars, *A Vision of the Use of Technology in Medical Education After the COVID-19 Pandemic*, vol. 9 (MedEdPublish, 2020)
78. D. Raigoso, N. Céspedes, C.A. Cifuentes, A.J. del Ama, M. Múnera, A survey on social assistive robotics: clinicians' and patients' perception of a social robot within gait rehabilitation therapies. Brain Sci. **11**(6), 738 (2021). https://doi.org/10.3390/brainsci11060738
79. I. Leite, C. Martinho, A. Paiva, Social robots for long-term interaction: a survey. Int. J. Soc. Robot. **5**(2), 291–308 (2013)
80. C. Breazeal, K. Dautenhahn, T. Kanda, Social robotics, in *Springer Handbook of Robotics* (Springer, Berlin, 2016), pp. 1935–1971
81. V. Venkatesh, M.G. Morris, G.B. Davis, F.D. Davis, User acceptance of information technology: toward a unified view. MIS Quart. **27**(3), 425–478 (2003)
82. T. Vandemeulebroucke, B.D. de Casterlé, C. Gastmans, How do older adults experience and perceive socially assistive robots in aged care: a systematic review of qualitative evidence. Aging Mental Health **22**(2), 149–167 (2018)
83. C. Bartneck, T. Belpaeme, F. Eyssel, T. Kanda, M. Keijsers, S. Šabanović, *Human-Robot Interaction* (Cambridge University Press, Cambridge, 2020)
84. W.Y.G. Louie, D. McColl, G. Nejat, Acceptance and attitudes toward a human-like socially assistive robot by older adults. Assist. Technol. **26**(3), 140–150 (2014)
85. A. Weiss, R. Bernhaupt, M. Tscheligi, D. Wollherr, K. Kühnlenz, M. Buss, A methodological variation for acceptance evaluation of human-robot interaction in public places, in *Proceedings of the 17th IEEE International Symposium on Robot and Human Interactive Communication, RO-MAN* (2008), pp. 713–718
86. T. Bickmore, D. Schulman, Practical approaches to comforting, in *Proceedings of ACM CHI 2007: Conference on Human Factors in Computing Systems* (2007), pp. 2291–2296
87. S. Hennemann, M.E. Beutel, R. Zwerenz, Drivers and barriers to acceptance of web-based aftercare of patients in inpatient routine care: a cross-sectional survey. J. Med. Internet Res. **18**, e337 (2016)
88. M.M. de Graaf, S.B. Allouch, T. Klamer, Sharing a life with Harvey: exploring the acceptance of and relationship-building with a social robot. Comput. Hum. Behav. **43**, 1–14 (2015)

Index

C. A. Cifuentes, M. Múnera, *Interfacing Humans and Robots for Gait Assistance and Rehabilitation*, https://doi.org/10.1007/978-3-030-79630-3

T

Printed in the United States
by Baker & Taylor Publisher Services